W0042849

28 Springer Series in Solid-State Sciences
Edited by Peter Fulde

Springer Series in Solid-State Sciences

Editors: M. Cardona P. Fulde H.-J. Queisser

Volumes 1 – 39 are listed on the back inside cover

The Structure and Properties of Matter

Editor: T. Matsubara

With Contributions by
T. Matsubara   H. Matsuda   T. Murao
T. Tsuneto   F. Yonezawa

With 229 Figures

Springer-Verlag Berlin Heidelberg New York 1982

Professor Dr. Takeo Matsubara
Professor Dr. Toshihiko Tsuneto

Department of Physics, Faculty of Science, Kyoto University, Kyoto 606, Japan

Professor Dr. Tsuyoshi Murao

Department of Chemistry, Faculty of Science, Kyoto University, Kyoto 606, Japan

Professor Dr. Hirotsugu Matsuda

Department of Biology, Faculty of Science, Kyushyu University, Fukuoka 812, Japan

Professor Dr. Fumiko Yonezawa

Department of Physics, Faculty of Science and Engineering, Keio University, Hiyoshi, Yokohama, 223, Japan

Series Editors:

Professor Dr. Manuel Cardona
Professor Dr. Peter-Fulde
Professor Dr. Hans-Joachim Queisser

Max-Planck-Institut für Festkörperforschung, Heisenbergstraße 1
D-7000 Stuttgart 80, Fed. Rep. of Germany

Revised Translation of the original Japanese Edition:
Bussei I edited by Takeo Matsubara
© by Takeo Matsubara
Originally published by Iwanami Shoten Publishers, Tokyo (1978).
English Translation by Takeo Matsubara, Toshihiko Tsuneto, Tsuyoshi Murao,
Fumiko Yonezawa, Takashi Odagaki, and Kazushige Machida

ISBN-13: 978-3-642-81731-1   e-ISBN-13: 978-3-642-81729-8
DOI: 10.1007/ 978-3-642-81729-8

Library of Congress Cataloging in Publication Data. Bussei. 1. English. The structure and properties of matter. (Springer series in solid-state sciences ; 28) Revised translation of: Bussei; part 1. rev. 2nd ed. Bibliography: p. Includes index. 1. Matter—Constitution. 2. Matter—Properties. 3. Condensed matter. I. Matsubara, T. (Takeo), 1921- II. Title. III. Series. QC173.B97613 1982 530 81-16629 AACR2

2153/3130-543210

Preface

This book originally appeared in Japanese in 1973 in the *Iwanami Series of Fundamental Physics* supervised by Professor Hideki Yukawa and published by Iwanami-Shoten. A revised second edition was published in 1978. The task we set ourselves was to grasp the properties of matter as a whole in a unified scheme and to present a general view of matter incorporating the results of modern physics.

To achieve this goal we have tried to explore the laws which describe the structure of macroscopic matter, namely, to ask in what kinds of physical states matter can, in principle, exist and why. Thus, using the methods of statistical physics and quantum mechanics, we have tried to systematically describe the properties of matter from a unified point of view.

Of course, we do not believe that such a standpoint can give an exhaustive description of condensed matter. One of the important viewpoints which obviously is omitted in such a unified approach is the historical one, which follows the development of physics in the course of time. Indeed, the physics of condensed matter has been characterized since the revolution in experimental techniques by the fact that it has developed into a scientific construction, which has been made possible by the accumulation of exact information gained through the close interaction between theory and experiment. No one can deny the important role played by ingeniously designed experiments in understanding the properties of matter. Here, however, we do not enter deeply into the interrelationship between theory and experiment.

Looking through the contents, one might believe that, contrary to the original intention of the editor, many gaps remain in a book which is supposed to cover the structures and properties of matter. For instance, plasma, which is important as the fourth state of matter, is almost untouched in this volume. Also, little attention has been paid to the surprisingly magnificent diversity in the complex structures of compounds. Even within the inorganic substances, we should have to mention, for instance, a huge manifold of silicates and other mineral substances. Furthermore, in contrast to other standard textbooks on solid-state physics, several fundamental items are omitted; crystallography including x-ray and neutron diffraction, thermal properties including lattice vibrations, and the general background of transport phenomena are some examples. For these problems, readers should refer to the reference books cited at the end of this volume.

Finally we should like to thank Professor P. Fulde for his kind suggestions and Dr. H. Lotsch of Springer-Verlag for his patient cooperation in making this English edition a reality.

Kyoto, January 1982   *Takeo Matsubara*

Contents

1. Atoms as Constituents of Matter

In this chapter, a brief review of atoms as the building blocks of matter is given. Since the starting point from which we discuss the structures and properties of matter is an atom, the aim of this chapter is to clarify how the main properties of atoms, the primary inputs of subsequent discussions, depend on atomic species.

1.1 Introduction

Materials which we treat, observe and measure in our daily life or in a laboratory are an agglomerate of an enormous number of particles. A macroscopic measurement of properties of matter senses nothing but the physical quantities which are exhibited by the group of all particles in the system. In principle, quantum mechanics will deduce the behavior of the group from the characteristics of constituent units such as mass, charge, spin, etc., and the interaction potential among particles. And, if the discussion is confined to a thermal equilibrium and its vicinity, statistical mechanics will derive possible condensed states of the group of particles under various macroscopic conditions exerted on the system. Therefore, one can theoretically predict the structure and the properties of matter on the basis of the intrinsic traits of the constituents. For example, the equilibrium state at a fixed temperature and a fixed pressure is the state which minimizes the Gibbs free energy of the system, and the Gibbs free energy can, in principle, be, evaluated provided that the characteristics of the constituents are known. Unfortunately, because of mathematical difficulties in handling interacting many-body systems, it is quite troublesome, even at present, to predict the complete structure and behavior of matter from the microscopic point of view stated above. Nonetheless, we can now grasp and understand various phases of matter—Why are some materials metallic, while others are nonmetallic under macroscopic conditions? How does the magnetism of matter appear? What is the essential difference between a liquid and a solid? and so on—in a unified manner based on the atomic viewpoint.

To understand changing properties on the basis of unchanging properfies is a powerful strategy for understanding in a unified fashion. Therefore, it might be reasonable to choose elementary particles as a basic constituent unit of our system. All elementary particles, however, are not stable. Moreover, we will be concerned mainly with materials which exist under conditions attainable on the earth. Therefore, it is sufficient for us to choose electrons and nuclei as the basic particles of our discussion. To be explicit, an electron is a stable elementary particle, while a nucleus is itself composed of the elementary particles protons and neutrons. The nucleus is not a single point, but has an internal structure whose length is about 10^{-13} cm. Yet even the change in structure of the nucleus of a radioisotope, which is substantial, is scarcely governed by the temperature, pressure, etc., on the earth surface. Consequently, we can regard a nucleus *a priori* as an unchangeable particle. The properties of a nucleus are characterized by an atomic number Z (the number of protons in the nucleus) and a mass number A. The latter is defined by $Z + n$, n being the number of neutrons in the nucleus. Nuclei which have a common Z and different n are called isotopes. For stable nuclei, Z and A obey an approximate relation,

$$Z \approx \frac{A}{2 + 0.015\, A^{2/3}}. \tag{1.1.1}$$

The nucleus of atomic number Z carries a charge Ze; e denotes the magnitude of electronic charge, $e = (4.80298 \pm 0.00020) \times 10^{-10}$ esu. The mass of a nucleus is in proportion to the mass number A within an error less than one percent. The rest mass per mass number is about 1,800 times as heavy as that of an electron. Since a nucleus is much heavier than the electron, the nucleus is sometimes assumed to rest when we discuss the motion of electrons.

In general, electrostatic force and gravity act between two charged particles with the mass and the strength of both forces inversely proportional to the square of the distance between the particles. The ratio of the proportional constants is about $2 \times 10^{39}\, Z/A$ for forces between electron and nucleus. Thus, the gravity is negligible compared with the electrostatic force. This is also the case for interactions between nuclei. Consequently the strength of the interaction between nuclei depends on Z but is almost independent of A.

Thus, the gross properpty of matter is determined by the atomic number of constituent nuclei. Each material of different atomic number has its own element name. At present, 103 elements are known from hydrogen H, $Z = 1$, to lawrencium Lr, $Z = 103$. Nuclei with higher Z than of Lr are extremely unstable even if they exist, and they are beyond our present consideration. There are some phenomena where we must take account of the finiteness of nuclear mass and hence the motion of the nucleus. Such phenomena are affected by the number of neutron in the nucleus. This is known as an isotope effect.

A nucleus can keep some electrons around it to become an atom (neutral atom or ion). Materials are also considered to be an assembly of such atoms. When an atom is brought close to other atoms, the state of electrons bound to the nucleus is no longer that of the isolated atom because of electron-electron and electron-nucleus interactions. This change of electronic state yields various condensed states of atoms and a variety of matter. In most situations, however, only a few electrons change their state considerably, and the change is closely related to the trait of the electronic state of an isolated atom. Therefore, we shall first discuss the properties of an isolated atom.

1.2 Atomic Scale

In order to describe the electronic states of an atom, we must refer to quantum mechanics. In fact, if the electrons in an atom would strictly obey the laws of classical mechanics and electromagnetism, they would lose their energy by radiation and the atom would eventually collapse. Before proceeding to a quantitative discussion of the steady state of an atom, let us first grasp the extent of physical quantities in the atomic scale.

According to quantum mechanics, the position and its conjugate momentum are subject to Heisenberg's uncertainty principle. We denote the uncertainty of position and momentum by Δx and Δp, respectively. Then Δx and Δp obey a relation

$$\Delta x \Delta p \approx \hbar \left(\equiv \frac{h}{2\pi} \right) . \tag{1.2.1}$$

Here, h is the Planck constant: $h = (6.6256 \pm 0.0005) \times 10^{-27}$ erg·s. Now, suppose an electron moves in a sphere of radius R centered on an atomic nucleus with a charge Ze. The kinetic energy E_k and the potential energy E_p of the electron in the ground state (the state of minimum energy) are roughly given by setting $\Delta x \approx R$ in (1.2.1), i.e.,

$$E_k \approx \frac{(\Delta p)^2}{2m_0} \approx \frac{1}{2m_0} \left(\frac{\hbar}{R} \right)^2 , \tag{1.2.2}$$

$$E_p \approx - \frac{Ze^2}{R} . \tag{1.2.3}$$

Here, m_0 denotes the electron mass: $m_0 = (9.1091 \pm 0.0004) \times 10^{-28}$ g. It is evident from (1.2.2,3) that the total energy

$$E = E_k + E_p \tag{1.2.4}$$

is minimized for the following R:

$$R \approx \frac{a_0}{Z} , \quad a_0 = \frac{\hbar^2}{m_0 e^2} = (5.29167 \pm 0.00007) \times 10^{-9} \text{ [cm] .} \tag{1.2.5}$$

The quantity a_0 is known as the Bohr radius. Although the radius R is quite small on a macroscopic scale, the radius of the nucleus is about 10^{-13} cm and much smaller than R. Therefore, we can regard a nucleus as a mass point without any structure on the atomic scale.

Inserting (1.2.5) into (1.2.2–4), we have

$$E \approx -E_0 Z^2, \quad E_0 = \frac{e^4 m_0}{2\hbar^2} = 2.154 \times 10^{-11} \ \text{[erg]}. \tag{1.2.6}$$

The energy E_0 is known as the Rydberg unit of energy, which is denoted hereafter by Ry. In passing, E_0 corresponds to a temperature $T_0 \approx 3.12 \times 10^5$ K through $E_0 = k_B T_0$ (k_B is the Boltzmann constant).

Using $\Delta p \approx m_0 v$ in (1.2.1, 5), the velocity v of electrons in an atom is estimated to be

$$v \approx \frac{\hbar}{m_0 R} \approx \frac{e^2}{\hbar} Z \approx 1.44 \times 10^8 Z \ \text{[cm} \cdot \text{s}^{-1}]. \tag{1.2.7}$$

The velocity v for relatively light atoms is much smaller than the velocity of light $c = (2.997925 \pm 0.000003) \times 10^{10}$ cm s^{-1}, and a nonrelativistic theory of motion applies to electrons in these lighter atoms as a good approximation. For heavier atoms, however, Z can take a value up to 100 and the electron velocity in these atoms approaches the velocity of light. This means that a relativistic correction is necessary for heavier atoms.

Now, let us examine a magnetic field produced by electric currents in atoms. Suppose that the current density at a position r is j. Then, according to Biot-Savart's law the magnetic field at the origin is given by

$$H = \frac{1}{c} \int \frac{j \times r}{r^3} \, dV. \tag{1.2.8}$$

For our system, we note that the magnitude of j is roughly equal to $ev/R^3 \approx e\hbar/(m_0 R^4)$ and the current density j is distributed over a volume of order R^3. Putting these values into (1.2.8), we find

$$H \approx \frac{jR}{c} \approx \frac{e\hbar}{m_0 c R^3} \approx \frac{2\mu_B}{R^3} \approx 1.25 \times 10^5 Z^3 \ \text{[G]}. \tag{1.2.9}$$

Here, $\mu_B = e\hbar/2m_0 c = (9.2732 \pm 0.0006) \times 10^{-21}$ erg$\cdot$G^{-1} is known as the Bohr magneton.

In general, the strength of the magnetic field produced by the magnetic moment μ at the origin is of the order of μ/r^3 at a distance r. The magnetic field in (1.2.9) is regarded to be produced by a magnetic moment μ_B at a distance R, and μ_B is of the order of the magnetic moment produced by an electron in atoms. In a coordinate system fixed on the electron, the nucleus carries a current of order Zev/R^3 and produces a magnetic field at the position of the electron, whose strength is about Z times larger than that of (1.2.9), i.e., about 1.25×10^5

the Rydberg energy $E_0 = m_0 e^4 / 2\hbar^2$ given in (1.2.6) as a unit of energy. Using these units, (1.3.4, 5) are reduced to

$$[-\Delta + V(r)]\, \Psi(r) = E\Psi(r)\,, \tag{1.3.6}$$

$$V(r) = -\frac{2Z}{r}\,. \tag{1.3.7}$$

To discuss phenomena in the atomic scale, it is convenient to choose the electron mass m_0 as a mass unit and

$$t_0 = \frac{\hbar^3}{m_0 e^4} \approx 2.419 \times 10^{-17}\ [\mathrm{s}]\,. \tag{1.3.8}$$

as a time unit. In fact, if an electron moves a distance a_0 during a time interval t_0, its velocity is a_0/t_0 and its kinetic energy is $(m_0/2)(a_0/t_0)^2 = E_0$. Thus, this energy coincides with the unit energy in the atomic scale. The unit system using a_0, m_0, t_0 as fundamental units is called the atomic unit.

Now, the Schrödinger equation (1.3.6) can be written in spherical coordinates as

$$\frac{1}{r^2}\frac{\partial}{\partial r}\left(r^2\frac{\partial\Psi}{\partial r}\right) + \frac{1}{r^2\sin\theta}\frac{\partial}{\partial\theta}\left(\sin\theta\frac{\partial\Psi}{\partial\theta}\right) + \frac{1}{r^2\sin^2\theta}\frac{\partial^2\Psi}{\partial\phi^2} + [E - V(r)]\,\Psi = 0\,. \tag{1.3.9}$$

We look for a solution written in a product form such that

$$\Psi(\mathbf{r}) = \Psi(r, \theta, \phi) = \mathrm{R}(r)\,\mathrm{Y}(\theta, \phi)\,, \tag{1.3.10}$$

where $\mathrm{R}(r)$ is a function of only radius r and $\mathrm{Y}(\theta, \phi)$ is a function of only angular variables θ and ϕ. If (1.3.10) is substituted into (1.3.9) and the results are multiplied by r^2, (1.3.9) is rearranged into a form

$$\frac{1}{\mathrm{R}}\frac{d}{dr}\left(r^2\frac{d\mathrm{R}}{dr}\right) + r^2[E - V(r)] = -\frac{1}{\mathrm{Y}}\left[\frac{1}{\sin\theta}\frac{\partial}{\partial\theta}\left(\sin\theta\frac{\partial\mathrm{Y}}{\partial\theta}\right) + \frac{1}{\sin^2\theta}\frac{\partial^2\mathrm{Y}}{\partial\phi^2}\right]. \tag{1.3.11}$$

Since the left-hand side of (1.3.11) depends only on r, and the right-hand side is a function of only θ and ϕ, both sides must be equal to a constant which is denoted by C. Then, the right-hand side of (1.3.11) becomes

$$-\left[\frac{1}{\sin\theta}\frac{\partial}{\partial\theta}\left(\sin\theta\frac{\partial\mathrm{Y}}{\partial\theta}\right) + \frac{1}{\sin^2\theta}\frac{\partial^2\mathrm{Y}}{\partial\phi^2}\right] = C\mathrm{Y}\,. \tag{1.3.12}$$

It is well known that the solution of the differential equation (1.3.12) is given by spherical harmonics $P_l^{|m|}(\cos\theta)\exp(im\phi)$ and the eigenvalue is expressed as

$$C = l(l+1). \tag{1.3.13}$$

Here, $P_l^{|m|}(\cos\theta)$ is the associated Legendre function, and m is an integer satisfying $|m| \leqslant l$.

Among the significant properties of the associated Legendre function, the following orthogonality relation is very important;

$$\int_{-1}^{1} P_l^m(x)\, P_{l'}^m(x)\, dx = \int_0^\pi P_{l'}^m(\cos\theta)\, P_l^m(\cos\theta) \sin\theta\, d\theta ,$$

$$= \begin{cases} 0 & \text{if } l \neq l' \\[2mm] \dfrac{2}{2l+1} \dfrac{(l+|m|)!}{(l-|m|)!} & \text{if } l = l' . \end{cases} \tag{1.3.14}$$

It follows immediately from the orthogonality relation that eigenfunctions belonging to different sets of (l, m) are orthogonal to each other. If we define

$$Y_l^m(\theta, \phi) = \sqrt{\frac{(2l+1)}{2} \frac{(l-|m|)!}{(l+|m|)!}}\, P_l^{|m|}(\cos\theta)\, e^{im\phi} , \tag{1.3.15}$$

$Y_l^m(\theta, \phi)$ is an orthonormalized angular wave function. Incidentally, the whole set of such eigenfunctions has been proved to form a complete set.

Substitution of the eigenfunction (1.3.15) into (1.3.11) yields

$$\frac{1}{r^2} \frac{d}{dr}\left(r^2 \frac{dR}{dr}\right) + \left[E - V(r) - \frac{l(l+1)}{r^2}\right] R = 0. \tag{1.3.16}$$

If we define $P(r) = rR(r)$, $P(r)$ satisfies

$$\frac{d^2P(r)}{dr^2} + \left[E - V(r) - \frac{l(l+1)}{r^2}\right] P(r) = 0. \tag{1.3.17}$$

Equation (1.3.17) is similar to the one-dimensional Schrödinger equation for a particle in a potential $V(r) + l(l+1)/r^2$.

In order to understand the physical significance of the term $l(l+1)/r^2$, let us now introduce an angular momentum operator L of an electron with respect to the nucleus

$$L = r \times p = -i\hbar(r \times \nabla) , \tag{1.3.18}$$

p being the momentum, and note that the z-component L of L and L^2 take the following form in the spherical coordinates:

$$L_z = -i\hbar \frac{\partial}{\partial\phi} , \tag{1.3.19}$$

$$L^2 = -\hbar^2 \left[\frac{1}{\sin\theta} \frac{\partial}{\partial\theta}\left(\sin\theta \frac{\partial}{\partial\theta}\right) + \frac{1}{\sin^2\theta} \frac{\partial^2}{\partial\varphi^2}\right]. \tag{1.3.20}$$

From a comparison of (1.3.20) with (1.3.12, 13), we find that $R(r)P_l^{|m|}(\cos\theta)$ $\exp(im\phi)$ is an eigenfunction of L_z and L^2 with eigenvalues $L_z = m\hbar$ and $L^2 = \hbar^2 l$ $(l+1)$.

Now, the total energy is expressed in terms of the radial momentum p_r and the azimuthal momentum p_l as

$$E = \frac{p_r^2 + p_l^2}{2m_0} + V(r).$$ (1.3.21)

Note that

$$\frac{p_l^2}{2m_0} = \frac{L^2}{2m_0 r^2} = \frac{\hbar^2 l(l+1)}{2m_0 r^2}$$ (1.3.22)

and the centrifugal potential is given by $L^2/2m_0 r^2$ since the magnitude of the centrifugal force is $L^2/m_0 r^3$. A comparison of (1.3.21, 22) with (1.3.17) written in the atomic units shows that (1.3.17) is the Schrödinger equation for the total energy (1.3.21), and thus the term $l(l+1)/r^2$ is regarded as the virtual centrifugal potential which appears as the problem is reduced to a one-dimensional case. The effective potential $V(r) + l(l+1)/r^2$ for the hydrogen atom is shown in Fig. 1.1 for various values of l.

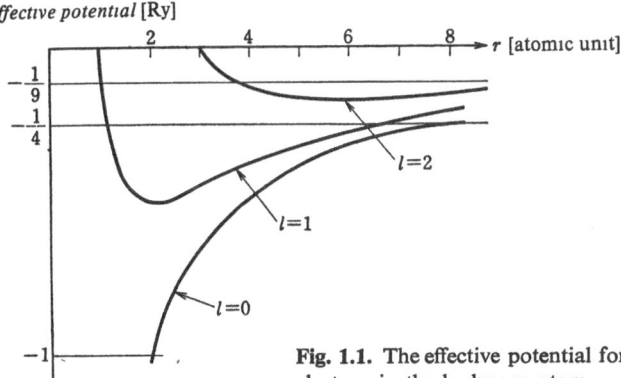

Fig. 1.1. The effective potential for the radial motion of the electron in the hydrogen atom

As one can see in Fig. 1.1, the effective potential except for $l = 0$ becomes positively infinite as r approaches zero. Consequently, the wave function $P(r)$ for those values of l will satisfy $\lim_{r\to 0} P(r) = 0$. In fact, if $l \neq 0$ and r is small, $V(r)$ and E in (1.3.17) are negligible compared with $l(l+1)/r^2$, then (1.3.17) may be rednced to

$$\frac{d^2 P(r)}{dr^2} - \frac{l(l+1)}{r^2} P(r) = 0.$$ (1.3.23)

The two independent solutions of the differential equation (1.3.23) are

$$P(r) = r^{l+1} \quad \text{or} \quad r^{-l}, \tag{1.3.24}$$

but the latter solution does not satisfy the normalization condition (1.3.3), for the integral

$$\int_0^\infty |R(r)|^2 r^2 dr \quad \text{diverges at} \quad r = 0 \quad \text{if} \quad l \geqslant 1.$$

For larger values of r, (1.3.17) is approximated by

$$\frac{d^2 P(r)}{dr^2} + EP(r) = 0. \tag{1.3.25}$$

This equation has independent particular solutions $\sin \sqrt{E}r$ and $\cos \sqrt{E}r$, for $E > 0$ and $P(r) = \text{constant}$ and $P(r) = r$ for $E = 0$, but both sets do not satisfy the condition (1.3.3). Therefore, E must be negative to describe the state of the atom of $N = 1$. For $E < 0$,

$$P(r) = \exp\left(\pm \sqrt{-E}r\right) \tag{1.3.26}$$

are two independent particular solutions of (1.3.25), and only the lower solution is allowed because of the condition (1.3.3).

Now, in order to interpolate two solutions for smaller and larger values of r, we write the solution as

$$P(r) = r^{l+1} \exp\left(-\sqrt{-E}r\right) \sum_{k=0}^\infty A_k r^k \quad (A_0 \neq 0). \tag{1.3.27}$$

Inserting this into (1.3.17), we find that A_k obeys the recurrence relation

$$A_k = -2A_{k-1} \frac{Z - (l+k)\sqrt{-E}}{(l+k)(l+k+1) - l(l+1)}. \tag{1.3.28}$$

If $\sqrt{-E} \neq Z/(l+k)$ $(k = 1, 2, \ldots)$, it is evident from (1.3.28) that the radius of convergence of the infinite series $\sum_{k=0}^\infty A_k r^k$ is infinite. Thus, the function $P(r)$ (1.3.27) is indeed a solution of (1.3.17), but we can prove from (1.3.28) that it is not a stationary state. In fact, if we choose A_0 such that $A_{k_0} > 0$ for a sufficiently large value of k_0, we can easily derive inequalities for $k > k_0$:

$$A_k \geqslant 2A_{k-1} \frac{\sqrt{-E}}{l+k+1}\left(1 - \frac{Z/\sqrt{-E}}{l+k}\right) > \frac{2\sqrt{-E}}{k} \frac{1 - \frac{Z/\sqrt{-E}}{l+k_0}}{1 + \frac{l+1}{k_0}} A_{k-1},$$

hence

$$A_k \geqslant A_{k_0} \frac{\gamma^k}{k!} \left(\frac{k_0!}{\gamma^{k_0}}\right), \quad \gamma = 2\sqrt{-E}\left(1 - \frac{Z/\sqrt{-E}}{l+k_0}\right) \Big/ \left(1 + \frac{l+1}{k_0}\right).$$

It follows that

$$\sum_{k=0}^{\infty} A_k r^k \geqslant A_{k_0} \left(\frac{k_0!}{\gamma^{k_0}}\right) \exp(\gamma r) + (k_0 \text{th order polynomial of } r)$$

and, since k_0 can take any large number and $\gamma \to 2\sqrt{-E}$ as $k_0 \to \infty$, P(r) diverges as $\exp(\sqrt{-E}r)$ when $r \to \infty$. Therefore, the solution (1.3.27) can be a stationary state only if a positive integer h exists such that

$$\sqrt{-E} = \frac{Z}{(l+h)}. \tag{1.3.29}$$

Under this condition, (1.3.27) for an arbitrary $l \geqslant 0$ is a stationary solution of (1.3.3). Thus, if $l \neq 0$, the linearly independent solutions of (1.3.3) are given uniquely by (1.3.27, 29).

When $l = 0$, the function (1.3.27) with (1.3.29) is again a solution of (1.3.3) and the other independent solution is represented by a similar series of the form

$$P(r) = \exp(-\sqrt{-E}r) \left\{\sum_{k=0}^{\infty} A_k r^k + \ln r \sum_{k=1}^{\infty} B_k r^k\right\} (A_0 \neq 0), \tag{1.3.30}$$

where A_k and B_k are constants. The similar consideration to the above, however, shows that the latter cannot be a stationary solution for any values of E since $\int_0^\infty |R(r)|^2 r^2 dr$ diverges at the upper limit for this solution. Thus, if we put

$$n = l + h \tag{1.3.31}$$

in (1.3.29), n must be a positive integer and the energy of a possible sationary state is written in the form

$$E = -\frac{Z^2}{n^2} \text{ [Ry]}. \tag{1.3.32}$$

The above argument of the stationary state of an electron in the one-electron atom is summarized by the following:

1) A stationary state is characterized by a set of three quantum numbers (n, l, m). n is a positive integer; for a given n, l can take n values, $l = 0, 1, 2, \ldots, n-1$; for a given l, m can take $(2l + 1)$ values, $m = -l, -l+1, \ldots, 0, 1, \ldots, l$.

2) The quantum number l specifies the angular momentum of the electron with respect to the nucleus and is called the azimuthal quantum number.
3) The quantum number m specifies the z-component of the angular momentum and is called the magnetic quantum number. States with the same set of (n, l) and a different value of m have equal energy if external fields are absent.
4) The energy of stationary states is determined only by the value of n. n is called the principal quantum number.

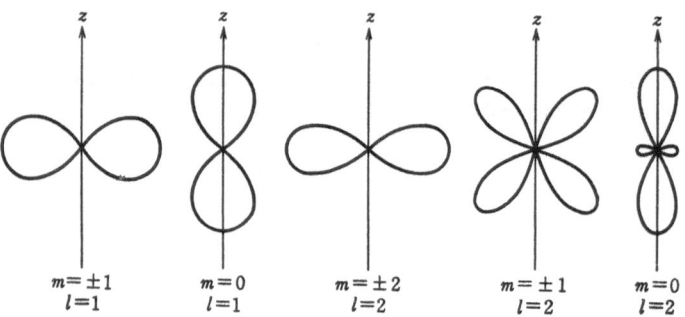

Fig. 1.2. Angular wave functions $(P_l^{|m|})^2$ for various values of l and m, plotted against the polar angle

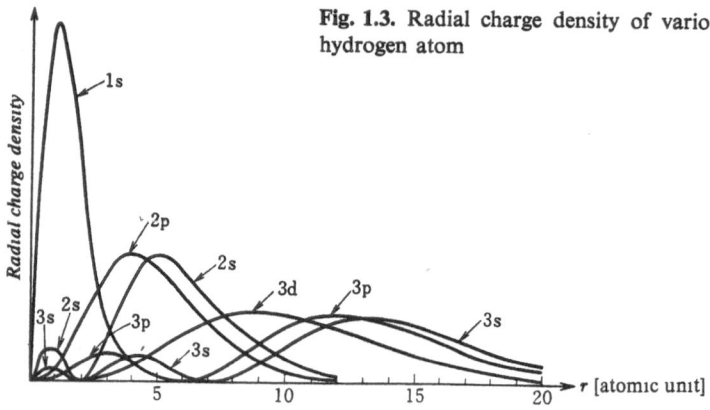

Fig. 1.3. Radial charge density of various orbitals of the hydrogen atom

In view of our derivation of these results, we note that properties 1)–3) are free from the explicit form of the potential $V(r)$ and are deduced from the fact that the potential $V(r)$ is a function only of distance r, in other words, the electron is put in a spherically symmetric field. On the other hand, property 4) is derived from the specific form of the Coulomb potential. Since the radial wave equation (1.3.17) depends on the angular momentum, each value of l might generally give a different set of energy eigenvalues. For the Coulomb potential, however, those eigenvalues for a given n are independent of l, though the lowest energy for a given l depends on the value of l. On the other hand, the validity of property 3) for any form of $V(r)$ is quite natural, since the z-axis of the coordinate system can be chosen in an arbitrary direction. As we shall see in Sect. 1.7, an external magnetic field, for example, breaks the spherical symmetry of the system and the energy eigenvalue under the external field depends on the value of m. However, as far as the field is spherically symmetric, the angular wave function has a common form, independent of $V(r)$. The eigenstates corresponding to $l = 0, 1, 2, \ldots$ are called $s, p, d, f, g, h, i, k, \ldots$ state or orbital, respectively. The angular wave functions are shown in Fig. 1.2 for a several values of l and m. If the radial wave function is normalized such that $4\pi\int_0^\infty |R(r)|^2 r^2 dr = 1$, $|R(r)|^2$ multiplied by the volume element $4\pi r^2 dr$ of a spherical shell between radii r and $r + dr$ is the probability of finding the electron in the shell, then, $P^2(r) = r^2 R^2(r)$ is proportional to the radial charge density of the electron, which is shown in Fig. 1.3 for various sets of quantum numbers.

1.4 Central-Field Approximation

In Sect. 1.3, we investigated the electronic states of a one-electron atom. For $N \geqslant 2$, however, we cannot solve the Schrödinger equation (1.3.1) exactly because of electron-electron interactions. In order to understand various electronic states of these atoms, a concept of electron orbitals has been introduced; each electron is assumed to occupy its own orbital and to move in the orbital independently of other electrons. In the quantum mechanical language, this concept corresponds to the assumption that each electron is expressed by its own one-electron wave function $\psi(r)$ and the entire wave function Ψ of all the electrons is given by a simple product of one-electron wave functions

$$\Psi(r_1, r_2, \ldots, r_N) = \psi_1(r_1)\,\psi_2(r_2) \ldots \psi_N(r_N), \tag{1.4.1}$$

where $r_1, r_2, \ldots, r_N$ denote the position of each electron. The one-electron wave functions $\psi_1(r_1), \psi_2(r_2), \ldots$ are the quantum analogue to the orbitals in the classical mechanics and are called the orbital functions.

Now, electrons in an atom move, being acted upon by the Coulombic repulsion from other electrons along with the attractive Coulombic force from the nucleus. The repulsive forces are determined by all of the coordinates of other electrons. Within the approximation (1.4.1), each electron can be treated as an independent particle, any correlation among their positions being neglected. Thus, each electron can be dealt with as if it moves in an average field produced by other electrons. In fact, to obtain the energy of the ground state on assumption (1.4.1), we write the potential energy $U_j(r_j)$ acting on the jth electron as

$$U_j(r_j) = \int V(r_1, r_2, \ldots, r_N)|\psi_1(r_1)|^2 |\psi_2(r_2)|^2 \ldots$$
$$|\psi_{J-1}(r_{J-1})|^2 |\psi_{J+1}(r_{J+1})|^2 \ldots |\psi_N(r_N)|^2 dr_1 dr_2 \ldots dr_{J-1} dr_{J+1} \ldots dr_N. \tag{1.4.2}$$

Then, it has been proved that the best approximation to the ground-state energy (the least approximate energy larger than exact value) is attained by a self-consistent eigensolution $\psi_j(r_j)$ which obeys the Schrödinger equation

$$-\frac{\hbar^2}{2m_0}\Delta\psi_j(r_j) + U_j(r_j)\,\psi_j(r_j) = \varepsilon_j\psi_j(r_j) \qquad (1.4.3)$$

with $U_j(r_j)$ given by (1.4.2). Here, $\psi_j(r)$ is normalized such that $\int|\psi_j(r)|^2dr = 1$. This procedure is known as the Hartree approximation.

The total energy of the system in this approximation scheme is expressed as

$$\int \ldots \int \Psi^*\left[-\frac{\hbar^2}{2m_0}\sum_{i=1}^{N}\nabla_i^2 + V(r_1, r_2, \ldots, r_N)\right]\Psi dr_1\,dr_2\ldots dr_N$$

$$= \sum_{j=1}^{N}\varepsilon_j - \sum_{j>k}\sum\int\int|\psi_j(r_j)|^2|\psi_k(r_k)|^2\frac{e^2}{|r_j - r_k|}\,dr_j dr_k\,. \qquad (1.4.4)$$

The second term on the right-hand side represents the compensation to the interaction energy between electrons since the first term counts it twice.

In general, the potential $U_j(r_j)$ will depend on the angular variables (θ_j, ϕ_j) as well as the radial variable $|r_j|$ in the spherical coordinates around the origin. For the sake of simplicity, we approximate $U_j(r_j)$ by its average over the angular variables and denote it by the same $U_j(r_j)$ $(r_j = |r_j|)$. Then, (1.4.3) takes a form identical to the Schrödinger equation (1.3.4) for an electron in the spherically symmetric field. Hence, as described in Sect. 1.3, the solution is expressed in a similar way to the hydrogen atom, i.e., using the spherical coordinates (r, θ, ϕ) whose origin is on the nucleus, the wave function is written as $\psi_j(r) = R_j(r)Y_j(\theta, \phi)$ and the angular part $Y_j(\theta, \phi)$ is characterized by the azimuthal quantum number l and the magnetic quantum number m. The radial wave function $R_j(r)$ obeys (1.3.16) with $U_j(r)$ instead of $V(r)$. Therefore, the eigenenergy ε_j (the orbital energy) of orbital j belonging to an azimuthal quantum number l is given by the eigenvalue E of (1.3.16) or (1.3.17). Using the same method as for the hydrogen atom, we denote the kth eigenenergy from the lowest by $\varepsilon_{nl}(n = k + l)$ and call n the principal quantum number. If $U_j(r)$ behaves as $1/r$, ε_{nl} is independent of l as we have seen in Sect. 1.3. Actually, as r becomes small, $U_j(r)$ asymptotically approaches the bare potential $-Ze^2/r$ of the nucleus, while as r is increased infinitely, $U_j(r)$ looks like $-e^2/r$ since $(N-1)$ electrons screen the nuclear charge. In addition, the orbitals for larger azimuthal quantum numbers have a smaller probability amplitude in the vicinity of $r = 0$ $[R(r) \approx r^l,$ as $r \to 0]$ since a stronger centrifugal force acts on them. Therefore, electrons of a common n and a larger l have less chance to feel the bare attraction of the nucleus compared with the case where the screening is absent, and the orbital energy ε_{nl} is expected to increase with l. In fact, from the numerical analysis and spectroscopic evidence we find the following sequence of the orbital energies in the increasing order

$$1s; 2s, 2p; 3s, 3p; [4s, 3d], 4p; [5s, 4d], 5p; [6s, 4f, 5d], 6p; [7s, 5f, 6d]. \quad (1.4.5)$$

Here, s, p, d, f stand for $l = 0, 1, 2, 3$, respectively. Orbitals in the square brackets have almost equal energy and a large energy gap exists between orbitals on both sides of the semicolons.

In the numerical analysis to obtain ϵ_{nl} of an electron, the orbitals of all other electrons must be specified because ϵ_{nl} really depends on these orbitals. The spectroscopic measurements of emission or absorption spectra of the atom provide information about the steady states of atoms. Experiments show that the ground state of atoms completely differs from a state in which all electrons occupy the orbital of the lowest energy. To obtain an agreement with experiments, we must take it into account that an electron has a spin and obeys the Pauli principle, as we will see in the following.

It is well known that an electron has an additional freedom called spin besides the three-dimensional space coordinates. In other words, an electron has an intrinsic angular momentum regardless of the states of motion in space. Hence, an electron carries an observable magnetic moment. In Sect. 1.3, we have shown that the magnitude of the angular momentum of orbital motion with respect to the nucleus is given by $\hbar\sqrt{l(l+1)}(l = 0, 1, 2, ...)$. The spin-angular momentum takes a value $\hbar\sqrt{l(l+1)}$ with $l = 1/2$. The z-component of angular momentum $\hbar m$ can take $(2l + 1)$ values $(m = -l, -l + 1, ..., l)$. Similarly, the z-component of spin-angular momentum takes two values $\hbar\sigma/2(\sigma = \pm 1)$, for $l = 1/2$. Correspondingly, two independent spin-wave functions exist. For example, we can define $\alpha(\sigma)$ and $\beta(\sigma)$ such as

$$\alpha(1) = 1, \alpha(-1) = 0; \quad \beta(1) = 0, \beta(-1) = 1. \quad (1.4.6)$$

Then, $\alpha(\sigma)$ and $\beta(\sigma)$ are understood to be the independent spin-wave functions and to specify the up-spin and down-spin state, respectively.

If the degree of spin freedom is taken into account, two orthogonal functions $\psi(\mathbf{r})\alpha(\sigma)$ and $\psi(\mathbf{r})\beta(\sigma)$ are derived from a single orbital function $\psi(\mathbf{r})$ describing a state of electron motion. These functions are called the spin-orbital functions. Since a given state (n, l) has $(2l + 1)$ independent orbital functions as seen above, $2(2l + 1)$ spin-orbital functions belong to a state (n, l).

The Schrödinger equation (1.3.1) does not contain any explicit spin variables. Therefore, the simple substitution of the spin-orbital functions $\psi_j(\mathbf{r}_j)\alpha(\sigma_j)$ or $\psi_j(\mathbf{r}_j)\beta(\sigma_j)$ for $\psi_j(\mathbf{r}_j)$ in (1.4.1) gives a similar approximation for the total wave function with spin variables. In other words, this corresponds to the approximation that each electron is assigned to a spin-orbital function. The Pauli exclusion principle plays an essential role in this approximation. According to the Pauli principle, a single independent spin orbital can be occupied by at most one electron. A state of a given (n, l) is completely occupied by $2(2l + 1)$ electrons and accommodates no more electrons. Thus, in the ground state of atoms, each (n, l) state has a definite number of electrons between 0 and $2(2l + 1)$. The

(n, l) orbital is called a closed shell of an (n, l) electron, if it is occupied by the maximum number $2(2l + 1)$ of electrons. If the orbital has less electrons than $2(2l + 1)$, it is called incomplete or open shell. The Pauli principle has originally been deduced from the empirical periodic table of elements. As we shall see in the next section, we can now derive the periodic table on the basis of the central-field approximation.

1.5 Shell Structure of Atoms and Periodicity of the Elements

To carry out the central-field approximation described in the preceding section, we must numerically solve the differential equation (1.4.3) for a given potential $U_j(r)$. The task is not so hard, if one uses a high speed computer. However, since each orbital feels a different potential $U_j(r)$ according to the original form of the Hartree approximation, an eigenfunction $\psi_j(r)$ of (1.4.3) is a solution of the j-dependent differential equation, and hence $\psi_j(r)$ may not satisfy the orthonormality condition

$$\int \psi_j{}^*(r) \cdot \psi_k(r) \, d^3r = \delta_{jk}. \tag{1.5.1}$$

Therefore, it is still difficult to discuss steady states other than the ground state (i.e. excited states) of atoms by this approximation. However, if we replace all $U_j(r)$ by a common potential $U(r)$, the orthonormality condition (1.5.1) is preserved. This approximation may be reasonable, because $U_j(r)$ is an effective potential produced by $N–1$ electrons other than the jth electron, and the long-range nature of the Coulomb force reduces the j-dependence of $U_j(r)$ originated by the removal of the jth electron, if $N–1 \gg 1$.

Suppose that $U_0(r)$ denotes an average potential at r exerted by all electrons. If r is much larger than every radii of charge density determined by electron orbitals, $U_0(r)$ for a neutral atom of $Z = N$ vanishes, for the nuclear charge is completely screened by electrons. When an electron moves around, however, it is the remaining $N–1$ electrons that actually screen the nuclear charge. Hence, the potential energy for an electron at r should approach asymptotically $-e^2/r$, which is significantly different from the above case. Consequently, we must eliminate the contribution of one out of N electrons to the potential for any value of r using some method. The following example is familiar as the Fermi-hole method.

As we shall see later, an electron with a definite spin does not attract other electrons of the same spin due to the Pauli exclusion principle. Now, an electron is assumed to carry a sphere of radius R around it from which other electrons are excluded and the radius R is chosen such that

$$4/3 \pi R^3 \rho(r) = 1, \tag{1.5.2}$$

where $\rho(r)$ denotes the total electron density at r. If we add positive charges with the same density $\rho(r)$ to the sphere, then the positive charges just compensate the contribution of the electron in the sphere to the potential energy. The potential energy at the origin produced by a uniform charge in the sphere of radius R is in proportion to the reciprocal of R, hence, from (1.5.2), to $\rho(r)^{1/3}$. The proportionality constant can be evaluated by the Hartree-Fock method and the potential energy acting on an electron is finally shown to be given by

$$U_0(r) - 3e^2 \left[\frac{3}{8\pi} \rho(r) \right]^{1/3} . \tag{1.5.3}$$

The sphere is called the Fermi hole after the fact that the Fermi statistics is deduced from the Pauli principle.

If r is large, however, (1.5.3) does not give a correct screening of nuclear charge mentioned previously, since $\rho(r) \approx 0$ for large values of r. This is simply corrected by

$$U(r) = \text{Min} \left\{ U_0(r) - 3e^2 \left[\frac{3}{8\pi} \rho(r) \right]^{1/3} , \quad -\frac{e^2}{r} \right\}, \tag{1.5.4}$$

where Min denotes the smaller in the braces.

Thus, by following the definition (1.4.2), $U(r)$ can be obtained from the orbital function $\psi_j(r)$ of orbitals occupied by electrons. Therefore, the prescription is summarized as follows. First we assume a proper $U(r)$. Next we find the eigenvalue ϵ and the eigenfunction $\psi(r)$ of the differential equation

$$-\frac{\hbar^2}{2m_0} \Delta\psi(r) + U(r)\psi(r) = \epsilon\psi(r) \tag{1.5.5}$$

to get the first approximation of the orbital function and orbital energy. Then, we distribute N electrons into each orbital from the lowest level in succession, taking account of the spin freedom and the Pauli principle, and we calculate $U(r)$ as a first approximation. Inserting the obtained $U(r)$ into (1.5.5), we proceed with the successive approximation.

Following this prescription, we can find all orbitals and their energy for each atom. In Table 1.1, we list the orbital energy given by *Herman* and *Skillman* [1.1]. We can observe the sequence of the orbital energy (1.4.5) from this table. Figure 1.4 shows the radius of each orbital at which the charge density of the orbital function reaches its maximum. This radius can be regarded as an extent of orbitals, and we will call it the radius of orbital from now on.

Now, we can compare the energies shown in Table 1.1 with the ionization energy of an atom measured by x-ray or electron-beam irradiation. For example, in the x-ray experiment, the absorption spectra are classified into several series such as K, L, M-series. The K-series is expected to be produced by the emission of K-shell electrons which are bound most tightly. The L and M-series are supposed to be due to the emission of L-shell and M-shell electrons, respectively. Already in 1922, Bohr had assigned the K-shell to $1s$ electrons, the L-

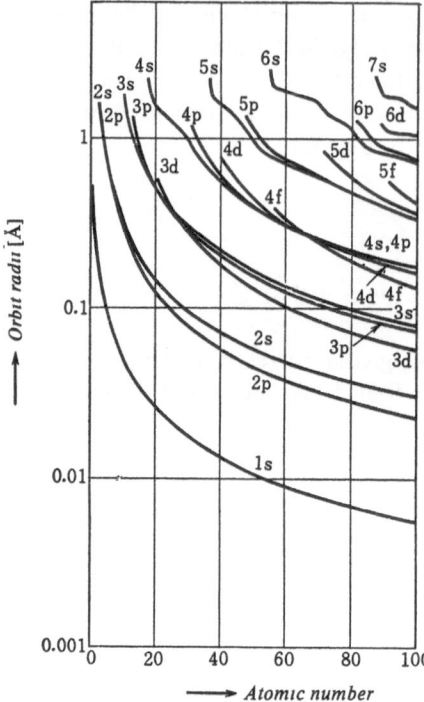

Fig. 1.4. Orbital radii of atoms [1.1]

shell to $2s$ and $2p$ electrons, and the M-shell to $3s$, $3p$ and $3d$ electrons. If the absolute value of orbital energy in Table 1.1 is regarded as the ionization potential of each element, the theoretical value can be compared with experiments as shown in Fig. 1.5. Here, to obtain a direct comparison between theory and experiments, the effect of electron spin on the orbital energy, which we shall discuss later, and the relativistc effect have been subtracted from the observed ionization potential. This comparison shows that the central-field approximation works very well as a first approximation for discussing the stationary states of atoms.

Now, if orbitals carry the same value of (n, l), the radial-wave functions and energy of these orbitals are the same. These orbitals are said to belong to a shell (n, l). In a given shell (n, l), there are $(2l + 1)$ independent orbital functions with a different angular part. The independent angular function can be chosen as $Y_l^m(\theta, \phi)$ $(m = -l, -l+1, ..., l)$. The angular function $Y_l^m(\theta, \phi)$ satisfies Unsöld's theorem

$$\sum_{m=-l}^{l} Y_l^{m*}(\theta, \phi) \, Y_l^m(\theta, \phi) = 2l + 1 \,. \tag{1.5.6}$$

It follows that the charge density of a closed shell does not depend on θ and ϕ,

Table 1.1. Configuration of outer electrons and the orbital energy (in Ry) of various atoms [1.1]

		1s	2s	2p	3s	3p	3d	4s	4p
H	1s	1.0000							
He	1s²	1.7210							
Li	2s	4.3980	0.4039						
Be	2s²	8.6980	0.6012						
B	2s²2p	14.3730	0.9239	0.4898					
C	2s²2p²	21.3780	1.2895	0.6603					
N	2s²2p³	29.7370	1.6959	0.8445					
O	2s²2p⁴	39.4560	2.1440	1.0409					
F	2s²2p⁵	50.5380	2.6349	1.2502					
Ne	2s²2p⁶	62.9900	3.1680	1.4710					
Na	3s	78.0500	4.7230	2.6690	0.3777				
Mg	3s²	94.9500	6.5520	4.1440	0.5051				
Al	3s²3p	113.6600	8.7150	5.9470	0.7245	0.3582			
Si	3s²3p²	134.0400	11.0870	7.9540	0.9975	0.4802			
P	3s²3p³	156.1100	13.6840	10.1800	1.2588	0.6139			
S	3s²3p⁴	179.8900	16.5100	12.6280	1.5300	0.7562			
Cl	3s²3p⁵	205.3600	19.5700	15.3030	1.8124	0.9067			
Ar	3s²3p⁶	232.5400	22.8650	18.2070	2.1068	1.0653			
K	4s	262.0900	27.0600	22.0080	2.9521	1.7328		0.3086	
Ca	4s²	293.5200	31.6270	26.1800	3.8750	2.4823		0.3987	
Sc	3d4s²	326.2900	35.9730	30.1300	4.4308	2.8831	0.5308	0.4309	
Ti	3d²4s²	360.7700	40.5210	34.2790	4.9813	3.2760	0.6279	0.4578	
V	3d³4s	396.9500	45.2890	38.6460	5.5371	3.7040	0.7183	0.4819	
Cr	3d⁵4s	434.4500	49.8230	42.7790	5.7114	3.6910	0.4789	0.4312	
Mn	3d⁵4s²	474.4700	55.5090	48.0500	6.6804	4.4776	0.8859	0.5253	
Fe	3d⁶4s²	515.8100	60.9570	53.0840	7.2691	4.8912	0.9625	0.5451	
Co	3d⁷4s²	558.8900	66.6480	58.3580	7.8785	5.3200	1.0415	0.5647	
Ni	3d⁸4s²	603.7000	72.5700	63.8600	8.5000	5.7560	1.1151	0.5831	
Cu	3d¹⁰4s	649.7000	78.1500	69.0200	8.6340	5.7090	0.7431	0.5091	
Zn	4s²	698.4000	85.1200	75.5500	9.7930	6.6610	1.2582	0.6185	
Ga	4s²4p	749.2000	92.6400	82.6300	11.2380	7.8920	2.0400	0.8377	0.3619
Ge	4s²4p²	801.8000	100.5100	90.0500	12.7620	9.1970	2.8901	1.0571	0.4683
As	4s²4p³	856.3000	108.7300	97.8200	14.3750	10.5870	3.8229	1.2754	0.5826
Se	4s²4p⁴	912.5000	117.3000	105.9500	16.0740	12.0580	4.8340	1.4953	0.7015
Br	4s²4p⁵	970.7000	126.2400	114.4300	17.8570	13.6110	5.9245	1.7185	0.8247
Kr	4s²4p⁶	1030.6000	135.3000	123.2500	19.7280	15.2490	7.0975	1.9456	0.9519

that is, the charge density is spherically symmetric. In other words, the average potential exerted by the electrons in the closed shell depends only on r. For atoms with larger values of N, almost all electrons belong to the closed shells, and this lends support to the central-field approximation.

The ground state of an atom is that in which there are no vacant spin orbitals that have lower energy than any that are occupied. Therefore, as the atomic number increases, the closed shells appear successively one after another in the order of (1.4.5). The configuration of the electrons is described by specifying the

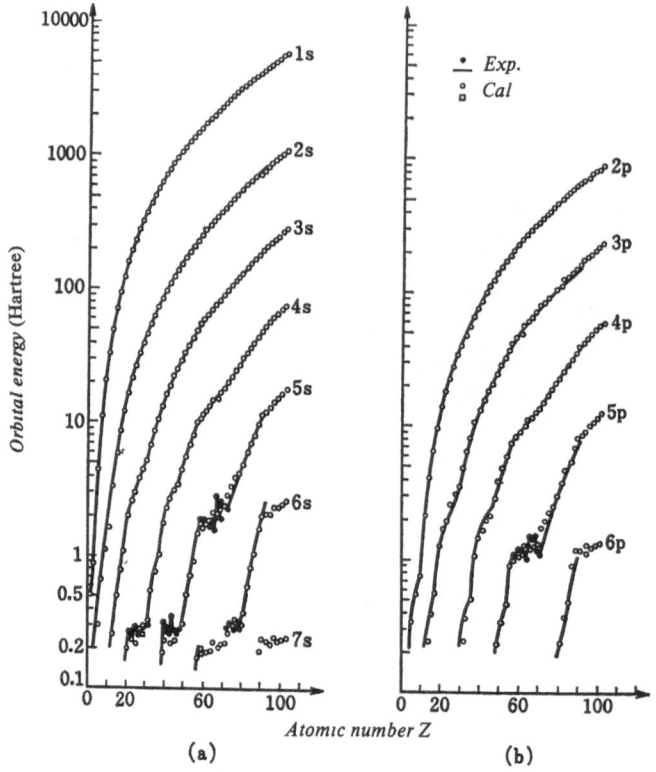

Fig. 1.5. Comparison between the theoretical and experimental values of orbital energy (1 Hartree = 2 Ry)

number of electrons assigned to each spin orbital. For instance, the ground-state electron configurations of Na and of Au are

Na: $1s^2 2s^2 2p^6 3s$ ($Z = 11$),

Au: $1s^2 2s^2 2p^6 3s^2 3p^6 4s^2 3d^{10} 4p^6 5s^2 4d^{10} 5p^6 6s\ 4f^{14} 5d^{10}$ ($Z = 79$).

Among elements whose ground state is composed only of the closed shells, those in which all closed shells left of a semicolon in (1.4.5) are filled and others unfilled are extremely stable, because the energy of the lowest shell of excited states is well separated from the highest closed shell occupied in the ground state. These elements are known as the inert elements. For general elements, the set of stable closed shells is called an inner shell and other shells an outer shell. Thus, the ground state of atoms except for the inert elements consists of an inner shell and some electrons occupying the outer shell. The electron distribution of the inner shell is spherically symmetric according to Unsöld's theorem and, as is shown in Fig. 1.4, outer electrons extend further outward compared with inner electrons. Therefore, outer electrons can be considered to move subject mainly to an effect of the effective nuclear charge that is reduced by the charges of inner electrons (the effective nuclear charge is equal to the total charge of outer electrons, but has the opposite sign). This speculation accounts for observations that those elements which have a common number of s and p electrons in the outer shell show similar chemical properties and that chemically similar elements appear periodically in a certain order. These similar elements are classified into family I, II, ... and VII according to the number of s and p electrons in the outer shell. This number determines the maximum valence of the element.

Table 1.2 gives the ground-state electron configuration of the outer shell which have been determined by spectroscopic experiments for each of the elements. The electron configurations of the inert gases, which have no outer shell, denote the orbital of the highest energy. The inert element is assumed to belong to the 0 family. Then, as the family number increases, outer electrons see a larger effective charge of the nucleus, which has an increasing tendency to absorb excess electrons from outside just to fill up the outer shell, and hence, become a negative ion. In other words, the electronegativity, or the energy released when a neutral atom becomes a negative ion by absorbing electrons from a place infinitely far away, is increased with the family number. At the same time, the ionization potential, or the energy required to take an electron from the neutral atom to infinity, increases with the family number. Consequently, when two atoms belonging to different families are brought together, a charge transfer takes place in general from the atom with lower family number to that with higher family number, and an electric dipole moment is produced. According to L. Pauling, let us define the electronegativity of each element in the following method: First, we define an ionic contribution Δ_{AB} to the binding energy between elements A and B by

$$\Delta_{AB} = D_{A-B} - \tfrac{1}{2}\left(D_{A-A} + D_{B-B}\right),$$

where D_{A-A}, D_{A-B} and D_{B-B} (in eV) are, respectively, the binding energy of the diatomic molecules A-A, A-B and B-B. The electronegativity x_A and x_B ($x_A > x_B$) of elements A and B are defined so as to satisfy

$$x_A - x_B = \sqrt{\Delta_{AB}/23.06}\,.$$

Using experimental values of D_{A-A}, D_{A-B} and D_{B-B}, we assign the electronegativity to each element such that the above relation holds for as many pairs of atoms as possible. Table 1.3 lists the electronegativities thus obtained. The difference $x_A - x_B$ roughly represents the dipole moment of molecule A-B in the Debye unit (10^{-18} cgs). The electronegativity of an atom is roughly in proportion to a mean of the ionization potential and electron affinity.

Table 1.2. Ground-state electron configuration of the elements [1.2]

		s	s^2	p	p^2	p^3	p^4	p^5	p^6	d	d^2	d^3	d^4	d^5	d^6	d^7	d^8	d^9	d^{10}
1s		H 1	He 2																
2s		Li 3	Be 4																
2p				B 5	C 6	N 7	O 8	F 9	Ne 10										
3s		Na 11	Mg 12																
3p				Al 13	Si 14	P 15	S 16	Cl 17	Ar 18										
4s, 3d	$4s^0$																		
	$4s$	K 19	Ca 20											Cr 24					Cu 29
	$4s^2$									Sc 21	Ti 22	V 23		Mn 25	Fe 26	Co 27	Ni 28		Zn 30
4p				Ga 31	Ge 32	As 33	Se 34	Br 35	Kr 36										
5s, 4d	$5s^0$																		Pd 46
	$5s$	Rb 37	Sr 38										Nb 41	Mo 42	Tc 43	Ru 44	Rh 45		Ag 47
	$5s^2$									Y 39	Zr 40								Cd 48
5p				In 49	Sn 50	Sb 51	Te 52	I 53	Xe 54										
6s, 4f, 5d	$6s^0$																Ir 77		
	$6s$	Cs 55	Ba 56															Pt 78	Au 79
	$6s^2$									La* 57	Hf 72	Ta 73	W 74	Re 75	Os 76				Hg 80
6p				Tl 81	Pb 82	Bi 83	Po 84	At 85	Rn 86										
7s, 5f, 6d	$7s^0$																		
	$7s$	Fr 87	Ra 88																
	$7s^2$									Ac 89	Th† 90								

		f	f^2	f^3	f^4	f^5	f^6	f^7	f^8	f^9	f^{10}	f^{11}	f^{12}	f^{13}	f^{14}
* 4f	$5d^0$	Ce (58)	Pr (59)	Nd 60	Pm (61)	Sm 62	Eu 63		Tb (65)	Dy (66)	Ho (67)	Er (68)	Tm 69	Yb 70	
	$5d$							Gd 64							Lu 71

		f	f^2	f^3	f^4	f^5	f^6	f^7	f^8	f^9	f^{10}	f^{11}	f^{12}	f^{13}	f^{14}
† 5f	$6d^0$				Np 93	Pu 94	Am 95		Bk 97	Cf 98					
	$6d$		Pa 91	U 92				Cm 96							

Table 1.3. Electro-negativity of the elements

H						
2.1						

Li	Be	B	C	N	O	F
1.0	1.5	2.0	2.5	3.0	3.5	4.0

Na	Mg	Al	Si	P	S	Cl
0.8	1.2	1.5	1.8	2.1	2.5	3.0

K	Ca	Sc	Ti	Ge	As	Se	Br
0.8	1.0	1.3	1.5	1.8	2.0	2.4	2.8

Rb	Sr	Y	Zr	Sn	Sb	Te	I
0.8	1.0	1.2	1.4	.18	1.9	2.1	2.5

Cs	Ba
0.7	0.9

It will be noted that the so-called transition elements and rare-earth elements form groups, which, respectively, have an open d shell and f shell in the ground state. As we can see in (1.4.5), the orbital energy of these d or f orbitals are almost the same as the s-orbital energy of a larger principal quantum number by 1 or 2, and the central-field approximation is not sufficient to determine a correct sequence of the electron shells. For these atoms, we must evaluate the total energy including the effect of spin, in addition to the orbital energy. At present, however, a complete prediction of the electron configurations of these atoms has not been accomplished even for their ground state.

In the above discussion, we defined the orbital radius. Similarly, we may define the atomic radius by the maximum radius of orbitals occupied by electrons in the ground state. For, in most crystals, the distance of adjacent atoms is roughly equal to the sum of their atomic radii. In fact, Fig. 1.6 given by

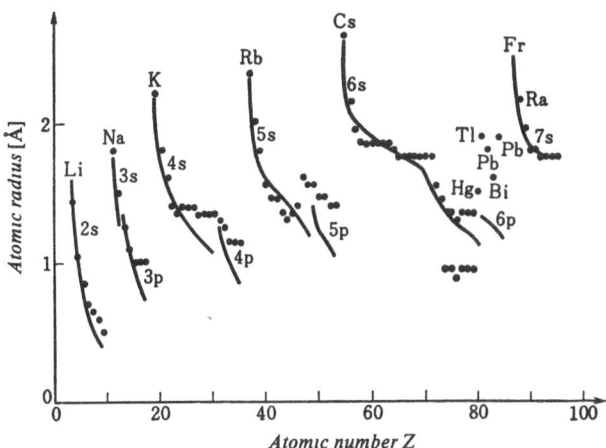

Fig. 1.6. Atomic radii. Comparison between theory and experiments [1.3]

Slater [1.3] shows a comparison between the maximum radius of electron orbitals and the atomc radius defined so as to give the proper atomic distance in crystals. The agreement between these two quantities is fairly good.

Generally speaking, the orbital radius of the closed shell is smaller than that of the open shell. Therefore, the atomic radius of atoms with an open shell is determined by the open-shell orbital. However, the atoms that have *d* or *f*-open shells are again exceptions. Namely, the ground state of these atoms except for Ir have open or closed *s* shells of a higher principal quantum number, and the radius of *d* and *f* shells are much smaller than that of the *s* orbital. Therefore, the *d* and *f* electrons in crystals move around the deep inside of the atom, compared with the distance between adjacent atoms, and are less affected by other atoms.

In addition to this property, *d* and *f* shells are composed of 10 and 14 independent spin orbitals, respectively. As far as the sum of orbital energies is concerned, all these orbitals are degenerated. By virtue of these facts, the transition elements and the rare-earth elements form important groups of magnetic elements which are closely related to each other, as we shall see in later chapters.

Meanwhile, elements Zn, Cd and Hg in which the *d* shell is just filled show somewhat similar chemical properties though they are not as chemically stable as the inert elements. Elements Cu, Ag and Au have one *s* electron outside the closed *d* shell and behave similarly to noble metals. This fact can be also understood if one notices that the radius of a *d* orbital is very small and *d* electrons greatly screen the nuclear charge for the motion of the outer electrons.

1.6 Stationary States of Atoms I

In Sect. 1.5, it has been shown that the central-field approximation describes the stationary states of atoms extraordinarily well in a simple manner. Within the same approximation, we can obtain the electron configurations whose energy are higher than that of the ground state energy. These states correspond to excited states because of the orthonormality condition (1.5.1). The atomic spectra can then be derived by considering a transition of electrons between the ground and excited states.

Contrary to the hydrogen and monovalent metals such as the alkali metals, copper, silver and gold, however, the spectra of atoms with more than two electrons in its open shell is quite complicated and cannot be explained within the approximation introduced in the previous section. In the central-field approximation, a spin orbital has been assigned to each electron. The argument based on the orbital energy is not sufficient precisely to specify the electron configurations in an open shell, for all orbital energies are equal in a single open shell. In general, eigenfunctions of the total Hamiltonian are approximately expressed by a proper linear combination of wave functions which correspond to each of the electron configurations of the open shell. If an open shell is occupied by more than two electrons, the expectation value of the total energy is supposed to be different from linear combination to linear combination because of the electron-electron interaction. Therefore, we must consider the subsequent problems: which linear combination is the best and how the energy levels are classified.

To investigate these problems, the Pauli principle given in Sect. 1.4 must be reformulated in a more general form for many electron systems. Using a symmetry analysis of the total Hamiltonian together with this reformulated Pauli principle, we can classify the steady states of atoms and hence understand the complicated structure of the atomic spectra.

As have been discussed in Sect. 1.4, an electron is specified by four coordinates, i.e., three position variables $r = (x, y, z)$ and a spin variable σ, which are denoted by a single symbol q for simplicity. Then, the total wave function of N electrons is expressed as $\Psi(q_1, q_2, ..., q_N)$. The Pauli principle is stated as: "The wave function for a many-electron system must be an antisymmetric function of $q_1, q_2, ..., q_N$". In other words, Ψ changes its sign for the interchange of two arbitrary arguments q_1 and q_k:

$$\Psi(q_1, ..., q_k, ..., q_l, ..., q_N) = - \Psi(q_1, ..., q_l, ..., q_k, ..., q_N) . \tag{1.6.1}$$

Therefore, when a permutation P is applied to $q_1, q_2, ... q_N$, Ψ does not change for even permutations and changes to $-\Psi$ for odd permutations:

$$P\Psi = \eta_P \Psi \tag{1.6.2}$$

$$\eta_P = \begin{cases} +1 \text{ for even permutations,} \\ -1 \text{ for odd permutations.} \end{cases} \tag{1.6.3}$$

The Pauli principle in this formulation can be shown to imply that in the form described in Sect. 1.4. Actually, in Sect. 1.4, N spin orbital functions $\phi_1, \phi_2, ..., \phi_N$ were introduced. Each of the coordinates of N electrons were attributed to a different function, and

$$\phi_1(q_1) \phi_2(q_2) ... \phi_N(q_N) \tag{1.6.4}$$

has been assumed to approximately represent the stationary state of the N-electron atom. But, since electrons can not be distinguished at all, wave functions which are derived by arbitrary permutations of $q_1, q_2, ..., q_N$ such as $\phi_1(q_2) \phi_2(q_1) ... \phi_N(q_N)$ are also expected to describe the same state. There are $N!$ kinds of these wave functions, corresponding to all possible permutations. Consequently, in order to construct a function satisfying (1.6.2) within the present approximation, we may take a linear combination of these $N!$ orbitals to get an antisymmetric function. As is well known, the only possible way to make the function antisymmetric is to write a determinant

$$\Phi = C \begin{vmatrix} \phi_1(q_1) \; \phi_2(q_1) \; \cdots \; \phi_N(q_1) \\ \phi_1(q_2) \; \phi_2(q_2) \; \cdots \; \phi_N(q_2) \\ \cdots\cdots\cdots\cdots\cdots \\ \phi_1(q_N)\phi_2(q_N) \; \cdots \; \phi_N(q_N) \end{vmatrix}, \tag{1.6.5}$$

where C is a constant. This determinant is known as the Slater determinant. The function Φ vanishes unless $\phi_1, \phi_2, ..., \phi_N$ are linearly independent, and thus Φ is meaningful only if $\phi_1, \phi_2, ..., \phi_N$ are linearly independent. In particular, if two of the ϕ's are identical, then Φ is identically zero. Therefore, not more than one electron can occupy the same spin orbital. The Slater determinent is uniquely determined by a set of spin orbital functions and hence it will be abbreviated by $\Phi [\phi_1, \phi_2, ..., \phi_N]$ in the following discussion.

It should be noted here that the state $\Phi [\phi_1, \phi_2, ..., \phi_N]$ is qualitatively different from those described by $\phi_1(q_1) \phi_2(q_2) ... \phi_N(q_N)$ and its $N!$ alterations derived by permutations. For example, in (1.6.5), Φ is identically zero if any pair of q are equal, while this is not the case for the state described by $\phi_1(q_1) \phi_2(q_2) ... \phi_N(q_N)$. Therefore, the Pauli principle formulated above implies not only the exclusion principle described in Sect. 1.4, but also restricts the properties of allowed stationary states.

Now, each shell (n, l) contains $2(2l + 1)$ spin orbitals. When $n[\leqslant 2(2l + 1)]$ electrons belong to this shell, $\binom{2(2l + 1)}{n}$ independent Slater determinants exist. Hence, only one stationary state is allowed for a closed shell. On the other hand, an open shell $[n < 2(2l + 1)]$ can take several stationary states depending on which orbitals are occupied by electrons. These states have an equal orbital energy, but the total energy of each state is generally different due to the electro-static repulsion between electrons. In addition, a better approximation to the stationary states of atoms will be realized by taking a linear combination of these allowed states.

It is useful then to note that "the eigenvalue and eigenfunction of the total energy are classified by a set of constants of the motion of the system". Here, the set of constants of the motion means the set of eigenvalues $(C_1', C_2', ... C_s')$ of the mutually commutative dynamical operators $(C_1, C_2, ..., C_s)$ including the Hamiltonian. Since $C_1, C_2, ..., C_s$ commute mutually, simultaneous eigenfunctions of $C_1, C_2, ..., C_s$ exist which are designated by eigenvalues $(C_1', C_2', ..., C_s')$, and the eigenfunctions can be chosen as an eigenstate of the Hamiltonian.

Now, it is clear from the starting equation (1.3.1) that the corresponding Hamiltonian (i) is invariant under a rotation around any axis passing through the origin, or spherically symmetric and (ii) does not contain the spin variables explicitly.

It follows from (i) that the total orbital angular momentum with respect to the origin

$$L = l_1 + l_2 + ... + l_N \tag{1.6.6}$$

commutes with the Hamiltonian. Here, l_j $(j = 1, 2, ..., N)$ is the orbital angular momentum operator of the jth electron. However, three cartesian components L_x, L_y, L_z of L do not mutually commute. The z component L_z and the square of the total angular momentum

$$L^2 = L_x^2 + L_y^2 + L_z^2 \tag{1.6.7}$$

are chosen as the commutative operators (dynamical variables).

The property (ii) tells us that the spin angular momentum s_j of the jth electron $(j = 1, 2, ..., N)$ commutes with the Hamiltonian and L, and hence they are also dynamic variables which are constants of the motion. Noting that the Pauli principle implies (1.6.2, 3), an arbitrary permutation P of electrons is also a dynamical variable with a constant of the motion η_P. Therefore, the dynamical variables which distinguish the electronic states must also commute with P. Since s_j itself does not commute with an arbitrary P, we must take the total spin angular momentum

$$S = s_1 + s_2 + \cdots + s_N \tag{1.6.8}$$

to be such a dynamical variable. Though three components S_x, S_y, S_z of S as well as the components of L do not commute mutually, we can choose S_z and

$$S^2 = S_x^2 + S_y^2 + S_z^2 \tag{1.6.9}$$

as the mutually commutative operators. Thus, the stationary states of atoms are classified by a set of eigenvalues of dynamical variables L^2, L_z, S^2 and S_z.

The eigenvalues of L^2 and S^2 are given by

$$L^2 = \hbar^2 L(L + 1) \quad (L = 0, 1, 2, 3, ...) \tag{1.6.10}$$

$$S^2 = \hbar^2 S(S + 1) \quad (S = 0, 1/2, 1, 3/2, ...) \tag{1.6.11}$$

and, for a given value of L, L_z can take $(2L + 1)$ eigenvalues

$$L_z = \hbar M_L \quad (M_L = -L, -L + 1, ..., L) \tag{1.6.12}$$

and, for a given value of S, S_z can take $(2S + 1)$ eigenvalues

$$S_z = \hbar M_S \quad (M_S = -S, -S + 1, ..., S). \tag{1.6.13}$$

The variables M_L and M_S are called the total orbital-magnetic quantum number and the total spin-magnetic quantum number, respectively. In the Schrödinger equation (1.3.1), the z-axis of the coordinate system can be chosen in an arbitrary direction, and hence the states with common L and S have equal energy.

The stationary states corresponding to $L = 0, 1, 2, 3, 4, 5, 6, 7, 8$ are designated, respectively, by S, P, D, F, G, H, I, K, L. For a given value of S, $(2S + 1)$, the number of possible eigenstates of S_z is called the multiplicity, the corresponding eigenstates are called the $(2S + 1)$-plet state or the $(2S + 1)$-plet term. The multiplicity is denoted by putting the number on the left shoulder of the term symbol. For example, $L = 1$, $S = 0$ is a singlet state denoted by ^1P and $L = 2$, $S = 1$ is a triplet state denoted by ^3D.

As we have seen before, only one Slater determinant is possible for the closed shell. Since L_z and S_z are the sum of l_z and s_z, respectively, of each electron, $L_z = 0$, $S_z = 0$ holds for the closed shell. This implies $L = S = 0$, for there is only one independent state. Therefore, the stationary state of the closed shell is ^1S and, so far as the stationary states of atoms are concerned, the stationary state does not contribute to the orbital and spin angular momentum.

As an illustrative example of an open shell, let us consider a system composed of two np ($n = 2, 3, ...$) electrons. The spin orbital functions allowed in an np-shell are identified by $m+$ and $m-$, where $m(= -1, 0, 1)$ is the magnetic quantum number of the orbital and $\pm$ denotes the up and down-spin states. Since two electrons are distributed into six independent spin orbital functions in this example, there are $\binom{6}{2} = 15$ independent Slater determinants whose M_L and M_S are tabulated in Table 1.4. Figure 1.7 shows the number of the Slater determinants as a function of M_L and M_S. From this figure, we can see that a state of $M_L = 2$ and $M_S = 0$ exists, but neither of $M_L = 3$, $M_S = 0$ nor of $M_L = 2$, $M_S = \pm 1$. This means that there must be a term $^1D(L = 2, S = 0)$. Similarly, a term $^3P(L = 1, S = 1)$ must exist, since a state of $M_L = 1$ and $M_S = 1$ exists, but neither of $M_L = 2$, $M_S = 1$ nor of $M_L = 1$, $M_S = 2$. As is shown in Fig. 1.8, ^1D has five independent states distinguished by M_L and ^3P nine (3×3) independent states. Subtraction of these states from Fig. 1.7 leads to a remaining term $M_L = M_S = 0$ which belongs to $^1S(L = 0, S = 0)$. Eventually, the stationary states of two np electron systems are classified into three terms $^3P + {}^1D + {}^1S$.

Table 1.4.

Slater determinant	M_L	M_S
$\Phi(1+, 1-)$	2	0
$\Phi(1+, 0+)$	1	1
$\Phi(1+, 0-)$	1	0
$\Phi(1+, -1+)$	0	1
$\Phi(1+, -1-)$	0	0
$\Phi(1-, 0+)$	1	0
$\Phi(1-, 0-)$	1	-1
$\Phi(1-, -1+)$	0	0
$\Phi(1-, -1-)$	0	-1
$\Phi(0+, 0-)$	0	0
$\Phi(0+, -1+)$	-1	1
$\Phi(0+, -1-)$	-1	0
$\Phi(0-, -1+)$	-1	0
$\Phi(0-, -1-)$	-1	-1
$\Phi(-1+, -1-)$	-2	0

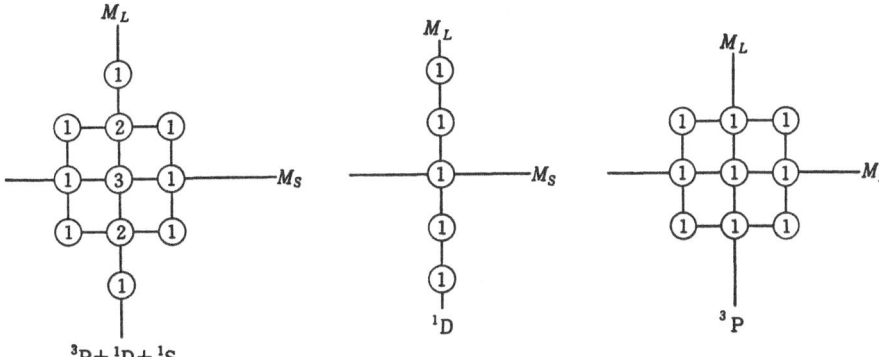

Fig. 1.7.

Fig. 1.8.

If each of two p electrons belongs to different shells, six independent spin orbital functions exist in each shell, and an assignment of each electron to these orbitals yields $36(= 6 \times 6)$ independent Slater determinants. Using the same argument as before, all of the states are classified by M_L and M_S as shown in Fig. 1.9a. It is easy to see that the stationary state includes a ^3D term. Subtracting this term from Fig. 1.9a produces Fig. 1.9b, from which we can deduce the existence of terms ^1D and ^3P. From Fig. 1.9c where terms ^1D and ^3P have been subtracted from Fig. 1.9b, we notice the existence of three terms ^1P, ^3S and ^1S.

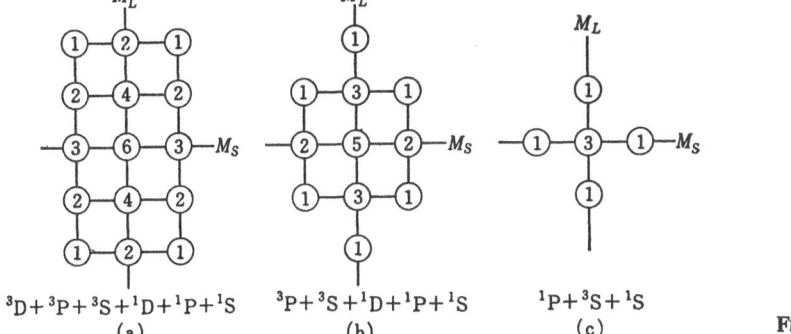

Fig. 1.9a-c.

In general, when each of two electrons belongs to different shells as in the latter example, the Pauli principle does not restrict the allowed state contrary to the former example, and hence the above result can be easily derived by a separate addition of the orbital and spin angular momenta.

Now, it is of interest to see which energy level among others minimizes the expectation value of the total energy of the atom. First, it can be proved that every levels have an equal expectation value for the sum of the total kinetic energy of electrons and potential energy due to the nucleus, since all of these energy levels have a common electronic configuration and all orbital functions are mutually orthogonal. Therefore, the difference in energy of each level is totally attributed to the different electro-static potential energy of electron-electron repulsion. Now, recall that the Slater determinant Φ (1.6.5) is zero, if $q_i = q_j$. This implies that if the ith and jth electrons have a common spin direction, the probability of both electrons approaching each other will be small. Therefore, it is expected that the larger the number of spins having a common direction, namely the larger S, the lower the energy will be. The fact that the energy of the term of maximum S is the minimum was empirically found by F. Hund through an analysis of atomic spectrum and is known as the First Hund Law. Hund also found empirically that the term of maximum L among terms with maximum S has the minimum energy; this is called the Second Hund Law.

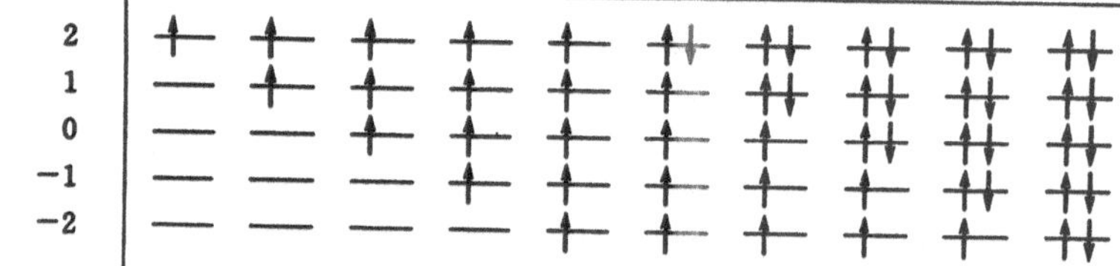

m_s \\ n	1	2	3	4	5	6	7	8	9	10
2	↑	↑	↑	↑	↑	↑↓	↑↓	↑↓	↑↓	↑↓
1	—	↑	↑	↑	↑	↑	↑↓	↑↓	↑↓	↑↓
0	—	—	↑	↑	↑	↑	↑	↑↓	↑↓	↑↓
−1	—	—	—	↑	↑	↑	↑	↑	↑↓	↑↓
−2	—	—	—	—	↑	↑	↑	↑	↑	↑↓
S	1/2	1	3/2	2	5/2	2	3/2	1	1/2	0
L	2	3	3	2	0	2	3	3	2	0
	^2D	^3F	^4F	^5D	^6S	^5D	^4F	^3F	^2D	^1S
Ion	Ti^{3+}	V^{3+}	Cr^{3+}	Cr^{2+}	Mn^{2+} Fe^{3+}	Fe^{2+}	Co^{2+}	Ni^{2+}	Cu^{2+}	Cu^+
A [cm^{-1}]	154	104	87	57		-100	-180	-335	-850	

Fig. 1.10. Electron configuration of $(3d)^n$

Thus, we can determine the state of the minimum energy. For example, Fig. 1.10 shows the spin configurations for the lowest energy state of $(3d)^n$ ($n \leqslant 10$). The parameter A denotes the magnitude of the spin-orbit interaction which we will discuss in the next section. Note that the spin configuration in Fig. 1.10 shows only one of the $(2L + 1)(2S + 1)$ independent states for a given set of L and S.

1.7 Stationary States of Atoms II

In Sect. 1.6, the stationary states of atoms were discussed on the assumption that the electronic energy is the sum of the kinetic and electro-static potential energies. However, as we have seen in (1.2.9), the orbital motion of an electron is transformed into a nuclear motion in the coordinate system fixed on the electron, and the magnetic field at the electron position produced by the nuclear motion interacts with the magnetic moment associated with the electron spin. The interaction energy is derived automatically from the Dirac equation which governs the behaviors of relativistic electrons and has the form

$$\sum_k \xi(r_k)\, l_k \cdot s_k \,, \qquad (1.7.1)$$

where r_k is the position vector of the kth electron and l_k and s_k are its orbital and spin angular momentum operators, respectively, For electrons in the central-field potential $U(r)$, the coupling constant $\xi(r)$ is given by

$$\xi(r) = \xi(r) = \frac{1}{2m^2c^2}\frac{1}{r}\frac{dU}{dr} \,. \qquad (1.7.2)$$

The interaction (1.7.1) is called the spin-orbit coupling. If this interaction is included, L and S are no longer constants of the motion. However, the total angular momentum

$$J = L + S = \sum_k (l_k + s_k) \qquad (1.7.3)$$

still commutes with the interaction energy (1.7.1), and hence with the total Hamiltonian of the system. Thus J is a constant of the motion. Therefore the energy levels are labeled by the quantum number J specifying the magnitude of the total angular momentum and the quantum number M characterizing the z component of J. Because of the interaction (1.7.1), these levels are supposed to depend on J. The quantum numbers J and M satisfy the following relations:

$$J^2 = \hbar^2 J(J+1) \quad (J = 0, 1, 2, \ldots), \qquad (1.7.4)$$

$$J_z = \hbar M \quad (M = -J, -J+1, \ldots, J). \qquad (1.7.5)$$

Except for atoms with larger atomic number, the magnitude of the spin-orbit coupling is less than the energy separation between terms with different sets of (L, S). Therefore, the labeling of energy levels according to (L, S) is still valid as in Sect. 1.6 and the wave functions characterized by (L, S) describe the stationary states well. This weak interaction is known as the Russell-Saunders coupling. Thus, in the Russell-Saunders coupling scheme, $(2L + 1)(2S + 1)$ terms labeled by (L, S) are classified further according to J. Since the z-axis is chosen in an arbitrary direction, $(2J + 1)$ levels which have the same J and different M are degenerate. From the addition rule of angular momenta, the composition of two angular momenta L and S characterized by L and S yields

$$J = L + S, L + S - 1, \ldots, |L - S| \,, \qquad (1.7.6)$$

and clearly the total number of levels is

$$(2L + 1)(2S + 1) = \sum_{J=|L-S|}^{L+S} (2J + 1), \tag{1.7.7}$$

as expected.

Now, for the $(2L + 1)(2S + 1)$ wave functions that belong to an (L, S) level, the interaction energy (1.7.1) can be written as

$$A\boldsymbol{L} \cdot \boldsymbol{S}, \tag{1.7.8}$$

where A is a constant determined by the electron configuration and L and S. Although this formula will be obtained rigorously by making use of the Wigner-Eckert theorem, we can understand it qualitatively as follows. If we regard $\boldsymbol{l}_k$ and $\boldsymbol{s}_k$ in (1.7.1) as vectors which precess about the resultant vectors $\boldsymbol{L}$ and $\boldsymbol{S}$, respectively, a time average of $\boldsymbol{l}_k \cdot \boldsymbol{s}_k$ eliminates all components of $\boldsymbol{l}_k$ and $\boldsymbol{s}_k$ except those along $\boldsymbol{L}$ and $\boldsymbol{S}$.

Using (1.7.3), (1.7.8) can be written as

$$A\boldsymbol{L} \cdot \boldsymbol{S} = \frac{A}{2} (\boldsymbol{J}^2 - \boldsymbol{L}^2 - \boldsymbol{S}^2)$$

$$= \frac{A}{2} \{J(J + 1) - L(L + 1) + S(S + 1)\}. \tag{1.7.9}$$

Therefore, each (L, S) level is split into several levels according to the total angular momentum J. For the example of an open shell with two p electrons, we have shown in Sect. 1.6 that terms ^3D and ^3P exist. Each of these terms is split into $J = 3, 2, 1$ and $J = 2, 1, 0$. These split levels are denoted by attaching the value of J to the term symbol such that $^3D_3, {}^3D_2, {}^3D_1; {}^3P_2, {}^3P_1, {}^3P_0$.

Thus, if $L \geqslant S$, J can take $2S + 1$ different values given by (1.7.6) and each (L, S) term is split into $(2S + 1)$ levels which lie closely to each other, hence $(2S + 1)$ is called the multiplicity. If $L < S$, the actual number of split levels is less than $(2S + 1)$. The number $(2S + 1)$ is called the multiplicity for this case as well. As an example of stationary states of atoms, Fig. 1.11 shows the energy levels of $3d\,4p$ electrons of Ca.

The coupling constant A increases sharply with the atomic number Z; for $3p$ orbital, A is about a several cm^{-1}, while for $4f$ and $5d$ orbitals, it goes up to several thousand cm^{-1} and exceeds the energy separation between two (L, S) levels (normally, this separation is about $10^3\ \mathrm{cm}^{-1}$). Therefore, in contrast with the Russell-Saunders coupling scheme, we cannot treat (1.7.1) as a perturbation in the subspace labeled by (L, S). In this case, it is a relevent approximation to consider the coupling energy (1.7.1) prior to the electron-electron interaction. The states of each electron is assumed to be described by the quantum numbers j and m, specifying the resultant of spin and orbital angular momenta of the electron,

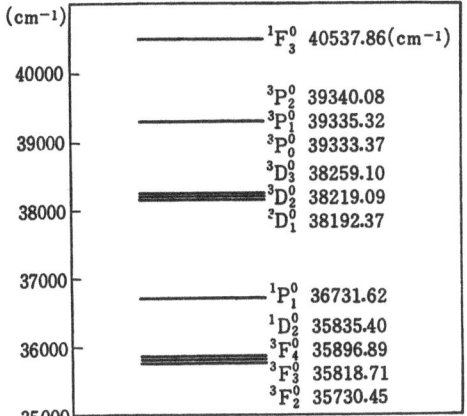

Fig. 1.11. Energy levels of $3d\,4p$ electron configurations of Ca

besides the principal quantum number n and the auxiliary quantum number l, and then energy levels are classified according to the total angular momentum of the atom. This case is known as the j–j coupling scheme.

As a final example, we consider a system under a weak external magnetic field H_z in the z direction. The Dirac equation implies that the interaction energy between an electron and the external field is written as

$$\mu_{\mathrm{B}}(L_z + 2S_z)\, H_z . \tag{1.7.10}$$

If the external magnetic field is sufficiently weak, we can regard (1.7.10) as a perturbation. A correction to each unperturbed level due to (1.7.10) is evaluated by taking an expectation value of (1.7.10) in the unperturbed states.

When the Russel-Saunders coupling scheme is relevent, the expectation values of L_z and S_z are easily calculated by assuming that the vectors L and S are precessing about the resultant vector J:

$$\langle L_z \rangle = \frac{J_z(J\cdot L)}{J^2}, \quad \langle S_z \rangle = \frac{J_z(J\cdot S)}{J^2}. \tag{1.7.11}$$

If we define a constant g by

$$\langle \mu_{\mathrm{B}}(L_z + 2S_z)\, H_z \rangle = g\mu_{\mathrm{B}}MH_z , \tag{1.7.12}$$

we have from (1.7.11)

$$g = \frac{(L + 2S)\cdot J}{J^2} = \frac{(J + S)\cdot J}{J^2}$$

$$= 1 + \frac{S\cdot J}{J^2} = 1 + \frac{J^2 + S^2 - (J - S)^2}{2J^2}$$

$$= 1 + \frac{J(J+1) + S(S+1) - L(L+1)}{2J(J+1)}. \tag{1.7.13}$$

The parameter g is called the Landé g factor. The relations (1.7.12, 13) derived from the vector model are also verified by means of the Wigner-Eckert theorem. Thus, $(2J+1)$ levels which are degenerated under no external field are split into levels which are separated from each other by an equidistance under a weak external magnetic field. This splitting is known as the Zeeman effect.

In this chapter, we have briefly discussed the atoms which are the fundamental constituent units of materials. In the succeeding chapters, we shall describe the various phases which the materials show as an agglomerate of these units.

2. System of Protons and Electrons

With the background material of Chap. 1, we begin to examine the possible physical states of matter having macroscopic size. In Chap. 2 a system composed of protons and electrons is considered as the simplest case. This is a system of hydrogen atoms with atomic number $Z = 1$. Although theoretically it is not more than a typical many-particle system interacting through a Coulomb force, a close examination reveals that this simple system will exhibit magnificently diverse states, giving a nice introduction to further studies into more complex systems.

2.1 General Comments

To begin with, let us make some general but very qualitative remarks concerning a system made up of nuclei and electrons, interacting with each other by Coulomb force. As we have noted in Chap. 1, nuclei can be regarded as a given entity, their internal structure being of little importance to our present purpose. We will also neglect the relativistic effects for simplicity. The only interaction we have to consider then is the static Coulomb force. Let us consider a system consisting of electrons and nuclei with charge Ze and mass M. Since we are usually interested in an electrically neutral system, we suppose there are N nuclei and ZN electrons. The Hamiltonian of our system is

$$
\mathcal{H} = \sum_i \frac{p_i^2}{2m} + \frac{1}{2} \sum_{i \neq j} \frac{e^2}{|r_i - r_j|} - \sum_{i,l} \frac{Ze^2}{|r_i - R_l|}
$$
$$
+ \sum_l \frac{P_l^2}{2M} + \frac{1}{2} \sum_{l \neq n} \frac{Z^2 e^2}{|R_l - R_n|},
$$

(2.1.1)

where r_i and R_l are the coordinates of the ith electron and the lth nucleus and p_i, P_l their momenta, respectively. When the system is made up of several kinds of atoms the above expression has to be generalized only slightly. Thus, depending on the choice of the parameters Z and M in it, we can describe a wide variety of matter such as metals, insulators and organic materials. This is indeed the gist of atomism.

First we consider the lowest energy state of the system which should be realized at a temperature 0 K and under 0 pressure. The reason why we begin our discussion with it will become clear in the following. Our system is governed by quantum mechanics. It was already clear in the case of the hydrogen atom that the classical theory encountered grave difficulties dealing with any system described by (2.1.1). When an electron is localized in a region of radius r around a proton in a hydrogen atom, its kinetic energy is $\hbar^2/2mr^2$ and the Coulomb energy $-e^2/r$. Hence, the total energy as a function of r has the minimum value $-me^4/2\hbar^2 = -\mathrm{Ry}$ (Rydberg constant $= 13.6\,\mathrm{eV}$)[1] at r equal to the Bohr radius $a = \hbar^2/me^2$ ($= 0.529\,\text{Å}$). What keeps atoms from collapsing is the quantum mechanical zero point motions of electrons. If the fundamental force were an attractive force behaving like r^3 in the limit of $r \to 0$, we would have the instability even in quantum mechanics. That such a force does not exist and that the Coulomb force is proportional to r^{-2} can be explained from more basic reasons, which are, however, not of concern here.

In the following we will consider a system consisting of a very large number of particles N. In studying such a system we have a new kind of problem not encountered in the theory of atoms and molecules. Since in a large system one particle interacts with many other particles simultaneously, we must see how the energy and other properties of the system vary as the total number of particles N is increased. Ordinarily, physical properties of a part of a homogeneous body do not depend on the size of the entire body provided it is sufficiently large. More precisely, when we express, for example, the total energy and the volume occupied by the system as

$$E = N\varepsilon, \quad V = Nv, \tag{2.1.2}$$

ε and v become independent of N in the limit of large N. It is not true that any system possesses this property. As an example of such a system that does not, we replace in the Hamiltonian (2.1.1) the Coulomb interaction by the Newtonian gravitational force. If all particles are localized in a region of radius R due to the mutual attraction, the potential energy of the system will be of the order of $-N^2(GM^2/R)$. We have neglected the contribution from electrons since the mass of electrons is small. For the same reason we consider only electrons in estimating the total kinetic energy. Since an electron is a fermion and obeys the Pauli principle, the volume occupied by an electron is of the order of R^2/N. When each electron is localized in such volume, the total kinetic energy is roughly equal to $\hbar^2 N^{5/3}/mR^2$. Hence the total energy of the system E is estimated to be

$$\frac{\hbar^2 N^{5/3}}{mR^2} - \frac{GM^2 N^2}{R}$$

and minimizing this we find

---

[1] Usually the Rydberg constant is a unit of wave number of a photon, namely, $R_\infty = me^4/4\pi c\hbar^3 = 1.097 \times 10^5\,\mathrm{cm}^{-1}$. Strictly speaking, R_∞ is the value when the mass of a proton is assumed infinite and is different from the value for a hydrogen atom $R_H = R_\infty M/(M + m)$.

$$R \approx \frac{\hbar^2}{GmM^2}\frac{1}{N^{1/3}}, \quad E \approx -\frac{G^2mM^4}{\hbar^2}N^{7/3}. \tag{2.1.3}$$

Consequently, the spatial extension of the system diminishes as the number of particles N is increased. Though the gravitational attraction is an extremely weak interaction, it acts between all particles because of its long range and the absence of screening.[2]

Our system consisting of nuclei and electrons with the Coulomb interaction does possess the property expressed by (2.1.2). Needless to say, it differs from the gravitational system in that positive and negative charges exist and the system is neutral as a whole. To lower the Coulomb energy, charge neutrality must also be kept locally as much as possible so that each nucleus is likely to be surrounded by Z electrons. Therefore, we expect interaction energies of any one particle with many other particles distant from it to add up to zero, the repulsive and attractive forces cancelling each other. In other words, one particle can effectively interact only with a finite number of particles within a finite distance from it. This is the fundamental property of the Coulomb interaction called screening. The screening distance, though different in individual cases, is of the order of the mean distance between nuclei. If a particle can effectively interact only with a finite number of particles, the potential energy of the total system should become proportional to the particle number N in the limit of large N. For this to be true, however, it is also necessary that the screening distance, and hence the mean distance between particles, be independent of N. As remarked before, there are enough electrons around each nucleus to keep the neutrality. To keep the energy low, the electronic state in the neighbourhood of individual nuclei should be almost the same as the ground state of a corresponding atom. Furthermore, because of the Pauli principle it is unlikely that these atoms overlap with each other appreciably. Thus, the ground state of our system should not be too different from an aggregate of the atoms. In fact, we know that the basic unit or building block of macroscopic bodies is atoms and molecules, and that the latters' properties are essential for an the understanding of the former. The appropriate unit in measuring the atomic distance in solids, for example, is also the Bohr radius and the unit of energy the Rydberg. If the mean distance between atoms is determined by the atomic size, it is quite natural that our system possesses the property (2.1.2). The above argument is, of course, only the zeroth approximation and more detailed discussions are necessary to understand specific mechanisms such as chemical bonds or metallic binding responsible for actual constructions of matter.

When we say atoms or molecules are the basic unit in building up large systems, another important fact is involved, namely that the ratio of electron

---

[2]When we substitute the number of nucleons in the sun 10^{57} for N and the nucleon mass for M in (2.1.3), we get $R \approx 5 \times 10^8$ cm, roughly the radius of a white dwarf. In a white dwarf, a contracted state of some stars after nuclear reactions have ceased, the pressure of the degenerate electron gas balances the gravitational force.

mass to nuclear mass is very small; for a nucleus of mass number A the ratio is approximately 1/1870 A. This fact is quite important even for the structures of atoms other than hydrogen (it is instructive to ask what the ground state of a helium atom would be if the nuclear mass were equal to the electron mass). When localized in a certain volume, kinetic energy of a nucleus is smaller by this ratio than that of an electron. Hence we can regard nuclei to have definite positions. As we will see in the next section, it is for this reason that in the theory of molecules as well as condensed bodies we first treat problems of electronic states with fixed nuclei. This basic approximation is called the Born-Oppenheimer approximation. The cloud of electrons is extended over the space between nuclei. That the negative charge is distributed in between well-localized nuclei with positive charge is indeed the basic mechanism of cohesion by which many atoms are bound together to constitute molecules or solids with complex spatial structures.

In reality we have to consider properties of matter at finite temperatures. To determine equilibrium states of a system at finite temperatures, we must take into account not only the energy of our system but also the degree of disorder due to thermal motion, namely the entropy. At temperature T each particle has a mean energy of thermal motion of the order of $k_B T$ (k_B is the Boltzmann constant). The basic unit of energy in our system Ry corresponds to the temperature of order 10^5 K. At temperatures much higher than this we have the so-called plasma state in which both nuclei and electrons are in a gaseous state and various matters almost lose their individual characteristics. It is naturally in a temperature region much lower than this that a milliard of molecules exist and various states such as gas, liquid and solid appear with diverse properties. Ordinarily we are interested in the region of temperature below 1000 K which is two orders of magnitude lower than 10^5 K (≈ 1 Ry/k_B). The reason for this is, as we will see in the next section, that spatial structures of matter are destroyed by thermal motion at a temperature lower by a factor of $\sqrt{m/M}$ than the temperature corresponding to our unit of electronic energy.

2.2 Electronic State of a Hydrogen Molecule

In this section[3] we consider the matter composed of protons and electrons and discuss what properties it will have under various conditions. In other words we are concerned with a macroscopic system of N hydrogen atoms. Although we are interested in the large N limit, we have to study a system consisting of a few hydrogen atoms as a preliminary discussion.

Since the case of a hydrogen atom has been thoroughly discussed as the exemplary application of quantum mechanics, we only recall here that its energy levels are given by

---

[3]For additional details see [2.1]

$$E_n = -\frac{me^4}{2\hbar^2 n^2} = -\frac{1}{n^2} \text{[Ry]} \quad (n = 1, 2, \ldots) \tag{2.2.1}$$

and that the wave function of the ground state $1s$ is

$$\varphi_{1s} = \frac{1}{\sqrt{\pi}} \left(\frac{1}{a}\right)^{3/2} e^{-r/a}, \tag{2.2.2}$$

where a is the Bohr radius (Fig. 2.1).

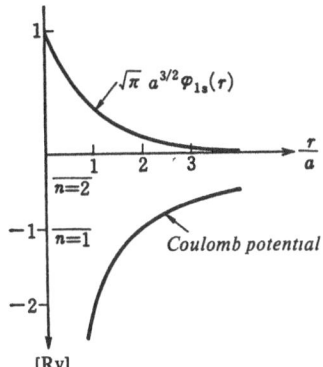

Fig. 2.1. The $1s$ wave function and energy levels of a hydrogen atom

The problem becomes a great deal more complicated when we proceed to the case $N = 2$, a hydrogen molecule. Since the treatment of a hydrogen molecule is not only the basis of the theory of chemical bonds but also involves problems fundamental to the electron theory of solids, we must discuss it in some detail here. Let us write the Hamiltonian of this system as

$$\mathscr{H} = \mathscr{H}_e + \mathscr{H}_p,$$

where

$$\mathscr{H}_e = \sum_{i=1,2} \frac{P_i^2}{2m} + \frac{e^2}{|r_1 - r_2|} - \sum_{i,l=1,2} \frac{e^2}{|r_i - R_l|} \tag{2.2.3a}$$

and

$$\mathscr{H}_p = \sum_{i=1,2} \frac{P_i^2}{2M} + \frac{e^2}{|R_1 - R_2|}. \tag{2.2.3b}$$

Here M is the proton mass and other notations need no explanation. First of all we want to know what the ground state of a hydrogen molecule looks like.

It is at present impossible to solve the Schrödinger equation $\mathscr{H}\Psi = E\Psi$ and obtain the exact wave function Ψ for the entire system which is a function of the coordinates of 4 particles and their spins. For our purpose of understanding the physics involved, however, it is perhaps more significant to proceed step by step, making appropriate approximations.

The first approximation to simplify the problem is, as we have remarked in the preceding section, to neglect the kinetic energy of protons. So we fix the positions of the protons and treat only the motion of electrons. Let the positions of the two protons be $R_1 = R/2$ and $R_2 = -R/2$. When R is fixed, the Hamiltonian (2.2.3a) describes 2 electrons moving in the Coulomb potential of the protons and interacting with each other through Coulomb repulsion. Suppose we can solve the Schrödinger equation and find the wave function $\psi_n(r_1, r_2, R)$ and the energy eigenvalue $E_n(R)$ where n denotes the set of quantum numbers specifying a state of the 2 electrons. Clearly, ψ_n and E_n depend on the distance R between the protons as a parameter. The total energy of the system in this approximation is given by the sum of E_n and the Coulomb energy between the protons

$$U_n(R) = E_n(R) + \frac{e^2}{R}. \tag{2.2.4}$$

Minimizing $U_n(R)$ with respect to R, we find the equilibrium distance and the total energy in the state n. In the next step, when we consider vibrations and rotations of the molecule in the electronic state n, $U_n(R)$ serves as potential energy for the protons. To improve the approximation further we have to find effects on the electronic state of the vibrational and rotational motions of the protons thus obtained. We note that this approximation scheme starting with the Born-Oppenheimer approximation is also used when we study electronic states in condensed matters such as solids.

Even if we fix the protons, it is extremely difficult to solve the Schrödinger equation for two electrons since they interact with each other. So we have to treat the electronic states approximately also, and there are two approaches to the problem.

The first approach is to adopt the same method as used in the problems of electronic structures of atoms. First, neglecting the interaction between the electrons, we solve the problem of a single electron moving in the Coulomb potential of the two protons and obtain its energy eigenstates (molecular orbitals). Then, letting electrons occupy these states in accordance with the Pauli principle, we obtain approximate eigenstates of the total system. This is called the molecular orbital method. It must be noted that the same approach is used for electronic states in crystals; there the Bloch functions appear as wave functions of an electron moving in a periodic potential of ions. Following this approach we first outline the ground-state wave function of an electron in the Coulomb potential of 2 protons. In the neighbourhood of each proton it should not be too different from the 1s wave function of a hydrogen atom provided R is not small compared to the Bohr radius. In this case, therefore, we try approximately to express the molecular wave function using the two $1s$ wave functions only. There are two possible wave functions of the form

$$\varphi_{\pm}(\mathbf{r}) = \frac{1}{C_{\pm}} [\varphi_a(\mathbf{r}) \pm \varphi_b(\mathbf{r})] , \qquad (2.2.5)$$

where $C_{\pm}$ are the normalization constants and we write $\varphi_{a,b}(\mathbf{r})$ in place of φ_{1s} $(\mathbf{r} \pm \mathbf{R}/2)$ for simplicity. In the symmetric function φ_+ the probability amplitude is large in the region between the two protons (Fig. 2.2). Since the electron feels the attractive force of both protons in this region, we expect φ_+ to have lower energy than φ_-. Figure 2.3 shows the energy of a H_2^+ ion in the state φ_+ and φ_- as a function of R. Since the total energy can be lower than the energy of a hydrogen atom, the symmetric state approximately described by φ_+ is called the bonding orbital, an electron binding two protons. On the other hand, the function φ_- has a node in between the two protons and its energy is always higher than that of a hydrogen atom. Therefore it is called the antibonding orbital and is an excited state (Fig. 2.3).

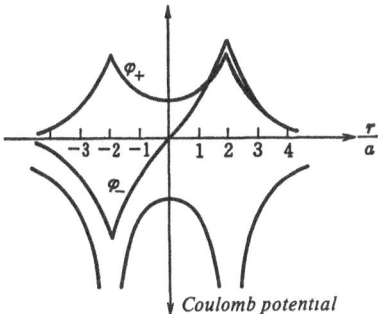

Fig. 2.2. The Coulomb potential and the wave functions on the axis connecting two protons

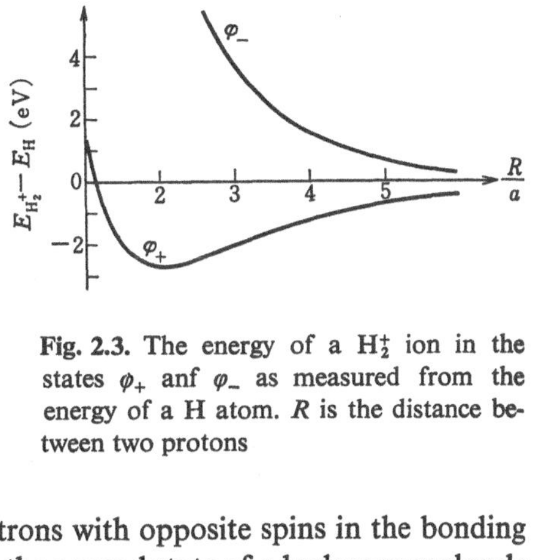

Fig. 2.3. The energy of a H_2^+ ion in the states φ_+ anf φ_- as measured from the energy of a H atom. R is the distance between two protons

Now we can accommodate two electrons with opposite spins in the bonding orbital φ_+. The resulting state should be the ground state of a hydrogen molecule in the present approximation. Because of the antisymmetry of wave functions with respect to the exchange of two electrons, it must be of the form

$$\varphi_+ = \frac{1}{\sqrt{2}} \begin{vmatrix} \varphi_+(\mathbf{r}_1)\alpha_1 & \varphi_+(\mathbf{r}_1)\beta_1 \\ \varphi_+(\mathbf{r}_2)\alpha_2 & \varphi_+(\mathbf{r}_2)\beta_2 \end{vmatrix} , \qquad (2.2.6)$$

where α and β denote two spin states of an electron. To obtain the energy we calculate the expectation value of $\mathscr{H}_e$ (2.2.3a) in this state. The total energy of a hydrogen molecule obtained in this way is shown in Fig. 2.4 [actually in this calculation the $1s$ function is replaced by $\exp(-r/\lambda a)$ with λ as a variational parameter; the minimum is at $\lambda = 1.197$].

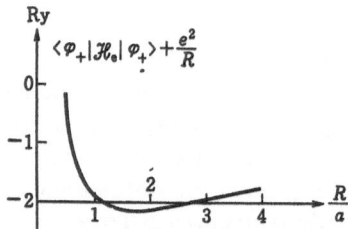

Fig. 2.4. The energy of a hydrogen molecule according to the molecular orbital method

As is clear from the figure, the molecular orbital method does not give the correct limiting state at large R, namely 2 hydrogen atoms. This is due to the fact that the state (2.2.6) always contains the ionized state $H^- + p$ with the same probability as the state of 2 hydrogen atoms $H + H$. Because of the neglect of the repulsive force between electrons, they move completely independently. In reality their motion is correlated due to the repulsion and the correlation obviously becomes more important as R increases.

This leads to the second approach, according to which we consider this correlation from the beginning and use the state of 2 hydrogen atoms $\varphi_a(r_1) \varphi_b(r_2)$ as the first approximation for the ground state of a hydrogen molecule. Since $\varphi_b(r_1) \varphi_a(r_2)$ is equally good bcause of thee indistinguishability of electrons, we must consider in accordance with the Pauli principle the following two wave functions:

$$\psi_{s,t} = \frac{1}{C_{s,t}} [\varphi_a(r_1) \varphi_b(r_2) \pm \varphi_b(r_1) \varphi_a(r_2)] \cdot \frac{1}{\sqrt{2}} (\alpha_1\beta_2 \mp \beta_1\alpha_2), \qquad (2.2.7)$$

where the indices s and t denote the spin singlet and the triplet state. The orbital part ψ_s is symmetric and ψ_t antisymmetric with respect to the interchange of electrons. To use only the $1s$ states seems to be a rather drastic approximation. When two hydrogen atoms are brought together, their wave function will surely be distorted because of the interaction. To distort the wave function, it is necessary to mix the excited states such as $2s$ and $2p$, etc.. Since, however, their energy levels are at least 10 eV higher than the $1s$ state, we expect that the degree of the mixing and hence the distortion is not too large. Incidentally, the approximation (2.2.6) was also based on this reason. The wave function (2.2.7) was first used by W. Heitler and F. London in their attempt to understand chemical bonds on the basis of quantum mechanics, and the theory based on this approximation is called the Heitler-London theory.

Contrary to the molecular orbital approach, this theory encounters difficulties when R is small. As $R \to 0$, the problem becomes identical to that of a helium atom as far as the electronic state is concerned. In this limit, while the molecular orbitals simply reduce to the helium states, the Heitler-London method leads to a wrong result. According to a more advanced treatment, the ratio of the probability amplitude of the ionized state $p + H^-$ to that of the state $H + H$ is about 0.18 in the ground state of a hydrogen molecule. This means that the correlation is rather strong and the Heitler-London theory is relatively good in this case. From the actual equilibrium distance between protons and the dissociation energy listed in Table 2.1, we see that the binding is very strong in the case of a H_2 molecule. As mentioned above, this binding is primarily due to the attraction electrons experience simultaneously from 2 protons. Note that the binding energy H_2^+ already amounts to 2.78 eV. In general, the binding which results from two ions sharing one or two electrons is called covalent binding.

Table 2.1.

	Ionization energy [eV]	Equilibrium nuclear distance [Å]
H_2^+	2.78	1.06
H_2	4.72	0.74
Li_2	1.08	2.67

Next we briefly consider the cases $N = 3$ or 4. What will happen if we bring together 3 or 4 hydrogen atoms? Will $H_2 + H$ or $H_2 + H_2$ bind tightly together to form H_3 or H_4 molecules? For instance, we can place 3 protons on vertices of an equilateral triangle and form an approximate molecular orbital in the same manner as before. Since, however, the symmetric orbital $\varphi_a + \varphi_b + \varphi_c$ holds only two electrons, the third one has to occupy an orbital whose wave function has a node. Therefore, it is unlikely that the energy of such a state is considerably lower than the sum of the energy of a H_2 molecule and a H atom. A similar argument also applies to the case of $N = 4$. In fact, as we will see in the next section, only a very weak attractive force acts between two hydrogen molecules. When they approach each other closely, the molecular orbitals of each molecule must be deformed because of the Pauli principle and this costs a large amount of energy, hence leading to the strong repulsion. This means that the covalent bond cannot hold together more than two protons, in other words, it has the property of saturation. Thus we arrive at the important conclusion that even in a system of very large N the basic building block is a hydrogen molecule.

Before closing this section, let us make a brief remark on the case $Z = 2$ and 3. The case of a helium atom $Z = 2$ with the closed 1s shell hardly needs any explanation. A Li atom $Z = 3$ has 2 electrons in 1s and 1 electron in 2s state. Though it has one valence electron like a hydrogen atom, we note as important differences that in Li there is the closed 1s shell and that there are 2p states which are almost degenerate with 2s states. With molecular orbitals made of 2s and 2p states two Li atoms can form a stable Li molecule. Because the 2s and 2p states are orthogonal to the 1s state, those molecular orbitals are considerably more extended in space than that of a H molecule. Moreover, three $2p$ states together with a $2s$ state can form almost degenerate 4 molecular orbitals. Hence it is possible that three or four Li atoms can strongly bind themselves to form Li_3 or Li_4 molecules. Indeed a large number of Li atoms can bind together to form a macroscopic body, that is, Li metal in which valence electrons are in states extending throughout the system. It must be added that the four molecular orbitals made of $2s$ and $2p$ states play a very important role in chemical bonds. One can see its importance from the fact that with 4-valence electrons carbon is the principal element in building up complicated structures of organic compounds.

2.3 Systems Made of Hydrogen Molecules

To be able to discuss the system made up of many hydrogen molecules, we must know more about the properties of the molecule, especially proton motion and intermolecular force. Putting aside translation of the molecule as a whole, we are concerned with its vibrations, that is, variations in the distance between protons and its rotations.

First, concerning the vibrations we consider only small oscillations of protons around their equilibrium position in the ground state of the molecule. In this case we can expand the potential energy for protons (2.2.4) around R_0 and approximate it by

$$U(R) \approx U(R_0) + \frac{k}{2} (R - R_0)^2 .$$

The problem then reduces to that of a harmonic oscillator of frequency $\omega = \sqrt{2k/M}$. Since the order of magnitude of the spring constant k is Ry/a^2, the energy quantum of the oscillation is roughly given by

$$\hbar\omega \approx \sqrt{\frac{m}{M}} \quad [Ry] .$$

The spread of the proton position due to zero-point oscillation is estimated from $k\overline{(R - R_0)^2}/2 \approx \hbar\omega$ to be about $(m/M)^{1/4} a$. Precisely because this is small compared to the spread of electron cloud, we could fix protons in the first approximation. The value of $\hbar\omega$ determined from the observed vibrational spectrum of H_2 is about 0.5 eV. That the vibrational energy of the ions is smaller by the factor $\sqrt{m/M}$ than the electronic energy is also true in the case of solids.

We turn now to the molecular rotations. If we neglect stretching of the molecule due to the centrifugal force, the rotational energy levels are given by

$$E_K = \frac{\hbar^2}{2I} (K + 1) K ,$$

where K is the rotational quantum number and I the moment of inertia of the molecule. For $K = 1$ this is about $k_B \times 170$ K. Let us recall that, protons being fermions with spin 1/2, the wave function of protons must be antisymmetric with respect to the interchange of 2 protons. When we use polar coordinates for the relative coordinate R of the protons, it can readily be seen that the rotational part of the proton wave function together with their spin state determines the symmetry. The total spin of the two protons can take values $S = 0$ (singlet) and $S = 1$ (triplet). Since the former is antisymmetric and the latter symmetric, the allowed rotational quantum numbers must be even for $S = 0$ and odd for $S = 1$. The molecule in the singlet state is called para-hydrogen and in the triplet state ortho-hydrogen. One might think that nuclear spins have little to do with macroscopic properties of matter, but the low temperature properties of solid hydrogen do depend on the spin states in an interesting way. Here we suppose for simplicity that all molecules are in the state of para-hydrogen. Below about 100 K they are in the lowest rotational state $K = 0$, so that the proton wave function is spherically symmetric. In this case we can regard the molecules as a spherical particle. In such low temperatures the vibrations of the molecules are also not thermally excited. In what follows, therefore, we need not consider the internal structure of the molecules.

Next we must know the interaction between two hydrogen molecules in its ground state. As said before, when they approach each other and their electronic wave functions start to overlap, very strong repulsion will act between them. Here we will study in some detail the force in the range where the overlap is negligible. Actually there are two kinds of long range force. The first is the quadrupole-quadrupole interaction. A H_2 molecule is not spherical and hence possesses a quadrupole moment. The second, more important force is the dispersion force (van der Waals force) which is due to quantum mechanical fluctuations. The dipole moment of the molecule is given by

$$d = e \sum_\alpha r_\alpha ,$$

where r_α is the position of the αth electron with respect to the center of the molecule. Why the contribution of protons is negligible will become clear in the following discussion. Although the expectation value of d in the ground state of a hydrogen molecule is 0, its fluctuations are always present. If at some moment the first molecule happens to have d_1 due to fluctuations, it will produce an electric field $- \nabla_2(d_1 \cdot R/R^3)$ at the second molecule. This induces a dipole moment in the second molecule given by

$$d_2 = - \alpha \nabla_2 \left(\frac{d_1 \cdot R}{R^3} \right), \tag{2.3.1}$$

where α is the molecular polarizability and $R = R_2 - R_1$, R_1 and R_2 being the centers of the molecules. Substituting (2.3.1) into the interaction energy between two dipoles

$$U_L = \frac{d_1 \cdot d_2}{R^3} - 3\frac{(d_1 \cdot R)(d_2 \cdot R)}{R^5},$$

we get

$$U_L = -\frac{\alpha}{R^6}\left[d_1^2 + 3\frac{(d_1 \cdot R)^2}{R^2}\right]. \qquad (2.3.2)$$

The expectation value of this does not vanish. One can obtain the polarizability in the ground state by perturbation theory as

$$\alpha = -e^2 \sum_n \frac{|\langle n|\sum_\alpha r_\alpha|0\rangle|^2}{E_0 - E_n}, \qquad (2.3.3)$$

where n denotes the excited states (we have assumed the molecule to be spherically symmetric). In the case of H_2 molecules, U_L depends on the relative orientation of two molecules. The isotropic part is equal to 10.9 $(a/R)^6$ Ry.

The dispersion force is quite universal as an intermolecular force and is a main force acting among rare gas atoms and most nonionic molecules. Its important features are, as can be seen from (2.3.2), firstly that it is attractive, secondly that it is a long range force proportional to $1/R^6$, and thirdly that unlike covalent bonding it does not have the property of saturation. When many particles interact simultaneously, the dispersion force between a pair of particles is hardly affected by the presence of other particles. Because of these features the dispersion force is important as a cohesive force for particles which do not interact with stronger forces of other kinds.

The repulsive force at a short distance and the long-range force together make up the interaction $U(R)$ between two hydrogen molecules depicted in Fig. 2.5. Note that the minimum value of $U(R)$ is smaller by three orders of magnitude than that of the interaction between hydrogen atoms and that the distance at the minimum is about 3Å, about 4 times the proton distance in the molecule. These facts justify our approach of regarding H_2 molecules as the basic building block.

What structure and properties do we expect for systems made up of hydrogen molecules? To simplify the problem we regard the molecules to be spherical and suppose that the interaction energy between them is a function only of the

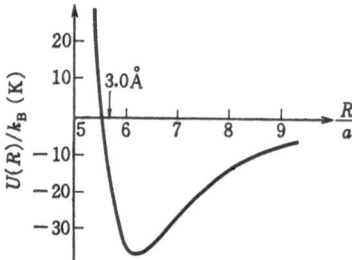

Fig. 2.5. The interaction potential between two hydrogen molecules

distance R. For ease of calculation, the potential depicted in Fig. 2.5 is often approximated by the following simple function

$$U(R) = 4\varepsilon \left[\left(\frac{\sigma}{R} \right)^{12} - \left(\frac{\sigma}{R} \right)^{6} \right] \tag{2.3.4}$$

called the Lenard-Jones potential, which we also adopt here. Note that $U = 0$ at $R = \sigma$ and that U takes the minimum value $-\varepsilon$ at $R = 2^{1/6} \sigma$. As a physical picture one can imagine hard spheres of radius $\sigma/2$ exerting a weak attractive force on each other. In the case of hydrogen molecules we have $\sigma = 2.93$ Å and $\varepsilon = 5.1 \times 10^{-15}$ erg $= k_B \cdot 37$ K. Thus we are now dealing with a system which consists of one kind of particles interacting with the potential U. This is far simpler than the original system consisting of protons and electrons.

The Hamiltonian of this simplified system is

$$\mathcal{H} = \sum_{i=1}^{N} \frac{p_i^2}{2M} + \frac{1}{2} \sum_{i \neq j} U(|x_i - x_j|), \tag{2.3.5}$$

where M is now the mass of the molecule and x_i is the coordinates of the ith molecule, p_i its momentum.

First, let us consider the lowest energy state of this system. In contrast to the original system with Coulomb interaction, an equilibrium state of this system exists even in the classical theory since $\mathcal{H}$ has a minimum. In the classical theory the kinetic energy vanishes at $T = 0$ K, so that the problem is simply to find the configuration of particle positions that minimizes the total potential energy

$$\frac{1}{2} \sum_{i \neq j} U(|x_i - x_j|).$$

In the case of potential such as (2.3.4) the configuration must be such that all nearest neighbor pairs of particles are in the bottom of the potential well as much as they can. This means that the mean distance between particles is of the order of σ. In the limit of a large particle number N, then, every particle will find the identical configuration of the surrounding particles except those near the surface. This implies that particles are arranged regularly in space with definite periodicity. Identical units placed on each lattice site of a lattice which has definite periodicity is called a crystal. In the classical theory, therefore, the lowest energy state of our system must be a crystal. For the Lenard-Jones potential of the form (2.3.4) the potential energy of various types of lattices per particle can easily be calculated. It is given by

$$
\begin{aligned}
U_0 &= 2 \sum_{i \neq 0} \varepsilon \left[\left(\frac{\sigma}{r_i} \right)^{12} - \left(\frac{\sigma}{r_i} \right)^{6} \right] \\
&= 2\varepsilon \left[\left(\frac{\sigma}{r_{nn}} \right)^{12} C_{12} - \left(\frac{\sigma}{r_{nn}} \right)^{6} C_6 \right],
\end{aligned} \tag{2.3.6}
$$

where r_i is the distance of the ith lattice site from the 0th site and $r_{n,n}$ is the nearest neighbor distance of a given lattice. The constants C_{12} and C_6 depend only on the types of lattices. The most stable one is the face-centered cubic lattice (fcc) with $C_{12} = 12.132$ and $C_6 = 14.454$.

Let us summarize our discussion so far. As long as we remain in the frame work of the classical theory, we expect any system of particles interacting with potential of the form of Fig. 2.5 to be in a condensed state at the absolute zero of temperature, which is moreover a crystal with the basic building blocks arranged in a regular manner.

This then leads us to the next question, whether or not the classical treatment is really valid in the present problem, or in other words, the quantum effect is indeed negligible. When a particle is localized in a region of radius R due to potential walls of surrounding particles, its kinetic energy is roughly equal to $K \approx (\hbar^2/2M)(\pi/R)^2$ according to the uncertainty principle. When we substitute $\sigma(2^{1/6} - 1)$ for R, we find a large value for K, which means that the quantum effect is quite important in this case. In fact, hydrogen molecules make up a crystal of hexagonal-close-packed (hcp) structure with the nearest neighbor distance equal to 3.75 Å, which is appreciably larger than the distance $2^{1/6}\,\sigma = 3.2$ Å of the potential minimum. The actual solid is considerably more expanded, due to zero point oscillations of particles, than the equilibrium density according to the classical theory. The case where the quantum effect is most important is helium which we will discuss in the next chapter. In all materials other than hydrogen and helium the quantum effect is rather small and hence they all solidify at sufficiently low temperatures.

So far we have been discussing the ground state of the system. In reality, of course, we are concerned with properties of matter at finite temperatures and under finite pressure. When both T and P are not too high, the basic building block remains to be H_2 molecules. In Fig. 2.6 we show schematically how the molecular solid, liquid and gas phase appear in the P-T diagram. Within the solid phase, phase transitions with changes in crystal structures occur which are omitted in this phase diagram. The way the three phases (solid, liquid and gas) appear is similar for most materials. It is one of the fundamental problems of statistical mechanics to see if the model system described by (2.3.5) can exhibit these phase transformations.

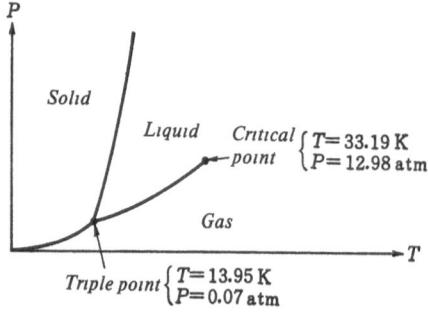

Fig. 2.6. The phase diagram of a system of hydrogen molecules. The critical and the triple point values are measured for ordinary hydrogen (75% ortho-hydrogen)

Before concluding this section we briefly comment on the electrical properties of a crystal of H_2 molecules. In general, the important characteristics of most molecular crystals is that they are insulators. To conduct electricity electrons must be moved in a crystal by an applied electric field. As remarked before, the electronic state of H_2 molecules in the crystal is hardly changed from that of the isolated molecule, a pair of electrons in spin singlet state occupying the bonding orbital. In order to move electrons, therefore, we have to remove an electron from this paired state of one molecule and place it onto another molecule. Then it can hop from one to adjacent molecules. As a result, there will be a H_2^+ and a H_2^- ion in the crystal. The energy of this state is almost 2 eV higher than before. This means that to excite electrons into conducting states we need finite energy, ΔE. Compared to this energy, the energy an electron acquires from the electric field when it moves from one molecule to its neighbor is negligible unless the field is extremely strong. Consequently it is hardly likely that the conduction takes place due to electrons excited by the applied field. Although thermally excited electrons can lead to conduction, their density is proportional to the exp $(-\Delta E/k_B T)$, which is extremely small in the case of solid hydrogen at low temperature. It must be added that ΔE is relatively small in semiconductors and zero in metals.

2.4 Metallic State at High Density

In the preceding section we have discussed properties of matter at moderate temperature and pressure, in which we are usually interested. As long as the electric force is the only interaction, high density states with mean distance between particles far smaller than the Bohr radius a would not be realized (at least as an equilibrium state). This is well testified by the difficulty of achieving static pressure of even one million atmosphere in a laboratory. Since extremely high density states exist within stars due to the gravitational force, however, properties of matter under high pressure are also of interest. Another, no less important reason for studying high density states is that often our problems become simpler and physical interpretation easier in the extreme cases. As a matter of fact, the subject of this section[4] will serve as the standard model when we discuss properties of metals.

Our system under low pressure was shown to be a molecular solid with H_2 molecules as the basic unit. What will happen to it when we compress it? Inside the planet Jupiter whose main constituent is hydrogen, for example, the pressure can reach as high as a few tens of megabars. What state of hydrogen matter do we expect to find there? Here we are not concerned with this very interesting but difficult problem, but are rather interested in a more extreme, hence simpler situation where the matter is compressed to such high density that the mean dis-

---

[4]For additional details see [2.1]

tance between protons is comparable to the proton distance in a H_2 molecule, 0.78 Å. Since in such states the molecules no longer have any significance as the building block, we must go back to our original components, protons and electrons.

Let there be N protons and N electrons in a volume Ω. We introduce a dimensionless parameter r_s defined by

$$\frac{4\pi r_s^3}{3} = \frac{1}{na^3} , \tag{2.4.1}$$

where n is the average density of electrons $n = N/\Omega$. Since the Coulomb energy per electron is of the order of e^2/ar_s, and the kinetic energy of an electron $\hbar^2/ma^2 r_s^2$ according to the Sect. 2.1, their ratio is

$$\frac{e^2/ar_s}{\hbar^2/ma^2 r_s^2} = r_s . \tag{2.4.2}$$

This means that in high density states with r_s much smaller than 1 the kinetic energy is dominant. As the first approximation, therefore, we can regard electrons to be freely moving, neglecting the Coulomb interaction. A free electron is described by a plane-wave state with momentum $\hbar k$ and energy $\xi_k = \hbar^2 k^2/2m$. If we use the periodic boundary condition on a cube of volume $L^3 = \Omega$, the allowed wave vectors are given by $k = (2\pi n_x/L, 2\pi n_y/L, 2\pi n_z/L)$ where n_x, n_y, n_z are integers. Since L is macroscopic, we can regard k to take continuous values.

To find the ground state of N free electrons we let them occupy, according to the Pauli principle, the plane-wave states in the order of increasing energy ξ_k. As a result, the occupied states will form a sphere in the momentum space, the so-called Fermi sphere (Fig. 2.7). Since the total number of states in the sphere including the spin states must be N, its radius called the Fermi momentum p_F is determined by

$$N = \sum_{\hbar|k|\leq p_F} 2 = 2 \frac{\Omega}{(2\pi\hbar)^3} \frac{4\pi}{3} p_F^3 , \tag{2.4.3}$$

The electrons occupying the states on the Fermi surface have momentum p_F and energy $\xi_F = p_F^2/2m$. The energy of the ground state of a free electron gas represented by the Fermi sphere is

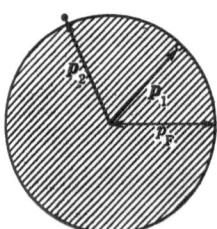

Fig. 2.7. Fermi sphere

$$E_\bullet^{(0)} = \frac{2}{N} \sum_{\hbar|k|\leq p_F} \frac{(\hbar k)^2}{2m} = \frac{\Omega}{N} \frac{p_F^5}{10\pi^2\hbar^3 m} = \frac{2.21}{r_s^2} \quad \text{[Ry per electron].} \quad (2.4.4)$$

We next consider excited states of the free-electron gas. When an electron in a state p_1 inside the Fermi sphere is displaced to a state p_2 outside it, the energy of the system increases by $(p_2^2 - p_1^2)/2m$ (Fig. 2.7). In such a way we can obtain excited states of this system simply by lifting electrons from inside the Fermi sphere to outside without conflicting with the Pauli principle. Obviously we need only a small energy to lift an electron from a state just below to just above the Fermi surface, but a larger energy which is comparable to the Fermi energy to lift one from deep inside the sphere. From this we can guess to a certain extent the physical properties of a system in which electrons can be regarded as almost freely moving. First of all it can conduct electricity, in other words it is metallic. A state in which electrons as a whole are moving with velocity V is represented by the Fermi sphere displaced by mV in the momentum space. Since we can reach such a state by displacing those electrons near the original Fermi surface, we expect an arbitrarily weak electric field to be able to excite electrons and lead to states with a net flow of electrons. Secondly, the characteristic energy of this system is the Fermi energy ζ_F which is now of the order of Rydberg (we have assumed that the interparticle distance is about the Bohr radius; as we will see later, ζ_F is a few eV even in actual metals). At temperatures much smaller than this only those electrons near the Fermi surface can be thermally excited. Hence their number is roughly $k_B T/\zeta_F$ times the total number of electrons. Since each of them carries thermal energy $k_B T$, the total thermal energy is proportional to $N k_B^2 T^2/\zeta_F$ so that the specific heat is proportional to $N k_B^2 T/\zeta_F$. That the observed specific heat of metals at low temperature is proportional to T was a puzzle for the classical theory.

Granted that electrons are almost free, we must next consider how protons behave in this case. If we calculate the same ratio as (2.4.2) for protons, we find $r_s M/m$. Hence we can still neglect kinetic energy of protons as in the case of molecules (in Sect. 2.5 we will comment on such extremely high density states that this rato is small compared to 1). The electron density is almost uniform in space, providing a uniform background of negative charge for protons. If their kinetic energy is negligible, we expect protons to arrange themselves in a lattice because of their mutual Coulomb repulsion. Let the position vector of a proton at the nth lattice site be R_n. Then the Coulomb interaction energy of our system is given by the sum of the following three terms:

electron-electron: $\quad E_{ee} = \dfrac{e^2}{2} \dfrac{N}{\Omega} \int d^3x \, \dfrac{1}{r}$

electron-proton: $\quad E_{ep} = -\dfrac{e^2}{\Omega} \sum_n \int d^3x \, \dfrac{1}{|x - R_n|} = -e^2 \dfrac{N}{\Omega} \int d^3x \, \dfrac{1}{r}$

proton-proton: $\quad E_{pp} = \dfrac{e^2}{2N} \sum_{n\neq n'} \dfrac{1}{|R_n - R_{n'}|}$

$$(2.4.5)$$

Though each of them diverges, the sum

$$E_C = \frac{e^2}{2} \left(\frac{1}{N} \sum_{n \neq n'} \frac{1}{|\mathbf{R}_n - \mathbf{R}_{n'}|} - \frac{N}{\Omega} \int d^3x \frac{1}{r} \right) \tag{2.4.6}$$

is finite. This is just the Coulomb energy calculated according to the classical theory. We list its values in the atomic unit calculated for the three types of lattices:

$$E_C = -\frac{1}{r_s} \times \begin{cases} 1.79186 & \text{(body-centered cubic)} \\ 1.79172 & \text{(face-centered cubic)} \\ 1.760 & \text{(hexagonal close packed)}. \end{cases} \tag{2.4.7}$$

This negative energy together with the exchange energy which we will discuss shortly is the main cause of the cohesion in metals. The electron cloud of negative charge spread in the space between localized positive ions holds the ions together just as in a hydrogen molecule. If protons were also distributed uniformly, the Coulomb energy (2.4.7) would obviously be equal to zero.

As the most important correction to the classical expression (2.4.7) for the Coulomb energy, we have the so-called exchange energy due to an essentially quantum-mechanical effect. In the calculation of the electron-electron interaction energy we have to take acount of the Fermi statistics. When we do this, we get, besides the first term of (2.4.6), the exchange energy

$$E_{ex}^{(1)} = -\frac{\Omega}{N} \frac{e^2 p_F^4}{4\pi^3 \hbar^4} = -0.916 \frac{1}{r_s} \quad \text{[Ry]}. \tag{2.4.8}$$

One can interpret this in the following way. Since two electrons with parallel spin avoid each other according to the Pauli principle, they exert less Coulomb repulsion on each other. Hence in (2.4.7) we overestimated the repulsive energy and so we now have a negative correction. To see this more explicitly it is instructive to look at the probability $g(|\mathbf{r}_2 - \mathbf{r}_1|)$ of finding another electron at $\mathbf{r}_2$ when there is one at $\mathbf{r}_1$. When we calculate this probability in the ground state of an electron gas represented by the Fermi sphere, we get the result shown in Fig. 2.8. This shows the correlation between electrons of parallel spins due to the Fermi statistics.

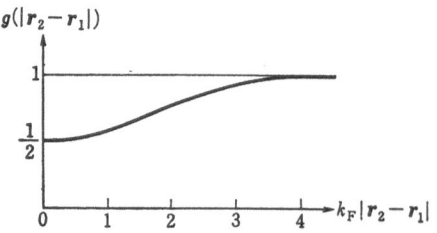

Fig. 2.8. The electron correlation function according to Fermi statistics

In the above approximation of free electrons we have neglected (i) correlation due to the interaction between electrons, (ii) the effect of the Coulomb potential of the proton lattice on electron states, and (iii) the motion of protons. Let us now briefly discuss the importance of these three effects as an introduction to the later chapters.

2.4.1 Electron-Electron Interaction

Being proportional to $1/r^2$ the Coulomb force is long-ranged and leads to unique effects. To simplify the problem, we replace protons by an uniform background of positive charge with density $e\bar{n} = Ne/\Omega$ and consider a gas of electrons moving in this medium. When we are interested in the motion of this gas which varies sufficiently slowly in space, we can use hydrodynamic equations. The number density of the particles $n(x, t)$ and the velocity field $v(x, t)$ satisfy the continuity equation

$$\frac{\partial n}{\partial t} + \text{div } (nv) = 0 \tag{2.4.9}$$

and the Euler equation

$$\frac{\partial v}{\partial t} + (v \cdot \nabla) v = -\frac{1}{mn} \nabla P + \frac{e}{m} \nabla\phi . \tag{2.4.10}$$

In the latter equation P is the local pressure, ϕ the scalar potential for an electric field, and all dissipative processes are neglected. The potential is related through the Poisson equation to the charge density

$$\nabla^2\phi = -4\pi(\rho_{ext} + \rho) , \tag{2.4.11}$$

where $\rho = e(n - \bar{n})$ describes fluctuations of the charge density of the electron gas and ρ_{ext} denotes charges brought in from outside. The pressure P is related to the energy density $\varepsilon = E/V$ as

$$P = \tfrac{2}{3}\varepsilon . \tag{2.4.12}$$

If we use as E the energy corresponding to the Fermi sphere of a free electron gas, we can easily derive an expression from (2.4.4, 5) for the pressure

$$P = \frac{(3\pi^2)^{2/3}}{5} \frac{\hbar^2}{m} n^{5/3} . \tag{2.4.13}$$

Note that in the above formulation we describe the Coulomb interaction between electrons through the field ϕ which contributes to the force in (2.4.10) and is generated by the density in (2.4.11).

As the first application we consider a static problem in which we place a point charge z at the origin, that is, $\rho_{\text{ext}} = z\delta(\mathbf{x})$. In (2.4.11) ρ is then the charge induced by ρ_{ext} around itself. Substituting (2.4.13) in (2.4.10) we get

$$\frac{2}{3}\frac{\xi_F}{\bar{n}}\nabla n = e\nabla\phi,$$

where we have replaced n in the denominator of the left-hand side by $\bar{n}$, assuming the deviation ρ to be small. By Fourier transforming ϕ and n, we can easily solve this equation together with (2.4.11) and obtain

$$\phi_k = \frac{4\pi z}{k^2\epsilon(k)}, \qquad (2.4.14)$$

where

$$\epsilon(k) = 1 + (\lambda_{\text{T-F}}k)^{-2}, \quad \lambda_{\text{T-F}}^2 \equiv \frac{v_F^2}{3\omega_P^2}, \quad \omega_P^2 \equiv \frac{4\pi e^2\bar{n}}{m} \qquad (2.4.15)$$

and $v_F = p_F/m$ is the Fermi velocity. The reason we have introduced $\lambda_{\text{T.F}}$ and ω_P will become clear shortly. When we transform back from ϕ_k, we find the familiar expression

$$\phi(r) = \frac{z}{r}\exp\left(-\frac{r}{\lambda_{\text{T.F}}}\right) \qquad (2.4.16)$$

for the screened Coulomb potential. This means that because of the Coulomb interaction between electrons, the electric field of the test charge is screened within a distance characterized by $\lambda_{\text{T.F}} = ar_s^{1/2}(\pi/12)^{1/3}$, the Thomas-Fermi screening length. From the way it enters into (2.4.14) it is natural to call $\epsilon(k)$ the wave-number-dependent dielectric constant of the medium which is now an electron gas. Though we have shown that the electric field of a test charge brought in from outside is screened, the field of electrons themselves will also be screened. Thus, the long-range Coulomb interaction between a pair of electrons is replaced by the screened interaction due to the presence of other electrons.

Where, then, do we find a special feature of the long-range force? To answer this question we now consider dynamical fluctuations of the electron density with long wavelength. Assuming that all quantities ρ, v, ϕ associated with the fluctuations are infinitesimally small, we make the approximation of keeping only terms linear in these quantities in (2.4.9–12). Then we readily obtain

$$\frac{\partial^2\rho}{\partial t^2} + \omega_p^2\rho = 0, \qquad (2.4.17)$$

where we have neglected a term with $\nabla^2 P$ because of long wavelength. This result means that a collective motion of electrons corresponding to the density fluctuations is a wave motion with frequency ω_p, called plasma oscillations. When interaction between particles is short ranged such as the one we discussed in Sect. 2.4, frequencies of oscillations with long wavelength are always proportional to the inverse wavelength as in sound waves. In contrast, charge density fluctuations can exert the long-range Coulomb force on each other, however far apart they appear. As a consequence the frequency remains finite in the limit of long wavelength. Since the corresponding quantum is $\hbar\omega_p \approx r_s^{-3/2}$ Ry, rather a large energy is necessary to excite plasma oscillations. Therefore, the plasma oscillations do not appear as a real process in most phenomena we observe in condensed matters. In other words, the degrees of freedom of the system corresponding to the density fluctuations with long wavelength compared to the interparticle distance are suppressed by the long-range Coulomb force. To disturb even local neutrality costs energy. As a result, the effective interaction between electrons becomes screened and short ranged, so that its effects are often not very important. This explains why the drastic approximation of entirely neglecting the interaction between electrons works surprisingly well in accounting for various properties of metals.

2.4.2 Interaction Between Protons and Electrons

When the electron-electron interaction is neglected, each electron moves independently in the Coulomb field of protons. We must then solve the problem of a single-electron state similar to that of molecular orbitals in the case of a molecule. Here protons are arranged in a lattice so that the Coulomb field also possesses the same periodicity as the lattice. To find electronic states in such periodic potential is the central problem of the electron theory of solids. The band theory developed for this problem provides the basis for understanding properties of metals, semiconductors, as well as insulators. This will be the subject of Chap. 5.

2.4.3 Motion of Protons

The most important modes of proton motion are lattice waves. Here we discuss only the longitudinal modes which are the proton density waves. We take into account the presence of electrons as a uniform medium with the dielectric constant $\epsilon(k)$, capable of the screening effect. Due to this medium protons interact with each other via the screened Coulomb force. As we did for electrons, we now replace the proton lattice by a continuum and use the hydrodynamical equations, again restricting our discussion to the long wavelength limit. This time denoted by ρ the proton density, we can derive in the same way as before

$$\frac{\partial^2 \rho}{\partial t^2} - \frac{e}{M}\bar{n}\nabla^2\phi = 0 . \tag{2.4.18}$$

Since protons are assumed to be at rest in the equilibrium state, we have no term coming from the pressure P. The electric field ϕ is produced by ρ and is screened by electrons. For the wave with wave number k we find from (2.4.14) $k^2\phi_k = 4\pi e \rho_k/\epsilon(k)$ which, when combined with (2.4.18), gives the dispersion relation for the density waves

$$\omega_k = \sqrt{\frac{4\pi\bar{n}e^2}{M(k^2 + k_{\text{T-F}}^2)}} \cdot k \,, \tag{2.4.19}$$

where $k_{\text{T-F}} = \lambda_{\text{T-F}}^{-1}$. Therefore, this mode is a sound wave with frequency ω proportional to k in the limit of long wavelength. The sound velocity is equal to $s = \sqrt{4\pi\bar{n}e^2/Mk_{\text{T-F}}^2}$. In the region $k \gg k_{\text{T-F}}$, though the above approximation is no longer valid, we find $\omega_k = \sqrt{4\pi\bar{n}e^2/M}$ which is just the plasma frequency of protons. For the lattice the shortest wavelength is the lattice constant. Thus we expect the dispersion relation of the lattice wave to be of the form shown in Fig. 2.9. The quantum of lattice oscillations is called phonons.

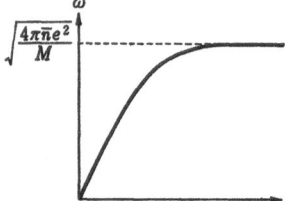

Fig. 2.9. The dispersion relation of proton density waves

The important results obtained are that long-wavelength phonons are sound waves with sound velocity $s \approx v_F\sqrt{m/M}$ and that the frequency of short-wavelength phonons is roughly equal to $\hbar\omega_p\sqrt{m/M}$. Note the presence of the factor $\sqrt{m/M}$. These conclusions remain valid for real metals. Phonons are basic elementary excitations in solids in general, and electron-phonon interaction plays a crucial role in some phenomena in metals. For instance, scattering of electrons by phonons is a principal mechanism of electrical resistance, and superconductivity is induced by attractive interaction between electrons mediated by phonons. These problems will also be discussed in Chap. 5.

In this section we have given qualitative discussions on metallic states which are valid, as noted at the beginning, only at very high density characterized by the condition $r_s \ll 1$. If estimated from (2.4.13), the pressure corresponding to $r_s = 1$ turns out to be 10^7 atm. So, pressure of at least 10^8 atm is perhaps necessary in the case of hydrogen which we have been discussing. Under high enough pressure of this magnitude we expect, not just hydrogen, but all matter to become metallic. On the other hand, we know that many elements are metals at 0 pressure. With the exception of hydrogen, all elements of single valence from Li to Cs are metals. As we remarked at the end of Sect. 2.3, atoms of these elements have the closed shell besides the valence electron, the presence of which affects the appearance of the metallic state strongly. Under 0 pressure the values of r_s for alkali metals are Li: 3.22, Na: 3.96, K: 4.87 etc., thus considerably larger than 1. Nevertheless, the simple theory we developed for the high density state is very useful for the study of these and other real metals.

2.5 Phase Diagram of the Proton-Electron System

Figure 2.10 is a highly schematic phase diagram showing various possible states the proton-electron system is expected to be in a wide range of pressure and temperature. The scale for pressure P is arbitrarily varied. It should also be noted that states under high pressure are only a theoretical guess and the numbers can only give the order of magnitude at best. In the solid states we have omitted possible phases with different crystal structures. In the following we will briefly comment on this phase diagram.

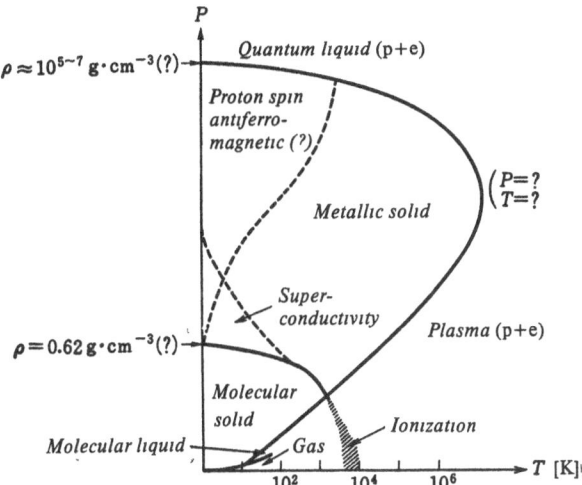

Fig. 2.10. The phase diagram of a $p + e$ system

As discussed in the preceding section, the molecular crystal becomes metallic when pressure is increased. Limiting ourselves to the case $T = 0$ K, we denote the energy of the molecular crystal and that of the metallic state with the same number of particles by E_i and E_m. Since $P = -\partial E/\partial V$, V being the volume of the system, the pressure under which both states coexist is given by the common tangent

$$P = -\left.\frac{\partial E_i}{\partial V}\right|_{V=V_i} = -\left.\frac{\partial E_m}{\partial V}\right|_{V=V_m}.$$

The volumes V_i and V_m of the system at this pressure determine the respective densities. In order to find the pressure at which the transition takes place, therefore, we have to know E_i and E_m separately as a function of density, which is an extremely difficult task. In the molecular crystal the basic unit at each lattice site is a H_2 molecule in which electrons are localized, so that it is an insulator. On the other hand, in the metallic crystal it is most likely that protons are the basic unit and electrons are not localized. Hence, the change of crystal structure and that of electronic state are closely related in this insulator-metal transition. In the former the structure is such as to lower the Coulomb energy as much as possible and in the latter to lower the kinetic energy of electrons instead. Although the insulator-metal transition in hydrogen has been studied since the beginning of the electron theory of solids, there is still no reliable theory. Experimentally too, it is outside the pressure range attainable at present. According to recent theoretical studies, the transition to metalic hydrogen is expected to occur at pressure in the neighborhood of 2.6×10^6 atm, and the mean proton distance then is about 1.15 Å. In general, the insulator-metal transition is one of the most interesting problems in solid-state physics and various approaches to this problem are actively being tried. For example, similar molecular crystal-to-metal transitions are expected in halogen elements. High pressure generated by shock waves is used in the experimental study and such transitions were reported to be observed in Br at about 3×10^5 atm. Theoretically one can think of a simpler problem in which only the density is changed with no change in the crystal structure. Suppose, for instance, we fix H atoms at lattice sites and change the lattice constant. When the lattice constant is large, it should be an insulator, electrons being localized near protons. When it becomes small, the system will surely be metallic. As suggested by the Heitler-London theory of a hydrogen molecule, the correlation between electrons due to the Coulomb interaction should be essential for the transition of this case.

As to properties at 0 K of metallic hydrogen we only note that it would probably be superconducting and that proton spins should be in an ordered state. Even in the crystalline state, protons at neighboring sites exchange their position due to quantum-mechanical tunneling, and as a consequence the exchange energy which depends on spin configurations of protons is finite. It is expected that proton spin ordering would be antiferromagnetic at high density.

What will happen to our system if the pressure is further increased? The free electron gas approximation becomes even better for electrons. At high enough density what we have said about electrons will apply to protons; the kinetic energy of zero point oscillations of protons will exceed the Coulomb energy so that at some point protons will also become liquid. In other words, the lattice of protons formed due to the Coulomb repulsion in the uniform background of negative charges will melt because of the zero-point oscillations. When the density is $\bar{n}$, the majority of the proton lattice waves have frequencies approximately equal to the proton plasma frequency $\omega_p' = \sqrt{4\pi\bar{n}e^2/m_p}$. Then, we can estimate the amplitude $\bar{x}$ of zero-point oscillations of protons around their lattice sites from the relation $m_p\omega_p'^2 \tilde{x}^2/2 \approx \hbar\omega_p'/2$ to be $\tilde{x} \approx \sqrt{\hbar/m_p\omega_p'}$. The ratio δ of $\tilde{x}$ to the lattice constant is, if we substitute the mean particle distance $r_0 = \bar{n}^{-1/3}$ for the latter, given by

$$\delta = \left(\frac{m_e}{m_p}\frac{a}{4\pi r_0}\right)^{1/4}.$$

When this exceeds a certain value, we expect the lattice to melt. Now, melting of solids at high temperatures is, of course, due to thermal motions. If we define the same ratio for the amplitude of thermal oscillations to the lattice constant, it is known empirically for alkali metals that the melting takes place when $\delta \approx 1/4$. If we tentatively use this value for the present case, though there is no justification for this, we get $r_0 = 5 \times 10^{-3}$ Å at the melting. Since the melting is due to quantum-mechanical zero-point oscillations rather than thermal motion, this liquid state may be called quantum liquid.

Since we have been concerned with the states at $T \approx 0$ K so far, we add a few remarks on finite temperature properties. The solid phase is thought to occur in a closed region in the P–T plane bounded by the melting curve as shown in Fig. 2.10. Since solid phases in general possess definite crystal symmetry while liquid or gas phases do not, the transition from one to the other can not be continuous. Hence the solid phase is bounded by the line of the first-order phase transition. From the figure it is clear that the melting curve must have a maximum. According to a recent study the maximum of the melting temperature is estimated to be $10^4 \sim 10^5$ K and at the density $\rho = 10^3 \sim 10^4$ g/cm^3.

At low pressure we have a gas of hydrogen molecules. When the temperature is raised, dissociation of the molecules starts, and when $k_B T$ exceeds the ionization energy of a hydrogen atom the gas becomes a plasma consisting of free electrons and protons. As shown in the figure, the temperature at which ionization occurs decreases as the pressure and hence the density are increased. When there are free electrons around atoms, they shield the Coulomb field of ions so that the ionization energy of the atoms is lowered. If we take this effect into account, the number of free electrons at a given temperature should be larger. This in turn enhances the screening effect. The process of ionization proceeds, therefore, as a cooperative phenomenon. The higher the density, the lower the temperature at which ionization starts and proceeds more rapidly. It is reasonable to expect that the boundary zone between the ordinary gas and the ionized plasma would join the insulator-metal transition line in the solid. The pressure effect on the ionization is an interesting problem which is in practice important in the study of stellar atmosphere.

In any material various phases like gas, liquid and plasma appear, apart from difference in scale, in the same way as in the case of hydrogen discussed here. At the same time systems composed of different elements possess an infinite variety of physical properties. The case of helium to be discussed in the next chapter will illuminate this point quite clearly.

3. Helium

In this chapter an assembly of helium atoms with $Z = 2$ is treated. Since there are two kinds of isotopes, ^{3}He and ^{4}He, which behave as a Fermi particle and a Bose particle, respectively, two different fluids, Fermi liquid and Bose liquid, appear. The new ingredient here is "superfluidity".

3.1 Characteristics of Helium

In this chapter we will consider a system made of $Z = 2$ nuclei and electrons, that is, an aggregate of He atoms. The rare gas elements which constitute the zeroth family of the periodic table have atoms with a closed shell of electrons and are hence chemically inactive (though Xe and Kr can form compounds with halogens). Although one might think systems made of such atoms could not be very interesting, they are quite important for physics of condensed matters, precisely because these atoms have a rather simple structure. Moreover, helium, the lightest of the rare gases, possesses unique properties not shared by any other materials.

A helium atom has a pair of electrons which occupy the $1s$ orbital in spin-singlet state. Energy necessary to excite it from the ground state to the lowest excited state is about 20 eV and the first ionization energy is 24.56 eV, both of which are largest among all atoms. Since, as we remarked in Sect. 2.2, He atoms do not form a stable diatomic molecule, they are the elementary building block. The interaction between two He atoms is qualitatively the same as that between two hydrogen molecules. Thus it is strongly repulsive when the $1s$ orbitals of two atoms overlap and becomes attractive due to the dispersion force outside. Since the excitation energy of a He atom is large, its polarizability α_0 is quite small, $\alpha_0 = 0.2 \times 10^{-24}$ cm^3, and in consequence the dispersion force is very weak. When the interaction between He atoms is expressed in the form of the Lenard-Jones potential (2.3.4), its parameters are $\varepsilon/k_B = 10.8$ K and $2^{1/6}\,\sigma = 2.88$ Å (see Table 3.1, where for comparison the parameters for other rare gases are included). When discussing properties of condensed states, it may be helpful to visualize He atoms as rigid balls of diameter 2.7 Å attracting each other weakly.

It is very fortunate for physics that besides the most abundant helium of mass number 4 a stable isotope ^{3}He exists. Since the electronic structure is identical in both isotopes, the interaction between ^{3}He atoms must be the same as in the case

Table 3.1. The parameters of rare gas elements

	ε/k_B [K]	$2^{1/6}\sigma$ [Å]	r_{nn} [Å]	T_B [K]	T_c [K]	Λ
³He				3.19	3.32	3.08
⁴He	10.8	2.88	(3.2)ᵃ	4.22	5.20	2.68
Ne	35.8	3.09	3.15	27.07	44.5	0.59
Ar	119	3.83	3.76	87.29	150.9	0.19
Kr	167	4.13	3.99	119.81	209.4	0.10
Xe	225	4.57	4.34	165.04	289.8	0.06

ᵃLiquid under saturated vapour pressure

of ⁴He. There is, however, an important difference in addition to that of mass. Composed of 3 nucleons, a nucleus of ³He possesses spin 1/2 and as a particle it is composed of an odd number of fermions. Suppose we interchange any two atoms in the wave function of a system consisting of many atoms. When doing so, we actually interchange 6 fermions (4 nucleons and 2 electrons) in the case of ⁴He and 5 in the case of ³He. As a result, the wave function should change its sign in the latter case but not in the former. When the internal structure of atoms stays unchanged and is hence irrelevant, we can regard ⁴He atoms as bosons and ³He as fermions. In any physical situations of interest to us here, the probability of two nuclei overlapping with each other is completely negligible. Nevertheless, this difference in statistics due to the difference in the nuclei manifests itself in a remarkable way in macroscopic properties of ⁴He and ³He at low temperatures.

Let us first ask if helium will condense into a solid or liquid state at a low enough temperature. This question was answered by H. Kamerlingh-Onnes in 1908, who initiated the modern low-temperature laboratory in Leiden. The leading principle of his research was provided by the theory of J. van der Waals, also of Leiden. In general, when a gas is compressed isothermally, its pressure and volume change as shown in Fig. 3.1. At high temperatures the ideal gas law $PV = RT$ holds, but as T is lowered, deviation from this law appears due to interaction between molecules. At a definite temperature characteristic of individual gases (called critical temperature T_c), the $P-V$ curve has an inflection point, called the critical point. Below this critical temperature the gas starts to

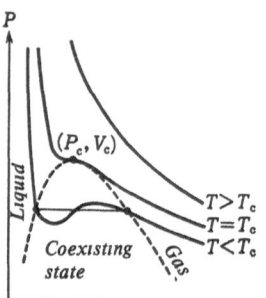

Fig. 3.1. The P-V diagram

liquefy when compressed as a certain specific volume is reached. After going through a coexistent phase of gas and liquid, it becomes entirely liquid, as shown in Fig. 3.1. According to van der Waals, the state of a real gas is approximately given by

$$\left(P + \frac{a}{V^2}\right)(V - b) = RT \tag{3.1.1}$$

(for derivation of this equation from statistical mechanics and discussions on gas-liquid transitions, see [3.1]). Since at the critical point (P_c, V_c, T_c) relations $\partial P/\partial V = 0$ and $\partial^2 P/\partial^2 V = 0$ hold simultaneously, the constants in (3.1.1) can be expressed as

$$a = \tfrac{9}{8} RT_c V_c, \qquad b = \frac{V_c}{3}.$$

Conversely, if we can accurately determine the constants a and b from measurements of the equation of states above T_c, we can predict the critical point. Following this idea, Kamerlingh-Onnes pursued his research and became convinced of the possibility of liquefying helium. Finally in 1908 he succeeded. In Table 3.1 the critical points of helium and other rare gases are listed. Note that except for helium, T_c is not appreciably different from the depth of the interaction potential ε/k_B. Under 1 atm helium 4 liquefies at 4.2 K and 3 at 3.2 K.

Does helium become a solid when we cool it further? Kamerlingh-Onnes thought that it would not solidify down to 0 K under low pressure. His guess was right. To solidify liquid ^{4}He and ^{3}He we must apply pressure higher than about 25 atm and 30 atm, respectively. This is due to the quantum effect we have mentioned in the study of a molecular solid of hydrogen. To express magnitude of this effect we usually use a quantity Λ defined as

$$\Lambda \equiv \frac{h}{\sigma\sqrt{m\varepsilon}}, \tag{3.1.2}$$

where ε and σ are the parameters appearing in the interaction potential between atoms of the form (2.3.4). It is readily seen that Λ is a measure of relative magnitude of the kinetic energy of the zero-point motion of atoms and of their interaction energy in the condensed states. A system with small value of Λ is classical and, as we saw in the preceding chapter, solidifies at a sufficiently low temperature. As shown in Table 3.1, Λ is small compared with 1 for all rare gases other than helium, for which the quantum effect is quite significant. In fact, helium is the only matter which stays liquid down to the absolute zero of temperature. One can say that the quantum fluctuations of atoms melt it even in the absence of thermal fluctuations. In ordinary liquids atomic positions are disordered due to thermal motion, and this disorder, of course, contributes to entropy. So to speak, a solid melts at high temperatures to increase its entropy. If, then, helium remains liquid down to 0 K, would it not be in conflict with the third law of thermodynamics? Liquid helium is, however, not an ordinary liquid. Since the entropy has to vanish as $T \to 0$ K, we expect it should be in some kind of ordered state at a sufficiently low temperature. Actually liquid ^{4}He becomes a superfluid and ^{3}He a Fermi liquid (which also becomes a superfluid of a different type). In this chapter we will mainly be concerned with the physics of these quantum liquids.

We can calculate the binding energy per atom at 0 K, that is, the energy necessary to remove an atom from the system in its ground state, from the internal energy of liquid helium, which can be determined experimentally from the heat of evaporation. The results are $k_B \times 7.15$ K for ^{4}He and $k_B \times 2.47$ K for ^{3}He. This should be compared to the potential energy per atom which is roughly $k_B \times 30 \sim 40$ K, implying that a good part of it is cancelled by the kinetic energy of zero-point motion.

Figures 3.2a, b are the phase diagrams of ^{4}He and ^{3}He. These and Fig. 3.3 showing the temperature dependence of the specific heat, should be enough to

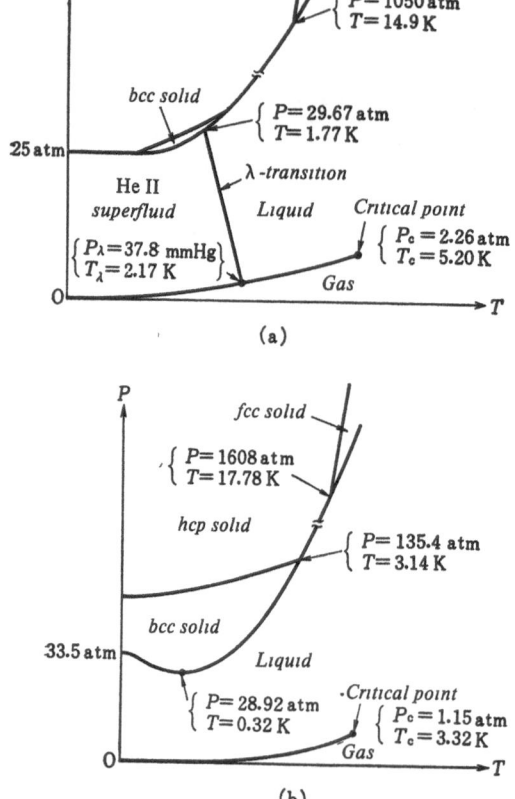

(a)

(b)

Fig. 3.2.(a) The phase diagram of ^{4}He (b) The phase diagram of ^{3}He

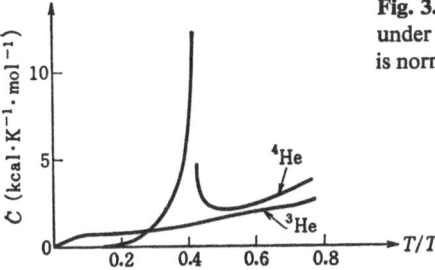

Fig. 3.3. The specific heat of liquid ^{3}He and ^{4}He under the saturated vapour pressure. Temperature T is normalized by the respective critical temperature T_c

convince us of remarkable differences in physical properties of the condensed states of the two isotopes. Although in the following we will only study the liquid states, it should be remarked that solid helium, being a typical quantum solid, also presents many interesting problems.

3.2 Superfluidity

When we cool liquid ^{4}He under saturated vapour pressure, we observe in the neighborhood of 2.172 K pronounced anomalies in temperature variations of its specific heat (Fig. 3.3), density or sound attenuation. Actually, Kamerlingh-Onnes himself had noted the anomaly of the density already in 1911 and later in 1926 discovered the anomaly in the specific heat. Since it seemed too extraordinary at that time, he thought there might be problems in his measurements and did not publish his findings. Only after 1938 was it established experimentally that liquid helium undergoes a transition to a new state with superfluidity.

Because of the shape of the anomaly in the temperature dependence of specific heat, this phase transition of liquid ^{4}He is called λ transition and the transition temperature is usually denoted by T_λ. Above T_λ liquid helium is in a normal state behaving like ordinary liquids, but below T_λ it is in superfluid state and is sometimes called He II. Let us first describe several experiments that show basic properties of the superfluid state.

1) He II can flow through extremely narrow channels without resistance. For example, it can flow through a capillary with an inner diameter of about 10^{-5} cm or a tube filled with fine powder of size less than 10^{-6} cm. Most remarkably, it flows without pressure difference at both ends of the channel provided the flow velocity is sufficiently small (Fig. 3.4). This means that energy of the flow does not dissipate as heat. Such a phenomenon can never be seen in ordinary liquids with viscosity.

We also observe superfluidity in thin films of liquid He II. For example, He II in the containers a and b in Fig. 3.5 can creep up the walls and flow into c. Under suitable conditions this flow takes place with no difference in the heights of the

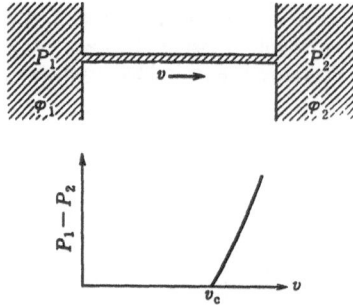

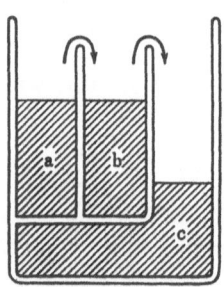

Fig. 3.4. The relationship between the flow velocity v and the pressure head $P_1 - P_2$

Fig. 3.5. The double beaker experiment

liquid levels in a and b. Such film flow is not possible in ordinary liquids because of the viscous drag of the walls. Superfluidity is observed even in extremely thin films with a thickness of about 10 Å, that is, films consisting of a few layers of He atoms. This means that in the He II phase it is a liquid with completely no viscosity.

2) On the other hand if we measure the viscosity of He II by one of the common methods of measuring liquid viscosity, we observe finite viscosity. In this method a stack of circular disks is suspended in a liquid and damping of its torsional oscillation due to viscosity is measured. The viscosity of He II observed in this way is strongly temperature dependent and vanishes at $T = 0$ K.

3) When we let He II flow out of an opening filled with fine powders, as shown in Fig. 3.6, the temperature inside the container rises. The so-called fountain effect is the reverse of this; when we raise the inside temperature by a heater, the liquid rises in the container.

4) At temperatures below T_λ, heat conduction becomes extremely good. Moreover, the conduction is not proportional to the temperature gradient as in ordinary cases. In fact the temperature gradient exists only in regions very close to walls. How well He II can carry heat is illustrated in the following experiment: when we immerse a thin, incandescent tungsten wire in the liquid, a thin

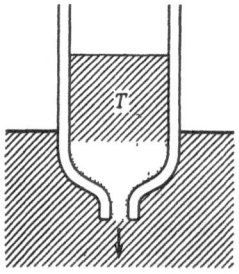

Fig. 3.6. Temperature rise due to a superfluid leak

vapour zone is formed around the wire but the liquid surrounding it does not boil and stays quiescent.

How can we understand these striking properties such as those described above? When the superfluidity was first discovered, F. London suggested that it might be related to the Bose-Einstein condensation. In an ideal Bose gas consisting of N noninteracting Bose particles in a volume V, the condensation occurs at temperature T_c given by

$$k_B T_c = \frac{2\pi\hbar^2}{m} \left(\frac{N}{2.6V}\right)^{2/3}. \tag{3.2.1}$$

Below T_c the single particle state of lowest energy (the state with momentum $p = 0$ in the case of a uniform system) is occupied by N_0 particles where

$$N_0(T) = N\left[1 - \left(\frac{T}{T_c}\right)^{3/2}\right], \tag{3.2.2}$$

that is, by a macroscopic number of particles. The remaining $N - N_0$ particles are thermally excited into higher single particle states. Obviously all particles fall into the lowest state at $T = 0$ K, $N_0(0) = N$. A macroscopic number of particles occupying a single quantum state is sometimes called a condensate. When a gas condenses into a liquid or a solid, it does so in real space, that is, the condensed phase separates itself into a region of space. In contrast to this, the condensation in the present case takes place in momentum space. Since in the condensate all particles occupy one state, it does not contribute to entropy. The entropy of the system is due to thermally excited particles.

If we substitute the mass of a ^{4}He atom as m and the number density of liquid ^{4}He as N/V in (3.2.2), we get $T_c = 3.13$ K. The entropy of the ideal Bose gas at T_c is equal to 1.28 R mol^{-1} where R is the gas constant. These values are not too different from $T_\lambda = 2.17$ K and the entropy of the liquid at T_λ, 0.8 R mol^{-1}.

After the suggestion by F. London, the two-fluid model was prodosed, according to which the superfluid and the normal fluid component coexist in He II just as in the ideal Bose gas below T_c where there are two components separated in the momentum space, namely the condensate and the thermally excited particles. The superfluid has no viscosity and does not contribute to entropy, whereas the normal fluid with viscosity behaves like ordinary liquids and contributes to entropy. If we denote the densities of the two fluids by ρ_s and ρ_n, their sum must of course be the density of the liquid ρ;

$$\rho = \rho_s + \rho_n.$$

The densities ρ_s and ρ_n depend on temperature, as in the ideal Bose gas, $\rho_s = \rho$ at $T = 0$ K and $\rho_s = 0$ at $T = T_\lambda$ (Fig. 3.7). Since the two fluids are separated in the momentum space, we can suppose them to move independently of each other. If we adopt such a two-fluid model, we can nicely explain the various phenomena mentioned above which seem to contradict each other. The super-flow phenomenon (i) is, of course, due to flow of the superfluid alone. The decay of disk oscillations (ii) is caused by the viscosity of the normal fluid. Since the normal fluid stays in the container, the entropy density of the liquid inside increases and hence the temperature should rise if the process is adiabatic, (iii). The unusual heat conduction (iv) can be explained as due to convective flows of the normal fluid moving from a high to a low temperature region and of the superfluid in the reverse direction.

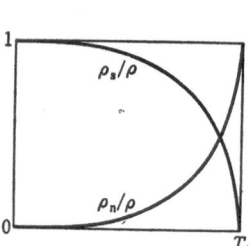

Fig. 3.7. Temperature variation of ρ_s and ρ_n

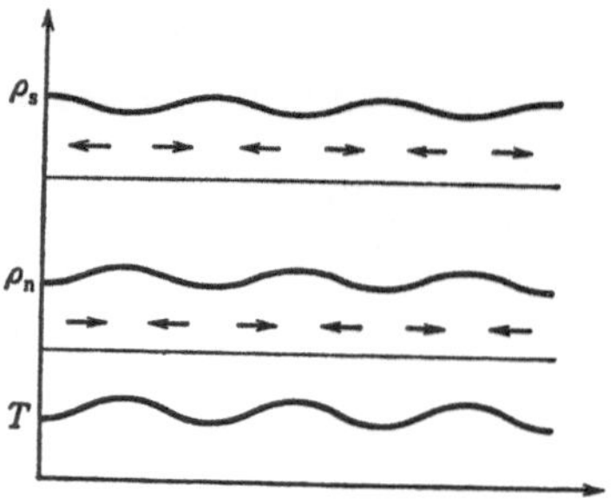

Fig. 3.8. Fluctuation of ρ_s, ρ_n and T associated with the second sound

The two-fluid model was also able to make predictions such as the possibility of the second sound. Ordinary sounds (referred to as the first sound in the case of He II), are fluctuations of the density ρ. Since the two fluids can move independently, fluctuations of ρ_s and ρ_n are possible with the total density $\rho = \rho_s + \rho_n$ kept constant (strictly speaking it does change slightly due to thermal expansion). Because ρ_n and hence the entropy density fluctuate, the second sound is accompanied by temperature fluctuation; it is a temperature wave. This is indicated in Fig. 3.8. Therefore, we can produce and observe the second sound by a heater and a thermometer. According to the two-fluid model the velocity of the second sound is given by

$$s_2^2 = \frac{\rho_s}{\rho_n} \frac{T\sigma^2}{C_v},$$

(3.2.3)

where σ is the entropy per unit mass and C_v the specific heat. By measuring the second sound velocity s_2 and the specific heat we can determine $\rho_n(T)$.

Thus, the low temperature phase of liquid ⁴He is characterized by the presence of the superfluid, namely, by finite ρ_s. Although the two-fluid model was initially conceived through an analogy with the ideal Bose gas, the model itself is independent of the latter. In the real liquid ⁴He, the interaction is rather strong and cannot be neglected. To get some idea about possible effects of the interaction, let us now discuss the ground state of a weakly interacting Bose gas. The ground state corresponds to $T = 0$ K where there is only the superfluid.

The Hamiltonian of this system has the same form as (2.3.5),

$$\mathcal{H} = \sum_{i=1}^{N} \frac{p_i^2}{2m} + \tfrac{1}{2}\sum_{i \neq j} U(x_i - x_j),$$

where m is now the mass of a ^{4}He atom. In Chap. 2 we have remarked that, when the classical approximation is valid, a system in its ground state is a crystal which minimizes the interaction energy. The fact that helium stays liquid at 0 K means that the kinetic energy is just as or rather more important than the interaction energy. With this in mind we will consider an extremely simple model with weak repulsive interaction between particles. We assume it to be purely repulsive because, if there is an attractive part as in the actual interaction, we expect the condensation into liquid state to take place, which makes our theoretical treatment rather difficult. Since the interaction is assumed to be weak, we include its effects by means of the simplest approximation, namely the Hartree approximation, sometimes called the mean-field approximation. This consists of approximating a wave function of a N-particle system by a product of single particle wave functions. Since we have a gas of Bose particles, its ground-state wave functon is then approximated by

$$\Psi(x_1, \ldots, x_N) = \prod_{i=1}^{N} \psi(x_i) \tag{3.2.4}$$

as in the ideal Bose gas. We assume $\psi(x)$ to be normalized as

$$\int dx \, \psi^*(x) \, \psi(x) = N \tag{3.2.5}$$

With this normalization the number density is given by

$$n(x) = \psi^*(x) \, \psi(x). \tag{3.2.6}$$

We determine the wave function $\psi(x)$ so as to minimize the expectation value of the total energy

$$E = \int \cdots \int dx_1 \ldots dx_N \Psi^*(x_1, \ldots, x_N) \mathcal{H} \Psi(x_1, \ldots, x_N)$$

under the subsidiary condition (3.2.5). With the Lagrange multiplier μ, we take the variation of $E - \mu N$ with respect to $\psi^*(x)$ and find an equation for $\psi(x)$;

$$-\frac{\hbar^2}{2m} \nabla^2 \psi(x) + \int |\psi(x')|^2 U(x - x') \, dx' \psi(x) = \mu \psi(x). \tag{3.2.7}$$

Since this is difficult to solve in general, we specialize in the case of point interaction, i.e.

$$U(x) = g\delta(x) .$$

This form is used to approximately express the interaction between hard spheres when their radius a is sufficiently small. It is known that the corresponding coupling constant is $g = 4\pi a\hbar^2/m$. Now (3.2.7) reduces to

$$-\frac{\hbar^2}{2m} \Delta^2\psi(x) + g|\psi(x)|^2\psi(x) = \mu\psi(x) . \tag{3.2.8}$$

This equation, sometimes called the nonlinear Schrödinger equation, has many interesting solutions. First of all, when the system is uniform and the mean particle density is n_0, we readily get $\mu = gn_0$. In terms of this we rewrite (3.2.8) as

$$-\tfrac{1}{2} \xi^2\Delta^2\psi(x) + \left(\frac{|\psi(x)|^2}{n_0} - 1\right) \psi(x) = 0 , \tag{3.2.9}$$

where ξ, a quantity with dimension of length, is defined by

$$\xi^{-2} \equiv \frac{mgn_0}{\hbar^2} \tag{3.2.10}$$

and is called the healing distance. If we impose a boundary condition $\psi = 0$ at walls, for example, ψ recovers its uniform value $\sqrt{n_0}$ at a distance of order ξ from the wall. This is due to the repulsive force between particles which keeps the density as uniform as possible. Since this effect is incorporated, our model is more realistic than the Bose gas. In the case of liquid ^{4}He, ξ is estimated to be a few Å.

Let us next consider a state with a flow of particles. When we calculate $\partial n(x)/\partial t$ using $(i/\hbar)\partial\Psi/\partial t = \mathcal{H}\Psi$ and the expression for the number density

$$n(x) = \int \cdots \int dx_1 \ldots dx_N |\Psi(x_1, \ldots, x_N)|^2 \sum_{i=1}^{N} \delta(x - x_i) , \tag{3.2.11}$$

we find the expression of the momentum density $j(x)$ as

$$j(x) = \tfrac{1}{2} \sum_{i=1}^{N} [p_i\delta(x - x_i) + \delta(x - x_i) p_i] . \tag{3.2.12}$$

Hence, for our wave function (3.2.4), its expectation value becomes

$$j(x) = -i \left(\frac{\hbar}{2}\right) [\psi^*(x)\nabla\psi(x) - \nabla\psi^*(x) \cdot \psi(x)] . \tag{3.2.13}$$

Let us write $\psi(x)$ in the form

$$\psi(x) = \sqrt{n(x)} \, e^{im\phi(x)/\hbar} , \tag{3.2.14}$$

where $n(x)$ is the particle density and $\phi(x)$ is the phase. Then, the velocity field $j(x)/mn(x)$ of the flow is simply given by

$$v(x) = \nabla\phi(x), \tag{3.2.15}$$

hence $\phi(x)$ plays the role of velocity potential. Thus we have an important conclusion that any flow described by $\psi(x)$ is potentially flow satisfying

$$\text{rot } v(x) = 0. \tag{3.2.16}$$

Since, furthermore, $\psi(x)$ is a wave function, its single-valuedness, or rather the multivaluedness of its phase, imposes a quantum condition on the flow. An integral of the velocity $v(x)$ taken along an arbitrary closed line C within the fluid which is called circulation, is now given by

$$\begin{aligned} \Gamma &\equiv \oint_C v \, dl = \oint_C \nabla\phi \, dl \\ &= nh/m \quad (n \text{ is an integer}), \end{aligned} \tag{3.2.17}$$

because the phase $m\phi(x)/\hbar$ can change by $2\pi n$ as x moves around C. The circulation is therefore quantized in the unit h/m (which is equal to 1.0×10^{-3} cm^2 s^{-1} for ^{4}He).

As an example we consider a rectilinear vortex line. It can easily be seen that (3.2.9) has a solution of the form $\psi = f(r) \exp(in\varphi)$ in the cylindrical coordinates. This solution represents a vortex line with velocity field

$$v_\varphi = \frac{n(h/m)}{r}$$

around the z axis (Figs. 3.9, 10). Because v_φ becomes infinite at $r = 0$, the amplitude $f(r)$ vanishes there. The central region where f is small is called the core of the vortex. Except at $r = 0$, rot $v = 0$ everywhere. Note that this, as well as the uniform flow state, corresponds to stationary states in quantum mechanics. What will happen if we rotate our fluid described by the wave function ψ? When we rotate an ordinary liquid in a container with a constant angular velocity ω, it will finally rotate like a rigid body. But in rigid body rotation we have rot $v = 2\omega \neq 0$, which is not allowed in our fluid. Instead, many vortex lines of $n = 1$ (which have lowest energy) with the density $2\pi m\omega/h$ per unit area appear and, the fluid as a whole mimicks rigid body rotation.

In the actual superfluid ^{4}He, it is confirmed by various experiments that vortex lines are generated when it is rotated. In one such experiment a cylindrical container with a thin wire at its center is used. When rotated, quantized circulation occurs around the wire, which is observed as a shift in vibrational frequencies of the wire. Very recently, images of vortex lines were taken by a method utilizing electrons trapped in the core of vortices.

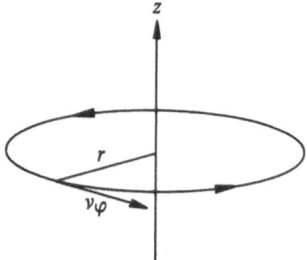

Fig. 3.9. Illustrating a vortex line

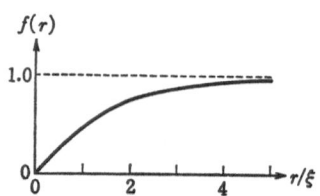

Fig. 3.10. The vortex-line solution $f(r)$

The superflow without a pressure head can also be understood in a natural manner as flow associated with the spatial variation of the phase of the wave function. When two vessels are connected by a capillary tube as in Fig. 3.4, we suppose the phase of ψ in vessel 1 to be φ_1 and that in 2 to be φ_2. If the phase difference $\varphi_1 - \varphi_2$ is not 0, there must be phase variation along the tube, which means there is flow in it.

Thus, we can understand the basic phenomena involving the superfluid alone in terms of the quantum-mechanical wave function ψ. Since ψ represents a state of a macroscopic number of particles, it may be called a macroscopic wave function. Such classical waves or classical fields are of course not unique to the superfluidity, the most familiar example being electromagnetic fields. Classical electromagnetic fields can be regarded as such, precisely because each wave mode is always occupied by a macroscopic number of photons. As in electromagnetic waves, the phase plays an essential role in superfluidity. In both cases phase relations are kept over a macroscopic scale.

In real liquid ^{4}He, as remarked already, the interaction between the atoms is not weak and the approximation used above is not applicable. This is perhaps best expressed in the condensate fraction. The number of particles N_0 in the lowest single particle state $p = 0$ has been estimated by various methods to be only about 10 % of the total number of particles. Because of the interaction, especially of its hard core part, He atoms in the liquid have to avoid each other in real space, which necessarily means spread in the distribution in momentum space. It is most likely, however, that even in the real superfluid, a macroscopic number of particles exists occupying a single state, in other words, the condensate exists. What is not permissible is to identify the condensate with the superfluid component. At 0 K we must have $\rho_s = \rho$, whereas even at 0 K the condensate density is only a tenth or so of ρ. Why, then, do the rest of the particles behave in the same way as the condensate? For simplicity let us consider the fluid at $T = 0$ K, i.e., in its ground state. Because of the interaction, quantum-mechanical motions of particles are correlated in such a way as to minimize the total energy. Suppose now the condensate is in a uniform flow state. When we look at the fluid in the frame of reference moving with it, the lowest energy state should be the same ground state. This means that all the particles have to move with the condensate; otherwise the correlation in the ground state can obviously not be maintained.

Can we then understand all the properties of the superfluid on the basis of the condensate described by the macroscopic wave function? Why, for example, doesn't superflow decay due to interaction with the walls? What constitutes the normal fluid at a finite temperature? To answer these questions we must consider excited states of our system.

3.3 Phonons and Rotons

To understand the thermodynamic properties of a physical system as well as its response to external fields, we must know the excited states of the system. In general, low-lying excited states of a system consisting of many particles can be described by so-called elementary excitations. For example, an excited state of a lattice which appears at a temperature low compared to its melting temperature, can be expressed by specifying how many phonons of each wave number and polarization there are in this state. The excitation energy is given approximately by the sum of the energy of individual phonons. Thus phonons are the unit of excitations, that is, the elementary excitations in this case. In a uniform system we can always choose elementary excitations carrying a definite momentum since the excited states can also be eigenstates of the total momentum operator. We denote by ε_k the energy of an excited state consisting of a single elementary excitation with momentum k (and, of course, with other quantum numbers such as spin if necessary) as measured from the energy of the ground state. A general excited state is specified by the number of elementary excitations with k present in it, i.e., by the distribution function n_k; its energy is then given by

$$E \approx E_0 + \sum_k{}' \varepsilon_k n_k ,$$ (3.3.1)

This is only approximate in most physical systems since interactions exist between elementary excitations. In a solid with anharmonic forces between atoms, a phonon can scatter with other phonons or decay into two phonons. Because of this, a state with two or more phonons present is in general no longer an exact eigenstate of energy even though a state with one phonon is. In many cases, however, the approximation is quite good as long as the density of elementary excitations is sufficiently small. At a finite temperature, there is a gas of thermally excited elementary excitations which determines thermodynamic as well as transport properties of the system. Once we have knowledge of elementary excitations, therefore, we can understand these properties at low temperatures to a considerable extent even without knowing microscopic structures of the system (more detailed discussions on elementary excitations can be found in [3.3]). It was L. D. Landau who, exploiting the concept of elementary excitations, developed an extremely effective theory of superfluid ⁴He, a Bose liquid, as well as of liquid ³He, a fermi liquid. In the following we will sketch the results of the Landau theory.

Landau assumed the elementary excitations in the superfluid ⁴He to have the dispersion relation of the form shown in Fig. 3.11, which is now called the Landau spectrum. It reduces to the phonon type,

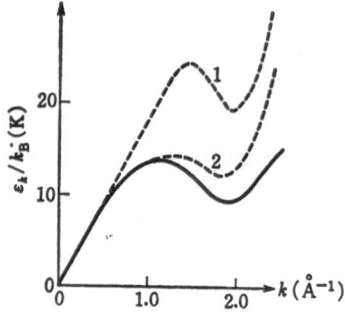

Fig. 3.11. Landau spectrum. See the text for dotted lines 1 and 2

$$\varepsilon_k = sk \tag{3.3.2}$$

in the region of small k, that is, of long wavelength. The velocity of sound s here is about 240 m s⁻¹. The excitations near the bottom of the valley in the spectrum are called rotons and have energy given approximately by

$$\varepsilon_k = \Delta + \frac{(k - p_0)^2}{2\mu}. \tag{3.3.3}$$

The phonon-roton spectrum Landau assumed was later confirmed experimentally by neutron inelastic scattering (Fig. 3.12). According to the experiments the

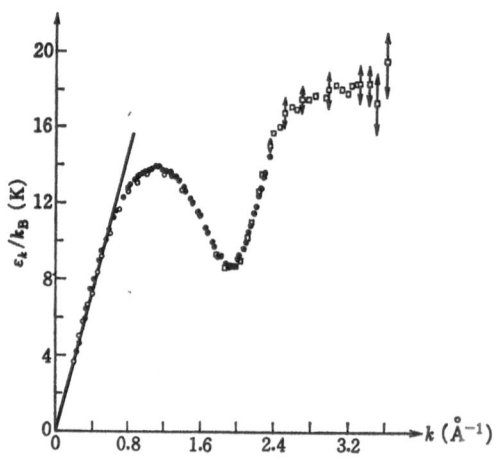

Fig. 3.12. The phonon-roton spectrum observed by the neutron inelastic scattering [3.1]

parameters appearing in (3.3.3) have the following values (at $T = 1$ K and under saturated vapour pressure):

$$\left. \begin{array}{l} \dfrac{\Delta}{k_B} = 8.65 \pm 0.04 \quad [\text{K}] \\[2mm] \dfrac{p_0}{\hbar} = 1.91 \pm 0.01 \quad [\text{Å}^{-1}] \\[2mm] \mu = 0.16m \end{array} \right\} . \tag{3.3.4}$$

Leaving aside for a while discussions on the microscopic picture of phonons and rotons, we first try to see what physical consequences we can draw from Landau's assumption. Since liquid ^{4}He is a boson system, elementary excitations should also obey Bose statistics. Hence their distribution in an equilibrium state at temperature T is given by

$$n(\varepsilon_k) = \left[\exp \left(\frac{\varepsilon_k}{k_B T} \right) - 1 \right]^{-1}. \tag{3.3.5}$$

Substituting here the energy ε_k shown in Fig. 3.11 and using (3.3.1), we can calculate the specific heat, for instance. The result agrees quite well with experimental data over a wide range of temperatures except near the λ point. Below about 0.9 K a dominant contribution to the specific heat is from phonons and above that from rotons.

Landau thought the gas of thermally excited phonons and rotons to be the normal fluid component. The normal fluid has the density ρ_n, in other words its flow carries mass. How do we determine ρ_n if the normal fluid is the gas of elementary excitations? If we had a microscopic theory of phonons and rotons in terms of dynamical variables of ^{4}He atoms, we would be able to calculate ρ_n from it. Since we do not yet have such a theory, however, we must follow Landau who used the following method.

Consider one elementary excitation in the presence of superfluid flow with uniform velocity v_s. In the frame of reference moving with v_s, the excitation with momentum k has obviously the energy ε_k because the superfluid is at rest in this frame. When we make a Galilean transformation from this frame to the rest frame, the total momentum k and the energy $E_0 + \varepsilon_k$ of our system will be transformed to

$$P = k + Nmv_s, $$
$$E = E_0 + \varepsilon_k + kv_s + \tfrac{1}{2} Nmv_s^2, \tag{3.3.6}$$

where E_0 is the ground state energy. Therefore, we must suppose this elementary excitation to have momentum k and energy $\varepsilon_k + k \cdot v_s$ in the presence of the superflow. Now we consider a case where the normal fluid is also moving with velocity v_n. This means that the gas of elementary excitations should have the equilibrium distribution in the frame of reference moving with the velocity v_n. Since the excitations have energy $\varepsilon_k + k \cdot (v_s - v_n)$ in this frame according to (3.3.6), their distribution is then given by $n(\varepsilon_k + k \cdot v_s + k \cdot v_n)$. In the frame moving with v_s, the momentum density is simply $\sum kn(\varepsilon_k + k \cdot v_s - k \cdot v_n)$. On the other hand, from the Galilean transformation of the momentum density j we have $j - \rho v_s = \rho_s v_s + \rho_n v_n - (\rho_s + \rho_n)v_s = \rho_n(v_n - v_s)$. Hence we arrive at the expression for ρ_n. When $v_s - v_n$ is small (compared to the velocity of sound, which is usually the case), we can expand the right-hand side and obtain

$$\rho_n = -\tfrac{1}{3} \sum_k k^2 \frac{\partial n}{\partial \varepsilon_k}. \tag{3.3.7}$$

When we use ε_k provided by neutron inelastic scattering, we get the results shown in Fig. 3.13. Except for the region near T_λ, the calculated ρ_n is in excellent agreement with the observed values obtained by various methods such as measurements of the second sound velocity.

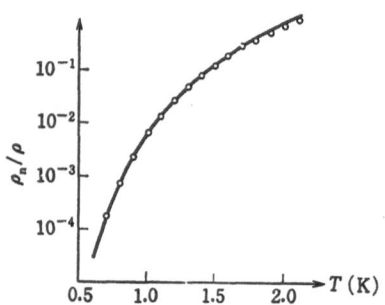

Fig. 3.13. Temperature dependence of ρ_n/ρ. The bold line is calculated from (3.3.7) with ε_k obtained by the neutron scattering experiments; (ooo) the values obtained from the velocity of the second sound

It is entirely natural that, being the gas of elementary excitations, the normal fluid can move independently of the superfluid and that it carries entropy. When spatial and temporal variation exist, one can set up the Boltzmann equation for the local distribution function $n_k(x, t)$ of elementary excitations, including the collision term due to interaction between the excitations. Here we just remark that the theory based on the Boltzmann equation for phonons and rotons is indeed very successful in quantitatively accounting for various transport phenomena such as viscosity and sound attenuation.

Some remarks are due here on the relationship between the elementary excitations and superfluidity. Firstly one might ask if a state with superflow is not also an excited state. The answer is yes, but it is a kind of excited state entirely different from phonons and rotons in the sense that the macroscopic number of particles are involved in the transitions between the ground state and a state with superflow or between states with different superflow. It is more appropriate to regard them as the different ground states in which the condensate has finite momentum corresponding to the flow. What corresponds to superflow in a crystal is its motion as a whole, that is, its center of mass motion. Since the very characteristic of a crystal is its spatial structure, phonons in a crystal cannot carry true momentum (only crystal momentum; see Chap. 5). As a consequence the gas of thermal phonons in a crystal does not behave like the normal fluid.

Our second remark is that the dispersion relation of the elementary excitations, namely, the Landau spectrum has important bearing on the superfluidity. We have already asked at the end of the preceding section why superflow does not decay due to friction with walls. Ordinarily fluid flow is dissipated because fluid particles are scattered or excitations like phonons are created by the interaction with walls. When an elementary excitation with momentum k is produced in the presence of superflow with velocity v_s, the energy of the system changes by $\varepsilon_k + kv_s$ according to (3.3.6). Therefore, the creation of such an excitation is now allowed by the conservation law unless

$$\varepsilon_k + k \cdot v_s \leqq 0 . \tag{3.3.8}$$

Clearly, if the spectrum is of the form shown in Fig. 3.11, this condition cannot be satisfied and hence the superflow is stable for $|v_s|$ smaller than the certain critical velocity v_c, which should be determined by the roton minimum (actually the critical velocity is determined by the creation of vortex lines rather than of rotons in most cases). In the case of the ideal Bose gas with $\varepsilon_k \propto k^2$, there are always excitations satisfying (3.3.8) for arbitrarily small v_s. It is because of the interaction between particles that the elementary excitations in liquid ^{4}He become phonons and rotons. In this sense the particle interaction keeps the coherent motion of superflow stable and is indispensable for superfluidity.

In the remaining part of this section we will briefly touch on the microscopic picture of phonons and rotons. The Landau theory is able to deal quite effectively with various phenomena in the superfluid without detailed knowledge of the microscopic structure of the ground state as well as the excitations. The next step in our understanding is naturally to construct a many-body theory of ^{4}He atoms which can reveal, among other things, the nature of phonons and rotons.

Longitudinal sound waves are density fluctuations and can propagate in liquids, in other words, they are good normal modes. There is, therefore, little doubt that the excitations of long wavelength which we have called phonons, are density fluctuations. Here we will show this, starting with the particle picture. Consider an excited state of our system in which one atom is moving with momentum k. A single-particle state with momentum k is described by a plane wave $\exp(i k \cdot x)$. Here we are considering a many-particle system in which particle motion is correlated because of their interaction. In particular, the hard-core repulsion does not permit wave-packets of particles to overlap with each other. We expect this correlation to be the same in the excited state as in the ground state. So, when the ith atom is moving with k, the wave function of our system should be given approximately by the $\exp(i\boldsymbol{k}\cdot\boldsymbol{x})\Psi_0$, where $\Psi_0(\boldsymbol{x}_1, ..., \boldsymbol{x}_N)$ is the wave function of the ground state. But wave functions of identical Bose particles must in general be symmetric with respect to the interchange of particles. Hence, we symmetrize the above expression as

$$\Psi_k(\boldsymbol{x}_1, ..., \boldsymbol{x}_N) = \sum_{i=1}^{N} \exp(i\boldsymbol{k}\cdot\boldsymbol{x}_i)\Psi_0(\boldsymbol{x}_1, ..., \boldsymbol{x}_N). \tag{3.3.9}$$

Here $\sum_{i=1}^{N} \exp(i\boldsymbol{k}\cdot\boldsymbol{x}_i)$ is just the Fourier component of the particle density $\rho(\boldsymbol{x})$ $= \sum_{i=1}^{N} \delta(\boldsymbol{x} - \boldsymbol{x}_i)$. Thus, starting with particle motion, we have arrived at the state with density fluctuation as the low-lying excited state. The energy of this state is given by

$$E - E_0 = \frac{1}{A} \int \cdots \int d\boldsymbol{x}_1 ... d\boldsymbol{x}_N \Psi_0^* \rho_k^* [\mathcal{H}, \rho_k] \Psi_0, \tag{3.3.10}$$

where A is the normalization constant. This can easily be calculated and the results for the excitation energy is simply

$$\varepsilon_k = E - E_0 = \frac{Nk^2}{2m} \frac{1}{A}. \tag{3.3.11}$$

The normalization constant can be expressed as

$$A = \iint d\boldsymbol{x}\, d\boldsymbol{y}\, e^{i\boldsymbol{k}(\boldsymbol{x}-\boldsymbol{y})} \rho_2(\boldsymbol{x} - \boldsymbol{y}),$$

where

$$\rho_2(\boldsymbol{x} - \boldsymbol{y}) \equiv \int \cdots \int d\boldsymbol{x}_1 ... d\boldsymbol{x}_N\, \rho(\boldsymbol{x})\, \rho(\boldsymbol{y}) |\Psi_0|^2. \tag{3.3.12}$$

This function $\rho_2(\boldsymbol{x} - \boldsymbol{y})$ is the density correlation function in the ground state. Because $\int d\boldsymbol{x}\, \rho_2(\boldsymbol{x}) = N\bar{\rho}$ ($\bar{\rho}$ is the mean particle density), the function P defined by

$$P(\boldsymbol{x} - \boldsymbol{y}) = (N\bar{\rho})^{-1} \rho_2(\boldsymbol{x} - \boldsymbol{y}) \tag{3.3.13}$$

is the probability of finding a particle at $\boldsymbol{y}$ when there is a particle at $\boldsymbol{x}$. If we denote the Fourier component of this function P by $S(k)$, the excitation energy ε_k becomes

$$\varepsilon_k = \frac{k^2}{2mS(k)}. \tag{3.3.14}$$

This important result, first obtained by R. Feynman, relates the excitation energy to the property of the ground state. The function $S(k)$ is called the structure factor and can be obtained by means of x-ray or neutron scattering experiments (Fig. 3.14). For intuitive understanding it is more convenient to use the pair correlation function defined by

$$g(r) = P(r) - \delta(r).$$

This is the probability of finding another particle at r when there is a particle at the origin. This function $g(r)$ given in Fig. 3.15 shows that He atoms cannot come too close to each other because of the hard core and that the nearest neighbors are about 3.4 Å apart.

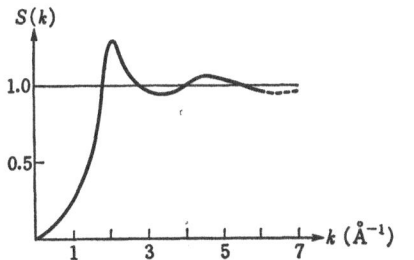

Fig. 3.14. The structure factor of liquid ⁴He

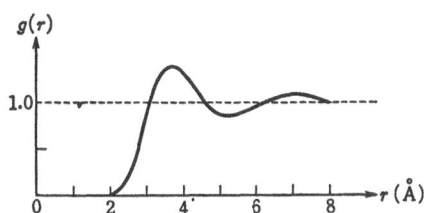

Fig. 3.15. Pair correlation function

Substituting $S(k)$ obtained from experiments, we find ε_k as shown by the dotted line 1 in Fig. 3.11. In the region of long wavelength it agrees well with the observed phonon spectrum. It also shows the valley corresponding to the roton spectrum, although the quantitative agreement is poor. This valley is caused by the peak in $S(k)$, which in turn expresses the spatial correlation between particles. The wave number corresponding to $r = 3.4$ Å at which $g(r)$ has the peak is equal to $2\pi/3.4 = 1.85$ Å⁻¹, close to the roton minimum, $p_0 = 1.91$ Å⁻¹. This suggests that the roton part of the spectrum appears due to a mechanism similar to the Bragg reflection in solids. When the wave number is close to p_0, neighboring particles move with almost the same phase, hence costing little energy.

To improve the result (3.3.14), we must take into account the correlation of particle motion in the excited state. The most important is the back-flow effect. For example, when a sphere is moving in a fluid, the fluid flows around the sphere as shown in Fig. 3.16, which is called the back-flow. When a sphere of radius a is moving in an incompressible fluid of density ρ, the effective mass of the sphere is larger by $(1/2)\rho(4\pi a^3/3)$ than its own mass (the fluid is assumed inviscid). It is clear that similar back-flow can occur also in a system of identical particles; in the case of particles with the hard core interaction, those in a path of a moving particle must give way to it. In the wave function (3.3.9) this effect is not incorporated. The theory which takes account of this effect gives the spectrum, as shown by the dotted line 2 in Fig. 3.11, considerably closer to the Landau spectrum.

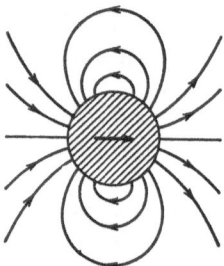

Fig. 3.16. Back-flow around a sphere

The effect of spatial correlation mentioned above and the back-flow effect are closely related. In this context it is interesting to refer to phonons in crystals. For instance, in the solid helium of bcc structure, the dispersion relation of lattice waves propagating in the (111) direction has a maximum and a minimum like the Landau spectrum. In Fig. 3.17 we schematically show the atomic motion in the lattice wave corresponding to the minimum. This figure suggests the relationship between the two effects we have discussed; it is natural that the back-flow effect is specially important in the roton region. Also it must be mentioned that according to the neutron scattering experiments, density fluctuations in liquid argon or in some liquid metals have a roton-like minimum in the dispersion relation, as one would expect from the above interpretation. The unique feature in the superfluid helium is that a roton is a good elementary excitation and has an infinite lifetime at 0 K.

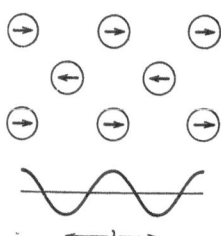

Fig. 3.17. Particle displacement associated with a lattice wave

The theory based on phonons and rotons becomes invalid near the λ point. As was remarked at the beginning, various quantities such as specific heat show conspicuous anomalies in the neighbourhood of the λ point. Such anomalies at phase transition have been studied intensively on the basis of the general theory of critical phenomena. The case of liquid ^{4}He has played an important role in this study because it is virtually an ideal system with little impurity or inhomogeneity, and also temperature control can be made extremely accurate (the specific heat measurements, for example, have been done as close to T_λ as $|T - T_\lambda|$ $\approx 10^{-6}$ K). The temperature dependence of the specfic heat in the neighbor-

Fig. 3.18. Specific heat near T_λ[3.2]

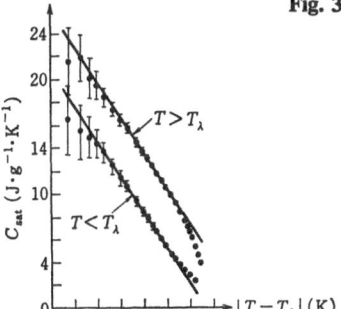

hood of T_λ is shown in Fig. 3.18; the specific heat diverges as $\ln|T - T_\lambda|$ on both sides of T_λ[1].

3.4 Liquid ³He; Fermi Liquid

Let us now turn to liquid ⁴He which is a typical fermion system. Its mass being 3/4 that of a ⁴He atom, a system of ³He atoms also remains liquid down to 0 K under low pressure. As we did for the electron gas in Chap. 2, the starting point of our discussion on properties of the liquid at low temperature is the ideal Fermi gas. Suppose there are N free Fermi particles in a volume V (which we take to be unit volume unless noted otherwise), and let $n_{p\alpha}$ be the number of particles in the state with momentum p, spin α and energy $\varepsilon_{p\alpha}$. In an equilibrium state at temperature T it is given by the Fermi distribution,

$$\bar{n}(\varepsilon_{p\alpha}) = \left[\exp\left(\frac{\varepsilon_{p\alpha} - \mu}{k_B T}\right) + 1\right]^{-1} \tag{3.4.1}$$

and especially at $T = 0$ K, that is, in the ground state, it is represented by a Fermi sphere. Here μ is the chemical potential determined by

$$\sum_{p\alpha} \bar{n}_{p\alpha} = N.$$

The total energy of the system is

$$E = \sum_{p\alpha} \varepsilon_{p\alpha} n_{p\alpha}. \tag{3.4.2}$$

Substituting (3.4.1), we can calculate the specific heat $C_v = (\partial E/\partial T)_v$, which at low temperature ($k_B T \ll \mu$) simplifies to

---

[1] More accurately, the specific heat anomaly is not logarithmic. If we express its temperature dependence as $(A/\alpha)\,[|1 - (T/T_\lambda)|^\alpha]$, the exponent α is $-0.04 \le \alpha \le 0.02$ according to recent measurements.

$$C_v = \frac{\pi^2}{3} N(0) k_B^2 T .$$

(3.4.3)

The linear dependence on T is characteristic of a degenerate Fermi gas in general. In this result $N(0)$ is the density of states at the Fermi surface

$$N(0) \equiv \sum_{p\alpha} \delta(\varepsilon_{p\alpha} - \mu) = \frac{m p_F}{\pi^2 \hbar^3} .$$

(3.4.4)

Because of the Pauli principle, particles occupying states within energy $k_B T$ below the Fermi surface can be thermally excited and acquire energy $k_B T$. Hence, $E(T) - E_0 \approx N(0) k_B T \cdot k_B T$, so that the specific heat is proportional to T as obtained above. When an external field H is applied, those particles with spin parallel to the field have their energy shifted by $-\hbar\gamma H/2$ ($\hbar\gamma/2$ being the magnetic moment of a ^{3}He nucleus in the present case), and those with spin antiparallel by $\hbar\gamma H/2$. Then the polarization of the system is given by

$$M = \tfrac{1}{2} \hbar\gamma \sum_p [\bar{n}(\varepsilon_{p\uparrow} - \tfrac{1}{2}\hbar\gamma H) - \bar{n}(\varepsilon_{p\downarrow} + \tfrac{1}{2}\hbar\gamma H)] .$$

In contrast to the $1/T$ dependence of the susceptibility at a high temperature when the Boltzmann distribution is valid, we find for the degenerate case

$$\chi = N(0) (\hbar\gamma/2)^2 .$$

(3.4.5)

Let us compare these results for the ideal Fermi gas with those observed for liquid ^{3}He. The specific heat data in Fig. 3.19 indeed show the linear dependence at a sufficiently low temperature. The susceptibility in Fig. 3.20 becomes inde-

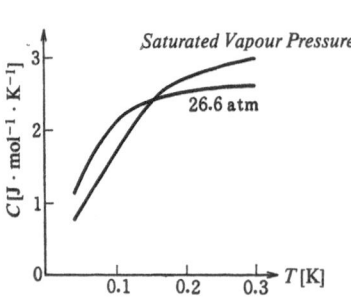

Fig. 3.19. Specific heat of liquid ^{3}He

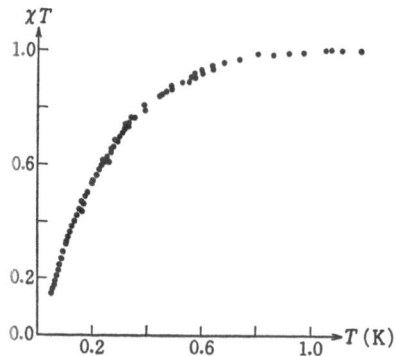

Fig. 3.20. Temperature variation of susceptibility; the ordinate is χ_T in arbitrary units [3.3]

pendent of T in the same region. From these and other evidence we can conclude that liquid ³He near 0 K shows the basic properties of the degenerate Fermi gas. In liquid ³He, however, we cannot neglect the interaction. The attractive part is necessary for the atoms to form a liquid, a condensed state. Even if we adopt for the attractive part a uniform potential well which holds the atoms together, a hard sphere model for the repulsive part requires a radius which is not small compared to the mean distance between atoms in liquid ³He. When we calculate the Fermi energy $\varepsilon_F = \hbar^2(3\pi^2 n)^{2/3}/2m$ from the density of liquid ³He at 0 K, 0.041 g cm⁻³, we find $\varepsilon_F/k_B \approx 5$ K. If liquid ³He were an ideal Fermi liquid we would then expect it to show the degenerate Fermi gas behaviour below 5 K. This is much higher than the actual value 0.1 K obtained from the specific heat and the susceptibility data. The coefficient $N(0)$ also does not agree with the empirical value. These facts are enough to show that the effect of the interaction is rather large in liquid ³He.

The theory of a fermion system with strong interaction such as liquid ³He was also presented by Landau. His theory, usually called the Fermi liquid theory, is able to describe in a coherent manner the equilibrium as well as the transport properties such as sound propagation, viscosity and spin diffusion. Since it has played an important role in the studies of electrons in metals and nuclear matter as well, we will discuss it in some detail.

3.4.1 Basic Ideas of the Fermi Liquid Theory

The Fermi liquid theory also starts with the question, "What are the elementary excitations of the system?". In the ideal Fermi gas, the ground state is the Fermi sphere and its excited states are obtained by raising a particle in a state $(\boldsymbol{p}, a)$ below the Fermi surface to a state $(\boldsymbol{p}', a')$ above, in other words, by creating a particle-hole pair. That the elementary excitation consists of a particle and a hole each of which is specified by momentum $\boldsymbol{p}$ and spin a and obeys the Fermi statistics, is the basic property we expect to find even in the presence of interaction between particles. Let us then make the basic assumption that there is a one to one correspondence between states of the ideal Fermi gas and states of the fermion system with interaction. As noted above, the ideal Fermi gas can be described by the occupation number n_{pa}. We therefore assume that this description is also valid for the interacting system. This assumption is obviously not valid when the system becomes a crystal, and even in a liquid state when it becomes a superfluid (as we will discuss in Chap. 4). The state for which this assumption is valid is called the normal Fermi liquid, with which we are concerned in the following.

Because of the interaction, that which is specified by $\boldsymbol{p}$, a is not a ³He atom itself but one with the effect of interaction incoporated, and hence its energy is different from the free particle value $p^2/2m$. We call it a "quasiparticle". According to the above basic assumption, the ground state of our interacting system is the Fermi sphere of the quasiparticles

$$n_0(\varepsilon_{p\alpha}) = \begin{cases} 1 & (\varepsilon_{p\alpha} \leq \mu) \\ 0 & (\varepsilon_{p\alpha} > \mu) \end{cases},$$ (3.4.6)

where μ is the chemical potential of the interacting system. The ground state energy is equal

$$E_0 = \sum_{p\alpha} n_0(\varepsilon_{p\alpha}) \, \varepsilon_{p\alpha}.$$ (3.4.7)

The excited states are described by the deviation of the quasiparticle distribution $n_{p\alpha}$ from $n_0(\varepsilon_{p\alpha})$

$$\delta n_{p\alpha} = n_{p\alpha} - n_0(\varepsilon_{p\alpha}).$$ (3.4.8)

We have to make the concept of quasiparticles more precise. Suppose we add a particle with momentum p (we omit mentioning its spin for a while) outside the Fermi sphere, in other words we create a state with a quasiparticle with momentum p and the Fermi sphere. This state is not an eigenstate of energy because of the interaction between quasiparticles. The quasiparticle above the Fermi surface can fall to lower (but empty) states, creating a particle-hole pair. Hence the quasiparticle has a finite lifetime τ. In other words, the state of the system described by the Fermi sphere plus the quasiparticle with p changes to other states in the time τ. But the lifetime of quasiparticles near the Fermi surface is very long due to the Pauli principle. If there were not the exclusion principle, τ would be given by the time between collisions of a particle with its neighbors. Since the velocity of particles near the Fermi surface is v_F and the mean distance between particles is $n^{-1/3}$, the lifetime would then be of the order of $n^{-1/3}/v_F$. Let us now take into account the Pauli principle. Suppose the energy of the quasiparticle is higher by Δ than the Fermi energy μ. Because of the energy conservation law, this quasiparticle can excite by collisions only particles which are within Δ below the Fermi surface. Moreover, since the quasiparticle cannot fall into occupied states and hence has to stay outside the Fermi sphere, it can transfer to colliding particles only an energy of the order of Δ and not of ε_F, consequently reducing their final states by a factor of Δ/ε_F. Thus the probability of allowed collisions is reduced by a factor $(\Delta/\varepsilon_F)^2$, leading to the estimate for the lifetime

$$\tau \approx \frac{n^{-1/3}}{v_F} \left(\frac{\varepsilon_F}{\Delta}\right)^2.$$

Since the quasiparticles we have to consider at temperature T are those with $\Delta \approx k_B T$, their lifetime is proportional to T^{-2}. It is known from the data of viscosity and thermal conductivity that at sufficiently low temperature that

$$\tau \approx 10^{-12}/T^2 [s]$$

(according to the above formula we find $10^{-11}/T^2$, which is not too bad). At

$T = 0.1$ K we have $\tau \approx 10^{-10}$ s, which corresponds to a mean free path of 50 Å, and at $T = 1$ mK, 5×10^{-3} cm. For the quasiparticle picture to be valid, its energy $\varepsilon_p - \varepsilon_F$ must be larger than the width due to the uncertainty $\hbar/\tau$. Hence the region of temperature where the Fermi liquid theory is valid is $k_B T \gg \hbar/\tau$, that is, $T \ll 0.1$ K.

Since we only have to consider a quasiparticle near the Fermi surface, we write its energy as approximately

$$\varepsilon_p = \mu + \frac{p_F(p - p_F)}{m^*}, \tag{3.4.9}$$

which amounts to keeping only the first term in the expansion of the energy in powers of $|p| - p_F$. Note that the effective mass of the quasiparticle is not equal to the mass m of the original particle due to the interaction with surrounding particles. One can say that the quasiparticle is dressed. The simplest mechanism of this is the back-flow effect we discussed in the preceding section.

We now consider the energy of an excited state specified by $\delta n_{p\alpha}$. The energy of individual quasiparticles depends on states of surrounding quasiparticles because of the interaction, so that it will change when there are other quasiparticle excitations. In other words, there is also interaction between quasiparticles, which should of course be distinguished from the original interaction between the bare particles (we have already mentioned such interaction between elementary excitations in the case of phonons). Thus we write the energy of the state with $\delta n_{p\alpha}$, including the effect of the interaction, as

$$E = E_0 + \sum_{p\alpha} \varepsilon_{p\alpha} \delta n_{p\alpha} + \frac{1}{2V} \sum_{pp',\alpha\alpha'} f_{p\alpha,p'\alpha'} \delta n_{p\alpha} \delta n_{p'\alpha'}$$
$$+ \text{(the 3rd order terms in } \delta n) + \dots. \tag{3.4.10}$$

When the total number of excited quasiparticles is small compared to N (at temperature T it is roughly $Nk_B T/\mu$), we can neglect the terms higher than the second order in δn. To write the energy in terms of quasiparticle distribution $\delta n_{p\alpha}$, as (3.4.10), may be considered as the mean field (Hartree-Fock) approximation applied to the interaction between quasiparticles. Consequently, the energy of a quasiparticle with p, α becomes

$$\tilde{\varepsilon}_{p\alpha} \equiv \frac{\delta E}{\delta n_{p\alpha}} = \varepsilon_{p\alpha} + \frac{1}{V} \sum_{p'\alpha'} f_{p\alpha,p'\alpha'} \delta n_{p'\alpha'} \tag{3.4.11}$$

and contains the effect of the potential field produced by other particles.

The coefficients $f_{p\alpha,p'\alpha'}$ characterize the effective interaction between quasiparticles. Near the Fermi surface they hardly depend on the magnitude of p and p', so that we regard them as functions of the angle between p and p' and of spin indices α, α'. We hence expand it in partial waves and write

$$f^{s,a}_{pp'} \equiv \tfrac{1}{2}(f_{p\uparrow p'\uparrow} \pm f_{p\uparrow p'\downarrow}) = \sum_l f^{s,a}_l P_l(\widehat{pp'}),$$ (3.4.12)

where the superscripts s and a correspond to $+$ and $-$, respectively, and P_l is the Legendre polynomial. Instead of f which has dimension energy $\times$ volume, we introduce, the following dimensionless parameters by multiplying f with the density of states per unit volume, $N(0) = m^*p_F/\pi^2\hbar^3$,

$$F^{s,a}_l = N(0) f^{s,a}_l$$ (3.4.13)

they are called Landau parameters. We note that the part of the interaction antisymmetric with respect to the spin indices, which has the superscript a, arises from the exchange effect in the case of liquid ³He.

3.4.2 The Equilibrium Properties

Let us first apply this theory to the equilibrium states. The equilibrium distribution at T is

$$\delta n_p(T, \mu) = \bar{n}(\bar{\varepsilon}_p) - n_0(\varepsilon_p),$$

where p stands for $p\alpha$ as in the following. The interaction between quasiparticles leads only to the correction of higher orders in T for the specific heat. The term linear in T is then given by the same expression as (3.4.3),[2]

$$C_v = \frac{m^*p_F}{\pi^2\hbar^3} k_B^2 T.$$ (3.4.14)

The compressibility κ may be calculated with the help of the expression

$$\kappa = -\frac{1}{V}\frac{\partial V}{\partial P} = \frac{1}{n}\frac{\partial n}{\partial P} = \frac{1}{n^2}\frac{\partial n}{\partial \mu},$$

where use has been made of the Gibbs-Duhem relation $Nd\mu - VdP = 0$ at $T = 0$ (since the leading term of κ is independent of T, we have put $T = 0$ here). Suppose a uniform external field U is applied so that the Hamiltonian of the system has an additional term $\int Un(x)dx = U\sum_p n_p$. Since $\partial n/\partial \mu = -\lim_{U\to 0} \partial n/\partial U$, we can find κ by calculating the response of the density n to the field U to the term linear in U. When U is present, the energy of the quasiparticle is given by

---

[2] The measurement of specific heat has been carried out down to about 50 mK, and the deviation from this formula is observed at a very low temperature. It is interesting that this deviation cannot be explained by the simple correction terms (αT^3). Recent studies have shown that it has a nonanalytic form $T^3\ln T$ and that such a correction can be derived from the Fermi liquid theory.

$$\varepsilon_p + U + \frac{1}{V}\sum_{p'} f_{pp'}\delta n_{p'}, \tag{3.4.15}$$

where we must note that δn_p is induced by the field U. Therefore,

$$\frac{\partial n_p}{\partial U} = \frac{\partial}{\partial U}\frac{1}{\exp[(\tilde{\varepsilon}_p + U - \mu)/k_{\mathrm{B}}T] + 1} = \frac{\partial n_p}{\partial \tilde{\varepsilon}_p} + \frac{\partial n_p}{\partial \tilde{\varepsilon}_p}\frac{\partial \tilde{\varepsilon}_p}{\partial U}$$

$$= \frac{\partial n_0(\varepsilon_p)}{\partial \varepsilon_p}\left(1 + \frac{1}{V}\sum_{p'} f_{pp'}\frac{\partial n_{p'}}{\partial U}\right). \tag{3.4.16}$$

This is an integral equation for the quantity $\partial n_p/\partial U$, and is one instance of a "self-consistency equation" that we always have in the mean-field approximation. Because of the isotropy of the system, $\partial n_p/\partial U$ does not depend on the direction of p, nor does the quantity $\Lambda = (1/V)\sum f_{pp'}\partial n_{p'}/\partial U$. We can regard $\partial n_0(\varepsilon_p)/\partial \varepsilon_p$ to be a δ function like at the Fermi surface, so that from (3.4.16) we have

$$\Lambda = -N(0)f_0^s(1 + \Lambda).$$

Hence we obtain

$$\kappa = -\frac{1}{n^2}\sum_p \frac{\partial n_p}{\partial U} = \frac{N(0)}{n^2}\frac{1}{1 + F_0^s}. \tag{3.4.17}$$

From this we find the velocity of ordinary sound $s_1 = (mn\kappa)^{-1/2}$ as

$$s_1 = \frac{p_{\mathrm{F}}}{m}\sqrt{\frac{m}{3m^*}}(1 + F_0^s). \tag{3.4.18}$$

Substituting the solution Λ into (3.4.15), we find the energy of the quasiparticle in the presence of the field U,

$$\varepsilon_p + \frac{U}{1 + F_0^s}$$

which means that the field U is replaced by the effective field $U/(1 + F_0^s)$. When F_0^s is positive, that is, repulsive, this becomes the screening effect and when it is negative, the effect is to enhance the external field U, physically a very reasonable result.

Next we turn to the susceptiblity which can be obtained by the same method as above if we introduce a uniform external field that changes its sign according to spin directions, $U = \pm \hbar\gamma H/2$. The result is

$$\chi = \frac{\hbar^2\gamma^2 N(0)}{4(1 + F_0^a)}. \tag{3.4.19}$$

Table 3.2 lists the Landau parameters and the effective mass of liquid ^{3}He determined from the experimental data of C_v, κ, and χ. Evidently the effect of the quasiparticle interaction is very important. The large values of F_0^s and its rapid increase with pressure reflect the hard-core repulsion between ^{3}He atoms. That F_0^a is negative means there is effectively a force which aligns the quasiparticle spins. One can interpret this as a pair of quasiparticles with the same spin avoiding each other due to the Pauli principle and hence experiencing less repulsive force between them. If F_0^a is less than -1, the liquid polarizes spontaneously, that is, becomes ferromagnetic. The large values of the Landau parameters mean, we want to emphasize, that the theory is valid for a system of fermions with strong interaction.

Table 3.2.

Pressure [atm]	0.28	27.0
$p_F/\hbar$ [cm^{-1}]	7.880×10^7	8.767×10^7
m^*/m	3.08	5.78
v_F [cm·s^{-1}]	5.38×10^3	3.19×10^3
F_0^s	10.77	75.63
F_1^s	6.25	14.35
F_0^a	-0.67	-0.72
F_1^a	-0.72	-0.66
s_0/v_F	3.60	12.2

The virtue of the Landau theory is that it is able to describe dynamical phenomena in the liquid ^{3}He as well as its equilibrium properties quite effectively. In this sense it also corresponds to the phonon-roton theory of liquid He II. In order to deal with dynamical phenomena we have to generalize the above theory to situations with spatial and temporal variations. Let us denote the characteristic length scale of variations by $1/q$ and the characteristic time scale by $1/\omega$. Considering that the characteristic energy of our system is the Fermi energy μ and that a particle-hole pair with total momentum q is qv_F, we expect the condition for spatially and temporally slow variation to be

$$\hbar q v_F \ll \mu, \quad \hbar \omega \ll \mu .$$

If these conditions are satisfied, we can speak of the local distribution $n_p(\boldsymbol{x}, t)$, which satisfies the Boltzmann equation

$$\frac{\partial n_p}{\partial t} + \boldsymbol{\nabla}_p \tilde{\varepsilon}_p \cdot \boldsymbol{\nabla}_x n_p - \boldsymbol{\nabla}_x \tilde{\varepsilon}_p \cdot \boldsymbol{\nabla}_p n_p = I(n_p) . \tag{3.4.20}$$

Here $\tilde{\varepsilon}_p$ is the local value of the quasiparticle energy

$$\tilde{\varepsilon}_p(\boldsymbol{x}, t) = \varepsilon_p + \sum_{p'} f_{pp'} \delta n_{p'}(\boldsymbol{x}, t) \tag{3.4.21}$$

and I is the collision term. The interaction between quasiparticles is responsible for the scattering as well as for the mean field we have been considering. When the deviation of the distribution function is small, we can linearize (3.4.20) with respect to $\delta n_p(x, t) = n_p(x, t) - n_0(\varepsilon_p)$ as

$$\frac{\partial \delta n_p}{\partial t} + v_p \cdot \nabla_x \delta n_p - \frac{\partial n_0}{\partial \varepsilon_p} v_p \cdot \nabla_x \sum_{p'} f_{pp'} \delta n_{p'} = I, \qquad (3.4.22)$$

where $v_p = \nabla_p \varepsilon_p = p/m^*$. When the sum over p and the spin is taken of both sides, the continuity equation for the number density of quasiparticles $n(x, t)$

$$\frac{\partial n}{\partial t} + \nabla_x \sum_p v_p \left(\delta n_p - \frac{\partial n_0}{\partial \varepsilon_p} \sum_{p'} f_{pp'} \delta n_{p'} \right) = 0$$

results because the collision term does not change the total number of particles. Hence we find the expression for the flow density as

$$j(x, t) = \sum_p \frac{p}{m^*} \left(1 + \frac{F_1^s}{3} \right) \delta n_p(x, t).$$

On the other hand, if there is one to one correspondence between the original particle and the quasiparticles, it must be that

$$j(x, t) = \sum_p \frac{p}{m} \delta n_p(x, t).$$

A state represented by the Fermi sphere displaced by q serves as a simple example for this problem. From the above two expressions we find that the effective mass m^* is related to the Landau parameter F_1^s as

$$\frac{m^*}{m} = 1 + \frac{F_1^s}{3}. \qquad (3.4.23)$$

The transport coefficients such as viscosity and thermal conductivity can be calculated with the help of (3.4.22). Instead of describing such calculations, let us here discuss sound propagation in the liquid ³He. Landau predicted that the nature of sound in the Fermi liquid would be different from the ordinary sound at a sufficiently low temperature. Such a sound mode, called the zero sound (as distinct from the ordinary one called the first sound), was later detected experimentally. In the case of the first sound we can regard the medium to be always in local thermal equilibrium, in other words, the period $1/\omega$ is much larger than the collision time $\tau(\omega\tau \ll 1)$. Thus the sound velocity is determined by the compressibility (3.4.18), an equilibrium property. In the opposite case, namely when $\omega\tau \gg 1$, we can neglect the collision term in (3.4.22). Let us consider density fluctuations in this collisionless regime. Dropping I and putting $\delta n_p \propto \exp(i\, q \cdot x - i\, wt)$ in (3.4.22), we get

$$(\omega - v_p \cdot q)\, \delta n_p = q \cdot v_p \frac{\partial n_0}{\partial \varepsilon_p} \sum f_{pp'} \delta n_{p'} \,. \tag{3.4.24}$$

Because of the factor $\partial n_0/\partial \varepsilon_p$ on the right-hand side, we can regard δn_p to be different from zero only at the Fermi surface. Taking a z direction along q and denoting polar angles of p by (θ, φ), we write

$$\delta n_p = \frac{\partial n_0}{\partial \varepsilon_p} \Lambda(\theta, \varphi) \,.$$

Restricting ourselves to waves independent of spin and taking a simple model with only F_0^s and F_1^s among the Landau parameters nonvanishing, (3.4.24) reduces to

$$\Lambda(\theta, \varphi) = \frac{\cos \theta}{(\omega/qv_F) - \cos \theta} \int \frac{d\Omega'}{4\pi} (F_0^s + F_1^s \cos \widehat{pp'})\, \Lambda(\theta', \varphi') \,.$$

If we specialize to longitudinal solutions independent of φ, we find for $\gamma = \omega/qv_F$

$$\frac{\gamma}{2} \ln \frac{\gamma + 1}{\gamma - 1} = 1 + \left(F_0 + \gamma^2 \frac{F_1}{1 + F_1/3} \right)^{-1} \,.$$

When F_0^s, $F_1^s \ll 1$, $\gamma \approx 1$, that is, the velocity of the sound s_0 becomes equal to the Fermi velocity v_F. In this limit the deformation of the Fermi sphere associated with the wave is concentrated in the direction $\theta = 0$, i.e., of q. In the opposite limit $F_0^s \gg 1$, $\gamma^2 = (1/3)F_0(1 + F_1/3)$ and the velocity s_0 approaches the velocity of the first sound. In this case the Fermi sphere has the deviation proportional to $\cos\theta$, that is, it is displaced as a whole. This means that locally, all particles flow uniformly as in the hydrodynamical sound.

To observe the zero sound it is necessary to use high frequency sound waves at a sufficiently low temperature so that $\omega\tau \gg 1$. Figure 3.21 shows the results of the measurements down to 2 mK of the sound velocity and the attenuation constant at 15.4 and 45.5 MHz. The transition from the first to the zero sound as T changes is clearly seen. If F_0^s, F_1^s and v_F given in Table 3.2 are used, we find the theoretical value for $(s_0 - s_1)/s_1 = 0.034$ which is in excellent agreement with the observed value 0.035.

To understand the sound attenuation we approximate the collision term in the Boltzmann equation by the simple form $\delta n_p/\tau$. This amounts to assuming that all quasiparticle excitations have a lifetime τ. When we do this, the above treatment of the zero sound remains valid if we simply replace ω by $\omega - i/\tau$ everywhere. Thus, if $\omega\tau \gg 1$, the relaxation time of the zero sound is also given by τ. The attenuation constant, defined as the inverse of the distance that the sound propagates before its amplitude decays by a factor of e^{-1}, is hence proportional to τ^{-1}. On the other hand, it is easy to show from the Navier-Stokes equation of fluid mechanics that the attenuation constant of the first sound is proportional to $\omega^2\eta$ and hence to τ because the shear viscosity coefficient η is proportional to τ. Recall that the lifetime τ depends on T as T^{-2}. Figure 3.21 beautifully shows the difference between the two sounds.

Another example to which the Fermi liquid theory has been successfully applied is the liquid mixture of ³He and ⁴He. Figure 3.22 is the phase diagram on a $T - x$ plane [x is the ³He concentration; $x = n_3/(n_3 + n_4)$]. Among several interesting facts one can notice in this phase diagram, it is particularly remarkable that even at $T = 0$ K ³He can dissolve in the superfluid ⁴He up to about 6 %. Ordinarily two components, especially isotopes, mix because of the mixing entropy. The unique example of mixing at $T = 0$ K here is evidently due to other reasons. Below about 0.1 K, the elementary excitations, phonons and rotons of the superfluid are hardly excited. When a single ³He atom is present in this superfluid, it will behave like a free fermion with an effective mass different from m_3 (already a quasiparticle due to the presence of the medium). When a finite concentration of ³He atoms is dissolved, they must become a degenerate Fermi liquid at a sufficiently low temperature. In this case the ³He quasiparticles interact with each other directly and indirectly by exchanging phonons. Thus at a sufficiently low temperature, we have again a Fermi liquid, this time in the medium of the superfluid ⁴He.

So far we have discussed the physics of the normal Fermi liquid. Will the liquid ³He remain normal when we cool it further, or will it make a phase

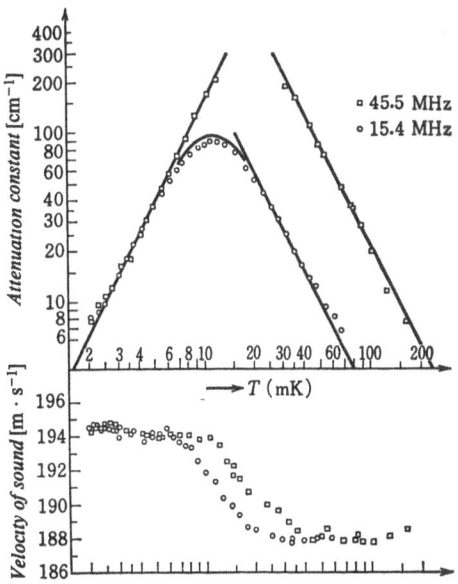

Fig. 3.21. Temperature variation of the velocity and the absorption coefficient of sound. The crossover from the first to the zero sound is clearly seen [3.4]

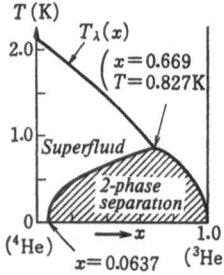

Fig. 3.22. The phase diagram of ^{3}He–^{4}He mixture. The liquid mixture exists except in the hatched region

transition? The order possessed by the normal Fermi liquid is entirely due to the Fermi statistics. If it undergoes some phase transition, the low temperature phase should be more highly ordered. Since it is liquid down to 0 K, the new order cannot be of spatial character like in crystals. There remains the possibility of a superfluid state. Being a fermion system, however, ^{3}He cannot become a superfluid of the same type as ^{4}He. It is known that conduction electrons in a metal compose another fermion system and we know that many metals undergo a phase transition at low temperature and become superconducting. In metals which become superconductors, the overall interaction between electrons near the Fermi surface is attractive, the phonon mediated attraction overcoming the Coulomb repulsion. At $T = 0$ K the electrons near the Fermi surface form pairs in spin-singlet state because of this attractive force and all the pairs formed are in a single-quantum state with respect to their center of mass motion. So one can again speak of a macroscopic wave function for the pairs which plays the role of order parameter in the superconducting state. Because of this, the superfluid state of liquid ^{4}He and the superconducting state have their essential properties in common, although microscopically they appear in entirely different mechanisms. In the former the fermions (even number of nucleons and electrons) first form a strongly bound state in real space, namely a ^{4}He atom, and then they make the Bose-Einstein condensation, while in the latter the fermions (conduction electrons) form pairs and condense simultaneously in momentum space because of the presence of the Fermi sphere. The microscopic theory of superconductivity based on the idea of pairing was presented by Bardeen, Cooper and Schrieffer (BCS) in 1957 (Chap. 5). The possibility that liquid ^{3}He also becomes a superfluid of the pairing type was pointed out in 1961. Since the interaction is strongly repulsive in short range and attractive in long range in this case, the possible pairing may be in a state with finite angular momentum, in contrast to the singlet s-state as in superconductors. Experimentally many efforts have been made since then, and in 1972 the phase transition was finally discovered at about 2 mK and under the melting pressure.

We mention here only the specific heat data. As shown in Fig. 3.23, the specific heat of the liquid under 33 atm makes a jump at 2.6 mK and then decreases more rapidly than that of the normal Fermi liquid as T is lowered. This

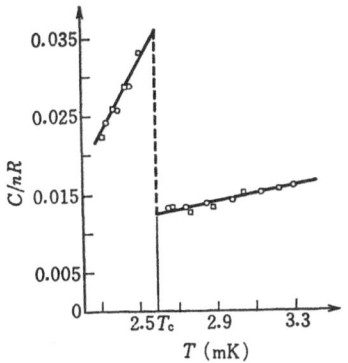

Fig. 3.23. Molar specific heat of liquid ^{3}He under 33 atm [3.5]

specific heat anomaly is similar to one we see in superconductors. This new phase has since been identified as the superfluid state with $^3P_2$ pairing (Chap. 4).

4. Superfluid Helium 3

As remarked at the end of Chap. 3, the new phase of liquid ³He was discovered in 1972. Since then the study of this new phase has made great progress both experimentally and theoretically. We now know that this phase is a superfluid state of the BCS type but with pairing in spin triplet and the orbital angular momentum $l = 1$ state. Since the pairs have internal degrees of freedom, in particular the spin, this superfluid exhibits new physical properties not seen in liquid ⁴He or in superconductors, and offers an exemplary case for the physics of "broken symmetry". The purpose of this chapter is to describe the basic experimental facts and to discuss some of the most characteristic phenomena of the new superfluid. For detailed discussions the readers are referred to the review articles listed in the bibliography.

4.1 Basic Properties

4.1.1 Thermodynamic properties

Liquid ³He becomes a typical degenerate Fermi liquid below about 0.1 K and then enters into a new phase in the mK region. As seen in Fig. 3.2, it remains liquid up to about 33.5 atm near 0 K, and its properties as a Fermi liquid vary greatly with pressure. The superfluid phase appears in the P-T plane, as shown in Fig. 4.1; the transition temperature T_c varies from 2.6 mK under the melting pressure to 0.9 mK under the saturated vapour pressure. We have already mentioned that the specific heat makes a jump at T_c as in superconductors (Fig. 3.23). It is extremely interesting that two distinct phases A and B exist, the former appearing in the high pressure, high temperature region. Moreover, the first-order transition line between the A and B phase, $T_{AB}(P)$, changes with the magnetic field, as shown in the figure. Note especially that when the field is weak, the T_{AB} line approaches the T_c line asymptotically at low temperatures. Actually, the normal-superfluid transition line T_c splits into two in the presence of the magnetic field (which is not shown in Fig. 4.2 because the splitting is too small to show for the field strength indicated). Thus the normal Fermi liquid first becomes another superfluid phase A_1 and then the A phase as it is cooled in the presence of the field. The superfluid is, therefore, quite active magnetically and responds to the field in an interesting manner as we will see.

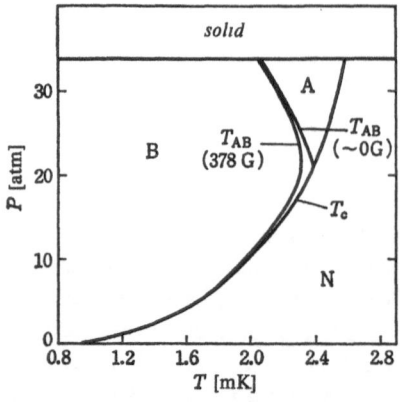

Fig. 4.1. The A and B superfluid phase in $P-T$ plane

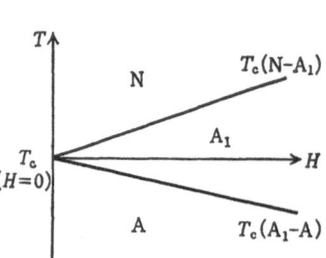

Fig. 4.2. The normal to the superfluid transition in the presence of a magnetic field

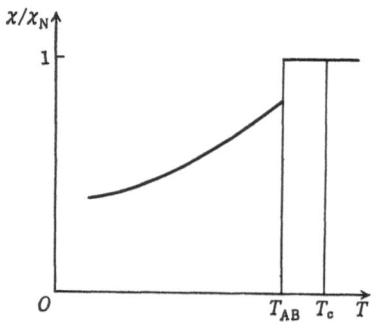

Fig. 4.3. Temperature dependence of susceptibility

Figure 4.3 shows the variation of the susceptibility χ when the liquid is cooled under appropriate pressure and undergoes the transitions from the normal to the A and then to the B phase. To very good accuracy the observed χ_A is equal to χ_N (we denote the normal phase by N). This fact means that in the A phase the spin of the ^{3}He atom (the nuclear spin) is not frozen by the pairing as in the spin-singlet pairing of superconductors. In the B phase the susceptibility χ_B shows the same temperature dependence as given by the BCS theory for superconductors but does not tend to zero as $T \to 0$ as in the latter. From these facts we must conclude that the superfluid phase of ^{3}He is due to the spin-triplet pairing instead of the singlet as in superconductors.

4.1.2 Transport Phenomena

What experimental evidence shows that in the new phase the liquid is indeed superfluid? There have been measurements of viscosity by means of an oscillating wire, the observation of unusually large thermal conduction due to convection currents, and the observation of the fourth sound, of which the last is decisive. As we mentioned concerning liquid ^{4}He, the superfluid but not the normal fluid can flow through a tube filled with fine powders. In such a "super-leak", sound waves can propagate only when the liquid becomes a superfluid. This was observed in liquid ^{4}He and called the 4th sound. Its velocity s_4 is given by $s_4^2 = (\rho_s/\rho)s_1^2$ where s_1 is the velocity of the 1st sound. The 4th sound was in-deed observed to propagate in the liquid ^{3}He below T_c in a container filled with CMN powder used for adiabatic demagnetization cooling. The velocity s_1 and the attenuation constant α of the zero sound in the bulk liquid are also observed to show anomalous temperature dependence when it enters the superfluid phase (Fig. 4.4). The superfluidity of ^{3}He, especially in its A phase, is qualitatively different from that of ^{4}He, as will be discussed in Sect. 4.6

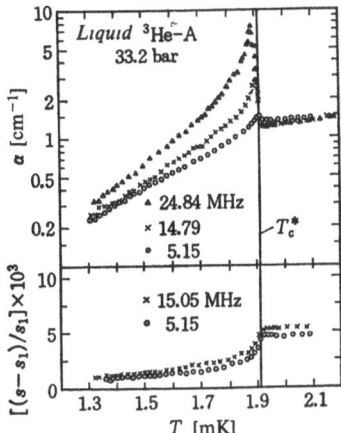

Fig. 4.4. Temperature dependence of the attenuation constant and the velocity s of the zero sound. s_1 is the velocity of the first sound [4.1]

4.1.3 Nuclear Magnetic Resonance (NMR) Experiments

One of the most interesting features of the ^{3}He superfluid is its magnetic behavior associated with the nuclear spin including the equilibrium properties mentioned in Sect. 4.1.1. Indeed the clear indication of the new phase was the observation of the shift in the resonance frequency of NMR. Figure 4.5 shows the resonant absorption of the ^{3}He sample on the melting curve in which the solid and the liquid phase coexist. The absorption due to the solid and the liquid above T_c has a peak at the Larmor frequency $\omega_0 \equiv \gamma H_0 (\hbar\gamma/2$ is the magnetic moment of ^{3}He nucleus[1], and H_0 the static field), whereas in the superfluid phase the peak clearly shows shift from it which depends on temperature (and pressure). This is quite surprising. The shift from ω_0 must be due to the magnetic dipole interaction be-

---

[1] In this chapter we will use the unit with $\hbar = 1$ and $k_B = 1$.

tween ^{3}He spins, but if configuration of ^{3}He atoms around any particular one is spherically symmetric (or cubic as in cubic crystals), the fields produced at its site by the surrounding spins cancel each other (on the average in liquids) and hence cannot lead to the shift. The appearance of the shift, therefore, means that in the new phase of the liquid, relative configration of ^{3}He atoms and direction of its spin are correlated in some coherent manner. We will discuss this problem of spin dynamics in Sect. 4.5.

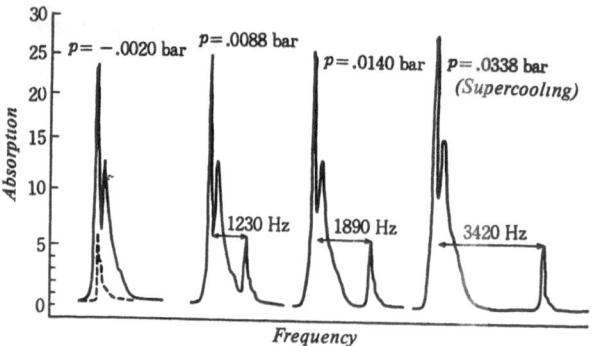

Fig. 4.5. NMR absorption versus frequency in the coexisting state of solid and liquid; $p = P - P_A$ where P_A is the pressure at which the liquid becomes the superfluid A phase. The small peak on the right side is due to the A phase liquid. The structure in the large peak due to the solid is of technical origin. The shift of the resonance frequency of the liquid is clearly seen [4.2]

Leaving the problem why liquid ^{3}He becomes a superfluid with spin triplet P-wave pairing to Sect. 4.4, we first explain its structure in more detail in the next section. Table 4.1 lists quantities relevant to the superfluid phase.

Table 4.1.

Pressure	0		34.36	at.
ϵ_F/k_B	1.64		1.02	K
T_c	0.9		2.6	10^{-3} K
$a = N^{-1/3}$ (mean inter-particle distance)	3.94		3.49	Å
ζ (coherence length)			~120	Å
$\gamma^2/k_B a^3$ (magnetic dipole-dipole interaction)	0.55		0.79	10^{-7} K
$\gamma H/k_B$ (Zeeman energy)		1.56 H [gauss]		10^{-7} K

4.2 Spin Triplet *P*-Wave Pairing

4.2.1 Cooper Pairs

As explained in Sect. 5.8 concerning superconductivity in metals, the ground state of a normal Fermi liquid is unstable for the formation of Cooper pairs if the interaction between particles near the Fermi surface is attractive. Here we briefly describe the generalized BCS theory, while in Sect. 5.8 we will discuss the original BCS theory with singlet S-wave pairing in more details.

Let us denote by $|k\alpha, -k\beta; F\rangle$ a state with a pair of particles with momentum k, $-k$ and spin α, β outside the Fermi sphere. Since it can make transitions to states $|k'\alpha, -k'\beta; F\rangle$ due to the interaction between particles, we construct a linear combination

$$\Phi = \sum_k \Psi_{\alpha\beta}(k) |k\alpha, -k\beta; F\rangle \qquad (4.2.1)$$

as an approximate eigenstate of the system (the Fermi sphere + a pair). The Schrödinger equation then becomes

$$(E - 2\xi_k)\left[\Psi_{\alpha\beta}(k) - \Psi_{\beta\alpha}(-k)\right]$$
$$= \sum_{|k'|>k_F} \left[V(k. k')\,\Psi_{\alpha\beta}(k') - V(-k, -k')\,\Psi_{\beta\alpha}(-k')\right], \qquad (4.2.2)$$

where $V(k, k')$ is the matrix element of the interaction for scattering the pair $(k', -k')$ to $(k, -k)$ and is assumed to be independent of spin. Also $\xi_k = k^2/2m - \varepsilon_F$. Because this is a pair of fermions, we have

$$\Psi_{\beta\alpha}(-k) = -\Psi_{\alpha\beta}(k). \qquad (4.2.3)$$

Also because of the symmetry $V(k, k') = V(-k, -k')$, the above equation simplifies to

$$(E - 2\xi_k)\Psi_{\alpha\beta}(k) = \sum_{|k|>k_F} V(k, k')\,\Psi_{\alpha\beta}(k'). \qquad (4.2.4)$$

As the phenomena under study occur at very low temperatures, only states close to the Fermi surface are involved. Hence we put $|k|, |k'| \approx k_F$ and $V(k, k') \approx V(\hat{k} \cdot \hat{k}')$ where $\hat{k} = k/|k|$, that is, V depends only on the scattering angle. Further we expand it as

$$V(k \cdot k') = \sum_l (2l + 1)\, V_l P_l(\hat{k} \cdot \hat{k}') \qquad (4i2.5)$$

and correspondingly, the wave function of the pair into partial waves

$$\Psi_{\alpha\beta}(k) \sim \sum_{lm} \psi_{\alpha\beta}^{(l,m)}(\xi_k)\, Y_{l,m}(\hat{k}), \qquad (4.2.6)$$

where $Y_{l,m}(\hat{k})$ is the spherical harmonics with angles of $\hat{k}$ as its variables. Substituting these into (4.2.4), we get

$$(E - 2\xi_k)\,\psi_{\alpha\beta}^{(l,m)}(\xi_k) = N(0)\,V_l \int_0^{\omega_c} d\xi_{k'}\psi_{\alpha\beta}^{(l,m)}(\xi_k')\,, \tag{4.2.7}$$

where $N(0) = mp_F/2\pi^2$ is the density of states at the Fermi surface of one spin state. We have also assumed that the interaction is cut off above ω_c by energy from the Fermi surface (in the BCS model of superconductivity it is of the order of the Debye frequency). Thus the equation for the Cooper pair is separated into each partial wave. If any one of V_l is attractive (< 0), a solution exists

$$\psi^{(l,m)}(k) = C(E - 2\xi_k)^{-1} \tag{4.2.8}$$

for which the energy eigenvalue is given by

$$1 = - N(0)\,V_l \int_0^{\omega_c} d\xi \frac{1}{2\xi - E}\,. \tag{4.2.9}$$

When $|E| \ll \omega_c$, we explicitly have $E \sim -2\omega_c \exp[-2/N(0)|V_l|]$.

4.2.2 Generalization of the BCS Theory

From the above consideration we expect the Fermi sphere to be unstable with respect to the formation of pairs with orbital angular momentum l if V_l is negative for particles near the Fermi surface. In the above treatment we restricted ourselves to states with only one pair plus the Fermi sphere. It is clear that in the many body system all particles near the Fermi surface participate in the pairing. Moreover, all the pairs formed should have the same total momentum (which is 0 in equilibrium states with no flow) in order to make the best use of the attractive interaction. This can be regarded as the Bose-Einstein condensation of the fermion pairs. The BCS theory treated this pair condensate in the mean field approximation. The essential point of the BCS theory is to assume that in the condensed state the expectation value (or the statistical average at finite temperature) of the pair wave function is finite, that is, the quantities

$$\left. \begin{aligned} \Psi_{\alpha\beta}(k) &\equiv \langle a_{k\alpha}a_{-k\beta}\rangle \\ \Psi_{\beta\alpha}^{\dagger}(k) &\equiv \langle a_{k\beta}^{\dagger}a_{k\alpha}^{\dagger}\rangle \end{aligned} \right\} \tag{4.2.10}$$

do not vanish, where $a_{k\alpha}$, $a_{k\alpha}^{\dagger}$ are the creation and annihilation operators for the single particle states (k, α). Clearly they correspond to the wave function (4.2.1) in the case of the Cooper pair. If the pair condensate described by $\Psi_{\alpha\beta}(k)$ exists, it produces a new mean field

$$\Delta_{\alpha\beta}(k) = -\sum_{k'} V(k, k')\, \Psi_{\alpha\beta}(k')$$
$$\Delta^{\dagger}_{\beta\alpha}(k) = -\sum_{k'} V(k', k)\, \Psi^{\dagger}_{\beta\alpha}(k')$$

$$(4.2.11)$$

In the mean field approximation the Hamiltonian of the system becomes

$$\mathscr{H} = \sum_{k\alpha} \xi_{k\alpha} a^{\dagger}_{k\alpha} a_{k\alpha} - \tfrac{1}{2}\sum [a_{-k\alpha}\Delta_{\alpha\beta}(k)\, a_{k\beta} + a^{\dagger}_{k\beta}\Delta_{\beta\alpha}(k)\, a_{-k\alpha}],$$
$$+ \text{const},$$

$$(4.2.12)$$

which can be diagonalized by the Bogoliubov transformation. To do it, it is convenient to write the Hamiltonian in the matrix form

$$\mathscr{H} = \tfrac{1}{2}\sum_{k}
\begin{pmatrix} a_{k\uparrow}^{\dagger} \\ a_{k\downarrow}^{\dagger} \\ a_{-k\uparrow} \\ a_{-k\downarrow} \end{pmatrix}^{\mathrm{T}}
\begin{pmatrix} \xi_k & 0 & & \Delta(k) \\ 0 & \xi_k & & \\ & & -\xi_k & 0 \\ \Delta^{\dagger}(k) & & 0 & -\xi_k \end{pmatrix}
\begin{pmatrix} a_{k\uparrow} \\ a_{k\downarrow} \\ a_{-k\uparrow}^{\dagger} \\ a_{-k\downarrow}^{\dagger} \end{pmatrix}$$

$$(4.2.13)$$

and diagonalize this 4×4 matrix. Here $\Delta(k)$ and $\Delta^{\dagger}(k)$ are the 2×2 matrices (4.2.11). Note that the square of the above 4×4 matrix is

$$\begin{bmatrix} \xi_k^2 \cdot 1 + \Delta\Delta^{\dagger} & 0 \\ 0 & \xi_k^2 \cdot 1 + \Delta^{\dagger}\Delta \end{bmatrix}$$

(**1** is the unit matrix). Therefore, if the 2×2 matrix $\Delta\Delta^{\dagger}$ is diagonal, the energy eigenvalues are given by

$$E_{\alpha}^{2} = \xi_k^2 + [\Delta(k)\Delta^{\dagger}(k)]_{\alpha}$$

$$(4.2.14)$$

as in the BCS theory. In this case the energy gap $\Delta(k)\, \Delta^{\dagger}(k)$ appears in the excitation spectrum.

4.2.3 Order Parameters of the Triplet P-Wave Superfluid

That which characterizes the superfluid state is the pair wave function (4.2.10) or the energy gap (4.2.11). In general, quantities which are nonvanishing only in an ordered state and represent its order are called order parameters. Let us now study properties of the order parameters (4.2.10) for our superfluid.

First of all $\Psi_{\alpha\beta}(k)$ has the same antisymmetry as (4.2.3). Next, we study how $\Psi_{\alpha\beta}(k)$ transform when the spin space is rotated (that is, when we rotate the quantization axis of the spin). When we make an infinitesimal rotation of angle $\delta\omega$ around an axis of direction $\hat{n}$ (a unit vector), the annihilation operator of spin 1/2 particles transforms as

$$a_{k\alpha} \rightarrow R_{\alpha\beta}a_{k\beta}$$

with

$$R = 1 + \frac{i}{2}\, \delta\omega\hat{n}\cdot\sigma\,,$$

where σ denotes the Pauli matrices

$$\sigma_x = \begin{bmatrix} 0 & 1 \\ 1 & 0 \end{bmatrix}, \quad \sigma_y = \begin{bmatrix} 0 & -i \\ i & 0 \end{bmatrix}, \quad \sigma_z = \begin{bmatrix} 1 & 0 \\ 0 & -1 \end{bmatrix}. \qquad (4.2.15)$$

Consequently the 2×2 matrix $\Psi(k)$ transforms as

$$\Psi(k) \rightarrow \Psi'(k) = R\Psi R^{\mathrm{T}}\,.$$

Now the operation of transposing a 2×2 matrix O can be written as $O^{\mathrm{T}} = -\sigma_2 O\sigma_2$. Hence,

$$\Psi'(k) = \Psi(k) + \frac{i\delta\omega}{2}\, \hat{n}\cdot(\sigma\Psi\sigma_2 - \Psi\sigma_2\sigma)\, \sigma_2\,. \qquad (4.2.16)$$

If $\Psi = iA\sigma_2$, it is invariant against rotation of the spin space and represents the spin-singlet pairing. In this case, because of the antisymmetry, the orbital part of the wave function $A = A(\hat{k})$ must satisfy $A(\hat{k}) = A(-\hat{k})$, that is, the orbital angular momentum of the pair l must be even. The simplest of such pairing is the S-wave pairing of the BCS theory, where the order parameter is just one complex constant A.

Next, let us assume the following form

$$\Psi = A\cdot(i\sigma\sigma_2)\,. \qquad (4.2.17)$$

Then (4.2.16) becomes

$$\Psi' = (A + \delta\omega\hat{n} \times A)\cdot(i\sigma\sigma_2) \qquad (4.2.18)$$

which means that A transforms like a vector in the spin space. When (4.2.17) is written in the form of a 2×2 matrix, it becomes clear that A_x, A_y, and A_z represent the components $\Psi_{\uparrow\uparrow}$, $\Psi_{\downarrow\downarrow}$, and $\Psi_{\uparrow\downarrow} = \Psi_{\downarrow\uparrow}$. This then is the order parameter for the spin triplet pairing. Obviously the orbital state of the pair should be with odd l.

With application to the liquid ³He in mind, we specialize in the P-wave pairing. The angular parts of the P-state wave functions with $l_z = \pm1$ and 0 are

$$Y_1^{\pm1} = \mp\sqrt{3/8\pi}(\hat{k}_x \pm i\hat{k}_y), \quad Y_1^0 = \sqrt{3/4\pi}\hat{k}_z\,,$$

where $\hat{k}_x$, $\hat{k}_y$, $\hat{k}_z$ are the components of the unit vector $\hat{k}$. For convenience the three states represented by $\hat{k}_x$, $\hat{k}_y$, and $\hat{k}_z$ are usually used instead of $Y_1^{\pm 1}$ and Y_1^0. So we write the vector A representing the orbital wave functions $\Psi_{\uparrow\uparrow}$, $\Psi_{\downarrow\downarrow}$ and $\Psi_{\uparrow\downarrow}$ as

$$A_i(\hat{k}) = A_{ij}\hat{k}_j \tag{4.2.19}$$

(in this chapter we understand that the sum is to be taken for repeated indices). Thus the order parameter can be represented as

$$\Psi(\hat{k}) = A_{ij}(i\sigma_i\sigma_2)\,\hat{k}_j\,. \tag{4.2.20}$$

Hence it is in general described by a set of 9 complex coefficients A_{ij}, which transform as a vector against rotation of both the spin space and the orbital space. Given a state specified by a set of A_{ij}, we can rotate it in both spaces to get a new state

$$A'_{ij} = R^{(s)}_{im}R^{(l)}_{jn}A_{mn}\,, \tag{4.2.21}$$

where $R^{(s)}$, $R^{(l)}$ are the corresponding rotation operators in each space. As the Hamiltonian of the system is invariant against the rotations (provided we neglect the interaction between the nuclear dipole moments which is very weak), all the states generated in this way are degenerate in energy. It must be noted that these states are physically different, just as states of an ideal ferromagnet with spontaneous polarization in different directions are all different ("broken symmetry").

4.2.4 ABM State and BW State

As we have seen above, the superfluid states with ^3P (the spin triplet *P*-wave pairing have large degrees of freedom and can be complicated, but in reality only a few types of states are of interest to us among all possible states. We focus now on the two types of states which are particularly important for the superfluid ^{3}He.

1) *ABM (Anderson-Brinkman-Morel) State*
In this class of states there are only $\Psi_{\uparrow\uparrow}$ and $\Psi_{\downarrow\downarrow}$ pairs if we take the axis of the spin space appropriately and both pairs are in the same orbital state. The order parameters are then expressed as

$$A_{ij} = Ad_i(\hat{n}_1 + i\hat{n}_2)_j\,, \tag{4.2.22}$$

where d, $\hat{n}_1$, $\hat{n}_2$ are all unit vectors and $\hat{n}_1 \cdot \hat{n}_2 = 0$. If we choose the *y*-axis of the spin space in the d-direction, $d = \hat{y}$, then there are only $\uparrow\uparrow$ and $\downarrow\downarrow$ pairs. If $\hat{n}_1 = \hat{x}$ and $\hat{n}_2 = \hat{y}$, both the $\uparrow\uparrow$ and $\downarrow\downarrow$ pairs are in the orbital state $k_x + ik_y$ with the *z*-component of the orbital angular momentum $m = 1$. Thus the unit vector $1 \equiv \hat{n}_1 \times \hat{n}_2$ is the symmetry axis of the orbital wave function with $m = 1$.

2) *BW (Balian-Werthamer) State*

Consider the states in which the total angular momentum of the ^3P pairs is zero. Since $J = 0$, the pair state must be invariant against simultaneous rotations of the spin and the orbital space, namely,

$$A_{ij} = A\delta_{ij} . \tag{4.2.23}$$

Then we have

$$\Psi = A \begin{bmatrix} -\hat{k}_x + i\hat{k}_y & \hat{k}_z \\ \hat{k}_z & \hat{k}_y + i\hat{k}_y \end{bmatrix} ,$$

the pairs with spin 1, 0, -1 being in the orbitals $-1, 0, 1$, respectively. Performing an arbitrary rotation on only the spin (or the orbital) part of this state, we generate a class of states called the BW states. If we rotate the spin state by an angle φ around the direction $\hat{n}$, we obtain

$$\left. \begin{aligned} A_{ij} &= AR_{ij}^{(s)}\delta_j \\ R_{ij}^{(s)} &= (\cos\varphi)\,\delta_{ij} + (1 - \cos\varphi)\,n_i n_j + (\sin\varphi)\,\varepsilon_{ijk}n_k \end{aligned} \right\} \tag{4.2.24}$$

which are the general form of the order parameter in the BW states.

4.3 Physical Properties of ABM and BW States

Since the mean fields $\Delta_{\alpha\beta}(k)$—(4.2.11)—which are related to the energy gap have the same structure as $\Psi_{\alpha\beta}(k)$, we can use the same representation as above,

$$\Delta(k) = \sum_{ij} d_{ij}(i\sigma_i\sigma_2)\,\hat{k}_j . \tag{4.3.1}$$

4.3.1 Energy Gap

In the ABM states, d_{ij} takes the same form as (4.2.22). We then find that the 2×2 matrix $\Delta^\dagger\Delta$ becomes diagonal as

$$\Delta^\dagger\Delta = |\Delta_A|^2[(\hat{n}_1 \cdot \hat{k})^2 + (\hat{n}_2 \cdot \hat{k})^2] \cdot 1 . \tag{4.3.2}$$

Because $\hat{n}_1$ and $\hat{n}_2$ are orthogonal, we can rewrite it as $|\Delta_A|^2[1 - (\boldsymbol{l} \cdot \hat{k})^2] \cdot 1$. Hence the energy gap in (4.2.14) of this state varies on the Fermi surface as $\sin^2\theta$ as shown in Fig. 4.6 where the z-axis is chosen in the direction of the symmetry axis **1** of the pair. Note that the gap vanishes at the poles. If we choose the $m = 0$ state k_z instead of $k_x + ik_y$, the gap vanishes on the equator. Because the energy gap is the binding energy of the pairs, we expect the choice of $m = 1$ in the ABM state to lead to lower free energy than that of $m = 0$, as can be con-

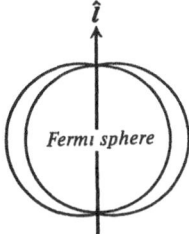

Fig. 4.6. Anisotropic energy gap

firmed by calculation. The anisotropy of the energy gap is reflected in various properties of the superfluid in this state, as we will see below.

In the BW states we have

$$\Delta\Delta^\dagger = |\Delta_B|^2 \hat{k}^2 \cdot 1 = |\Delta_B|^2 \cdot 1 \tag{4.3.3}$$

and the gap is isotropic as in the BCS theory. Many properties of this class of states is, therefore, similar to what the BCS theory predicts. Because of what we have said above, the condensation energy of pairs is larger in the BW state than in the ABM state. In fact, if the Fermi liquid effects to be discussed below are absent, the BW states with the isotropic gap have the lowest free energy below T_C, and hence should always be the one to appear.

The magnitude of the energy gap Δ_A and Δ_B in (4.3.2, 3) are determined by the self-consistency condition for the mean field, namely, by the gap equation. In the model with only $V_{l=1} = -V_1$ nonzero, the gap equation becomes

$$d_{\alpha i} = \tfrac{3}{2} V_1 \sum_{k'} d_{\alpha j} \hat{k}'_i \hat{k}'_j \frac{\tanh(\beta E_{k'}/2)}{E_{k'}} , \tag{4.3.4}$$

where we have assumed E_k of the form (4.2.14). It is clear from this equation that the energy gap $\Delta_B(T)$ of the BW states has the same dependence on temperature as the gap of the BCS theory. The gap $\Delta_A(0)$ of the ABM states at $T = 0$ is equal to

$$\Delta_A(0) = \Delta_B(0) \, e^{5/6 - \ln 2} .$$

Hence the average energy gap $\Delta_A^2 \sin^2\theta$ is 0.883 $\Delta_B(0)$. The difference in the condensation energy of the two classes of states is therefore about 10%. The transition temperature T_C is determined by the same equation (4.3.4) if we put the gap parameter in E_k equal to zero. Consequently T_C does not depend on the form of the order parameter $d_{\alpha i}$ so that T_C is common to all possible states. These facts are essential for understanding the phase diagram, as we will see in the next section.

4.3.2 Normal (Super) Fluid Component $\rho_n(\rho_s)$

The density of the normal fluid ρ_n is given according to (3.3.7) in Chap. 3 by

$$\rho_{nij} = -\sum_{k\sigma} k_i k_j \frac{\partial n}{\partial \varepsilon_{k\sigma}}. \tag{4.3.5}$$

In the present case, the excitation energy $\varepsilon_{k\sigma}$ is $E_{k\sigma} = \sqrt{\xi_k^2 + |\Delta_k|^2}$ which in general depends on the direction of k, so that ρ_n becomes a tensor. For actual calculations the expression used in the theory of superconductivity is convenient, which now becomes

$$\rho_{nij} = 4\pi T N(0) k_F^2 \int \cdot \frac{d\Omega_k}{4\pi} \hat{k}_i \hat{k}_j \sum_n \frac{|\Delta_k|^2}{\{[(2n+1)\pi T]^2 + |\Delta_k|^2\}^{3/2}}. \tag{4.3.6}$$

In the BW states with the isotropic gap, ρ_n and hence ρ_s are isotropic and identical to the BCS result. In the ABM states ρ_n is a diagonal tensor. When relative velocity of the normal and the superfluid is parallel to the 1 vector, it is $\rho_{n\parallel}$ and when perpendicular, $\rho_{n\perp}$, which are given near T_c by

$$\left.\begin{aligned}
\frac{\rho_{n\parallel}}{\rho} &= 1 - \frac{7\zeta(3)\Delta^2}{10\pi^2 T_c^2} \\
\frac{\rho_{n\perp}}{\rho} &= \frac{2\rho_{n\parallel}}{\rho} - 1.
\end{aligned}\right\} \tag{4.3.7}$$

respectively.

Since more quasiparticles are thermally excited near the poles with small energy gap (Fig. 4.6), it is natural that $\rho_{n\parallel}$ is greater than $\rho_{n\perp}$. This has been confirmed by the experiment in which $\rho_s = \rho - \rho_n$ is measured with fixed 1 (the direction of the 1 vector can be controlled by walls and a magnetic field, see Sects. 4.5, 6), as shown in Fig. 4.7. For quantitative comparisons with experimental data, the Fermi liquid effects must be taken into account.

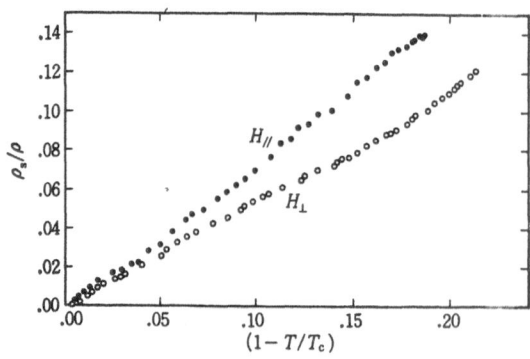

Fig. 4.7. Superfluid density $\rho_s = \rho - \rho_n$ in the A phase. $H_\parallel$ corresponds to $\rho_{s\perp}$ [4.3]

4.3.3 Susceptibility

In superconductors with the spin singlet pairing only thermally excited particles contribute to the susceptibility since energy larger than the gap is necessary to break up the pairs. Therefore, χ vanishes at $T = 0$ and its temperature dependence is given by

$$\chi = \chi_N \frac{1}{2k_B T} \int_0^\infty d\xi \, \text{sech}^2 \left\{ \frac{1}{2k_B T} [\xi^2 + \varDelta^2(T)] \right\} \equiv \chi_N Y(\varDelta). \tag{4.3.8}$$

In the ABM states, if the $\boldsymbol{d}$ vector is parallel to an external field $\boldsymbol{H}$, there are only pairs with spin $\uparrow\downarrow$ and $\downarrow\uparrow$ along the direction of $\boldsymbol{H}$ according to (4.2.20, 22). Hence the susceptibility for the field $\boldsymbol{H}\|\boldsymbol{d}$ is given by the above expression for the singlet pairs. If, however, $\boldsymbol{d} \perp \boldsymbol{H}$, for instance $\boldsymbol{d}\|\hat{\boldsymbol{x}}$ and $\boldsymbol{H}\|\hat{\boldsymbol{z}}$, there are only pairs with spin $\uparrow\uparrow$ and $\downarrow\downarrow$. Moreover, the $\uparrow\uparrow$ pairs and the $\downarrow\downarrow$ pairs can be formed independently of each other. For this reason the density of particles with spin $\uparrow$ and $\downarrow$ is changed by $\pm(1/2)N(0)\gamma H$ in the magnetic field just as in the normal state and then the pairing takes place on each Fermi sphere for particles with $\uparrow$ and $\downarrow$ spin. Hence, $\chi_\perp = \chi_N$ in this case. Thus χ is also a tensor in the ABM states. If there is no other effect, the field dependent energy is in general $-\chi_{ij}H_i$ H_j for a weak field, so that the $\boldsymbol{d}$ vector tends to lie in the plane perpendicular to the magnetic field. This is in accordance with the fact mentioned in Sect. 4.1 that the observed χ_A is equal to χ_N in the A phase.

In the BW states χ is also isotropic. According to (4.2.23), however, a fraction $\overline{k_z^2} = 1/3$ (average over the Fermi surface) of pairs are with $\uparrow\downarrow + \downarrow\uparrow$ spin at any temperature. These pairs behave like the singlet pairs so that their contribution to χ is again given by (4.3.8). The contribution of the rest of the pairs is the same as χ_N. Therefore we obtain

$$\chi_{BW} = \chi_N [\tfrac{2}{3} + \tfrac{1}{4} Y(\varDelta)]. \tag{4.3.9}$$

Unlike χ of the BCS theory, χ_{BW} is equal to $(2/3)\chi_N$ at $T = 0$. The temperature dependence of χ_B observed in the B phase agrees with this result. As in the case of the normal fluid density, the Fermi liquid correction is necessary for the quantitative agreement.

4.4 Fermi Liquid Effects

The theory sketched in the preceding section is the generalization of the weak coupling BCS theory to the ^3P pairing. According to it, among other things, the BW states with the isotropic gap is always the most stable one below T_c. For various reasons we identify the B phase to be the BW states. Why, then, does the A (and A_1) phase appear in the real liquid? Why do we have, to start with, the ^3P pairing? To answer these basic questions it is necessary to take account of the fact that liquid ^{3}He is the Fermi liquid with a strong interaction between particles so that the pairs are formed by the quasiparticles near the Fermi surface rather than the original bare particles. We have mentioned that the hard core repulsion prevents the ^1S pairing as in superconductors. This hard core part of the interaction also leads to the effect of aligning spins of neighboring particles as remarked in Sect. 3.4.2. This effect is best expressed by the susceptibility (3.4.19) or rather by the Landau parameter appearing in it, F_0^a, which is equal to -0.7 under saturated vapour pressure, not very far from the condition for ferromagnetism $F_0^a = 1$. Because of this we expect the probability of finding, say, up-spin particles near an up-spin particle, to be high, in other words, fluctuations of the spin density should be large in the liquid. These fluctuations are some-times called paramagnons. It is physically plausible that this effect is advantageous for the spin-triplet pairing.

The Landau parameters express the effective interaction between quasi-particles near the Fermi surface. More precisely they give only the magnitude of the forward scattering with small momentum transfer q (Fig. 4.8). On the other hand, all scattering processes with q as large as p_F are involved in the pair-ing. Hence it is not possible to find the pairing interaction V_l from the Landau parameters. Nevertheless, we can hope to get some crude estimate. For this purpose we rewrite f^s and f^a in (3.4.12) in terms of the parameters appropriate for spin singlet and triplet quasiparticle pairs, f^0 and f^1. Since $f^1 = f_{\uparrow\uparrow}$ $f^0 + f^1 = 2f_{\uparrow\downarrow}$ we have

$$\begin{aligned} f^0 &= f^s - 3f^a \\ f^1 &= f^s + f^a \end{aligned} \Big\} . \tag{4.4.1}$$

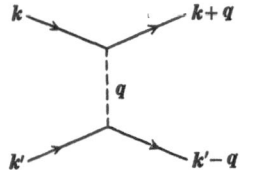

Fig. 4.8.

Take for example the $l = 1$ component of the partial wave expansion. As we know from the observed susceptibility that f^a is negative, the spin-dependent part of the interaction then acts as an attractive force for the triplet pairs and a repulsive one for the singlet pairs. Its magnitude is rather large, as remarked above.

Before explaining the A phase intrusion, we next consider the Fermi liquid effects on the various properties of the superfluid. It must first be emphasized that even in the present case the weak coupling BCS theory is valid once we know the effective interaction for the quasiparticle pairing, although the Fermi liquid effect is quite important in determining this interaction itself. Recall that the Fermi liquid effects are due to the mean field produced by the deviation $\delta n_{p\alpha}$ of quasiparticle distribution from the Fermi sphere. When the pairs are formed in the superfluid phase, a corresponding deviation $\delta n_{p\alpha}$ appears. The mean field, however, depends only on the integral over p, that is, over energy ξ_p, of $\delta n_{p\alpha}$ with fixed direction $p/|p|$. According to the BCS theory, the pairing takes place symmetrically; particle pairs above and hole pairs below the Fermi surface. Hence for any direction $p/|p|$ their contributions to the deviation of quasiparticle density cancel each other. As a result the pair formation itself does not lead to any Fermi liquid effect (when there is the particle-hole symmetry, that is, when we neglect change in the density of states with energy).

The Fermi liquid effects are, of course, very important when we consider responses in the superfluid states to external perturbations such as ρ_n, χ or the transport coefficients. Here we briefly comment on the corrections in ρ_n and χ.

The normal fluid density ρ_n is determined from the momentum density of the fluid when quasiparticle excitations as a whole move with velocity v_n and the superfluid is at rest. In this situation the energy of a quasiparticle is given by

$$\bar{\varepsilon}_{p\alpha} = \varepsilon_{p\alpha} - p\cdot v_n + \frac{1}{V}\sum f_{pp'}\delta n_{p'\alpha} \,.$$

We solve for $\delta n_{p\alpha}$ as we did for an external field U, replacing U by $-p\cdot v_n$ in (3.4.16), and then calculate the momentum density. In the present case we must remember $\varepsilon_{p\alpha}$ has the energy gap. The result is

$$\rho_n = \frac{\rho_n^0}{1 + (\rho_n^0/nm^*)(F_1^s/3)} \,, \tag{4.4.2}$$

where ρ_n^0 is what we get if we use the expression without the Fermi liquid correction (4.3.6) for quasiparticles with the effective mass m^* [$N(0) = m^*p_F/2\pi^2$ or in (4.3.7) we use $\rho = m^*n$]. This applies to both $\rho_{n//}$ and $\rho_{n\perp}$. One can interpret this result as follows. Since the real particles carry mass m, it seems that we only have to multiply ρ_n^0 by m/m^* to get ρ_n. However, we must recall that m^* is due to the backflow around the particle moving in the liquid; the backflow contributes to the momentum density also. Now, in the present case the superfluid as well as the normal fluid participate in the backflow around a moving quasiparticle. The superfluid contribution to the backflow should then make ρ_n greater than $\rho_n^0 m/m^*$ as in (4.4.2). Note also that $\rho_n \to mn$ when $T \to T_c$, as it should.

A similar consideration can be made for the susceptibility. In the normal state the correction to the free Fermi gas result $\gamma^2 N(0)$ is given by the factor $(1 + F_0^a)^{-1}$ in (3.4.19), which is due to the following mechanism: due to a magnetic field the density of particles with spin parallel to the field is increased which then produces the mean field enhancing the polarization. If the spin polarization is depressed by the pairing, this enhancement is also reduced proportionately. Hence we expect the factor $(1 + F_0^a)^{-1}$ to be replaced by $[1 + (\chi^0/\chi_N^0)F_0^a]^{-1}$, where χ^0 and χ_N^0 are the expressions without the Fermi liquid correction. Thus,

$$\chi = \frac{(1 + F_0^a)\chi_N}{1 + (\chi^0/\chi_N^0)F_0^a} . \tag{4.4.3}$$

Because of this correction χ_{BW}/χ_N at $T = 0$ is equal to $(2/3)(1 + F_0^a)/[1 + (2/3)F_0^a]$ instead of $2/3$ from (4.3.9). The comparison with experimental data is as shown in Fig. 4.9.

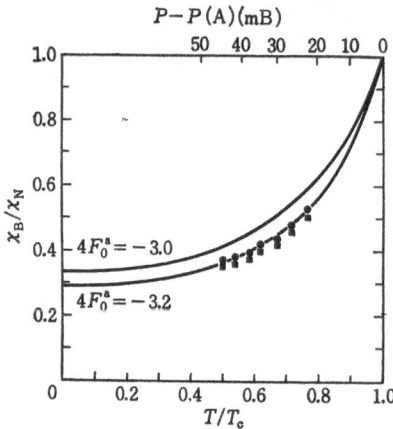

Fig. 4.9. Susceptibility in the B phase; the solid lines are theoretical with slightly different Landau parameters. The experimental values are obtained by NMR experiments [4.4]

As these examples show, the mean fields appearing in the responses to external fields are greatly changed when the liquid becomes the superfluid. Now, the interaction is the force field one particle produces on another. The first particle acts on the second directly as well as indirectly through surrounding particles, the latter effect is taken into account in the Fermi liquid theory as the mean field. The spin-dependent part of this mean field is what we discussed in the beginning of this section. Clearly it is related to the enhancement effect in the susceptibility. This means that, if spin fluctuations are suppressed by the pairing as in the BW states, this part of the mean field is reduced in the same way as χ. In other words, the effective interaction responsible for the spin-triplet pairing is itself affected by the pair formation. This feedback should be absent in the ABM states with $\chi = \chi_N$. Therefore, the pairing interaction can be larger in the ABM states than in the BW states. With the help of this consideration we can understand the phase diagram (Fig. 4.1) at least qualitatively.

From what we have said, we can identify the A phase and the B phase to be the ABM and the BW states, respectively. Near T_c where the condensation energy is still small, the A phase may be preferable because of the larger pairing interaction. At lower temperatures the B phase is more stable because of the larger condensation energy. That the A phase only appears in the high pressure region is plausible since the magnitude of negative F_0^a increases with pressure (Table 3.2). We can also explain the change of T_{AB} with a magnetic field. Recall that the transition temperature to the normal state is the same for all states. Hence, if we expand the free energy in powers of the magnitude of the energy gap Δ, the second-order term which is proportional to $T - T_c$ is the same for the A and the B phase, and the difference appears only in the fourth-order term. When a magnetic field is present, we have an additional term χH^2. Now χ in the ABM states is constant, $\chi_{ABM} = \chi_N$, while is the BW states χ_{BW} decreases as given by (4.3.9). Near T_c, $Y(\Delta) - 1 \propto - |\Delta|^2$ so that the magnetic energy is higher in the BW states by $|\Delta|^2 H^2$ than in the ABM states. In the presence of the field, therefore, the A phase should be the stable phase near T_c. The form of the transition line T_{AB} under the weak field (Fig. 4.1) can also be understood from the above considerations. This unique phase diagram is only possible for such phase transitions as the ^3P pair condensation where the order parameter has many degrees of freedom.

As remarked before, the A_1 phase appears before the A phase in the presence of a magnetic field. Since in the field the density of particles with spin parallel to it increases and so does the density of states for them [by an amount of order $\gamma H d N(0)/d\varepsilon_F$], they form pairs before particles with spin anti-parallel to the field. According to the experiments, the transition temperature T_{N-A_1} from the normal to the A_1 phase is higher than T_{A_1-A} at which the pairs of both spins appear by $4 \times 10^{-2} T_c(\gamma H/k_B T_c)$. We have stated above that the pairs with spin parallel to the field first appear. It must be noted that this is the case when there is no Fermi liquid effect and that which spin pairs first appear in the real liquid is still an unresolved problem, both theoretically and experimentally. Also not much is known about the properties of the A_1 phase under a strong magnetic field.

4.5 Nuclear Spin Dynamics

The phenomena in which the characteristics of the spin-triplet pairing appear most clearly are related to the nuclear spin dynamics, and the NMR experiment is a powerful probe for this reason. If the spin and the translational degrees of freedom are completely decoupled, nothing more than the Larmor precession can be expected. They are, however, coupled by the magnetic dipole interaction (which will be called the d-d interaction in the following) which is very weak compared to the pairing interaction but is essential in the spin dynamics.

The d-d interaction has the form

$$H_{d-d} = \tfrac{1}{2} (\tfrac{1}{2} \gamma)^2 \sum_{n \neq m} \left[\frac{\sigma_n \cdot \sigma_m}{|r_n - r_m|^3} - 3 \frac{\sigma_n \cdot (r_n - r_m) \, \sigma_m \cdot (r_n - r_m)}{|r_n - r_m|^5} \right], \qquad (4.5.1)$$

where $(1/2) \, \gamma \sigma_n$ is the magnetic moment of the nth particle and r_n its position. Rewriting it in the momentum representation,

$$H_{d-d} = -\frac{\pi}{3}\left(\frac{1}{2}\gamma\right)^2 \sum_{kk'q} \left(\frac{3q_i q_j}{q^2} - \delta_{ij}\right) a_{k\alpha}^\dagger \sigma_{i\alpha\beta} a_{k+q\beta} a_{k'\alpha'}^\dagger \sigma_{j\alpha'\beta'} a_{k'+q\beta'} \qquad (4.5.2)$$

(sums over the vector components i, j and the spin indices α, β, ... should be taken), we calculate its expectation value in the superfluid states in the mean field approximation described in Sect. 4.2. Then it is expressed in terms of the order parameters Ψ, $\Psi^\dagger$ (4.2.10),

$$E_{d-d} = -\frac{\pi}{3}\left(\frac{1}{2}\gamma\right)^2 \sum_{kk'} \left[\frac{3(\hat{k}-\hat{k}')_i(\hat{k}-\hat{k}')_j}{|\hat{k}-\hat{k}'|^2} - \delta_{ij}\right] \qquad (4.5.3)$$
$$\cdot \sigma_{i\alpha\beta}\sigma_{j\gamma\delta}[\Psi_{\gamma\alpha}^\dagger(\hat{k})\,\Psi_{\beta\delta}(\hat{k}')]\,,$$

where we have neglected the terms independent of the order parameters. When we use the representation $\Psi = A(k)\cdot(i\sigma\sigma_2)$ (4.2.17), the part containing the spin operators becomes

$$\mathrm{Tr}\{\sigma_i\Psi\sigma_j^T\Psi^\dagger\} = -2(A_iA_j^* + A_jA_i^* - \delta_{ij}A_kA_k^*)\,, \qquad (4.5.4)$$

If we further use (4.2.19), we get

$$E_{d-d} = -\frac{\pi}{3}\gamma^2 \sum_{kk'} \left[\frac{3(\hat{k}-\hat{k}')_i\hat{k}_l(\hat{k}-\hat{k}')_j\hat{k}_n'}{|\hat{k}-\hat{k}'|^2} - \delta_{ij}\hat{k}_l\hat{k}_n'\right] A_{ll}A_{jn}^*\,. \qquad (4.5.5)$$

If the first term is integrated over the directions of $\hat{k}$ and $\hat{k}'$, it must reduce to the form, $a\delta_{il}\delta_{jn} + b\delta_{ij}\delta_{ln} + c\delta_{in}\delta_{jl}$. The coefficients a, b, c can easily be determined by the integrals of the three possible contractions of the expression in the curly parenthesis. As a result we obtain

$$E_{d-d} = \frac{\pi}{20}\gamma^2\,(A_{ll}A_{jj}^* + A_{lj}A_{jl}^* - \tfrac{2}{3}A_{lj}A_{lj}^*)\,, \qquad (4.5.6)$$

where A_{ij} is the integral of the original A_{ij} over the energy ξ and is related to the energy gap as $\varDelta_{ij} = N(0)V_1A_{ij}$. Thus the dipole energy E_{d-d} depends on the form of the order parameters and hence on the pair states.

In the A phase (ABM states) we substitute (4.2.22) and get

$$E_{d-d} = \frac{\pi}{10}\gamma^2\frac{\varDelta^2}{V_1^2}[\tfrac{1}{2} - (l\cdot d)^2]\,. \qquad (4.5.7)$$

For the B phase (BW states) Eq. (4.2.24) is to be used. It is interesting that E_{d-d} depends only on the rotation angle θ but not on the direction of the unit vector n:

$$E_{d-d} = \frac{2\pi}{15}\gamma^2\frac{\varDelta^2}{V_1^2}[2\cos^2\theta + \cos\theta + \tfrac{3}{4}]\,. \qquad (4.5.8)$$

The energy of the pair condensation is independent of $1 \cdot d$ or θ. This degeneracy is lifted by the $d - d$ interaction. If we estimate its magnitude using

$$V_1^{-1} \sim N(0) \ln \frac{\omega_c}{\Delta} ,$$

we find

$$E_{d-d} \sim N \left(\frac{\Delta}{\varepsilon_F}\right)^2 \frac{\gamma^2}{a^3} ,$$

where $N = a^{-3}$ is the number density of particles.

In the NMR experiments we usually observe the total magnetic moment of the system. Let us, therefore, study the dynamics of the total spin of our system. For simplicity we consider the orbital state to be fixed. When we rotate the spin state by θ (that is, to rotate it by angle $|\theta|$ around the axis $\hat{n} = \theta/|\theta|$), we go into the new state in which E_{d-d} has a different value according to the above dependence on the angles (4.5.7) or (4.5.8). The operator for infinitesimal rotations of the spin space is just the total spin S (since we are interested in uniform rotations of all spins in the system). Therefore, each component of S and the corresponding component of θ are canonically conjugate with each other. This means that $\partial E_{d-d}/\partial\theta$ acts as a torque on S. Since the total spin S can be regarded as a classical quantity, its equation of motion can readily be written down:

$$\frac{dS}{dt} = \gamma S \times H - \frac{\partial E_{d-d}}{\partial \theta} . \tag{4.5.9}$$

We have assumed that the time variation is so slow that at each instant the system is in the equilibrium state with a given order parameter. For the same reason we further assume that the total spin S is always given by $\gamma S = \chi H_{\text{eff}}$, where the effective field H_{eff} in the rotating system is equal to

$$H + (1/\gamma)(d\theta/dt)$$

according to the Larmor theorem. Hence we have

$$\frac{d\theta}{dt} = \gamma \left(\frac{\gamma S}{\chi} - H\right). \tag{4.5.10}$$

From these coupled equations we can draw many interesting results. Eliminating S, we obtain

$$\frac{d^2\theta}{dt^2} - \gamma \frac{d\theta}{dt} \times H + \frac{\gamma^2}{\chi} \frac{\partial E_{d-d}}{\partial\theta} = -\gamma \frac{dH}{dt} . \tag{4.5.11}$$

Here we will only consider the A phase which is relatively simple. When there is a static field $H_0 // \hat{z}$, the d-vector tends to be perpendicular to it (Sect. 4.3). We choose the y-axis in its direction. To minimize E_{d-d}, the l-vector also points in the same direction. Let us study small oscillations of S and d in this state.

4.5.1 Small Oscillations

Clearly oscillations of (θ_x, θ_y) and θ_z separate in (4.5.11). What is usually observed in the NMR experiments are the transverse oscillations, which in our case obey

$$- \omega^2 \theta_x + i \omega \omega_0 \theta_y + \Omega_A^2 \theta_x = 0 ,$$
$$- \omega^2 \theta_y - i \omega \omega_0 \theta_x = 0 .$$

Hence the resonance frequency is determined by

$$\omega^2 = \omega_0^2 + \Omega_A^2 , \tag{4.5.12}$$

where $\omega_0 = \gamma H_0$ and the shift Ω_A is given by

$$\Omega_A^2 \equiv \frac{\pi}{10} \frac{\gamma^4 \Delta^2}{V_{1\chi}^2} . \tag{4.5.13}$$

We note that the magnitude of Ω_A^2 is roughly $(\gamma^2/a^3)(\Delta/\varepsilon_F)^2 \varepsilon_F$ and that it has the same temperature dependence as the energy gap. As remarked before, the observation of this shift was the first indication of the new superfluid phase. The theory was mainly worked out by A.J.Leggett. In this transverse oscillation S precesses around $H_0 || \hat{z}$ and d oscillates in the xz plane (Fig. 4.10). The longitudinal oscillations have a resonance frequency independent of H_0. In this case d oscillates in the xy plane. One can make a similar treatment of the B phase resonances.

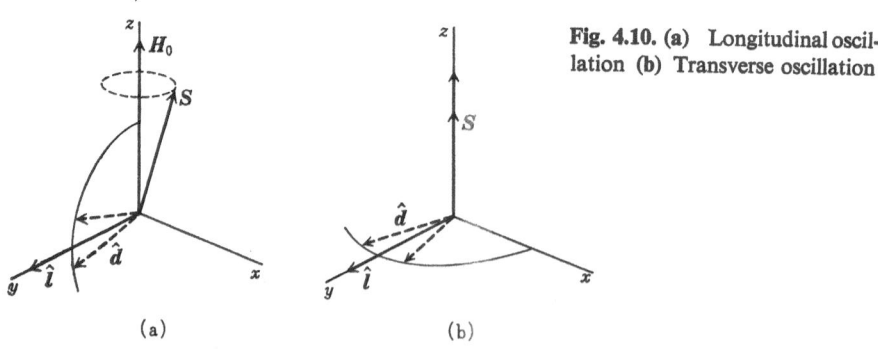

Fig. 4.10. (a) Longitudinal oscillation (b) Transverse oscillation

(a)   (b)

4.5.2 Ringing

The longitudinal oscillations can be solved even when the amplitude is not small. It is instructive to look at this mode as a kind of ac Josephson effect. The state in which d is rotated by θ_z from the y-direction around the z-axis has, according to (4.2.17, 22), the order parameter

$$
\begin{bmatrix}
f(\hat{k})\,e^{-i\theta_z} & 0 \\
0 & f(\hat{k})\,e^{i\theta_z}
\end{bmatrix},
$$

where $f(k) = k_x + ik_z$ is the orbital wave function. From this form it is clear that the phase difference between $\uparrow\uparrow$ pairs and $\downarrow\downarrow$ pairs is just $2\theta_z$. If the $d - d$ interaction is not included, the superfluid consisting of $\uparrow\uparrow$ pairs and that of $\downarrow\downarrow$ pairs are decoupled and the energy of the system is independent of the phase difference. The dipole interaction E_{d-d} now provides a weak coupling between the two superfluids. In the Josephson effect, two superconductors are weakly coupled by a tunnel junction and if there is a phase difference between the two superconductors, electrons flow without a potential difference. In the present case, the transfer of particles between the spin $\uparrow$ and $\downarrow$ superfluids occurs due to E_{d-d} when there is the phase difference. The total spin S_z is proportional to $n_\uparrow -n_\downarrow$, so that the z component of (4.5.9, 10) is the pair of canonical equations of motion for the number of particles and the phase angle:

$$
\left.
\begin{aligned}
\frac{dS_z}{dt} &= -\frac{\partial E_{d-d}}{\partial \theta_z} \\
\frac{d\theta_z}{dt} &= \frac{\gamma^2 S_z}{\chi} - \omega_0
\end{aligned}
\right\}.
\tag{4.5.14}
$$

The equilibrium state under a static magnetic field H_0 is obviously $\gamma S_z = \chi H_0$ according to the second equation. If we change the magnetic field suddenly by ΔH_0, θ_z and S_z will start oscillating. Substituting the explicit form of E_{d-d} and eliminating S_z, we obtain for θ_z just the equation of motion for a pendulum:

$$
\frac{d^2\theta_z}{dt^2} + \frac{1}{2}\,\Omega_A^2 \sin 2\theta_z = 0 .
\tag{4.5.15}
$$

In general, the solutions are given in terms of the elliptic functions. We only write the result for the frequency of the oscillation

$$
\left.
\begin{aligned}
\omega &= \frac{\pi\Delta\omega_0}{K(\Omega_A/\Delta\omega_0)}, & \Delta\omega_0 &> \Omega_A \\
\omega &= \frac{\pi\Omega_A}{2K(\Delta\omega_0/\Omega_A)}, & \Omega_A &> \Delta\omega_0
\end{aligned}
\right\},
\tag{4.5.16}
$$

where $\Delta\omega_0 = \gamma\Delta H_0$ and K is the complete elliptic integral. Since S_z is associated with the magnetic moment, this means that when we suddenly change the field by a certain value, the moment of the system starts to oscillate with the corresponding frequency (4.5.16). This is called ringing. The observed ringing frequency in the A phase is compared with the theory in Fig. 4.11. When $\Delta\omega_0 \rightarrow 0$, $\omega \rightarrow \Omega_A$, and as the pendulum swings just to the top, $\omega = 0$ when $\Delta\omega_0 = \Omega_A$. The same effect is seen in the B phase, but the oscillation is more complicated. The characteristics of the spin-triplet pairing is clearly shown in this phenomenon.

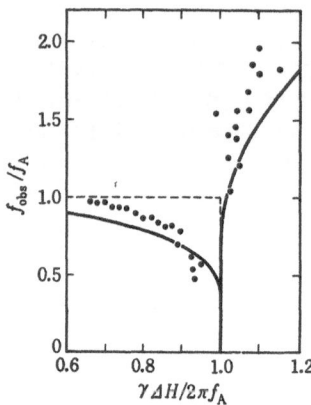

Fig. 4.11. Ringing frequencies [4.4]. The solid line illustrates theory [4.5]

Although we have discussed only spatially uniform motions, there are of course modes with finite wavelength. The simplest of those are spin waves, which are Goldstone bosons associated with the superfluid phase. Due to the $d - d$ energy, the gap Ω_A appears in the dispersion relations for the spin waves.

4.6 Texture and Superfluidity

As mentioned before, the superfluidity in liquid ³He shows features qualitatively different from that of ⁴He or superconductors. This section is devoted to this subject. In this limited space we have to focus our attention on the behaviour of the superfluid component to be described by spatial variations of the order parameter, and assume that the normal fluid is always at rest and in spatially uniform states.

In superconductors as well as in superfluid ⁴He, the order parameter is the condensate wave function Ψ and the condensation energy is a function of $|\Psi|^2$, independent of its phase. Consequently, spatial variations occur in such a way that change in $|\Psi|$ is kept as small as possible. So in the first approximation we only have to consider variations of the phase, the continuous degree of freedom with respect to which the energy of the ordered phase is degenerate. This approximation is valid in the so-called London limit where spatial variations associated with superfluid flow occur over a long distance compared to the coherence length ζ. In this approximation the phase plays the role of velocity potential for superfluid flow, and as a result, rot $v_s = 0$. Vorticity can exist only as a form of vortex lines. A vortex line has a core of size ζ where the magnitude $|\Psi|$ is greatly reduced (Sect. 3.2).

In the ^3P superfluid the condensation energy is degenerate with respect to rotations of the spin and the orbital state of pairs [(4.2.21); we neglect the $d - d$ interaction for simplicity]. In the London limit, to which our discussion is restricted, the order parameters A_{ij} always retain the form (4.2.22) for the ABM states in the A phase and (4.2.24) for the BW states in the B phase with constant coefficient A. The rotational degrees of freedom which still remain in each case correspond to the phase in the ordinary superfluid. Since in the B phase with the isotropic energy gap the superfluid properties not involving spin are similar to those of the ordinary superfluid, we will consider only the A phase in the following.

As the order parameters of the A phase we can use the d-vector and the orthogonal unit vectors $\hat{n}_1$, $\hat{n}_2$. In the London limit we only have to consider spatial variations of these vectors, that is, nonuniform states are represented by the fields of these vectors. For simplicity we assume that the d-vector related to the spin state is fixed (it is not difficult to genelarize our discussions to include its spatial variations). It is convenient to think of the orbital part of the order parameter as a triad of the unit vectors, $\hat{n}_1$, $\hat{n}_2$ and $l = \hat{n}_1 \times \hat{n}_2$. This is somewhat analogous to the triad of the basis vectors describing a crystal. Since among other things the superfluid density ρ_s is different depending on the relative direction of its flow and the l-vector, we can regard the triad field as describing the "texture" of this anisotropic superfluid. The best example of a liquid with texture is liquid crystals, whose order parameter (in the nematic phase) is the director indicating the direction of a rod-like molecule.

What actually happens if the triad field varies in space? If we change by ϕ the phase of the pair wave function $A_j = A(\hat{n}_1 + i\hat{n}_2)_j$ [we omit the vector d because we have assumed it to be fixed], we get the new state

$$Ae^{i\phi} = A[(\hat{n}_1 \cos \phi - \hat{n}_2 \sin \phi) + i(\hat{n}_2 \cos \phi + \hat{n}_1 \sin \phi)] . \qquad (4.6.1)$$

This means that to change the overall phase by ϕ is equivalent to rotating the triad $(\hat{n}_1, \hat{n}_2, l)$ around the l vector by ϕ. Now, the state of uniform superfluid flow with velocity v_s is associated with the spatial variation of the phase $\phi = 2m_3 v_s \cdot x$ as we learnt in the case of the ordinary superfluids (all pairs moving with momentum $2m_3 v_s$). For this phase we can easily derive from (4.6.1)

$$v_s = \frac{1}{2m_3} \hat{n}_{1j} \nabla \hat{n}_{2j} . \qquad (4.6.2)$$

This leads to the term in the momentum density $\rho_{sij} v_{sj}$. The phase variation we have just considered corresponds to rotation around the fixed l-vector. Of course, the l-vector can also vary in space, and it turns out that $\nabla \times l$ gives rise to additional terms in the momentum density. In the weak coupling theory the momentum density is given by

$$j_s = \rho_{s\parallel} [2v_s - l(l \cdot v_s) + \tfrac{1}{2} \nabla \times l - \tfrac{1}{2} l(l \cdot \nabla \times l)], \tag{4.6.3}$$

where $\rho_{s\parallel}$ corresponds to $\rho_{n\parallel}$ obtained in Sect. 4.3. Since l is the direction of the angular momentum of the pairs, it is conceivable that we have the terms related to $\nabla \times l$ (the general form of j_s can be found from the requirement that it is a vector made by contractions of $A_i^* \nabla_k A_j$ and its complex conjugate).

In the ordinary superfluids an integral $\oint v_s ds$ around a closed line is equal to $2\pi n$ (n is an integer). Do we have the same quantization of circulation in the present case? In order to see it, let us calculate $\nabla \times v_s$ from (4.6.2). Making use of the fact that we are dealing with the triad of orthogonal unit vectors, we can reduce it to the form

$$(\nabla \times v_s)_i = \frac{1}{4m_3} \epsilon_{ijk} l \cdot (\nabla_j l \times \nabla_k l), \tag{4.6.4}$$

where ϵ_{ijk} is the antisymmetric unit tensor. Hence it does not vanish when the direction of l changes, so that the circulation can also take an arbitrary value. This implies an important result that vorticity and hence angular momentum can be present in the superfluid diffusely rather than as vortex lines. Let us show it in concrete example.

Suppose there is the A phase superfluid in a long cylindrical container and it is uniform along its axis, which we take in the $\hat{z}$ direction. If l is pointing everywhere in $\hat{z}$ direction, we have $\nabla \times v_s = 0$. Then circular flow around the central axis is in general given by the vortex line solutions, namely $\phi = n\varphi$, where φ is the angle in the cylindrical coordinates. In this case the triad makes n rotations around l as it goes around the axis once. Hence the order parameter must be discontinuous on the central axis, that is, there must be the core of the vortex line. Now, remove the condition $l \parallel z$. Is it possible to eliminate the core by changing l? The answer is yes. Choose $n = 2$ and assume $l \parallel z$ at some radius R. Then the triad makes two complete rotations around l as we go around a circular path of radius R, because there is a vortex line with $n = 2$. Now, instead of having $l \parallel z$ everywhere, we rotate the triad around the axis φ as we move in from the circle of radius R to the center so that l points down at the center as in Fig. 4.12 (everything is uniform along the z direction). If we construct the triad field in this way, we have $l \parallel -\hat{z}$, $\hat{n}_1 \parallel \hat{x}$, $\hat{n}_2 \parallel \hat{y}$ at the center as we approach it from any direction. The field is, therefore, continuous. By escaping in the third dimension, we succeeded in avoiding the singularity. Near the center the order parameter is uniform so that there is no flow, and near $r = R$ we have the vortex flow with $n = 2$. For this texture (called the Anderson-Toulouse structure) we can find the

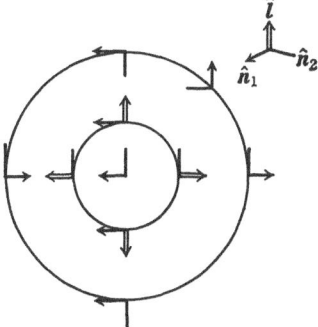

Fig. 4.12. The triad field in a cylinder

energy associated with the spatial variation and the total angular momentum around the central axis, which are listed in Table 4.2, where R is taken to be the radius of the container. In contrast to the energy of a vortex line proportional to $\ln(R/\xi)$ this structure has energy which is independent of R.

Table 4.2.

Structure	Free energy associated $/\pi A$	Angular momentum $/\pi R^2 m_3 A$
Vortex line ($l//z$)	$4 \ln(R/\xi)$	1
Radial disgyration	$\ln(R/\xi)$	0
M–H	5.85	0.885
A–T	23.1	1.70

$A = \dfrac{21\zeta(3)}{40} \dfrac{N}{m_3^*} \left(\dfrac{\Delta(T)}{2\pi T}\right)^2$; the energy is per unit length and for T near T_c

When we consider textures in a container, however, we must take into account the boundary condition at the wall. Let us for simplicity assume specular reflection at the wall, namely, that particles are reflected with its momentum component perpendicular to the wall reversed. If the l-vector, the symmetry axis of the pair wave function, is perpendicular to the wall, the pairs are not affected by the wall. If l is parallel, the pair wave function will surely be disturbed by the scattering, so that the magnitude of the order parameter will be reduced within a region of thickness comparable to the coherence length from the wall. Therefore, the boundary condition l perpendicular to the wall is imposed by the energy of the order of $\xi N(0) \Delta^2$ per unit area. In the case of a cylindrical container, then, l must be in the radial direction on the wall. If it is radial everywhere, $l \| \hat{\rho}$, we have a texture with singularity on the central axis (this texture is called radial disgyration). To avoid the singularity we just have to apply the procedure described above. The resulting texture is called the Mermin-Ho structure (which is just a central part of the A-T structure), the energy and the angular momentum of which are also listed in Table 4.2. In this structure too there is circular flow near the wall. Thus we have an interesting situation; when liquid ^{3}He becomes the A phase superfluid in a cylindrical container, it will spontaneously start to make circular flow even when the container is at rest. The experimental observation seems to be very difficult, unfortunately.

As an example of a three-dimensional structure let us consider the field with l pointing radially from the origin O (Fig. 4.13). Since the texture is not only the field of the l-vector but also of the triad, this seemingly point structure is actually associated with a singular vortex line. One can either attach two vortex lines with $n = 1$ and -1 to O, or one vortex line with $n = 2$ which then resembles the monopole. In general, structures in the triad field are characterized by quantum numbers such as an integral

$$N = \frac{m_3}{4\pi} \int dS \, \text{rot} \, v_s$$

taken over an arbitrary closed surface. If it contains the monopole, $N = 1$. In a container, the vortex line attached to the monopole must end at the wall. To lower the energy, then, the vortex line will contract and finally the monopole will touch the wall.

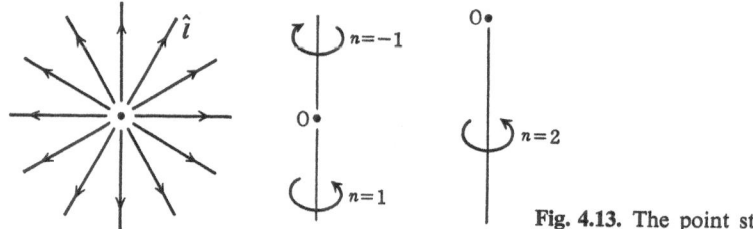

Fig. 4.13. The point structure

From the above examples it is clear that the superfluidity in the A phase is qualitatively different from that of the ordinary superfluids. In fact, it can be shown that the persistent flow, which is the symbol of superfluidity, is not stable without the boundary condition. If the l-vector is not fixed by the boundary condition at the wall or by a magnetic field (via the d-d interaction), one can reduce the quantum number of circulations along a closed path of persistent flow either to 0 or 1 by twisting the l-vector continuously. For discussions on these problems or for the classification of possible structures the homotopy theory is sometimes useful.

In this section we have considered textures concerning the orbital part only. It must be added that there have been many studies on the problems of textures including the d- vector or of the $\hat{n}$-vector in the B phase. In particular, the domain structures involving the d-vector have been studied in detail.

Because of the limitation of space we have left out many important problems such as (i) the transport phenomena, (ii) collective modes including spin waves and orbital waves, (iii) intrinsic orbital angular momentum in the A phase, (iv) properties of the A_1 phase under strong magnetic field and (v) two-fluid dynamics. For these subjects the readers are referred to the review articles listed in the bibliography.

5. Metals

In Chaps. 2, 3, we have considered materials of macroscopic scale composed of single element atoms with $Z = 1$ (hydrogen, Chap. 2) and $Z = 2$ (helium, Chap. 3), and examined what kinds of physical states they can take. It is rather surprising to observe that within the examples of only these two substances, many kinds of possible phases are already realized if suitable conditions are provided. Thus we have the phases of vapor, liquid, plasma, molecular crystal, metal and even superfluid and superconducting states. However, one should recognize that nature is far more complex, rich in variety and allows the existence of many characteristic states. Therefore, it would be necessary to proceed further with the consideration of the materials with $Z = 3, 4, \ldots$ so on and examine their possible schemes of existence. If we limit our consideration to within the states of absolute zero, $T = 0$, and no external pressure, $P = 0$, it turns out that Li with $Z = 3$ becomes a metal, as already mentioned, and Be with $Z = 4$ is also a metal. To know how many elements are metals in their normal states, let us look at the periodic table. As Table 5.1 shows, there are about 70 among 103 elements which are normally called metals. In other words, the metallic state is most common in the elements, and the inactive gases (Ne, Ar, Kr, etc.), the halogen family, nonmetallic elements such as O and N, and a few semiconducting or semimetallic elements occupying a narrow region in the periodic table between the metals and nonmetals belong to the minority group.

Therefore we should first ask why so many elements become metals. Postponing the discussion about the nonmetallic states until Chap. 6, we shall give in this chapter, a brief survey of the metallic states. It should, however, be noted that as mentioned in Chap. 2, a material can take many different phases with varying circumstances (temperature and pressure). For instance, the nonmetallic elements which will be discussed in Chap. 6 may become metals under sufficiently high pressure. Such a transition between different phases is one of the subjects to be discussed.

Table 5.1. Periodic table. (A*) Actinide elements; (L*) Lantanide elements

Element U (kcal·mol⁻¹) T_m (°C)											

Ia	IIa												Semimetals semiconductors		Non-metals			
													IIIb	IVb	Vb	VIb	VIIb	0
Li 38 180	**Be** 77 1278												**B** 131 2030	**C** 170 3727	**N** 1 50 −210	**O** 1 74 −219	**F** 0 52 −220	**Ne** 0 52 −249
Na 26 98	**Mg** 35 650						Metals						**Al** 77 660	**Si** 107 1410	**P** 75 44	**S** 66 119	**Cl** 6 −101	**Ar** 1 77 −189

Ia	IIa	IIIa	IVa	Va	VIa	VIIa	VIII			Ib	IIb	IIIb	IVb	Vb	VIb	VIIb	0
K 22 64	**Ca** 42 850	**Sc** 80 1539	**Ti** 112 1800	**V** 122 1715	**Cr** 94 2000	**Mn** 67 1250	**Fe** 99 1530	**Co** 102 1490	**Ni** 102 1452	**Cu** 80 1083	**Zn** 31 420	**Ga** 65 30	**Ge** 89 959	**As** 29 817	**Se** 49 217	**Br** −7	**Kr** 2 67 −157
Rb 20 39	**Sr** 39 800	**Y** 97 1509	**Zr** 146 1900	**Nb** 174 1950	**Mo** 151 2600	**Tc** (152) 2200	**Ru** 154 2000	**Rh** 133 1970	**Pd** 90 1553	**Ag** 68 961	**Cd** 27 321	**In** 57 155	**Sn** 72 232	**Sb** 62 631	**Te** 47 453	**I** 19 114	**Xe** 3 76 −112
Cs 19 29	**Ba** 43 850	**L***	**Hf** 146 2230	**Ta** 187 3027	**W** 200 3390	**Re** (186) 3180	**Os** 187 2500	**Ir** 159 2360	**Pt** 135 1771	**Au** 88 1063	**Hg** 15 −39	**Tl** 43 304	**Pb** 47 327	**Bi** 50 271	**Po** 35 254	**At** (300)	**Rn** −71
Fr (19) 27	**Ra** (42) 700	**A***															

La	Ce	Pr	Nd	Pm	Sm	Eu	Gd	Tb	Dy	Ho	Er	Tm	Yb	Lu
102 826	98 804	86 919	76 1019	(64) 1027	50 1072	43 826	83 1312	90 1356	67 1407	71 1461	71 1497	58 1545	40 824	99 1650

Ac	Th	Pa	U	Np	Pu	Am	Cm	Bk	Cf	Es	Fm	Md	No	Lr
(104) 1050	137 1751	(132) (1425)	125 1131	(113) 637	92 640									

5.1 Characteristics of Metals

The most eminent macroscopic properties of metals may be concisely characterized to be good electric and thermal conductors, opaque for visible light with luster and easily deformable under external stresses. It is natural to consider that such properties are to be attributed to the cohesive mechanism of metals. Let us first take the simplest alkaline metals, which are Li ($Z = 3$), Na ($Z = 11$), K ($Z = 19$), Rb ($Z = 37$) and Cs ($Z = 55$). The electronic configurations of these atoms consist of closed shells and a valence electron with ns state which is loosly bound to a closed shell and circulates around it. If we discard these closed shells, the electronic structures of these atoms are similar to that of H ($Z = 1$). In the ground state of a system composed of a large number of atoms, hydrogens form a molecular crystal in which the unit of the crystal structure is a hydrogen molecule. Why is the unit of the crystal structure of alkaline metals not diatomic molecules like Li_2 and Na_2? Actual crystals form a lattice consisting of positive ion cores and of valence electrons which move relatively freely. One reason for this difference from the hydrogen crystal case is attributed to the presence of ionic croes. Atomic nuclei are forced to keep each other apart beyond a certain distance due to the ionic cores. The wave functions of valence electrons are also pushed away from atomic nuclei to spread over a crystal so that they are orthogonalized to the wave functions of ionic core electrons. Moreover, contrary to the hydrogen case, since there are np orbitals having nearly equal energies, the formation of a large assembly of atoms does not give rise to an energy increase. Instead it offers a possiblity of forming molecular orbitals extending over the whole crystal in which all valence electrons may be accommodated. The valence electrons in such extended orbitals would gain energy from a negative Coulomb potential due to the ionic cores on the one hand, and would keep the increase of the kinetic energies to a minimum on the other hand, thus making the metallic state stable.

This situation does not change essentially for divalent metals and multivalent metals. It is clear that this picture explains qualitatively the characteristics of metals mentioned above. The presence of valence electrons which move freely throughout a crystal (simply called free electrons of metals hereafter) makes the system a good electric conductor. Since free electrons are also carriers of heat, metals are good thermal conductors as well. Direct evidence which shows that free electrons are carriers of both, electric and thermal currents is provided by the Wiedemann-Franz law for metals. That is, the ratio of electric and thermal conductivity is a constant depending only on temperature irrespective to materials. We have summarized the measured values of specific resitivity and thermal conductivity for several metals in Table 5.2. The values of specific resitivity of metals vary over the range of $10^{-8} - 10^{-6}$ in the unit of $\Omega \cdot m$, and the largest among them is for Bi which is called a semimetal. This is because the conductivity of Bi has an intermediate value between metals and semiconductors such as Si and Ge. We will discuss semimetals later in Chap. 6 along with semiconductors. To check the validity of the Wiedemann-Franz law, we also tabulate the Lorentz number L defined by

$$L = \frac{k\rho}{T} \tag{5.1.1}$$

for several selected metals. A simple transport theory gives the Lorentz number as

Table 5.2. Specific resitivity ρ and thermal conductivity k of metals. ρ is the value at 20°C $[10^{-8}\Omega m]$ k is the value at 0°C [J cm^{-1} s^{-1}k^{-1}]

	ρ	k		ρ	k		ρ	k
Li	9.4	—	As	35	—	Cs	21	—
Be	6.4	2.3	Rb	12.5	—	Ta	15	0.54
Na	4.6	1.35	Sr	30.3	—	W	5.5	1.70
Mg	4.5	1.5	Zr	49	—	Os	9.5	—
Al	2.75	2.38	Mo	5.6	1.43	Ir	6.5	1.48
K	6.9	0.99	Rh	5.1	—	Pt	10.6	0.69
Mn	4.2	0.22	Pd	10.8	0.67	Au	2.4	3.1
Fe	9.8	0.76	Ag	1.62	4.18	Hg	95.8	0.09
Co	6.37	—	Cd	7.4	0.92	Tl	19	0.41
Ni	7.24	0.125	In	8.2	0.25	Pb	21	0.35
Cu	1.72	3.85	Sn	11.4	0.64	Bi	120	0.085
Zn	5.9	1.13	Sb	38.7	0.18	Th	18	—

$$L = \frac{\pi^2}{3}\left(\frac{k_B}{e}\right)^2$$
$$= 2.7 \times 10^{-13}[\text{esu} \cdot \text{K}^{-2}] = 2.45 \times 10^{-8} \ [\text{W} \cdot \Omega \cdot \text{K}^{-2}] \tag{5.1.2}$$

which explains the experimentally observed values pretty well.

Optical properties of metals are attributed to the behaviour of free electrons. Bright metals such as Al and Ag do not absorb visible light but completely reflect it. This can be understood in the following way. High frequency parts of electromagnetic waves of light stimulate forced oscillation in free electrons which is the collective plasma oscillation stated in Sect. 2.4. Its eigenfrequency is given by

$$\omega_p^2 = \frac{4\pi n e^2}{m}, \tag{5.1.3}$$

where m is electron mass and n is electron density. A typical electron density yields the plasma frequencies in the ultraviolet region. Therefore the forced oscillation of free electrons which is antiphase with the electromagnetic oscillation almost cancels out the electric field inside a metal. As a result, an electromagnetic wave cannot propagate into a metal, and is completely reflected at the surface of a metal. This is the origin of metallic luster. Polished metals are good reflectors. There are metals having characteristic colors such as copper and gold. This is because of the presence of localized d electrons in addition to free electrons which absorb photons of particular wave length.

Metallic bonds are characterized by not having directional anisotropy. This can be roughly understood by the fact that the positive ions are embedded in a free electron gas and bonded through it to each other. Thus electrons do not localize to form strong covalent bonds which generally have preferred directions. However, this is not always the case if d electrons exist or when the population of p electrons increases as the number of valence is increased, as the possibility of hybridization of spd orbitals causes a tendency toward directional bonding. In fact, to be a metal, the number of np electrons per atom is limited, at the most, to two electrons (Table 5.1). The group V elements with the configuration $(ns)^2 \ (np)^3$ are either insulators (as N and P) or semimetals (as As, Sb

Table 5.3. Lorentz numer of metals in unit of 10^8 W$\cdot\Omega\cdot$K^{-3}

	$T = 0°C$	$T = 100°C$		$T = 0°C$	$T = 100°C$
Cu	2.23	2.33	Pb	2.47	2.56
Ag	2.31	2.37	Pt	2.51	2.60
Au	2.35	2.40	Ir	2.49	2.49
Zn	2.31	2.33	Mo	2.61	2.79
Cd	2.42	2.43	W	3.04	3.20
Sn	2.52	2.49			

and Bi). The group IV elements with the configuration $(ns)^2 (np)^2$ are insulators diamond), semiconductors (Si and Ge) or semimetals (graphite and grey tin). All these crystals have a strong directional anisotropy in their bonding. White tin, which does not have diamond structure and Pb which has a face-centered cubic structure are exceptionally metallic.

The metals which have incomplete d-electron shells are called transition metals. They exhibit different properties from other metals (metals of sp electrons are sometimes referred to as simple metals). For example, optical properties and magnetic properties are quite different from others as we will see in Chap. 8. The most direct evidence to show their singular nature is found in their cohesion. In Table 5.1 we have listed the cohesive energy U (i.e. the energy necessary to break up the crystal into individual neutral atoms) and the melting temperature T_m for all elements. As is easily seen from this table, the cohesive energies of transition metals are larger than those of other metals. This is due to the additional binding energy coming from d electrons.

By the way, let us compare the melting temperature of metals with that of nonmetallic elements. It turns out that the binding force of metals is stronger than that of molecular crystals, but weaker than that of diamond. The fact that metals have an intermediate melting temperature makes them easier to work with, but the easy deformability of metals is essentially based on the nondirectional metallic bonding. To consider this point further, it is important to take account of the dislocation theory of metals which is discussed in Chap. 10. The easy deformability of metals reveals itself in the elastic properties of metallic crystals. In Table 5.4 we show elastic constants for several selected crystals. All these crystals may be regarded as macroscopically isotropic. It should be noted that the rigidity μ of metals is small, and the Poisson ratio ν is large compared with nonmetals. According to the dislocation theory, the critical shearing stress σ to remove a dislocation is given by

$$\sigma = \frac{2\mu}{1-\nu} \exp\left(-\frac{2\pi c}{1-\nu}\right), \tag{5.1.4}$$

Table 5.4. Elastic constants of selected isotropic substances

	10^{11} dyn·cm^{-2}			Poisson ratio
	c_{11}	$c_{44} = \mu$	c_{12}	ν
Na	0.603	0.586	0.459	0.315
Ni	24.65	12.47	14.73	0.30
Al	10.82	2.85	6.13	0.34
Cu	16.84	7.54	12.14	0.345
Pb	4.66	1.44	3.92	0.44
diamond	110	44	33	0.18
NaCl	4.87	1.26	1.24	0.10
KBr	3.46	0.505	0.58	0.07

where c is a numerical constant depending on crystal lattices. As the Poisson ratio ν approaches $1/2$ and μ becomes small, then σ becomes also small. In other words, small shearing stress yields to permanent deformation. The fact that $\nu \cong 1/2$ and μ is small for most metals implies that cohesive force depends only on the volume, and deformation can take place only in such a way that the total volume is kept constant.

We have so far given preliminary discussions of metals in order to be able to understand their main properties qualitatively. Our elementary discussions build a simple image for metals. According to it metals can be described by a system consisting of a free electron gas and ionic cores embedded in it. On the basis of this image, we can construct a free electron model for metals, or a band theory of electrons moving through a periodic potential exerted by ionic cores, to be described in the next section. However, as we will see later, such a crude model exhausts the essence of metals and can successfully explain gross features of microscopic properties of simple metals.

5.2 Band Theory of Metals

The fundamental symmetry of a crystal is uniquely determined by three fundamental vectors a_1, a_2 and a_3. All lattice points in a crystal are expressed by

$$R_l = l_1 a_1 + l_2 a_2 + l_3 a_3 . \tag{5.2.1}$$

Due to this periodicity, the Hamiltonian should be invariant under a coordinate transformation such that

$$r \rightarrow r + R_l . \tag{5.2.2}$$

The averaged potential $V(r)$ exerted on electrons in a crystal should satisfy the following equation:

$$V(r + R_l) = V(r) . \tag{5.2.3}$$

Since the energy of electrons is given by the eigenvalues of the Schrödinger equation

$$-\frac{\hbar^2}{2m} \nabla^2 \psi(r) + V(r)\,\psi(r) = \varepsilon\psi(r) , \tag{5.2.4}$$

the problem is to solve this eigenvalue problem with a periodic potential $V(r)$ of (5.2.3).

In the free electron model, $V(r)$ is replaced by a constant averaged potential $-U_0$. Choosing the origin of the energy at the bottom of this negative potential, we obtain the solution of (5.2.4) as

$$\psi(r) \approx e^{ik\cdot r}, \quad \varepsilon = \frac{\hbar^2 k^2}{2m}.$$
(5.2.5)

For simplicity we use the periodic boundary condition of the Born-vou Karman type where a cube with side length L is chosen as a fundamental region. Since $\psi(x + L, y, z) = \psi(x, y, z)$,

$$k_x L = 2\pi n_1, \quad k_y L = 2\pi n_2, \quad k_z L = 2\pi n_3$$
(5.2.6)

(n_1, n_2 and n_3 being arbitrary integers) are held, and the quantized energy levels are given by

$$\varepsilon = \frac{\hbar^2}{2m} \left(\frac{2\pi}{L}\right)^2 (n_1^2 + n_2^2 + n_3^2).$$
(5.2.7)

Since $L = \Omega^{1/3}$. is of a macroscopic size, distribution of the energy levels of (5.2.7) is almost continuous even when n_i changes. Let us introduce the density of states $n(\varepsilon)$ which is the number of eigenstates having the energy within the range ε and $\varepsilon + d\varepsilon$. We can easily derive $n(\varepsilon)$ from (5.2.7) as

$$n(\varepsilon) = \frac{4\pi\Omega(2m)^{3/2}\sqrt{\varepsilon}}{h^3},$$
(5.2.8)

taking account of the electron spins. The free electron gas is obtained by filling up electrons into these states according to the Pauli exclusion principle. In the free electron model there are only two parameters involved: the depth U_0 of the potential and the free electron number N_e. At the absolute zero temperature, the Fermi distribution is such that pairs of electrons with up and down spins fill up the levels of the density of states given by (5.2.8) from the low-energy side. We define the maximum energy as the Fermi energy ε_F. For a given N_e,

$$\int_0^{\varepsilon_F} n(\varepsilon)\, d\varepsilon = N_e$$

determines ε_F, that is, explicitly

$$\varepsilon_F = \frac{\hbar^2}{2m}\left(\frac{3N_e}{8\pi\Omega}\right)^{3/2}.$$
(5.2.9)

The eigenfunction of free electrons is characterized by the quantized wave vectors of (5.2.6), and is written as

$$\psi_k(r) = \frac{1}{\sqrt{\Omega}} e^{ik\cdot r},$$
(5.2.10)

which is normalized in a volume Ω. In order to see the effects of the periodicity of a potential, it is convenient to introduce here the reciprocal lattice vectors in the space spanned by the wave vector k. Let us define three fundamental reciprocal vectors b_1, b_2 and b_3 such that they satisfy the following equations:

$$a_i \cdot b_j = 2\pi\delta_{ij} \quad (i, j = 1, 2, 3) . \tag{5.2.11}$$

The explicit form of the solution of (5.2.11) is given by

$$\left.\begin{aligned}
b_1 &= \frac{2\pi}{\Delta} a_2 \times a_3 \\
b_2 &= \frac{2\pi}{\Delta} a_3 \times a_1 \\
b_3 &= \frac{2\pi}{\Delta} a_1 \times a_2
\end{aligned}\right\}, \tag{5.2.12}$$

where $\Delta = a_1 \cdot (a_2 \times a_3)$ is equal to the volume of a unit cell. Therefore, the dimension of b_i is inverse to that of a_j, being the same dimension as the wave vector. In terms of a set (n_1, n_2, n_3) of arbitrary integers, we define the reciprocal lattice vectors as

$$B_n = n_1 b_1 + n_2 b_2 + n_3 b_3 . \tag{5.2.13}$$

Using the definition (5.2.11) for b_j, we can show that any lattice vector R_l of (5.2.1) satisfies

$$B_n \cdot R_l = 2\pi(n_1 l_1 + n_2 l_2 + n_3 l_3) = 2\pi \times \text{integer.}$$

Then it is evident that $\exp[i(B_n \cdot R_l)] = 1$. This means that the function $\exp(iB_n \cdot r)$ has the same periodicity as the crystal lattice, that is

$$\exp[iB_n \cdot (r + R_l)] = \exp(iB_n \cdot r) .$$

Thus the periodic potential $V(r)$ should be expanded as

$$V(r) = \sum_n V_n \exp(iB_n \cdot r) \tag{5.2.14}$$

where the sum $\sum_n$ is taken over all integers, or over all reciprocal lattice vectors. Equation (5.2.14) is nothing but the three-dimensional Fourier decomposition. By assuming that V_n's are small enough to treat $V(r)$ as a small perturbation, we shall examine energy band structures within the nearly free electron approximation (NFE model). We take a fundamental region of the volume $\Omega = N\Delta$ which contains $N(= N_1 N_2 N_3)$ unit cells. N_1, N_2 and N_3 are the numbers of lattice points in a_1, a_2 and a_3 directions, respectively. We also assume the following unperturbed wave function:

$$\psi_k = \frac{1}{\sqrt{N\Delta}} e^{ik \cdot r}, \tag{5.2.15}$$

where k is given in terms of a set of three integers (h_1, h_2, h_3) by

$$k = \frac{h_1}{N_1} b_1 + \frac{h_2}{N_2} b_2 + \frac{h_3}{N_3} b_3 . \tag{5.2.16}$$

As is seen from (5.2.16), the distribution of k, although taking N discrete values, is almost continuous in a unit cell of the reciprocal lattice space. The matrix element of the potential energy $V(r)$ with respect to the unperturbed states (5.2.15) has the following form:

$$(k|V|k') = \frac{1}{N\Delta} \sum_n V_n \int_\Omega \exp\left[i(k' - k + B_n) \cdot r\right] dr .$$

Note that the integral in the right-hand side is $N\Delta$ when

$$k' - k + B_n = 0 \tag{5.2.17}$$

and vanishes otherwise. Thus the nonvanishing matrix element can be written as

$$(k|V|k - B_n) = V_n . \tag{5.2.18}$$

In other words, the wavevector of the state which gives a nonvanishing matrix element mixed with the state k through V should have the form

$$K(n) = k - B_n . \tag{5.2.19}$$

If we introduce

$$[K(n)|V|K(m)] = V_{m-n}, \tag{5.2.20}$$

then the eigenvalue equation is reduced to

$$\det \left\| \left[\frac{\hbar^2}{2m} K(n)^2 - \varepsilon\right] \delta_{nm} + V_{m-n} \right\| = 0 . \tag{5.2.21}$$

The rank of this determinant is infinite. The solutions of this equation, if found, give the exact energy eigenvalues irrespective of the values of V_n. However, in practice we can solve it only when V_n is very small. If $(\hbar^2/2m) [K^2(n) - K^2(m)]$ $(n \neq m)$ were larger than any V_l, the eigenvalue is approximately determined by the well-known perturbation theory as

$$\varepsilon_n(k) = \frac{\hbar^2}{2m} (k + B_n)^2 + \frac{2m}{\hbar^2} \sum_{m \neq n} \frac{|V_{m-n}|^2}{K(n)^2 - K(m)^2}, \tag{5.2.22}$$

This is not valid when the two wave numbers $K(n)$ and $K(m)$ are equal, that is, when k and $k' = k - B_n$ satisfy

$$k^2 = (k - B_n)^2 . \tag{5.2.23}$$

This is the case where two energies of the unperturbed states are degenerate. We must re-examine the original secular equation in this case to have a correct solution:

$$\varepsilon(k) = \frac{\hbar^2}{2m} k^2 \pm |V_n|^2 . \tag{5.2.24}$$

For a given k there are two energies $\varepsilon(k)$. They are separated by a gap

$$\Delta\varepsilon = 2|V_n| . \tag{5.2.25}$$

Thus there exists an energy range which is forbidden for electrons. The condition (5.2.23) may be rewritten as

$$k \cdot B_n = \tfrac{1}{2} |B_n|^2 \tag{5.2.26}$$

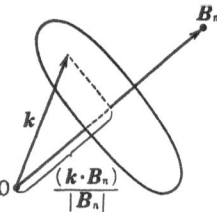

Fig. 5.1. Representation of the condition (5.2.26)

This is easily interpreted in geometrical terms. As shown in Fig. 5.1, we consider a vector which connects the origin and the lattice point B_n in the reciprocal space along with the plane which bisects this vector perpendicularly. Then it is evident that all the vectors which connect the origin to any point on this plane satisfy the condition (5.2.26). Therefore in the nearly free electron approximation, most energy eigenvalues do not alter much from the free electron case except for the states with the wave vectors which lie on the planes bisecting the reciprocal lattice vectors perpendicularly. On such planes, energy eigenvalues differ appreciably from corresponding free electron states, and generally, discontinuity of energy is expected. Now we shall introduce here the concept of Brillouin zones. They are the regions in the reciprocal space which contain the origin and are bounded by the planes which bisect the vectors connecting the origin and any one of reciprocal lattice points. We call the smallest volume zone the first Brillouin zone and similarly call the second, third and so on with increasing enclosed volumes. In this scheme, energy gaps generally appear on the Brillouin zone boundaries.

It will be instructive to demonstrate the above results by considering a one dimensional system. We depict in Fig. 5.2a the eigenvalues for free electrons when neglecting the second term of the right-hand side in (5.2.22). They are parabolas centered at the points corresponding to different n. Glancing at this figure, one immediately notices that the number of states is increased compared with the original free electron case (i.e. single parabola). The periodicity of the lattice defines the reciprocal lattice vectors. It is obvious that any wave vector k' in the reciprocal space may be put in a form of the sum $k - B_n$, where k is a vector limited within the first Brillouin zone and B_n is a reciprocal lattice vector suitably chosen for the given wave vector k'. Therefore, a state labeled by a wave vector k' can be labeled as well by a suitable n and a wave vector k within the first Brillouin zone. From this argument, we can see that the really independent states in Fig. 5.2a are either those included in the first Brillouin zone or the states described by a single parabola.

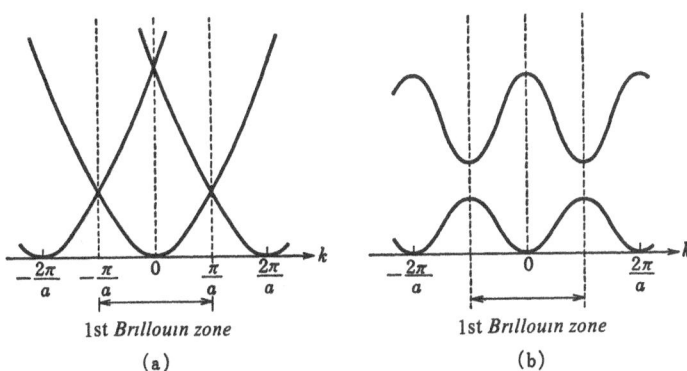

Fig. 5.2a, b. The first Brillouin zone of one-dimensional crystal. (a) Free electron; (b) with a periodic potential

The first choice of selecting the independent states is called the reduced zone scheme. The second choice is called the extended zone scheme. In addition to these two schemes, we sometimes use the third one called the periodic zone scheme in which all the states in Fig. 5.2 are taken into account. Although in this scheme the states are overcounted, it has a merit such that the energy eigenvalue is manifestedly a periodic function in a wave vector. When V_n is not zero, energy gaps appear at the points where the zeroth-order energies are degenerate as shown in Fig. 5.2b. In the reduced zone scheme a set of energy eigenvalues forms bands separated by gaps.

The three-dimensional case is essentially the same, but difficult to visualize. However, it will be useful for later discussion to explain here the cases of bcc and fcc lattices in some detail. We demonstrate the first Brillouin zone of a bcc lattice in Fig. 5.3. According to (5.2.12), the reciprocal lattice of a bcc lattice forms a

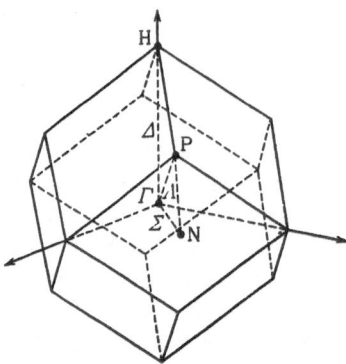

Fig. 5.3. The first Brillouin zone of a bcc lattice

face centered lattice, and hence there are twelve equivalent reciprocal lattice points nearest to the origin and twelve faces which bisect the vectors connecting these reciprocal lattice points and the origin. The letters in this figure are commonly used to denote main symmetry points. In the reduced zone scheme the eigenvalues and the wave functions are given by (within the free electron approximation)

$$\varepsilon_n(k) = \frac{\hbar^2}{2m} \, |k - B_n|^2 \tag{5.2.27}$$

$$\psi_k(r) = \exp{(ik \cdot r)} \exp{(-iB_n \cdot r)} \, . \tag{5.2.28}$$

Introducing (ξ, η, ζ) defined by

$$k = \frac{2\pi}{a} (\xi, \eta, \zeta) \, ,$$

where a is the length of a cube, we can express $\varepsilon_n(k)$ as

$$\frac{2ma^2}{\hbar^2} \, \varepsilon_n(k) = [\xi - (n_2 + n_3)]^2 + [\eta - (n_1 + n_3)]^2 + [\zeta - (n_1 + n_2)]^2 \, .$$

$$\tag{5.2.27'}$$

As an example we illustrate the curves of (5.2.27') from $\varGamma$ point to H point along $\varDelta$-line in Fig. 5.3. The numbers in this figure denote the degeneracies of a band, or the number of wave functions associated with each band. It is not hard to evaluate bands in other directions in k-space. We show the final results in Fig. 5.4b omitting all the detail of calculations.

For a fcc lattice, its reciprocal lattice is body centered. The first Brillouin zone is shown in Fig. 5.5. In Fig. 5.6, the bands $\varepsilon_n(k)$ along main symmetry lines are depicted.

When the periodic potential $V(r)$ is nonvanishing but small, the band degeneracies in Figs. 5.4, 5.6 will be removed and many band gaps will appear.

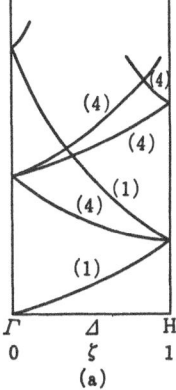

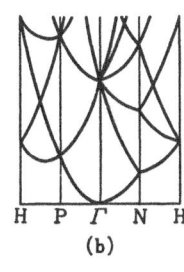

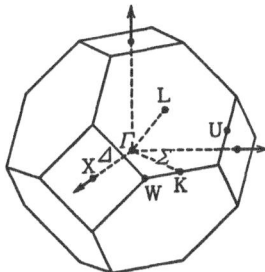

Fig. 5.4a,b. Energy bands of a free electron in a bcc lattice

Fig. 5.5. The first Brillouin zone of a fcc lattice

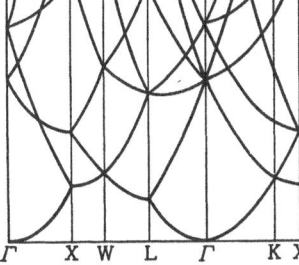

Fig. 5.6. Energy bands of a free electron in a fcc lattice

However, in contrast to the one-dimensional case, band structures of a three-dimensional crystal are different for each direction, and therefore the location and magnitude of gaps depend on the directions in k-space. Thus the forbidden bands do not always appear in the density of states as a function of energy. In the free-electron model it is impossible to describe detailed characteristics of various materials because the number of model parameters is too small. By taking account of a periodic potential, it becomes possible to explain qualitatively delicate differences between various properties depending on crystal structures and elements.

To solve the eigenvalue problem of the Schrödinger equation (4.24) for a given $V(r)$, without resorting to perturbation theory on the basis of free electrons, is a hard task. Fortunately, in virtue of high speed computers, it eventually becomes feasible to perform band calculations fairly accurately for many elements. There are several useful devices for this purpose, that is, the orthogonalized plane-wave method (OPW), argumented plane-wave method (APW) and Green's-function method which fit with computer calculations. Since the main references of the band calculations are listed in the end of this volume ([5.1–4]), we briefly review important facts and refer to the results which are concerned with later discussion, without going into details.

Because of the periodicity of $V(r)$ expressed by (5.2.3), the eigenvalues and eigenfunctions of the Schrödinger equation (5.2.4) are required to be

$$\varepsilon = \varepsilon_n(k) \tag{5.2.29}$$

$$\psi_{kn}(r) = e^{ik\cdot r}u_n(r), \tag{5.2.30}$$

where $\varepsilon_n(k)$ is a periodic function of k. Thus k can be limited to the fundamental region (for instance, in the first Brillouin zone) in the reciprocal space and n is a quantum number denoting the band index (reduced zone scheme). The function $v_n(r)$ is a periodic function satisfying

$$u_n(r + R_l) = u_n(r). \tag{4.2.31}$$

Equations (5.2.29.31) are the results of the so-called Bloch theorem. Since the original Schrödinger equation is invariant under the symmetrical operations which form a space group, eigenfunctions and eigenvalues are required to have these properties. Although their general verifications will not be given here, it should be pointed out that the properties expressed in (5.2.29–31) have already manifested themselves in the solution of the NFE approximation. Periodicity of the eigenvalues is apparent in the one-dimensional case from Fig. 5.2. Eigenfunctions have also a form of (5.2.30), since they were given by (5.2.28) or their linear combinations with different n. The reason why the NFE approximation is not useful in practical calculations is that the expansion in terms of plane waves does not rapidly converge. Since we are interested in energies of valence electrons in metals, the corresponding wave functions must be orthogonal to electron wave functions in inner shells, thus they oscillate rapidly in ion core regions. This means that in order to solve the secular equation (5.2.21) with fair accuracy, we must treat the determinant with large dimensions. To avoid this difficulty, in the OPW method we choose from the beginning the wave functions which are orthogonalized to those of electrons in inner shells as unperturbed wave functions. Let $\varphi_\alpha(r - R_l)$ and ε_α be the wave function and its eigenenergy of an inner electron in a level α at site R_l. If we construct from $\varphi_\alpha(r - R_l)$ the function

$$\chi_\alpha(r) = \frac{1}{\sqrt{N}} \sum_l \exp(ik \cdot R_l)\, \varphi_\alpha(r - R_l), \tag{5.2.32}$$

it satisfies the Bloch theorem (5.2.30). When $\varphi_\alpha(r - R_l)$ and $\varphi_\beta(r - R_l)$ in different inner shell orbits are orthogonal, it is easy to see that the functions $\chi_\alpha(r)$ are orthogonal too. Now we consider the linear combination

$$\mathrm{OPW}(k) = \frac{1}{\sqrt{\Omega}} e^{ik\cdot r} - \frac{1}{\sqrt{\Omega}} \sum_\alpha C_{\alpha k}\chi_\alpha(r) \tag{5.2.33}$$

so as to orthogonalize to arbitrary $\chi_\alpha(r)$ by choosing $C_{\alpha k}$ as follows:

$$C_{\alpha k} = \int \chi_\alpha^*(r)\, e^{ik\cdot r}\, dr \,. \tag{5.2.34}$$

Because of (5.2.32), $\chi_\alpha(r)$ is invariant under the replacement k by $k - B_n = K(n)$. Thus in the OPW(k) of (5.2.33) we can replace k by $K(n)$. By expanding in terms of OPW(k) instead of plane waves in the NFE approximation as

$$\psi_k(r) = \sum_n a_{kn}\, \text{OPW}(k - B_n)\,, \tag{5.2.35}$$

we obtain a secular equation to determine eigenvalues in the same form as (5.2.21)

$$\det \left\| \left[\frac{\hbar^2}{2m} K(n)^2 - \varepsilon \right] \delta_{nm} + \Gamma_{nm} \right\| = 0\,, \tag{5.2.36}$$

where Γ_{nm} is, however, written as

$$\Gamma_{nm} = V_{n-m} + \sum_\alpha (\varepsilon - \varepsilon_\alpha)\, C_{\alpha K(m)}{}^* C_{\alpha K(n)}\,. \tag{5.2.37}$$

Note that Γ_{nm} contains ε in contrast to (5.2.21). The convergence in the OPW method is much more improved than that of the NFE approximation. It is known that a few numbers of OPW(k) less than ten are enough to have a good result of band calculations for alkali metals.

We show in Fig. 5.7 a result of the band calculation for Na of a bcc lattice derived by the OPW method. Comparing it with the bands calculated in the NFE approximation (as shown in Figs. 5.4a,b), we can observe a clear correspondence between both calculations. By adding electrons to these bands for this alkali metal (taking into account the freedom of electron spins), the Fermi level is located at the position indicated in these figures. Up to this energy, the band structures calculated by the OPW method and the NFE approximation almost coincide with each other.

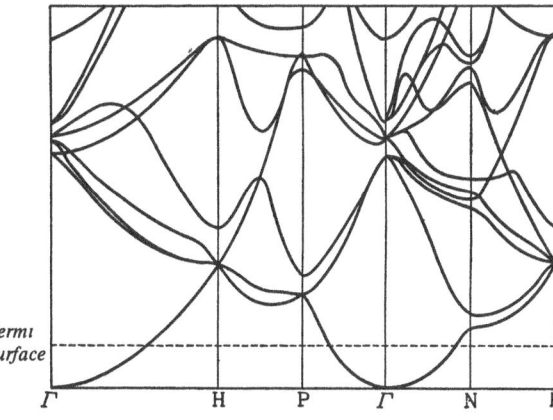

Fermi
surface

Γ H   P Γ N   H   Fig. 5.7. Band structure of Na

In multivalent metals, the Fermi surface is situated outside the first Brillouin zone because the free electron number increases. In the reduced zone scheme the Fermi surface is located at a band with high band index number. In such a case one might think that the NFE approximation will not be justified, but in practice the energy band structure calculated for fcc Al is very similar to that of the NFE approximation. Various band calculations by the OPW, APW and Green's-function method have been done for Al, all yielding almost identical results. In Fig. 5.8, a cross section of the Fermi surface by (110) plane of Al is shown.

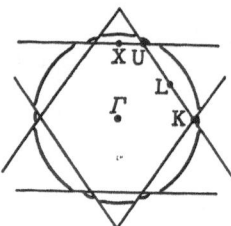

Fig. 5.8. A cross section of the Fermi surface of Al in the extended zone scheme

Now let us outline the APW method. A fundamental idea of this method is in an observation that electron wave functions in a crystal should resemble that of a free ion near an ion core region and behave like plane waves outside ion cores. We divide the crystal potential into two parts, one consisting of spheres centered at each atomic nucleus and the other part outside these spheres. We further assume that inside a sphere with radius R, a spherically symmetric potential is felt and outside the spheres the potential may be taken as constant. Then outside the spheres, the wave function can be approximated by a linear combination of plane waves, that is, by

$$\psi_k^{\text{out}}(r) = \sum_n C_n(k) \, e^{i\boldsymbol{K}(n)\cdot r} \, . \tag{5.2.38}$$

Inside the sphere, the wave function of an electron in a spherically symmetric potential takes the following form:

$$\psi_k^{\text{in}}(r) = \sum_n \sum_{lm} A_n(l, m) \, Y_{lm}(\theta, \varphi) \, R_l(r, \varepsilon) \, , \tag{5.2.39}$$

where the radial wave function $R_l(r, \varepsilon)$ satisfies

$$-\frac{1}{r^2} \frac{d}{dr} \left(r^2 \frac{dR_l}{dr} \right) + \left[\frac{l(l+1)}{r^2} + v(r) \right] R_l = \varepsilon R_l \, . \tag{5.2.40}$$

Since these two wave functions should be matched smoothly at $r = R$,

$$A_n(l, m) = 4\pi(\mathrm{i})^l C_n(k) \frac{j_l[\boldsymbol{K}(n), R]}{R_l(R, \varepsilon)} Y_{lm}(\theta_{\boldsymbol{K}(n)}, \varphi_{\boldsymbol{K}(n)}) \tag{5.2.41}$$

is held. The coefficient $C_n(k)$ is determined so as to minimize the energy expectation value calculated by these wave functions. The linear equations for $C_n(k)$ lead to the following secular equation for eigenvalues:

$$\det \left\| \left[\frac{\hbar^2}{2m} K(n)^2 - \varepsilon \right] \delta_{nm} + \Gamma_{nm}^{\text{APW}} \right\| = 0 \tag{5.2.42}$$

which is similar to (5.2.21), where Γ_{nm}^{APW} is given by

$$\Gamma_{nm}^{\text{APW}} = \frac{4\pi R^2}{\Omega} \left[\left\{ -\frac{\hbar^2}{2m} [K(n) \cdot K(m)] - \varepsilon \right\} \frac{j_1(|B_n - B_m|R)}{|B_n - B_m|} \right.$$
$$\left. + \sum_{l=0}^{\infty} (2l + 1) P_l(\cos \theta_{nm}) j_l(|K(n)|R) j_l(|K(m)|R) \frac{R_l'(R, \varepsilon)}{R_l(R, \varepsilon)} \right] \tag{5.2.43}$$

and θ_{nm} is the angle between $K(n)$ and $K(m)$. In (5.2.43) the outside potential is chosen to be zero. If this is not zero, we simply add the matrix element V_{nm}^{out} to the right-hand side of (5.2.43). Since the APW method accurately takes into account the effect of a potential near nuclei, this method gives very accurate energy eigenvalues provided the assumed potential is a good approximation.

We now proceed to the Green's-function method (or KKR method after Korringer, Kohn and Rostoker who invented it). In order to solve the following Schrödinger equation

$$\left(\frac{\hbar^2}{2m} \nabla^2 + \varepsilon \right) \psi_k(r) = V(r) \psi_k(r) ,$$

we introduce the Green's function defined by

$$\left(\frac{\hbar^2}{2m} \nabla^2 + \varepsilon \right) G(\varepsilon, k, r - r') = \delta(r - r') \tag{5.2.44}$$

which satisfies the same boundary condition as $\psi_n(r)$. To be more precise, the Green's function should be a solution which satisfies the Bloch theorem (5.2.30, 31). It is not difficult to show that such a Green's function is given by

$$G(\varepsilon, k, r - r') = -(4\pi)^{-1} \sum_l \frac{\exp(ik|r - r' - R_l|)}{|r - r' - R_l|} \exp(ik \cdot R_l) , \tag{5.2.45}$$

where we put $(\hbar^2/2m)k^2 = |\varepsilon|$. In term of this Green's function, the original eigenvalue problem becomes equivalent to solve the following integral equation:

$$\psi_k(r) = \int G(\varepsilon, k, r - r') V(r') \psi_k(r') \, dr' . \tag{5.2.46}$$

This equation is also equivalent to the variation problem which minimizes the following functional:

$$\Lambda = \int \psi_k^*(r)\, V(r)\, \psi_k(r)\, dr$$
$$+ \iint \psi_k^*(r)\, V(r)\, G(\varepsilon, k, r - r')\, v(r')\, \psi_k(r')\, dr\, dr' \,. \tag{5.2.47}$$

If we take a trial function

$$\psi_k(r) = \sum_{l,m} C_{lm}(k)\, R_l(r)\, Y_{lm}(\theta, \varphi) \tag{5.2.48}$$

with the coefficients $C_{lm}(k)$ to be determined variationally, then we obtain the secular equation very similar to (5.2.21, 43), that is,

$$\det \| [K(n)^2 - \varepsilon]\, \delta_{nm} + \Gamma_{nm}^{KKR} \| = 0 \,, \tag{5.2.49}$$

where Γ_{nm}^{KKR} is given by

$$\Gamma_{nm}^{KKR} = \frac{4\pi R^2}{\Omega} \sum_{l} (2l + 1)\, P_l(\cos \theta_{nm})$$

$$\times j_l[|K(n)|R]\, j_l[|K(m)|R] \left[\frac{R_l'(\varepsilon, R)}{R_l(\varepsilon, R)} - \frac{j_l'(kR)}{j_l(kR)} \right]. \tag{5.2.50}$$

A merit of the Green's function method is that, since the function (5.2.45) depends only on specific crystal structures, once numerical tables of these functions for typical lattice structures are prepared, the same procedure is repeatedly applicable to other systems without essential changes.

When, as in transition metals, the d bands are partially filled, or as in noble metals, the filled d bands are near the Fermi surface, the OPW method is not successful. Because localized d levels heavily affect the sp elctronic states, a significant deviation from the NFE approximation is caused. However, even in such a case, the APW and KKR methods can be applied to yield reasonably good results. An example of such a band calculation is shown in Fig. 5.9a for fcc Cu. The shape of the Fermi surface shown in Fig. 5.9b which is deduced from the calculated band structure, is quite different from those of alkali metals.

5.3 Fermi Surface

Properties of an electron assembly in which each electron moves independently in an average potential with crystal periodicity are thought to be determined by the manner of filling electrons into energy bands $\varepsilon_n(k)$ as functions of wave vector k from a low energy side according to the Pauli principle. If we consider the ground state of the system, the maximum energy ε_F of the filled electron defines the Fermi surface in k-space through a relation

$$\varepsilon_n(k) = \varepsilon_F \,. \tag{5.3.1}$$

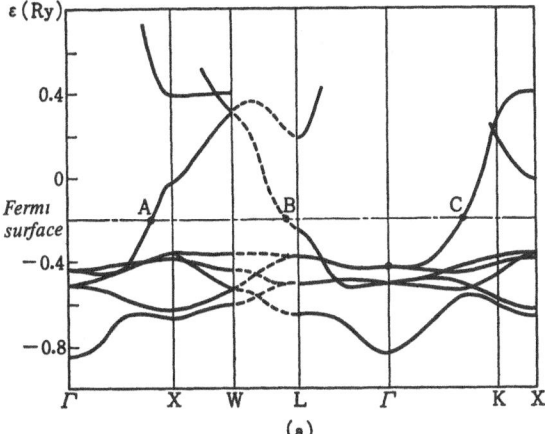

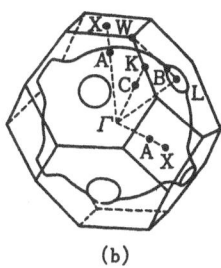

(b)

Fig. 5.9a,b. Band structure and Fermi surface of Cu

(a)

In the case of free electron gas, this is a sphere with a radius k_F. When a periodic potential is taken into account, the Fermi surface is not always of a simple shape. Since almost all properties of metals are determined by the behavior of electrons near the Fermi surface as long as the independent particle approximation is valid where the electron-electron interaction is replaced by its average, the relationship between metallic properties and the Fermi surface is very crucial. Theoretically, one can calculate in principle the energy bands $\varepsilon_n(k)$ by various methods discussed in the previous section, and then, once the number of electrons is given, the Fermi surface is determined from (5.3.1). As many reference papers cited at the end of this volume tell us, many calculations of band structures for various metals already exist. Examining these results, we observe that except for transition metals and noble metals, energy bands can be well described by the NFE approximation, and the Fermi surface, which may extend over several Brillouin zones in general, is nearly spherical (see Fig. 5.8 for Al). In other words, this means that the secular equation for energy eigenvalues $\varepsilon_n(k)$ in the OPW, APW and KKR methods can be transformed into a similar form (5.2.21) for the NFE approximation, that is, into a form

$$\det \left\| \left[\frac{\hbar^2}{2m} K(n)^2 - \varepsilon \right] \delta_{nm} + \Gamma_{nm} \right\| = 0 \tag{5.3.2}$$

and the matrix element Γ_{nm} is small (compared with ε_F). In the NFE approximation, Γ_{nm} is the matrix element of a periodic potential $V(r)$ with respect to the basis states described by the plane waves with wave vectors $K(n)$ and $K(m)$ [or Γ_{nm} may be said to give the matrix element for the scattering of plane waves $K(n) \rightarrow K(m)$ by a potential $V(r)$]. Therefore, the fact that Γ_{nm} is a small quantity may be interpreted that there is an effective potential exerted on the valence electrons which is weak. Let us call such an effective potential a pseudopotential, and denote it by V_{ps}. Then corresponding to (5.2.20) we can put

$$\Gamma_{nm} = (\boldsymbol{K}(n)|V_{\mathrm{ps}}|\boldsymbol{K}(m)) . \tag{5.3.3}$$

The pseudopotential is not unique as is clear from the fact that there are many Γ_{nm}, such as $\Gamma_{nm}^{\mathrm{OPW}}$, $\Gamma_{nm}^{\mathrm{APW}}$, $\Gamma_{nm}^{\mathrm{KKR}}$ depending on the computing method.

Before discussing details of the pseudopotential in the next section, let us examine the above theoretical consideration once again. Although much labor is needed to calculate $\varepsilon_n(\boldsymbol{k})$, answers thus obtained remain rigorously unchecked. Numerical calculations themselves have a limit in their accuracy, and there are several problems inherent in determining an averaged potential $V(\boldsymbol{r})$, although we have omitted its process. The best way to obtain an average electron-electron interaction in an independent particle approximation is a self-consistent Hartree-Fock approximation, but it is not feasible in practice for crystals. When we take these difficulties seriously, we are led to suspect to what extent big calculations of band structures from first principles are really meaningful. We need to know the experimental data in order to check the theoretical calculations, and the experimental methods should determine $\varepsilon_n(\boldsymbol{k})$ directly without resorting to theories. In principle this is possible by using spectroscopic methods, or various photon absorption and emission experiments in solids. In fact, as will be stated in the next chapter, such spectroscopic methods are very powerful for experimentally investigating band structures of insulators and semiconductors which have no Fermi surface. Fortunately the Fermi surface exists in metals which sharply divides the $\boldsymbol{k}$-space into two regions, that is, one is occupied by electrons and the other is unoccupied. Owing to the Pauli principle, electrons near the Fermi surface mainly respond to external perturbations in metals. Thus we can obtain useful information on the shape and size of Fermi surfaces from appropriately designed experiments. In fact, such a field of investigations is classified by the term Fermiology. Very accurate information on Fermi surfaces for various metals have been accumulated [5.5]. Comparing experimental data on Fermi surfaces with calculated results, we can check assumed periodic potentials and band structures based on it. Moreover, we can theoretically predict various metallic properties by using calculated band structures. If a Fermi surface inferred experimentally is not much different from that of a NFE approximation, Γ_{nm} (which can reproduce the Fermi surface) or matrix elements of the pseudopotential would be determined empirically. Thus a successful pseudopotential theory of metals can be established as long as we are dealing with simple metals.

We now introduce some experimental methods to determine the Fermi surface. As stated repeatedly, properties of metals often reflect electronic behavior near the Fermi surface. For instance, the low-temperature electronic specific heat of a metal which is proportional to the electron density of states at the Fermi energy ε_{F} gives information on the inverse of the velocity of an electron with Fermi momentum averaged over the Fermi surface. From the electrical conductivity, the electron relaxation time averaged over the Fermi surface with a certain weight is extracted. Since these methods only give information about electronic properties averaged over the Fermi surface, they are rather indirect for studying the shape of Fermi surfaces in detail. In order to investigate the geometrical shape of Fermi surfaces experimentally, we must resort to a method to detect anisotropy depending on directions. Moreover, relationships between observed quantities and the Fermi surface must be simple enough. Even if we need a theory to relate them, this theory should be reliable in its foundation. In such circumstances, it turns out that conduction electron motions in magnetic fields are most useful.

Let us list below the main experimental methods in Fermiology:

1) de Haas-van Alphen effect
2) Cyclotron resonance (Azbel-Kaner resonance)
3) Utilization of various size effects
4) Magneto-acoustic effect
5) Magneto-galvano effect.

All these experiments are performed under the presence of magnetic fields. Relevant methods giving information about the Fermi surface without using magnetic fields are the following:

6) Anomalous skin effect
7) Positron annihilation
8) Kohn effect.

Although information thus obtained depends on experiments used, the main items are:

(I) maximum and minimum value of cross section A of Fermi surface
(II) extremum value p of diameter of Fermi surface
(III) averaged value of radius of curvature of Fermi surface
(IV) topological informations on the Fermi surface,
(V) derivative $dA/d\varepsilon$ of the Fermi surface where cross section A is regarded as a function of energy.

We shall briefly review some of these methods without going into detail.

Within the approximation to treat electron motion under magnetic fields semi-classically, we may start with

$$\hbar \frac{dk}{dt} = - \frac{e}{c} v_k \times H \, . \tag{5.3.4}$$

This is the classical Newtonian equation of motion for an electron in a crystal with momentum $\hbar k = p$ and charge $-e$ which moves under the Lorentz force. The velocity vector v_k of the k electron is related to electronic energy $\varepsilon(k)$ as a function of k as follows:

$$v_k = \frac{1}{\hbar} \operatorname{grad}_k \varepsilon(k) \, . \tag{5.3.5}$$

Equation (5.3.4) can be also derived quantum-mechanically, but we shall omit the proof here. Equation (5.3.5) can be interpreted as giving a formula of the group velocity for the waves in a dispersive medium. This could also be derived by calculating the expectation value of the velocity operator p/m with respect to the Bloch wave function $\psi_k(r)$. From (5.3.4, 5) it immediately follows that the direction of the derivative dk/dt must be perpendicular both to the applied magnetic field H and to $\mathrm{grad}_k\ \varepsilon(k)$. Since the vector $\mathrm{grad}_k\ \varepsilon(k)$ is perpendicular to the plane $\varepsilon(k) = \mathrm{const}$, k should move on the plane $\varepsilon(k) = \mathrm{const}$ keeping its direction perpendicular to H. Choosing the direction of H as the z-axis, we obtain from (5.3.4)

$$\hbar \frac{dk_z}{dt} = 0 \qquad\qquad (5.3.6)$$

or $k_z = \mathrm{const}$. Therefore the motion of an electron under a magnetic field can be visualized as in Fig. 5.10a. That is, first we draw the Fermi surface in k-space as a surface with constant energy ε_F and find the cross section between the Fermi surface and a plane $k_z = \mathrm{const}$ which is perpendicular to H. Then the k-vector of the electron moves along this cross-sectional curve. This is the electron orbit in k-space. It should be noted that the directions of motions on the Fermi surface depend on the distribution of electrons in k-space. As shown in Fig. 5.10b when the energy at the center of k-space is lowest and electrons occupy the central region, each point on the Fermi surface is called electron-like. On the contrary, as shown in Fig. 5.10c when outside regions have low energy in k-space and the central region is unoccupied, each points on the Fermi surface are called hole-like. The moving directions of the k-vector are opposite for these two cases. Now, in order to look at an electron motion in real space, let us introduce the coordinate $r(x, y, z)$ of an electron and re write (5.3.5) as

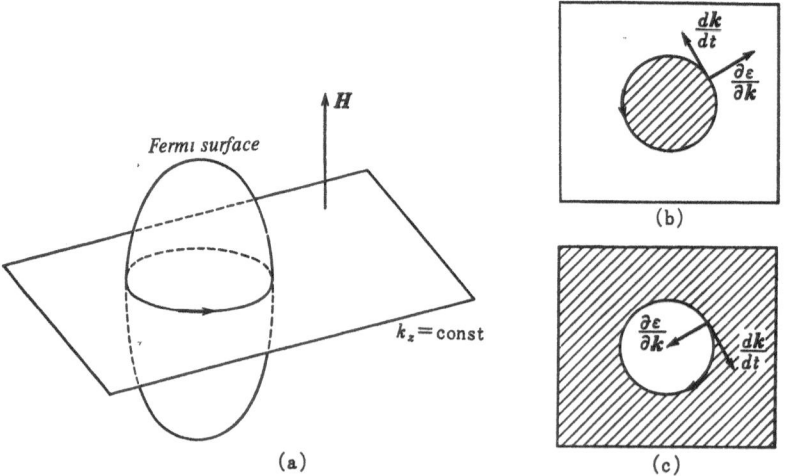

Fig. 5.10a-c. Electron orbit in the k-space in the presence of a magnetic field

$$\frac{dr}{dt} = v_k = \frac{1}{\hbar} \frac{\partial \varepsilon(k)}{\partial k}. \tag{5.3.7}$$

Upon using (5.3.7), the x and y components of the vector equation (5.3.4) may be put in the form

$$\left.\begin{aligned}
\hbar \frac{dk_x}{dt} &= -\frac{e}{\hbar c} \frac{\partial \varepsilon(k)}{\partial k_y} H = -\frac{e}{c} H \frac{dy}{dt} \\
\hbar \frac{dk_y}{dt} &= \frac{e}{\hbar c} \frac{\partial \varepsilon(k)}{\partial k_x} H = \frac{eH}{c} \frac{dx}{dt}
\end{aligned}\right\} \tag{5.3.8}$$

which are readily integrated to give the result

$$\left.\begin{aligned}
\frac{\hbar c}{eH} k_x &= -y + Y \\
\frac{\hbar c}{eH} k_y &= x + X
\end{aligned}\right\}, \tag{5.3.9}$$

where X and Y are constants of integration. It turns out from (5.3.9) that two orbits C in k-space and C' in real space are similar, that is, C' is obtained by rotating C by an angle $90°$ and multiplying a scale factor $\hbar c/eH$. At this point we now understand that the motions of conduction electrons under a magnetic field are closely related to the shape of the Fermi surface. When the orbit C (or C') is closed, as is the case in Fig. 5.10, electrons undergo a periodic motion which is called cyclotron motion. Its frequency ω_c or period T can be derived in the following way. We have from (5.3.8)

$$\left|\frac{dk}{dt}\right| = \frac{eH}{c\hbar^2} \sqrt{\left(\frac{\partial \varepsilon(k)}{\partial k_x}\right)^2 + \left(\frac{\partial \varepsilon(k)}{\partial k_y}\right)^2} \tag{5.3.10}$$

and hence

$$\begin{aligned}
T &= \frac{2\pi}{\omega} = \oint_C dt = \oint \frac{dk}{k} \\
&= \frac{c\hbar^2}{eH} \oint_C \frac{dk}{\sqrt{\left(\frac{\partial \varepsilon(k)}{\partial k_x}\right)^2 + \left(\frac{\partial \varepsilon(k)}{\partial k_y}\right)^2}} \\
&= \frac{c\hbar^2}{eH} \frac{\partial A(\varepsilon, k_z)}{\partial \varepsilon},
\end{aligned} \tag{5.3.11}$$

where $A(\varepsilon, k_z)$ is the area enclosed by the orbit C which is a function of energy ε and z component k_z of k. The cyclotron mass m_c^* is defined through the cyclotron frequency as

$$\omega_c = \frac{eH}{m_c^* c}.$$ (5.3.12)

In the free electron case m_c^* is the electron mass m. For the general case where electrons move in a periodic potential, it is related with the cross section $A(\varepsilon, k_z)$ of the Fermi surface through

$$m_c^* = \frac{\hbar^2}{2\pi} \frac{\partial A(\varepsilon, k_z)}{\partial \varepsilon},$$ (5.3.13)

as is easily proved from (5.3.11,12).

After the preliminary discussion so far, let us proceed to the explanation of the experimental methods to determine the Fermi surface with an applied magnetic field. The de Haas-van Alphen effect is closely related to quantization of the cyclotron motions. The mathematical details will be explained in Chap. 8 on magnetism. In short, electronic levels are modified according to the quantization of cyclotron motions affecting electron distribution in a complicated way. As a result, the electronic energy or the Fermi energy oscillates with the strength of a magnetic field. According to the theory given in Chap. 8, the magnetization of a system oscillates with H, the phase of this oscillation is given by

$$\phi = \frac{c\hbar \bar{A}}{2\pi eH}$$ (5.3.14)

and the frequency is proportional to $1/H$. In (5.3.14), $\bar{A}$ is the maximum or minimum value among the cross sections of the Fermi surface perpendicular to H. If there are several extremal values of the cross sections of the Fermi surface perpendicular to H, the observed quantity is a superposition of these oscillations in which the frequencies are determined by (5.3.14). This experiment is a useful tool for investigating the shape of the Fermi surface, because we can get the corresponding extremal values of the Fermi surface sliced perpendicularly to various directions of H. For example, if we observe the de Haas-van Alphen effect of Cu, the oscillation of magnetization when applying H in the [100] direction drastically differs from the case when the field is applied in the [111] direction. In the former case, only one frequency is observed, while in the latter case two frequenies are detected. This observation supports the shape of the Fermi surface shown in Fig. 5.9b. This figure indicates that, among the cross sections cut perpendicularly to the [111] direction, there are two extremals; one is a large circle and the other a small circle with its center at the L-point.

In the cyclotron resonance, the cyclotron motion of electrons is induced by applying a magnetic field H along the metal surface and resonance absorption of the impressed microwave is produced with a suitable microwave frequency ω. By keeping ω constant and varying H, surface impedance gets a minimum value whenever ω/ω_c takes an integer. Then from (5.3.12, 13)

$$\frac{\omega}{\omega_c} = \omega \frac{c\hbar^2}{2\pi e} \frac{(dA/d\varepsilon)}{H}$$ (5.3.15)

is derived by which we can determine $dA/d\varepsilon$ experimentally.

The utilization of the size effect is a method to determine the diameter of an electron orbit in k-space (or a cross section of the Fermi surface) by cleverly using the fact that the orbit of cyclotron motion in the coordinate space is equal to that of k-space rotated by an angle of 90° and multiplied by $c\hbar/eH$. If we perform an experiment of cyclotron resonance on a metallic sample with d thickness, the resonance phenomena are observed as far as the orbits of the cyclotron motion are confined within the sample thickness d. With decreasing H, the orbits in the coordinate space extend up to the point where electrons are scattered by the surface of the sample to cease the resonance phenomena. Thus, the critical magnetic field H_0 above which no resonance is observed is given by the relation

$$d = p \frac{c\hbar}{eH_0},$$ (5.3.16)

which yields the diameter p of the cross section of the Fermi surface in a certain direction. If the same experiments are repeated after changing the relative direction of the sample surface, we can know the diameters of the Fermi surface in various directions and hence the shape of the Fermi surface itself.

In the magneto-acoustic effect, we measure ultrasonic attenuation in metals as a function of the applied field H which produces the cyclotron oscillation in the sample. The attenuation of sound oscillates with a frequency which is proportional to $1/H$. This is because the electron orbit of cyclotron motion produces a geometrical resonance whenever the extension of the cyclotron orbits is changed by one wavelength of a sound wave. Therefore, if we denote the wavelength of the sound wave by λ, the oscillation in the sound attenuation has a phase which is related to the diameter p of the Fermi surface along the direction of sound propagation through

$$\psi = p \frac{c\hbar}{eH} \frac{1}{\lambda},$$ (5.3.17)

where $pc\hbar/eH$ is the extremal diameter of the cyclotron orbit in real space along the direction of sound propagation. In the same way as the size effect, the measurement of ψ gives rise to the value of p and we can use it to determine the shape of the Fermi surface.

Finally we consider the magneto-galvano effect, which is related to the phenomena in which the electric resistivity of metals in a magnetic field generally increases with the strength of the field. Since the magnetic-field dependence of electric resistivity changes with the direction of applied fields and the nature of orbital motions of electrons or positive holes, its measurement supplies useful topological information about how the Fermi surface extends to various directions in k-space. So far we have considered only the closed cyclotron orbit such as shown in Fig. 5.10. There are, however, the cases of the special shapes of Fermi surfaces where a sliced cross section along a certain direction is not closed in a Brillouin zone, extending without limit (in such a case the periodic zone scheme is most useful). For example, we take the Fermi surface of Cu shown in Fig. 5.9b. Since the narrow neck part of the Fermi surface extends towards L point and is connected periodically with the neighboring Fermi surface in the next Brillouin zone, the open orbit comes out when slicing parallel to the [111] direction in the k-space. Although we omit details here, according to the theory of magneto-resistance, the magnetic field dependence of magneto resistance is quite different for closed and open orbits. Therefore, if we measure resistivity under a strong magentic field in various directions, we can infer the topological structure of the Fermi surface.

We cite references in the Bibliography about other remaining experimental methods to elucidate the Fermi surface. We present only one typical example here of the experimental determination of the Fermi surface by means of the de Haas-van Alphen effect. It is the Fermi surface of Pb shown in Fig. 5.11, which provides us with a beautiful example indicating that even heavy metals can have a NFE-like nearly spherical Fermi surface.

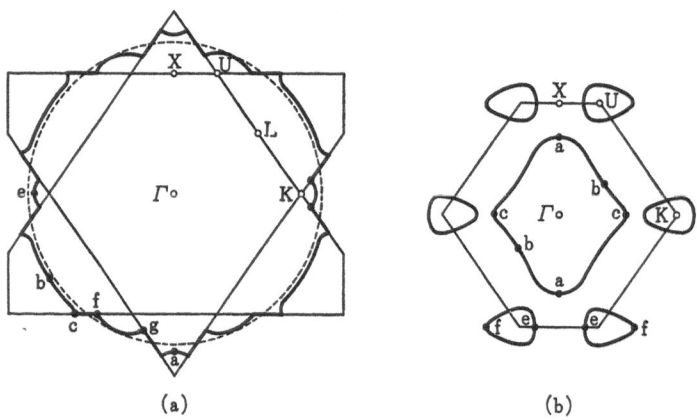

(a)   (b)

Fig. 5.11a, b. Fermi surface of Pb (fcc lattice). (a) Fermi surface in the extended zone scheme and free electron Fermi surface (- - - -); (b) Fermi surface in the reduced zone scheme

5.4 Pseudopotential of Simple Metals

In so-called simple metals in which conduction electrons come only from sp electrons, the Fermi surfaces derived experimentally are nearly spherical and the NFE approximation is valid, as we have seen in the previous section. It is the aim of this and the next sections, on the basis of these facts, and introducing a weak effective pseudopotential which acts on conduction electrons, to exploit a pseudopotential theory by which various properties of simple metals may be derived. It would be possible to establish a unified picture for simple metals, if we could choose pertinent pseudopotentials for all metallic elements (experimentally or theoretically), by using different characters of each element which may be discriminated and all properties which may be understood within a simple NFE approximation [5.6].

Before going into detail, let us consider first why a pseudopotential could be so weak. At the moment we define a pseudopotential V_{ps} by the matrix element:

$$\Gamma_{nm} = (\boldsymbol{K}(n) \,|\, V_{ps} \,|\, \boldsymbol{K}(m))$$

which appeared in the secular equation (5.3.3) of the NFE approximation. The reason why V_{ps} is effectively weak is easily understood by examining (5.2.37) for Γ_{nm} in the OPW method. The second term in the right-hand side almost cancels out the first term of the potential arising from bare ions which is negative and strong, especially in the vicinity of nuclei. Note that $\varepsilon - \varepsilon_a > 0$. That is, in order that the wave functions of conduction electrons be orthogonal to those of electrons bound to nuclei, the former should oscillate rapidly near nuclei yielding high kinetic energies which almost cancel the large negative potential energies due to electron-ion core attraction. On the other hand, in the intermediate region between ion cores, electronic states can be well described by the plane waves of a nearly free electron type. Thus the NFE approximation is valid in these regions. In fact, several examples of calculations show that the two terms in the right-hand side of (5.2.37) almost cancel out each other within the ion core region. It is hard to see analytically that Γ_{nm} in the APW and KKR method is effectively small. However we can still interpret it as follows. In the APW method we have divided the space into two regions; a spherical region (each of radius R) centered at the origins of spherical potentials and an intermediate region between these spheres. In the latter region the potential is almost a constant, and hence electronic wave functions are expressed by plane waves. This is modified by scattering due to the strong potential in the spherical region. In the APW and KKR method we try to find out coherent waves spreading throughout a crystal by taking into account the incident and scattered waves at each spherical region. As far as the wave function in the intermediate region is concerned, the important quantity which matters is not the strong potentials themselves but the phase shift η_l of scattered waves due to the potential. Although η_l may be very large in general, if we put

$$\eta_l = p_l \pi + \delta_l \,, \tag{5.4.1}$$

then the part of $p_l \pi$ does not affect the scattered plane wave. In other words, this is virtually equivalent to the potential which produces a phase shift η_l if we focus our attention on the scattered wave only in the intermediate region and neglect the change in the electronic wave function within the spherical region due to the strong attractive force. Thus we can replace it by a weak effective pseudo-potential.

Although introduction of a pseudopotential will simplify the description of the electronic state of simple metals, this merit is compensated for by a complication such that a pseudopotential properly chosen turns out to depend on energy and angular momentum in general, reflecting electronic states of ion cores. We need only the matrix element Γ_{nm} with respect to $K(n) = k - G_n$ and $K(m) = k - B_m$. However, we cannot generally expect that the matrix element of a selected pseudopotential is a function of the difference between $K(n)$ and $K(m)$. Therefore, when expressed in the real space representation, the potential becomes nonlocal and is difficult to write down in a simple form. Should it be expressed in real space, it would formally take on the following form:

$$V_{ps}(r) = \sum_l f_l(r, \varepsilon)\, \mathscr{P}_l ,\tag{5.4.2}$$

here $\mathscr{P}_l$ is a projection operator onto the state with the angular momentum l. However, since it is too complex to use as it is, we have to simplify it for practical purposes. As to the energy dependence, we may fix the energy at

$$\varepsilon = \varepsilon_F$$

because we are interested in the energies near the Fermi surface. To avoid the complexity due to the nonlocality, we restrict the matrix elments to those for the wave vectors on the Fermi surface, that is,

$$|K(n)| = k_F, \quad |K(m)| = k_F$$

and we further assume the matrix elements to be a function of $K(n) - K(m) = B_m - B_n = q$ only. Let us express the pseudopotential as the sum of the contribution of ion cores at each lattice sites

$$V_{ps}(r) = \sum_l v(r - R_l) .\tag{5.4.3}$$

Introducing $k = K(n)$, $k' = K(m)$ and $k - k' = q$, we decompose the matrix element (5.4.3) into two factors:

$$\langle k | V_{ps} | k' \rangle = S(q)\, v(q)\tag{5.4.4}$$

$$S(q) = \frac{1}{N} \sum_l \exp(-iq \cdot R_l)\tag{5.4.5}$$

$$v(q) = \frac{1}{\Omega} \int e^{-iq \cdot r} v(r)\, dr ,\tag{5.4.6}$$

where N is the total number of atoms, Ω is the volume per atom, $S(q)$ is the structure factor and $v(q)$ is the form factor of a potential. They are assumed to be a function of q only. For the determination of energy bands, we need only the values of the pseudopotential at the reciprocal lattice points. However, in order to calculate various properties of metals in terms of the pseudopotential, we sometimes need a knowledge of $v(q)$ for the values of q in a wider range. For instance, we must know the pseudopotential for all q when evaluating frequencies of lattice vibration of metals. Therefore it is desirable that $v(q)$ is given by an analytic function. There are several proposed model potentials for this purpose. We introduce here the simplest and most often used model potential among them. It is convenient to devide $v(q)$ into two parts: one is the pseudopotential $v^{\mathrm{ion}}(q)$ of bare ions and the other is the potential which contributes to the screening effect due to conduction electrons. We put

$$v(q) = \frac{v^{\mathrm{ion}}(q)}{\epsilon(q)}, \tag{5.4.7}$$

where $\epsilon(q)$ is the dielectric constant of a conduction electron system, which is assumed for simplicity to be given by a free-electron gas model. Then we can show that

$$\epsilon(q) = 1 - \frac{8\pi e^2}{\Omega q^2} \chi(q) \tag{5.4.8}$$

$$\chi(q) = \frac{1}{N} \sum_{|k|<k_{\mathrm{F}}} \frac{2m/\hbar^2}{k^2/2 - (k+q)^2/2}, \tag{5.4.9}$$

where $\chi(q)$ is a Fourier component of the density-density correlation function of free electron gas which often appears when discussing the linear response of electron gas to an external disturbance. This is given explicitly by

$$\chi(q) = -\frac{z}{2}\left(\frac{2}{3}\varepsilon_{\mathrm{F}}^0\right)^{-1}\left(\frac{1}{2} + \frac{4k_{\mathrm{F}}^2 - q^2}{8qk_{\mathrm{F}}} \ln\left|\frac{q+2k_{\mathrm{F}}}{q-2k_{\mathrm{F}}}\right|\right), \tag{5.4.10}$$

where z is the free electron number per atom, and $\varepsilon_{\mathrm{F}}^0$ is the Fermi energy of free-electron gas. The simplest form of bare ion potential $v^{\mathrm{ion}}(r)$ may be given by

$$v^{\mathrm{ion}}(r) = \begin{cases} 0 & (r \leq R_{\mathrm{e}}) \\ -\dfrac{2e}{r} & (r > R_{\mathrm{e}}) \end{cases}. \tag{5.4.11}$$

This represents a vanishing core potential inside the sphere with radius R_{e}, and outside $r > R_{\mathrm{e}}$, a Coulomb potential of a point charge $(-ze)$ at the origin. In this model potential R_{e} is the only parameter, which is thought to be of an order of an ion core radius. The form factor of the potential (5.4.11) can be easily calculated as

$$v(\boldsymbol{q}) = -\frac{4\pi z}{q^2 \epsilon(q)\Omega} \cos qR_{\text{e}} \, . \tag{5.4.12}$$

In Fig. 5.12 we depict $v(\boldsymbol{q})$ of (5.4.12) as a function of $\boldsymbol{q}$. In the limit $\boldsymbol{q} \to 0$, $v(\boldsymbol{q}) \to -2\,\varepsilon_{\text{F}}^0/3$, and $v(\boldsymbol{q}_0) = 0$ when

$$\boldsymbol{q}_0 = \frac{\pi}{2R_{\text{e}}} \, . \tag{5.4.13}$$

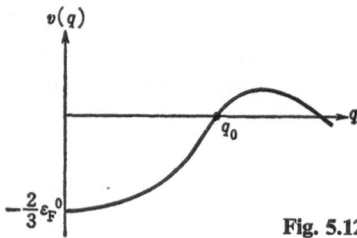

Fig. 5.12. A sketch of a structure factor of a pseudopotential

With increasing $\boldsymbol{q}$ this function takes a positive maximum and approaches zero in oscillations. This behavior of the form factor is typical and commonly found in other model potentials.

The more elaborate model potential which takes into account the detailed properties of ions, has the following form:

$$v^{\text{ion}}(\boldsymbol{r}) = \begin{cases} -\sum_l A_l \mathscr{P}_l & (r < R_M) \\ -\dfrac{ze}{r} & (r > R_M) \end{cases}, \tag{5.4.14}$$

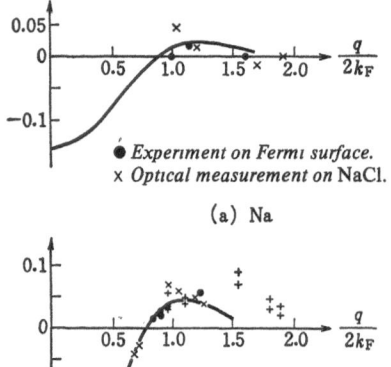

Experiment on Fermi surface.
x *Optical measurement on NaCl.*

(a) Na

Experiment on metallic Mg (Fermi surface and Optical measurement)
+,x *Optical measurement on MgO.*

(b) Mg

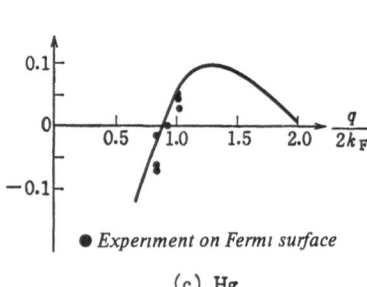

Experiment on Fermi surface

(c) Hg

Fig. 5.13a–c. Pseudopotentials for Na, Mg and Hg. Solid lines represent the model potential [5.8]

where R_M is a parameter which is chosen to be larger than the ion core radius R_c, and $\mathscr{P}_l$ is the operator which projects out the wave function with angular momentum l, and A_l is another parameter which is determined by comparing the energy levels of an electron trapped by the potential (5.4.14) with the measured values of spectroscopic energy levels for isolated ions.

Animalu and *Heine* [5.7] determined the values of A_0, A_1, A_2, and R_M and functional forms of the form factors for 24 different elements. *Cohen* and *Heine* [5.8] also published graphs in which comparisons between calculated form factors of the Animalu-Heine model potential and empirical $v(q)$ determined by various methods are given. A few examples of their results are shown in Fig. 5.13. Generally speaking, the empirically determined $v(q)$ seems to fit well with the theoretical curves. In Table 5.5 we show the values of various parameters relevant to the pseudopotential for simple metals.

Table 5.5. Parameters for the pseudopotential of simple metals from *Animalu* and *Heine* [5.7]

	A_0	A_1	A_2	R_M	Ω	z	q_0	k_F
Li	0.336	0.504	0.455	2.8	144.9	1	0.91	0.5890
Na	0.305	0.339	0.402	3.4	254.5	1	0.87	0.4882
K	0.240	0.256	0.368	4.2	481.4	1	—	0.3947
Rb	0.224	0.226	0.384	4.4	587.9	1	—	0.3693
Cs	0.205	0.207	0.366	4.8	745.5	1	—	0.3412
Be	1.01	1.22	1.48	2.0	54.4	2	1.44	1.0287
Mg	0.78	0.88	0.99	2.6	155.9	2	1.13	0.7242
Ca	0.54	0.50	1.49	2.6	293.5	2	—	0.5865
Ba	0.45	0.34	1.07	3.4	424.1	2	—	0 5188
Zn	0.99	1.14	0.98	2.2	102.0	2	1.42	0.8342
Cd	0.88	0.98	1.11	2.6	144.8	2	1.28	0.7423
Hg	0.97	1.11	0.85	2.6	157.8	2	1.33	0.7223
Al	1.38	1.64	1.92	2.0	111.3	3	1.35	0.9276
Ga	1.44	1.58	1.41	2.4	131.4	3	1.40	0.8776
In	1.32	1.46	1.10	2.4	175.3	3	1.32	0.7972
Tl	1.44	1.51	0.98	2.4	191.7	3	1.39	0.7738
Sn	1.84	2.04	1.62	2.0	181.5	4	1.42	0.8674
Pb	1.92	(2.00)	0.90	2.1	203.4	4	1.47	0.8350

5.5 Structure of Simple Metals

Among about 70 metallic elements, 25 metals belong to simple metals, another 25 metals are classified into transition metals, and the rest are rare-earth metals called lantanides or actinides (Table 5.1). For simple metals, the pseudopotential theory of the previous section is applicable and successful in explaining various properties qualitatively. In this section as an example of the application of the pseudopotential theory, we shall discuss the structure of simple metals [5.9].

In Table 5.6, crystal structures and atomic radii R_a determined from

$$\Omega = \frac{4\pi}{3} R_a^3 \tag{5.5.1}$$

Table 5.6. Structural data for simple metals: (tet.) tetragonal, (ort) orthorhombic, (tri) trigonal

	Structure	z	R_a(obs)	R_a(cal)	R_a(cal)−R_a(obs)	q_0	$R_a q_0\, 2\pi$
Li	hcp → bcc	1	3.26	3.76	0.50	0.91	0.472
Na	hcp → bcc		3.93	4.24	0.31	0.87	0.544
K	bcc		4.86	5.36	0.50	—	—
Be	hcp	2	2.35	3.07	0.72	1.44	0.539
Mg	hcp		3.34	3.70	0.36	1.13	0.601
Zn	hcp		2.90	3.09	0.19	1.42	0.655
Cd	hcp		3.26	3.30	0.04	1.28	0.664
Hg	tet. → tri. (A10)		3.35	2.88	−0.47	1.33	0.709
Ca	fcc → bcc		4.12	4.48	0.36	—	—
Ba	bcc		4.66	5.41	0.75	—	—
Al	fcc	3	2.98	3.26	0.28	1.35	0.640
Ga	ort.(A11)		3.15	3.09	−0.06	1.40	0.702
In	tet.(A6)		3.47	3.38	−0.09	1.32	0.729
Tl	hcp		3.58	3.09	−0.49	1.39	0.792
Sn	D → tet.(A5)	4	3.51	3.26	−0.25	1.42	0.795
Pb	fcc		3.65	3.18	−0.47	1.47	0.854

for selected simple metals are shown. It should be noted that the atomic radii of the simple metals do not change so much even if the crystal structures are different. Now let us first show that the order of magnitude of R_a for simple metals can be roughly explained on the basis of a free electron model. Assuming z free electrons per atom, we express the energy of a free electron as

$$\varepsilon^0(\mathbf{k}) = V_0 + \frac{\hbar^2}{2m} k^2, \tag{5.5.2}$$

where V_0 is the averaged potential calculated in the following way. Ions having positive charge $+ze$ are embedded in the electron gas. We choose a model ionic potential adopted in the previous section:

$$v^{\mathrm{ion}}(r) = \begin{cases} -A_0 & (r < R_M) \\ -\dfrac{ze^2}{r} & (r > R_M) \end{cases}. \tag{5.5.3}$$

For simplicity we retain only A_0. The values of A_0 and R_M are already tabulated in Table 5.5. If we assume that z electrons are distributed uniformly inside the atomic sphere (with volume $\Omega = 4\pi R_a^3/3$), the potential produced by this charge density at a position r distant from the center of the atomic sphere is given as

$$v_e(r) = \begin{cases} \dfrac{3ze^2}{2R_a}\left[1 - \dfrac{1}{3}\left(\dfrac{r}{R_a}\right)^2\right] & (r < R_a) \\[3mm] \dfrac{ze^2}{r} & (r > R_a). \end{cases} \tag{5.5.4}$$

For $r > R_a$[note $R_a > R_M$], (5.5.3, 4) cancel out to make the potential vanish. Therefore it is enough to calculate the average potential for $r < R_a$ which is given as

$$V_0 = \frac{1}{\frac{4\pi}{3}R_a^3}\left[\int_0^{R_a} v^{\mathrm{ion}}(r)\, 4\pi r^2\, dr\right] + \int_0^{R_a} \frac{3ze^2}{2R_a}\left[1 - \frac{1}{3}\left(\frac{r}{R_a}\right)^2\right] 4\pi r^2 dr . \tag{5.5.5}$$

Performing the integration by using (5.5.3) we obtain

$$V_0 = -\frac{0.3ze^2}{R_a} + 1.5\frac{ze^2 R_M^2}{R_a^3} - A_0\frac{R_M^3}{R_a^3}. \tag{5.5.6}$$

Then we substitute this V_0 into (5.5.2) and perform the sum of $\varepsilon^0(k)$ over occupied k-states to give the energy per atom

$$zV_0 + \tfrac{3}{5} z\varepsilon_F^0, \tag{5.5.7}$$

where $\varepsilon_F^0 = \hbar^2 k_F^2/2m$ is the Fermi energy of the free electron. However, in this result electron-electron Coulomb interaction energy is doubly counted so that a half of the Coulomb energy must be subtracted, that is we need to add a term

$$-\tfrac{1}{2}\,\frac{1.2z^2 e^2}{R_a}. \tag{5.5.8}$$

Furthermore, we add the exchange energy ε_{ex} and correlation energy ε_{cor} of the electron gas which were evaluated in (2.2.4). Thus, finally we have the total energy of electron gas per atom in the form

$$\begin{aligned} U_0 &= -\frac{0.9z^2 e^2}{R_a} + 1.5\frac{z^2 e^2 R_M^2}{R_a^3} - A_0\frac{zR_M^3}{R_a^3} + \frac{3}{5}z\varepsilon_B^0 + \varepsilon_{ex} + \varepsilon_{cor} \\ &= -\frac{0.9z^2 e^2}{R_a} + 1.5\frac{z^2 e^2 R_M^2}{R_a^3} - A_0\frac{zR_M^3}{R_a^3} + \frac{1.105z}{r_s^2} - \frac{0.458z}{r_s} + zU_c(r_s). \end{aligned} \tag{5.5.9}$$

Since the desired value of R_a should minimize this total energy, we determine R_a from the condition

$$\frac{\partial U_0}{\partial R_a} = 0 \qquad (5.5.10)$$

and tabulate the calculated value R_a(cal) in Table 5.6. Both R_a(obs) and R_a(cal) are in good agreement within an error of 10%. This result clearly indicates that the cohesive energy of simple metals comes mostly from the energy of free electrons moving in the averaged potential and depends solely on the volume (or on R_a). In other words, the manner of the packing of atoms in simple metals, or equivalently, the volume of metals, can be roughly described by a free electron gas model to zeroth approximation. In order to determine the structure of various metals we may therfore only consider the energy difference for different structures keeping the atomic volume approximately constant.

The discrepancies between R_a(cal) and R_a(obs) in Table 5.6 can be improved quantitatively if the band structure effect is taken into account perturbationally. So far we have neglected the periodic arrangement of ion cores. In the NFE approximation, energy gaps appear on the planes in the k-space which bisect the reciprocal vector G of the crystal lattice, and the magnitude of a gap is proportional to $v(G)$. As will be shown later quantitativey as an effect of band structure, an extra term proportional to the square of the energy gap

$$U_{bs} \propto - [v(G)]^2 \qquad (5.5.11)$$

is added in the cohesive energy of the system in addition to the structure independent term U_0 calculated above. This structure-dependent part is a decreasing function of $v(G)$. As stated in the previous section, since $v(G)$ vanishes at $G = q_0$, if the reciprocal lattice vector G (which is determined in the first approximation by packing spheres of atomic radius) happens to be larger than q_0, $|v(G)|$ increases with increasing G, and when $G < q_0$, $|v(G)|$ increases with decreasing $|G|$. That is, in real space when $G > q_0$, the lattice constant should become small to lower the energy, and when $G < q_0$, the lattice constant should become larger. Now the values of q_0 are known for simple metals, as tabulated in Table 5.5. If we read off the q_0 for each metal and calculate $R_a q_0/2\pi$, then we can draw the graphs $[R_a(\text{cal}) - R_a(\text{obs})]$ vs $R_a q_0/2\pi$ by using Table 5.5. The results are shown in Fig. 5.14. Clearly we can see that when $R_a q_0/2\pi$ is large (in this case it is probable that $q_0 > G$), the calculated values of R_a are smaller than the observed ones. In contrast, when $R_a q_0/2\pi$ is small the calculated values are larger than the observed values. It is amusing that these results are indeed as expected. This encourages us to go further into a quantitative discussion of the structures of simple metals on the basis of pseudopotential theory.

Now let us proceed to the more detailed formulation. We divide the total energy of a crystal into two parts, one is the structure-independent part U_0 and the other, the structure-dependent part U_s, is produced by the regular arrangement

Fig. 5.14. Plot of $R_a(\text{cal}) - R_a(\text{obs})$ vs $R_a q_0 / 2\pi$

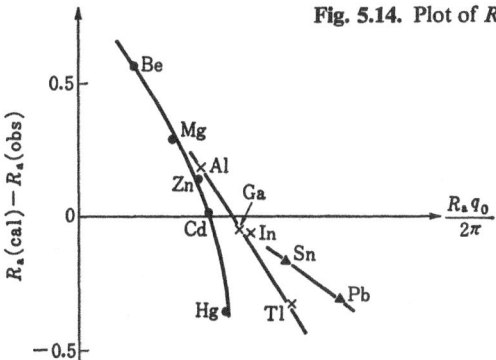

of ion cores. There are two main contributions to U_s. One is the electrostatic energy U_E due to the ion core lattice embedded in a uniform electron gas. This is the Ewald electrostatic energy given by

$$U_E = - \alpha \frac{z^2 e^2}{2R_a},\tag{5.5.12}$$

where α depends on the crystal structures and is tabulated in Table 5.7 The other is the contribution from the energy change of electrons perturbed by a periodic potential. This is called the band structural energy U_{bs}. The energy of an electron with wave vector k is evaluated in second-order perturbation theory, which leads to

$$\varepsilon(k) = \frac{\hbar^2}{2m} k^2 + \langle k | V | k \rangle + \frac{2m}{\hbar^2} \sum_G' \frac{|\langle k + G | V | k \rangle|^2}{\frac{1}{2} k^2 - \frac{1}{2} (k + G)^2}.\tag{5.5.13}$$

The first and second terms in the right-hand side have been already taken into account to give rise to U_0. Thus only the third term contributes to U_{bs}. Summing up (5.5.13) over all occupied k-states yields

$$U_{bs} = \frac{1}{N} \left(\frac{2m}{\hbar^2} \right) \sum_{k < k_F} \sum_G' \frac{|\langle k + G | V | k \rangle|^2}{\frac{1}{2} k^2 - \frac{1}{2} (k + G)^2}.\tag{5.5.14}$$

Table 5.7. Madelung constants

Structure	α
bcc	1.79186
fcc	1.79175
hcp	1.79168
sc	1.76012

If we assume that $\langle k + G | V | k \rangle$ is a function of $|G| = G$ only and use (5.4.4–6) we obtain

$$\langle k + G | V | k \rangle = S(G)\, v(G).$$

Upon using $\chi(q)$ defined in (5.4.9), U_{bs} may be rewritten as

$$U_{bs} = \sum_G{}' |S(G)|^2 [v(G)]^2 \chi(G). \tag{5.5.15}$$

However, this is not correct. As is well known in the Hartree-Fock approximation, we must avoid double counting of the electron-electron interaction when summing up the one-electron energy $\varepsilon(k)$. A correct procedure is the following: when an ionic potential is switched on, a uniform charge of free electron gas redistributes to screen it. The overcounted electronic self-energy is given by the product of the induced part of charge density and the potential produced by this induced charge density. We have to subtract half of this self-energy. According to linear response theory, the induced charge density is given by a product of the disturbing potential [which is $S(G)v(G)$ in the present case] and the density-density correlation function, and the result may be put in the form

$$\rho(G) = \frac{2}{\Omega} S(G)\, v(G)\, \chi(G). \tag{5.5.16}$$

On the other hand, the screened potential becomes

$$V^{sc}(G) = S(G)\, v(G) - S(G)\, v^{ion}(G)$$
$$= S(G)\, v(G)[1 - \epsilon(G)]. \tag{5.5.17}$$

Therefore the energy to be subtracted is

$$\tfrac{1}{2}\Omega \sum_G \rho(G)\, V^{sc}(G) = \sum_G |S(G)|^2 [v(G)]^2 \chi(G)[1 - \epsilon(G)]$$

and the correct result for U_{bs} is given by

$$U_{bs} = \sum_G{}' |S(G)|^2 [v(G)]^2 \chi(G)\, \epsilon(G). \tag{5.5.18}$$

Since the density-density correlation function (or its Fourier component) $\chi(G)$ is generally a negative quantity because of (5.4.10), U_{bs} is proportional to $-[v(G)]^2$ which is nothing but the starting point (5.5.11) of our discussion. U_{bs} depends on the product of the structure factor $[S(G)]^2$ and the form factor $[v(G)]^2$ of the pseudopotential. Since the latter depends on the metal itself, a discussion on possible structures of metals must be given for each element individually. However, there are some general conclusions which can be drawn from the form of U_{bs}, and hence we shall give a review of the structures of simple metals on the basis of (5.5.18).

We can discard the dielectric constant $\epsilon(G)$ for the present discussion because $\epsilon(G)$ is nearly equal to unity for large values of G. $\chi(G)$ in (5.4.10) is a function of G which is schematically depicted in Fig. 5.15. Since $\chi(G) = -1$ for small G and $\chi(G)$ tends to vanish for $|G| > 2k_F$, we can replace $\chi(G)$ by the dotted rectangular line in Fig. 5.15 as a rough approximation, or

$$U_{bs} \approx - \sum_{|G|<2k} |S(G)|^2 \, [v(G)]^2 . \qquad (5.5.19)$$

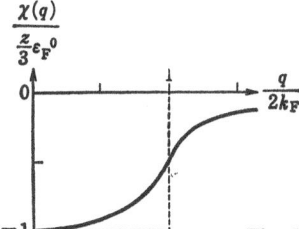

Fig. 5.15. Curve for $\chi(q)$

In this approximation, the structure is more stable as $[v(G)]$ becomes larger and the number of the reciprocal lattice vectors $|G|$ which give the same $[S(G)]^2$ increases. Now look back at Table 5.6 once again. It turns out that what we must explain about the structures of simple metals are the following points:

1) the reason why lighter elements assume hcp structure and heavier elements form bcc structure in alkaline metals,
2) the reason why divalent elements form hcp structure except for Ca, Ba and Hg,
3) the reason why Hg, Ga and In take especially low-symmetry structures,
4) the reason why the fcc structure appears in multivalent metals such as Al and Pb.

We first examine the contribution from the structure factor $[S(G)]^2$ for three kinds of typical crystal structures fcc, bcc and hcp. As is noted before, the volume per atom or atomic radius R_a is determined by minimizing U_0 which is structure independent. Thus we must compare different structures keeping R_a a fixed value. Hence it is convenient to take the unit of reciprocal vectors to be $2\pi/R_a$.

We take the fcc structure for example. Main reciprocal lattice vectors (which form bcc lattices) are $G_{100} = (4\pi/a)$ (100) and $G_{111} = (2\pi/a)$ (111). There are six of G_{100} and eight of G_{111}. Since the lattice constant a is related to R_a by

$$\frac{a^3}{4} = \frac{4\pi}{3} R_a^3 \quad \text{(fcc)}, \qquad (5.5.20)$$

the magnitudes of these reciprocal lattice vectors are given by

$$|G_{100}| = \frac{4\pi}{a} = \frac{2\pi}{R_a} 2\left(\frac{3}{16\pi}\right)^{1/3}$$

$$|G_{111}| = \frac{2\pi}{a}\sqrt{3} = \frac{2\pi}{R_a}\sqrt{3}\left(\frac{3}{16\pi}\right)^{1/3}. \tag{5.5.21}$$

These vectors satisfy $[S(G)] = 1$ and contribute to U_{bs} with the relative weight of six and eight respectively; that is

$$U_{bs} \approx -6[v(G_{100})]^2 - 8[v(G_{111})]^2 \quad \text{(fcc)}. \tag{5.5.22}$$

For a bcc structure, the most important reciprocal lattice vectors are $G_{110} = (2\pi/a)(110)$ and its equivalent eleven vectors. Using

$$\frac{a^3}{2} = \frac{4\pi}{3} R_a^3 \quad \text{(bcc)}, \tag{5.5.23}$$

we evaluate the magnitude of the reciprocal lattice vector as

$$|G_{110}| = \frac{2\pi}{a}\sqrt{2} = \frac{2\pi}{R_a}\sqrt{2}\left(\frac{3}{8\pi}\right)^{1/3} \tag{5.5.24}$$

which leads to

$$U_{bs} \approx -12[v(G_{110})]^2 \quad \text{(bcc)}. \tag{5.5.25}$$

A similar result can be derived for a hcp structure. We summarize in Fig. 5.16 the important reciprocal lattice vectors expressed in units of $2\pi/R_a$ and their weights for the three structures fcc, bcc and hcp. If we assume that $v(G)$ is not so different for main $|G|$, then the structure having reciprocal vectors which contribute to U_{bs} as much as possible, is the most stable. Therefore a crucial factor is $2k_F$ which may be related to the valence of metal z by

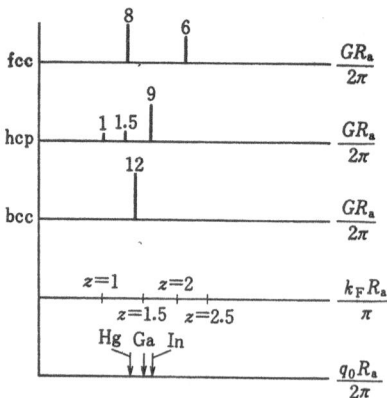

Fig. 5.16. Comparison of various lattice structures

$$2 \left(\frac{4\pi}{3} k_F^3\right) \frac{1}{(2\pi)^3} \left(\frac{4\pi}{3} R_a\right)^3 = z \tag{5.5.26}$$

or

$$\frac{k_F R_a}{2\pi} = \left(\frac{9z}{32\pi^2}\right)^{1/3} . \tag{5.5.27}$$

We also show the values of $k_F R_a/2\pi$ for several z in Fig. 5.16. From this figure we can deduce very interesting conclusions. According to (5.5.19), the structure having vectors $|G|$ smaller than $2k_F$ as many as possible, has the largest negative U_{bs}. Thus it is concluded that hcp is most stable for metals with $z = 1$. This is indeed the case for light alkaline metals, but not for heavy alkaline metals which crystallize in a bcc structure. This might indicate that simple nonlocal pseudopotential theory is not applicable to the latter. Figure 5.16 shows that a bcc structure becomes stable when $z = 1.5$. There are no simple metals with a half integer z. However, 50%–50% alloys of monovalent and divalent metals are known to form a bcc structure very often. This fact supports the above conclusion. When $z = 2$, hcp is again stabilized, explaining the structure of Be, Mg, Zn and Cd. For larger z, fcc becomes stable which corresponds to the structure of Al and Pb. Ca and Ba are exceptions for which, as in the case of heavy alkaline metals, simple pseudopotential theory may not be adequate. Besides these simple lattices, more complicated structures actually appear which can be also accounted for by examining special properties of $[v(G)]$. So far we have treated $[v(G)]$ as a constant, discarding the fine details of its G-dependence. If we look at the pseudopotentials of metals which take special structures such as Hg, Ga and In more carefully, it is discovered that the magnitudes of the main reciprocal lattice vectors of high-symmetry lattcies (fcc, bcc, hcp) are very close to q_0. In Fig. 5.16 the values of q_0 for these metals expressed in units of $2\pi/R_a$ are also shown. Now, since $|v(G)| \cong 0$ for these metals and for the high-symmetry structures, a structure having $|G|$ differing from q_0 makes U_{bs} largely negative by lowering the crystal symmetry. In fact, Hg takes a structure derived by compressing fcc along [111] at room temperatures. At low temperatures, another tetragonal form is obtained by strongly distorting fcc. Ga takes a complicated structure derived from fcc in which each atom has seven nearest neighbors, three atoms being on the same atomic plane and the remaining four atoms on two layers located at up and down levels. In also takes a similar structure. All these distortions lower U_{bs} by making $|G|$ deviate from q_0 so that the values of $|v(G)|$ become large.

These qualitative arguments based on (5.5.19) give an answer to the main part of the questions raised before about the structures of simple metals. We can discuss it even more quantitatively using the pseudopotential of each metal. In fact, there are many such calculations already. For instance, pseudopotential theory successfully explains why and how the axial ratios c/a of hcp lattices for divalent metals deviate from the ideal value $c/a = 1.633$.

5.6 Transition Metals

As briefly discussed in Sect. 5.2, in monovalent noble metals such as Cu, Au and Ag, the Fermi surface deviates from a sphere in the [111] direction as shown in Fig. 5.9b, and the NFE approximation is not applicable to these metals. This is because those filled d-bands located immediate below the Fermi surface mix with sp electron bands, resulting in a large deviation from the NFE approximation. It is another evidence of the inapplicablity of the pseudopotential theory that noble metals take fcc structures in contradiction to the theory, which predicts hcp structures for metals with $z = 1$.

The metals ranging from group IIIa to VIII in the left side of noble metals in Table 5.1 have an electronic configuration with incompletely filled d levels, and their energy bands resemble those of Cu depicted in Fig. 5.9a, but with a lower Fermi level leaving some unfilled parts in the d-bands. All these are called transition metals. They cannot be treated by simple pseudopotential theory and exhibit remarkable magnetic properties because of their special electronic structure. Since the magnetism in the transition metals will be discussed in Chap. 8, we shall give in this section only a short review of the cohesive mechanism and crystal structures of transition metals.

In view of the fact that among the transition metals in the first series, Fe, Co and Ni exhibit ferromagnetism and Mn and Cr antiferromagnetism, there has been some doubt about treating d electrons in these metals as itinerant particles in the frame of band electron theory. As will be discussed in Chap. 8, in order to explain the magnetic properties, it is essential to take into account the d-electron correlation, and hence simple picture of one-electron approximation seems to be far from reality. However, as the experimental investigations on the Fermi surfaces of transition metals progress, it becomes evident that d electrons participate in conduction, forming a part of the Fermi surfaces. Furthermore, the Fermi surfaces which are theoretically derived from energy-band calculations are often in good agreement with experimental data. Thus we can regard the band theory of transition metals as a reasonable starting point when discussing overall thermodynamical properties of materials except for their magnetism in which electron correlation plays an essential role.

We can apply the APW and KKR methods to calculate band structures for transition metals although OPW is not applicable. The essential difference from the simple metal case is that we cannot perturbationally treat the matrix elements Γ_{nm}^{APW} and Γ_{nm}^{KKR} as small quantities which appear in the secular equation to determine energy eigenvalues. Reflecting the presence of d levels, Γ_{nm}'s generally depend on energy. As cited at the end of the volume, many band calculations for transition metals by means of APW and KKR methods exist, and the band structures are very similar to each other among the same crystal structure of metals. As an example, we show in Fig. 5.17 the band structure for fcc Ni calculated by the KKR method. We can see the similarity between this and the band structure for Cu given in Fig. 5.9. Noting this similarity in the band struc-

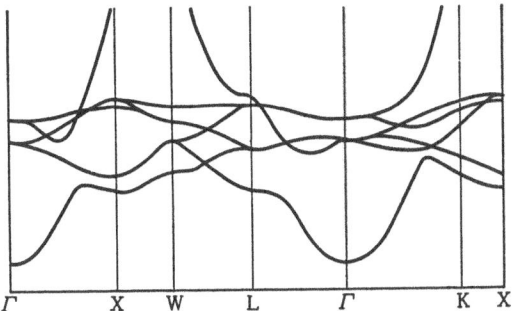

Fig. 5.17. Band structure of Ni

$$\Gamma \quad X \quad W \quad L \quad \Gamma \quad K \quad X$$

tures among metals with different numbers of d electrons, we can successfully explain overall properties of transition metals by assuming a common band structure for all transition metals and by changing the filling of bands which depends on the d-electron number. Especially in order to explain the empirical fact that properties of transition metal alloys depend only on the average atomic valence z, the common band model is very useful. Just as for simple metals, we can reproduce the band structures calculated from the first principle by introducing a pseudopotential; there are several investigations to reproduce band structures of transition metals by using a few parameters. One such method is that called the *combined interpolation scheme* in which we treat sp electrons by the NFE approximation and d electrons by Bloch functions which consist of five atomic d orbital functions, assuming that d electrons are localized. Then we consider the interaction between sp and d electrons to make hybridized orbitals. Under a given k vector, the secular equation has the following form:

$$\det \left| \begin{array}{c:c} \begin{array}{c} sp \text{ bands} \\ (\text{NFE approx.}) \end{array} & \begin{array}{c} sp\text{-}d \text{ hybrid} \\ \text{components} \end{array} \\ \hdashline \begin{array}{c} sp\text{-}d \text{ hybrid} \\ \text{components} \end{array} & \begin{array}{c} d \text{ bands} \\ (\text{LCAO}) \end{array} \\ \underbrace{}_{\text{4-dim.}} & \underbrace{}_{\text{5-dim.}} \end{array} \right| = 0 . \qquad (5.6.1)$$

Now one of the most eminent characteristics of transition metals is that the cohesive energy is larger than that of other metals and changes systematically within transition metals. In Fig. 5.18 we replot the cohesive energies U (kcal · mole^{-1}) of metals given in Table 5.1. As the number of d electrons increases, U increases and then decreases passing a maximum point. This evidently shows that d electrons contribute to metallic cohesion. The other characteristics of transition metals are in their crystal structures, which show a certain regularity as summarized in Table 5.8. The transition metals belonging to the same column in the periodic table exhibit a common regularity. This remarkable fact should be explained by a general principle. In the following we shall give a qualitative explanation for these characteristic properties of transition metals [5.10.11].

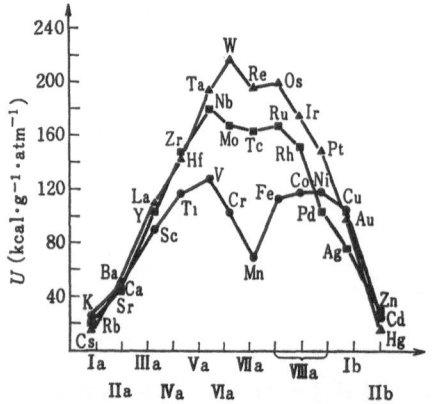

Fig. 5.18. Cohesive energy of transition metals

Table 5.8. Crystal structures of transition metals

x	3	4	5	6	7	8	9	10	11
$3d4s$ series	Sc	Ti	V	Cr	Mn	Fe	Co hcp	Ni	Cu
	hcp	hcp	bcc	bcc	complex	bcc	fcc	fcc	fcc
$4d5s$ series	Y	Zr	Nb	Mo	Tc	Ru	Rh	Pd	Ag
	hcp	hcp	bcc	bcc	hcp	hcp	fcc	fcc	fcc
$5d6s$ series	La	Hf	Ta	W	Re	Os	Ir	Pt	Au
	hcp	hcp	bcc	bcc	hcp	hcp	fcc	fcc	fcc
		hcp		bcc		hcp		fcc	

For this purpose we shall adopt the common band model as a basic assumption. To check the validity of the common band model, we show a part of the band structures for the first series of transition metals calculated by the APW method in Fig. 5.19. Energy dispersion curves along a high symmetry line in k-space are drawn for Ti of hcp, V, Cr and Fe of bcc, and Co, Ni and Cu of fcc structures. We can clearly see that almost the same curves for the same crystal structure are obtained except that the position of the Fermi surface is different. From these band structure curves we can calculate the electronic density of states which are shown schematically in Fig. 5.20 for bcc, fcc and hcp structures. The dotted lines are the integrated density of states to indicate the electron number accommodating the states below the given energy. These curves represent the manner of changes in s, p, d electron energies for different crystal symmetry, and the main features may be summarized in the following way. It is noted that for bcc structures there is a deep valley in the density of states around the energy below which six electrons per atom may be accommodated. On the other

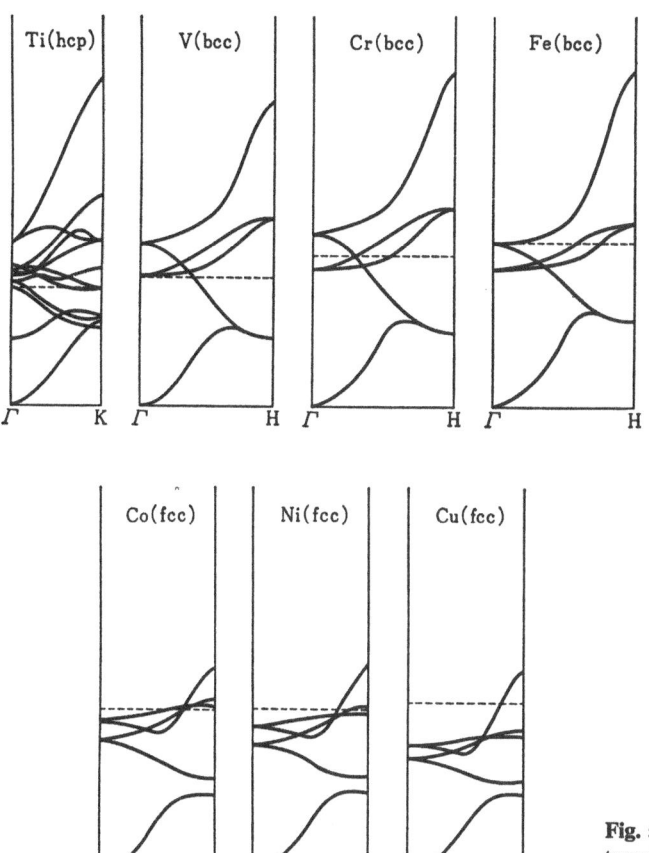

Fig. 5.19. **Fig. 5.19.** Band structures of transition metals

hand, for fcc structures there is no such valley in the density of states, but its highest peak shifts to the high energy side. The density of states for a hcp structure is somehow similar to that of a fcc structure, but close examination reveals that there is a valley in the high energy side which, in turn, makes the density of states in the low energy side slightly higher than that of bcc and fcc structures.

First let us explain qualitatively the characteristic point of the cohesive energy in transition metals (Fig. 5.18). Imagine that we make a transition metal crystal from an assembly of atoms infinitely separated from each other. With decreasing interatomic distance, we assume that the d levels of isolated atoms are broadened to form d-bands with a bandwidth W and the center of gravity of levels at $-\varepsilon$ relative to when at infinity. For simplicity, if we approximate this density of states of band per atom $n(\varepsilon)$ by a constant, we have

$$n(\varepsilon) = \frac{10}{W} \qquad (5.6.2)$$

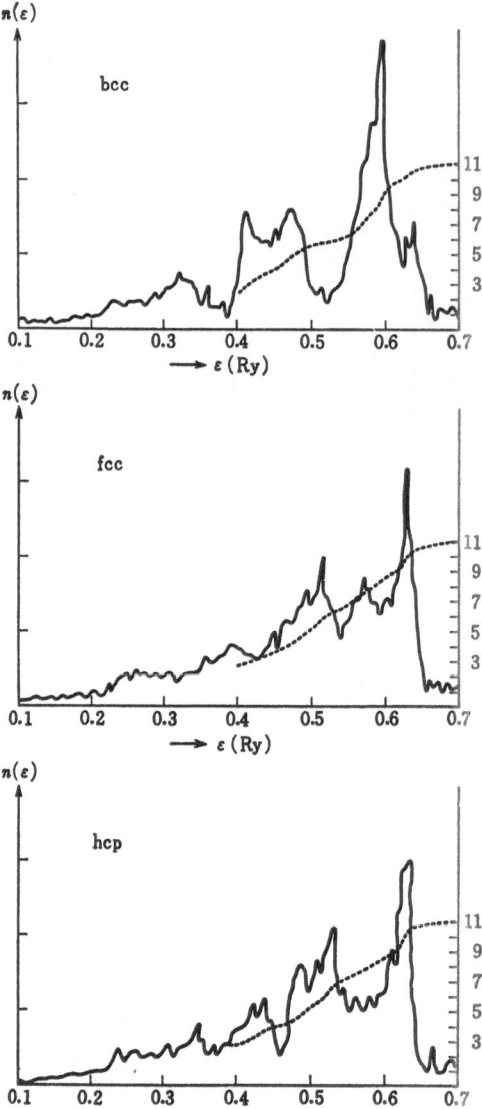

Fig. 5.20. Density of states of transition metals for three crystal structures. Dotted lines represent the integrated density of states (right scale) [5.10]

because the band should accommodate ten d electrons per atom. Let x be the number of d electrons per atom for given metal. Then it should hold

$$\int_{-s-W/2}^{\varepsilon_F} n(\varepsilon)\, d\varepsilon = \frac{10}{W}\left(\varepsilon_F + \varepsilon + \frac{W}{2}\right) = x \,. \tag{5.6.3}$$

The cohesive energy U is calculated as

$$U = - \int_{-\varepsilon-W/2}^{\varepsilon_F} \varepsilon n(\varepsilon)\, d\varepsilon = -\frac{10}{W}\frac{1}{2}\left[\varepsilon_F^2 - \left(\varepsilon + \frac{W}{2}\right)^2\right]$$

which, upon using (5.6.3), may be rewritten in terms of x as

$$U = x\varepsilon + \frac{W}{20} x(10 - x). \tag{5.6.4}$$

As shown in Fig. 5.21b, this function represents a curve with a maximum at $x = 5 + 10\ (\varepsilon/W)$. Although this curve is not so similar to the measured curve, it gives us an insight to the role played by d electrons in the cohesion mechanism of transition metals.

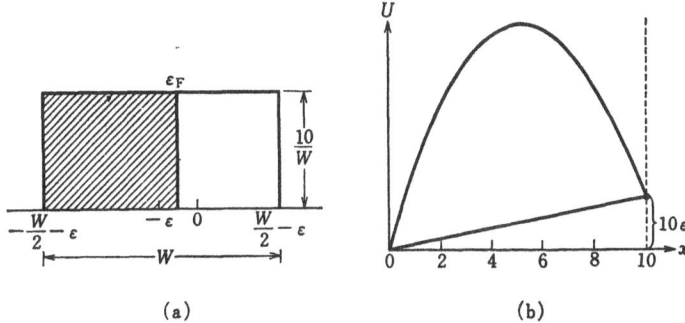

(a)   (b)

Fig. 5.21a, b. Model of d-band and cohesive energy

To discuss the d-electron number dependence of U, it is necessary to take into account the stability of different structures simultaneously. For $3d$-electron metals, we also have to take into account magnetic structures of crystals. This complicates the problem further. It is hard to explain by a simple theory why U becomes small for around Cr, Mn and Fe, and why Mn has an exceptionally complicated structure. On the other hand, situations are somewhat simpler for $4d$-and $5d$-electron metals, because they exhibit no magnetic ordering. Let us write the contribution to U in the form

$$U = x\varepsilon + U_{\mathrm{bs}} + U_{\mathrm{E}} + U_{\mathrm{core}} + U_{\mathrm{ex}}, \tag{5.6.5}$$

where the first term expresses the contribution from the shifted center of gravity of energy bands, the second term band structural energy given by

$$U_{\mathrm{bs}} = \int^{\varepsilon_F} \varepsilon n(\varepsilon)\, d\varepsilon. \tag{5.6.6}$$

U_E is the Ewald energy, U_{core} the repulsive interactions between ion cores, and U_{ex} is due to the exchange interaction. We must calculate U for different crystal structures to compare their stability. However, except for the first two terms in the right-hand side of (5.6.5), the remaining terms are known to be insensitive to crystal structures. Thus, the most crucial term is U_{bs}. Since in (5.6.5) the origin of energy is chosen so that the center of gravity of energy bands is zero, we should generally rewrite it as

$$U_{bs} = \int_{\varepsilon_{min}}^{\varepsilon_F} (\varepsilon_0 - \varepsilon) \, n(\varepsilon) \, d\varepsilon , \qquad (5.6.7)$$

where ε_0 is defined by

$$\varepsilon_0 = \frac{\displaystyle\int_{\varepsilon_{min}}^{\varepsilon_{max}} \varepsilon n(\varepsilon) \, d\varepsilon}{\displaystyle\int_{\varepsilon_{min}}^{\varepsilon_{max}} n(\varepsilon) \, d\varepsilon} . \qquad (5.6.8)$$

Using $n(\varepsilon)$ curves given in Fig. 5.20, we calculate U_{bs} for bcc, fcc and hcp as a function of

$$x = \int_{\varepsilon_{min}}^{\varepsilon_F} n(\varepsilon) \, d\varepsilon \qquad (5.6.9)$$

and plot $\Delta U = U_{bs}(\text{bcc}) - U_{bs}(\text{fcc})$ and $\Delta U = U_{bs}(\text{hcp}) - U_{bs}(\text{fcc})$ in Fig. 5.22a. As we include the sp bands in these figures, for transition metals the value of x starts with $x = 3$. Since the larger U is, the more stable the system, this figure tells us that bcc is found to be stable at $x = 5 \sim 6$ and $x = 10 \sim 11$ and hcp is stable at $x = 3 \sim 4$ and $x = 7 \sim 9$, but there is no place where fcc is stable. This result obviously contradicts the observed facts given in Table 5.6. Though the result that bcc is stable at $x = 5 \sim 6$ is in accordance with an empirical fact, fcc should have been stable for $x \sim 9$. We can partially remove this discrepancy by taking into account the effect of the first term in the right-hand side of (5.6.5). Since the coordination numbers of bcc and fcc are different,

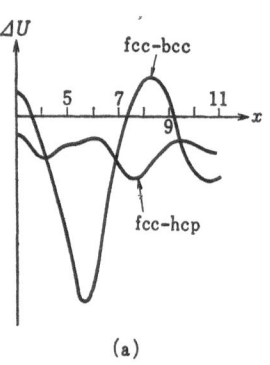

(a)

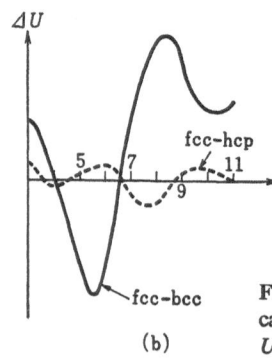

(b)

Fig. 5.22a, b. Comparison of calculated $U(\text{fcc})$, $U(\text{hcp})$ and $U(\text{bcc})$ [5.10]

there must be different contributions to the first term depending on electron number and coordination number which affect the repulsive interaction between atoms. Thus we try to assume

$$\Delta\varepsilon = \varepsilon_{fcc} - \varepsilon_{bcc} = 0.001 \quad [Ry/atom] \tag{5.6.10}$$

to explain the empirical fact. The curve for $U(fcc) - U(bcc)$ becomes that shown by the solid line in Fig. 5.22b. As to the difference between hcp and fcc, it is expected that, since the packing of atoms for both structures are almost identical, the energy differences are not appreciably large. Thus, to examine the stability between them is a very delicate problem. However, one difference should be pointed out, that is, the crystal potential of hcp exerted on electrons, which has hexagonal symmetry, is slightly lower than that of fcc with cubic symmetry. This can be indeed observed in Fig. 5.19 when we compare two band structures of Ti (hcp) and Ni (fcc). In the former case, the degeneracy of d levels are completely removed and hybridized strongly with sp levels. Therefore it might be possible that small differences in the parameter which controls the hybridization effect makes bands of hcp crystal alter significantly. Now we try once again to lower $U_{bs}(hcp)$ rather arbitrarily by 0.006 Ry/atom. The result is shown by the dotted line in Fig. 5.22b, indicating that hcp is stable at x observed in the real transition metals.

Generally speaking, it is hard to discuss the stability of different crystal structures theoretically and to deduce reliable conclusions because of our theoretical accuracy. However, it is interesting that the relatively simple argument mentioned above gives qualitatively a correct explanation about the appearance of characteristic structures of transition metals. A physical reason why a bcc structure is stabilized at $x = 5 \sim 6$ is as follows. As is seen from Fig. 5.20 the bands split into two parts in a bcc structure in which each band accommodates nearly the same number of electrons. This is because of the spatial extension of d orbitals and their atomic configuration. The lower band has a bonding character and the higher band an antibonding character. The center of gravity ε_0 of these bands is situated around the valley between the two bands. If we fill up electrons from the lower energy band, the contribution of the lower band to U_{bs} is positive and that of the upper band is negative, because of (5.6.7). Thus $U_{bs}(bcc)$ has a large positive value at $x = 5 \sim 7$. On the other hand, there is not such a deep valley in the density of states of a fcc structure, therefore the distribution is shifted toward the high energy side. Hence, the center of gravity of the bands is located at a higher energy position than that of a bcc structure. As far as the electron number is small, $U_{bs}(fcc)$ is greater than $U_{bs}(bcc)$ because of the presence of the large difference $\varepsilon - \varepsilon_0$. At about $x = 5 \sim 6$, this inequality is reversed. When the electron number exceeds $x = 6$, the sign of $U_{bs}(fcc) - U_{bs}(bcc)$ is reversed again because $\varepsilon_0 - \varepsilon < 0$ in a bcc structure becomes large. Similar qualitative arguments may be applied to $U_{bs}(fcc) - U_{bs}(hcp)$ as well.

5.7 Alloys

We can easily make solid solutions by mixing different metallic elements. This is one of the characteristics of metals. Roughly speaking, this is because valence electrons of individual metallic elements become free electrons by alloying, without bounded to particular ions. Thus, when different atoms are mixed up, the valence electrons of these atoms merge into conduction electrons, and even if the different atoms produce extra charges in the host metal, an increase of energy due to the presence of an internal field is avoided by a screening effect of conduction electrons. However, there are infinite combinations of alloying of different metals, and the appearance of various mixtures is not simple. From old times, the existence of special phases of alloys have been recognized by several empirical laws which have not been investigated systematically on the basis of the modern electron theory of metals. In this section we shall discuss some topics of alloys and give simple arguments for developing the electron theory of alloys, stressing the fundamental aspects of the problem.

Usually alloys mean substitutional mixtures, in which the lattice points of a host metal of bcc, fcc or hcp structure are partially replaced by different kinds of metal atoms. To form a uniform alloy some condition is required for constituent metallic atoms. It is often observed that the two metals which have the same crystal lattices and similar atomic volumes are easily solved to form a substitutional alloy for any concentration ratio of two metals. For particular concentration ranges, it sometimes happens that a regular arrangement of atoms takes place in an alloy, which we call an ordered alloy. There is a group of alloys called electronic compounds in which a definite crystal structure appears only when the ratio of the valence electron number to atom number gets a certain value. These are alloys which obey the Hume-Rothery law, and we shall shortly come back to these alloys to give a brief account on recent interpretations of this electronic compound from an electron theoretical point of view. Some alloys are called intermetallic compounds when they appear in special structures with a specific concentration ratio. Among them a particularly famous one is the so-called Laves phase which is a phase of alloys of AB_2 type, appearing only at a certain ratio of atomic radii. Contrary to substitutional alloys, interstitial alloys are such that atoms with radii quite different from that of host atoms penetrate into interstitial positions of the host lattice to form an alloy. In this case, Hägg's law exists which relates the ratio of atomic radii and crystal structures assumed by the interstitial alloys. If we include in a broad sense semimetallic or semiconducting elements as a constituent atom to form intermetallic compounds, we have a further abundance of alloys with many varieties in which some are metallic, others are semimetallic or semiconductive. We shall discuss some of the semiconductive alloys in Chapter 6.

Let us first consider the case in which a host metal contains a small amount of different kinds of atoms as an impurity. This is a dilute alloy. Since impurity atoms are far apart in this case, we can neglect the interaction between them.

Let z be the atomic valence of an impurity atom, then in the simplest model of a free electron gas, an alloying effect is just to increase the free electron number by z per impurity atom. If the atomic valencies of an impurity atom and host atom are different, an impurity atom brings into the electron gas an excess charge ze, around which free electrons tend to gather to screen it. This effect has been already discussed in the section on metallic hydrogen in Chap. 2 by using the Thomas-Fermi theory, which can be applied here also. According to the theory, the potential of the excess charge ze is modified from the Coulomb form $1/r$ to

$$V(r) = - ze \frac{e^{-r/\lambda}}{r}, \tag{5.7.1}$$

where the screening constant λ is given by

$$\lambda^2 = \frac{v_F^2}{3\omega_p^2} \tag{5.7.2}$$

[see (2.4.51)], v_F is the Fermi velocity of the free electron and ω_p is the plasma frequency of the electron gas. The charge distribution $\delta n(r)$ of the screening electron cloud around ze is determined by Poisson's equation

$$\nabla^2 V(r) = - 4\pi e \delta n(r) \tag{5.7.3}$$

which leads to

$$\delta n(r) = \left(\frac{z}{4\pi\lambda^2}\right) \frac{e^{-r/\lambda}}{r}, \tag{5.7.4}$$

where λ is of an order of 10^{-8} cm for ordinary metals. Therefore, the excess charge of an impurity atom is screened out within the region occupied by this atom. Although the above argument is based on classical mechanics, the more rigorous quantum mechanical treatment does not alter the conclusion except that in the latter treatment $\delta n(r)$ does not diverge at $r \to 0$ as in (5.7.4), and $\delta n(r)$ behaves asymptotically at $r \to \infty$ as

$$\delta n(r) \propto \frac{\cos(2k_F r + \varphi)}{r^3} \tag{5.7.5}$$

which is an oscillatory damping. This is known as the Friedel oscillation which is due to the presence of a sharp Fermi surface of free electrons in the momentum space. This solution satisfies (5.7.4), and moreover the following relation

$$\int_0^\infty \delta n(r)\, 4\pi r^2 dr = z \tag{5.7.6}$$

generally holds. This is Friedel's sum rule which ensures charge nentrality.

When the concentration of impurity atoms increases, the theory for dilute alloys is not applicable and at the same time, the band theory loses its foundation because the periodicity of the crystal lattice is destroyed. In such a case how do we treat electronic states? One of the approximations commonly used is such that the band structure of a host metal is kept unchanged and the electron number per atom and the bottom of the energy band are varied by alloying. This is called the rigid (or common) band model. An example in which this model proves successful is an explanation of the Hume-Rothery law based on it.

In the alloy of Cu and Zn, so-called brass, several special crystal structures appear when the concentration ratio takes certain values. For example, it assumes a bcc structure called β-phase for the composition CuZn, a γ-phase for Cu_5Zn_8 which is a complex cubic structure containing 52 atoms in a unit cell, and ε-phase with a hcp structure for $CuZn_3$. It turns out, however, that, even in many other alloys, the same structures as β, γ and ε-phases appear when the composition of an alloy satisfy certain rules. According to Hume-Rothery, let the atom number for a given composition be A (for instance for Cu_5Zn_8, $A = 5 + 8 = 13$) and the total number of the atomic valence be e (for instance for Cu_5Zn_8, $e = 5 + 2 \times 8 = 21$), then the β-phase appears when $e/A = 3/2$, the γ-phase when $e/A = 2/13$ and the ε-phase when $e/A = 7/4$. This empirical rule is in surprisingly good accordance with observations, provided that the atomic valence of transition metals such as Fe, Co and Ni is counted as 0, although for ordinary metals the valence electron number is taken to be equal to the atomic valence. If we use an abbreviation such that an alloy composed of a monovalent atom and a trivalent atom at an atomic ratio of 3:1 is expressed by $(I)_3 (III)$, then the composition of atoms to form the β, γ, and ε-phases are realized in practice in the combination as tabulated in Table 5.9.

Table 5.9. Various phases of electronic compounds (Hume-Rothery law)

	β phase $\dfrac{e}{A} = \dfrac{3}{2}$		γ phase $\dfrac{e}{A} = \dfrac{21}{13}$		ε phase $\dfrac{e}{A} = \dfrac{7}{4}$	
Example	$(I)(II)$	CuZn	$(I)_5(II)_8$	Cu_5Zn_8	$(I)(II)_3$	$CuZn_3$
	$(I)_3(III)$	Ag_3Al	$(I)_9(III)_4$	Cu_9Al_4	$(I)_5(III)_3$	Ag_5Al_3
	$(I)_5(IV)$	Cu_5Sn	$(I)_{31}(IV)_8$	$Cu_{31}Si_8$	$(I)_3(IV)$	Cu_3Sn
	$(III)(VIII)$	AlFe	$(II)_{21}(VIII)_5Zn_{21}Co_5$			

It is a very interesting problem to account for the Hume-Rothery law by electron theory. The common band model explains it in the following way. An approximate total energy of an alloy is obtained

$$U = \int^{\varepsilon_F} \varepsilon n_0(\varepsilon)\, d\varepsilon \qquad (5.7.7)$$

by filling valence electrons coming from each constituent atom in a common band up to the Fermi energy ε_F. In (5.7.7), $n_0(\varepsilon)$ is the density of states of the common band. Now let us consider an energy constant surface in the k-space. Since the volume enclosed by this surface should be proportional to the electron density of states, $n_0(\varepsilon)$ is determined by how this surface changes as a function of ε. At the boundary of the Brillouin zone, the energy generally changes discontinuously. Therefore it is expected that $n_0(\varepsilon)$ should show abrupt changes in the energy where the energy constant surface touches the Brillouin zone. This is schematically illustrated in Fig. 5.23a where the dotted line indicates the density of states for free electrons. Therefore, it is expected that U, which given by (5.7.7), varies as a function of ε_F as shown in Fig. 5.23b, and the total energy is lowered to stabilize the system when the Fermi surface touches the Brillouin zone (or more precisely, the Jones zone in the extended zone scheme). Since the Fermi surface is determined by an electron number only as a first approximation, it is expected that a system is relatively stable when it has a crystal structure such that the Fermi sphere determined by the electron number touches the Jones zone associated with the assumed crystal lattice.

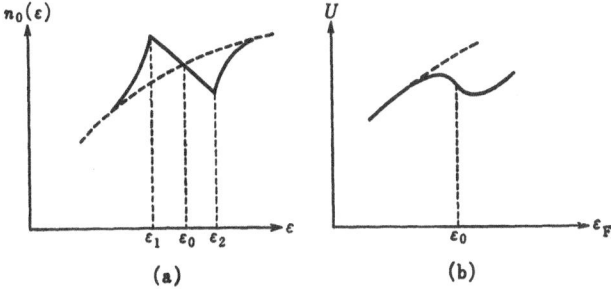

Fig. 5.23. (a) Density of states near zone boundary. (b) Crystal energy U as a function of ε_F

In fact, the three values of e/A cited in Table 5.9 nearly correspond to the electron numbers when the Fermi surface touches the Jones zone and are determined from the crystal structures of each phase. Since it is easy to prove this for bcc and hcp, we shall explain here only the case of the γ-phase, leaving the proof of the other cases to the reader's exercises. The γ-phase has a complex crystal structure which we will not show here, but the structure factor $S(G)$ of this lattice has been calculated to yield the following values:

G	110	200	211	220	300	222	321	400	330	411
$S(G)$	0.32	0	0.32	0	0.32	2.68	1.05	0	8.85	5.63

$S(G)$ is large for (330) and (411), at which large energy gaps are expected to occur. The Jones zone constructed by these two planes (330) and (411) and their equivalents are shown in Fig. 5.24. Thirty-six planes in total enclose the origin nearly spherically, and it is known that the number of states included in this volume is 90 per unit cell. Since there are 52 atoms in a unit cell, the number of states per atom in this Jones zone is given as

$$90/52 = 22.5/13$$

which is very near to $e/A = 21/13$.

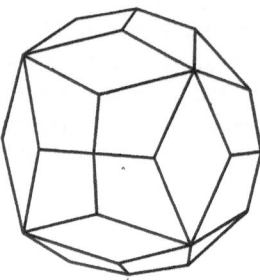

Fig. 5.24. Jones zone of γ phase

The above argument is one of the qualitative explanations for the Hume-Rothery law. It seems to need a theory based on more solid ground. In connection with this, we recall that the discussion on structures of simple metals given in Sect. 5.5 predicts a bcc lattice to be stable for $z = 1.5$. It is interesting to note that this precisely corresponds to $e/A = 1.5$.

When the periodicity of a lattice is perturbed drastically as in alloys, it would be better to consider the interactions between atoms in coordinate space rather than in k-space. For this aim, we go back to Sect. 5.5 and look at the band structure energy

$$U_{\mathrm{bs}} = \sum_{G}' |S(G)|^2 [v(G)]^2 \chi(G) \, \epsilon(G)$$

which is transcribed by putting

$$\Phi_{\mathrm{bs}}(q) = [v(q)]^2 \chi(q) \, \epsilon(q) \tag{5.7.8}$$

and using the definition of $S(G)$, as

$$U_{\mathrm{bs}} = \frac{1}{N^2} \sum_i \sum_j \sum_q \exp[iq(R_i - R_j)] \, \Phi_{\mathrm{bs}}(q) \,. \tag{5.7.9}$$

Performing the sum over q, we have

$$U_{\mathrm{bs}} = \frac{1}{2N} \sum_i \sum_j \Phi_{\mathrm{bs}}(|R_i - R_j|) \,, \tag{5.7.10}$$

where $\Phi_{bs}(R)$ is the Fourier transform of $\Phi_{bs}(q)$ defined by

$$\Phi_{bs}(R) = \frac{2\Omega}{(2\pi)^3} \int \Phi_{bs}(q) \exp{(iq \cdot R)} dq .\qquad (5.7.11)$$

It is seen from (5.7.10) that the band structure energy is apparently expressed as a sum of two-body potentials between ions. This potential $\Phi_{bs}(R)$ represents an effective interaction between ion cores mediated by electrons, and in the limit $R \to \infty$, it tends to an asymptotic form $-z^2e^2/R$ which exactly cancels the direct Coulomb repulsion. Therefore, if we put

$$\Phi(R) = \frac{z^2e^2}{R} + \Phi_{bs}(R) ,\qquad (5.7.12)$$

$\Phi(R)$ may be regarded as an effective interatomic potential acting between atoms in metals. This potential is of a central force. This is due to the second-order perturbation theory. A typical variation of $\Phi(R)$ as a function of R is depicted in Fig. 5.25, where the oscillation in the range of large R comes from the same reason as in the Friedel oscillation in (5.7.5). This oscillating long-range interaction is one of the characteristics of the interaction between ions mediated via free electrons, and this should be the case in alloys as well.

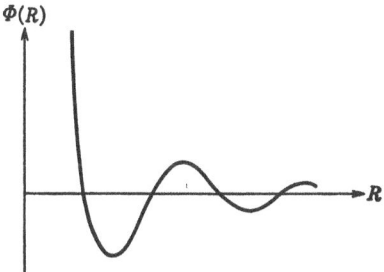

$\Phi(R)$

R

Fig. 5.25. Effective interatomic potential in metals

To extend it to the binary alloy consisting of A and B atoms, we first introduce two pseudopotentials screened by conduction electrons for two kinds of ions A and B:

$$v_i(q) = \frac{v_i^{ion}(q)}{\epsilon(q)} \quad (i = \text{A, B}) ,\qquad (5.7.13)$$

where the dielectric function $\epsilon(q)$ is assumed to be determined only by the density of the free electrons or the average valence number z, and takes the same value for the A and B ion. Then it is shown that the effective interaction potential between the ith and jth ions which are separated by a distance R is given by

$$\Phi_{ij}(R) = \frac{z_i z_j e^2}{R} + \frac{2\Omega}{(2\pi)^3} \int [v_i(q) \, v_j(q) \, \epsilon(q) \, \chi(q)] \, e^{i q \cdot R} dq \tag{5.7.14}$$

for the same reason that leads to (5.7.12). Now we assign at the ith lattice point the variables $x_i^{(A)}$ and $x_i^{(B)}$ such that $x_i^{(A)}$ $(x_i^{(B)})$ takes unity when the ith lattice point is occupied by the A (B) atom and zero otherwise. Then it is not difficult to show that the energy of an alloy can be expressed as

$$U = \frac{1}{2N} \sum_{i \neq j} \sum [\Phi_{AA}(R_{ij}) \, x_i^{(A)} x_j^{(A)} + \Phi_{BB}(R_{ij}) \, x_i^{(B)} x_j^{(B)}$$
$$+ \Phi_{AB}(R_{ij}) \, (x_i^{(A)} x_j^{(B)} + x_i^{(B)} x_j^{(A)}) \, . \tag{5.7.15}$$

In the most crude approximation, we can replace $x_i^{(A)}$ and $x_i^{(B)}$ by their averages

$$\bar{x}_i^{(A)} = C, \quad \bar{x}_i^{(B)} = 1 - C \, , \tag{5.7.16}$$

where C is the concentration of the A atom. In this approximation, U may be expressed as

$$\bar{U} = \frac{1}{2N} \sum_{i \neq j} \sum W(R_{ij}) \tag{5.7.17}$$

with

$$W(R) = C^2 \Phi_{AA}(R) + (1 - C)^2 \Phi_{BB}(R) + 2C(1 - C) \, \Phi_{AB}(R) \tag{5.7.18}$$

This result can be further put in a simpler form by introducing $W(q)$, the Fourier transform of $W(R)$, that is,

$$\bar{U} = \frac{1}{2} \left[\sum_G W(G) - \frac{\Omega}{(2\pi)^3} \int W(q) \, dq \right] . \tag{5.7.19}$$

In the first term, the sum is taken over all reciprocal lattice vectors corresponding to the given crystal structure. The second term comes from a fact that the term $i = j$ in the sum of (5.7.17) is excluded. Now $W(q)$ is given by the sum of the Fourier transforms of $\Phi_{AA}(R)$, etc., through (5.7.18) which contains the factor $\epsilon(q) \chi(q)$ as seen from (5.7.14). As we have seen in Sect. 5.5, this function has an infinite derivative at $q = 2k_F$. Therefore U as a function of the Fermi energy ε_F varies with infinite slope when one of the reciprocal lattice vector G satisfies the condition

$$|G| = 2k_F \, . \tag{5.7.20}$$

Equation (5.7.20) may be interpreted as expressing the condition that the Fermi surface touches the Brillouin zone (or Jones zone). Thus this is another way to explain the Hume-Rothery law.

A phenomenon in which the long range force as depicted in Fig. 5.25 might play an important role is observed in the so-called long period regular alloys. CuAu has a regular arrangement of a fcc structure as shown in Fig. 5.26a (CuAu I phase) from room temperature up to 385°C. Between 385°C and 410°C, a one-dimensional long periodic structure appears, as shown in Fig. 5.26b (CuAu II phase). In this phase, a regular arrangement with a period five times the

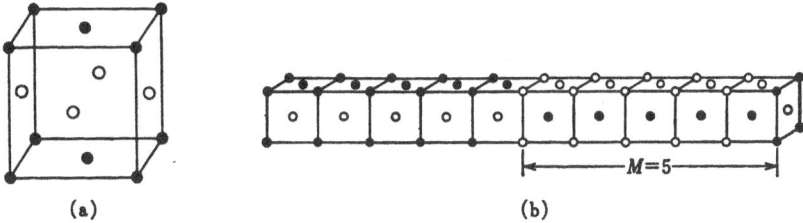

(a)   (b)

Fig. 5.26. (a) Crystal structure of CuAu I phase (b) Crystal structure of CuAu II phase

original lattice constant takes place along one of the cubic axes, as confirmed by observing the extra Bragg spots in x-ray diffraction patterns. There are many examples of such a long period structure observed in other alloys, and its existence as a stable thermodynamic phase seems to be established. It is hard, however, to understand it in terms of short range interaction only. We explain this phenomena on the basis of the band theory as follows [5.12]. In the disordered phase of CuAu which has a fcc structure on average, the Brillouin zones are of the form depicted in Fig. 5.27a by the dotted lines. When the system enters the CuAu I phase, it's symmetry becomes tetragonal and hence the Brillouin zone is modified to the form as shown by the solid lines in Fig. 5.27a. The volume of the Brillouin zone of this phase is just half of that of the fcc phase, which can accommodate all valence electrons. Therefore, if the regular atomic arrangement causes an energy gap near the boundary of the new Brillouin zone and all valence electrons are accommodated within it, a large energy gain would be expected. This is a simple interpretation of the ordering energy of the CuAu I phase. However, if the Fermi surface is nearly spherical, it cannot fit the new Brillouin zone of cubic shape, leaving holes in the corners and resulting in that part of the electrons occupy the second Brillouin zone, as shown in Fig. 5.27b. Now, as a long periodic structure with a period M along one direction is produced, some of the reciprocal points are split and the Brillouin zone will be changed as in Fig. 5.27c. When the Fermi surface fits itself into such a deformed Brillouin zone, the energy of the system will be lowered further. One can estimate the volume of a sphere which touches the new deformed Brillouin zone (as shown in Figs. 5.27c, 5.27d, there are two possibilities discriminated by $\pm$), by a short manipulation yielding a volume given by

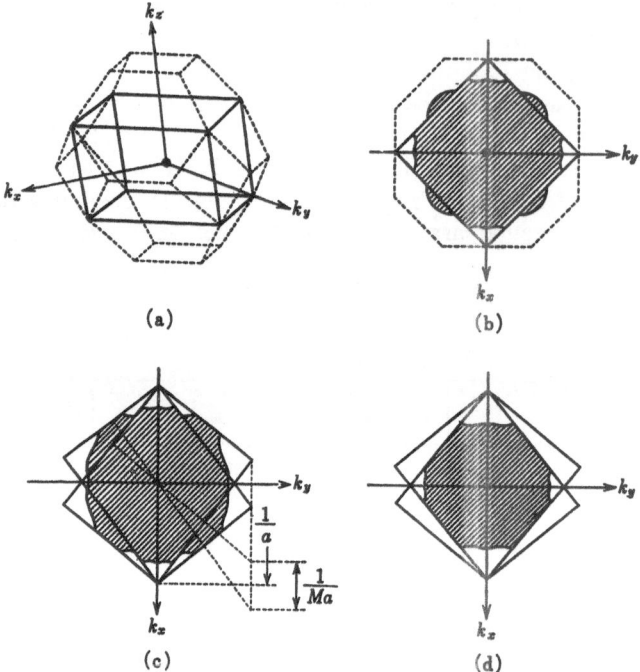

(a)

(b)

(c)

(d)

Fig. 5.27a-d. Change of Brillouin zone for different CuAu phases. (a) Brillouin zone for CuAu, (b) Fermi surface of CuAu I; (c, d) Fermi surface of CuAu II [5.12]

$$\Omega_0 = \frac{\pi}{6a^3} \left(2 \pm \frac{1}{M} + \frac{1}{4M^2} \right)^{3/2}. \tag{5.7.21}$$

Now we assume this is equal to the volume inside the Fermi surface, then $\Omega_0 a^3$ is a quantity proportional to the electron number per atom e/A. From this we can draw interesting conclusions. That is, if the number of electrons are changed in an alloy with long period structure, the long period M should change accord-

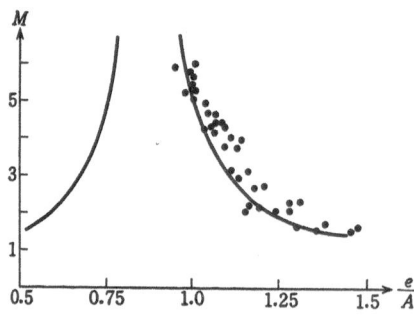

Fig. 5.28. Long period M vs e/A [5.12]

ing to (5.7.21). This prediction is indeed verified by experiment. By adding a few metal atoms with different valencies to the CuAu II phase, the number of electrons per atom can be changed, and at the same time the long period M varies as a function of e/A as shown in Fig. 5.28 by the dotted points. The solid lines are the theoretical curves calculated from (5.7.21) with an adjusted proportional coefficient. It would be possible to theoretically explain the phase transitions CuAu I → CuAu II → CuAu III (disorder) and accompanied changes in the properties by using the common band model and taking into account the lattice distortions and the entropy effect.

5.8 Superconductivity

In this last section, let us consider superconductivity which is one of the most striking phenomena among various properties of metals. Superconductivity can be observed in many metallic elements, alloys, intermetallic compounds and a few special semiconductors limited to appear only in very low temperatures. A certain rule seems to exist for the occurrence of superconductivity in these substances.

In 1911, H. Kamerlingh-Onnes first found this phenomenon when he measured extremely small electric resistivity in mercury below a certain transition T_c and coined the name of superconductivity. Since then this remarkable phenomenon has been found in many other materials. Experimental data on it have been accumulated. However, it is only in 1957 that J. Bardeen, L. N. Cooper and J. R. Schrieffer gave a complete theory (so-called BCS theory) of superconductivity on the basis of microscopic electron theory. It is a rare event in the history of physics that for about 50 years after experimental discovery, a correct theoretical interpretation has not been given.

We briefly summarize several well-known characteristics of superconductivity.

1) Superconductors undergo the second-order phase transition at T_c below which they become a superconducting state. The order of magnitude of T_c is around liquid helium temperatures. The maximum T_c so far known is 23 K for Nb_3Ge. The specific heat behaves exponentially near 0 K, that is,

$$\frac{C_e}{\gamma T_e} = 8.5 \exp\left(-1.5 \frac{T_e}{T}\right) \tag{5.8.1}$$

and varies as T^3 near the transition temperature. The specific heat jump of the λ shape is observed at T_c.

2) The superconducting state is destroyed by a strong magnetic field above the critical field H_c and returned to a normal state. The critical field H_c depends on temperature and the parabola law

$$H_c = H_0 \left[1 - \left(\frac{T}{T_c} \right)^2 \right] \tag{5.8.2}$$

holds approximately. Below H_c, superconductors are completely diamagnetic, that is, magnetic fields can penetrate only near the surface and are excluded from the bulk of superconductors. This is called the Meissner effect.

3) The current induced in a multiply-connected (for instance, a ring) superconductor persists without decay. This is called a persistent current. A magnetic flux which emerges across the closed circuit of a persistent current is quantized to take discrete values. The elementary unit of this quantized flux is given by

$$\phi_0 = \frac{\pi \hbar c}{e} \approx 2 \times 10^{-7} \ [\text{G} \cdot \text{cm}^2]. \tag{5.8.3}$$

4) There is an isotopic effect in the transition temperature T_c; T_c is approximately inversely proportional to the root square of an ionic mass M:

$$T_c \sqrt{M} = \text{const.} \tag{5.8.4}$$

5) In order that materials exhibit superconductivity, there are certain limitations and rules depending on the average valence electron number z per atom and on the electronic density. For instance, superconductors of metallic elements are limited to the valence number z such that

$$2 \leqq z \leqq 8 .$$

This rule also applies well to compounds except for very few cases. Ferromagnetic and antiferromagnetic elements never become superconductors. Table 5.10 shows the transion temperature T_c of several metals and alloys.

Table 5.10. Superconducting transition temperatures of metals and alloys

Metal	T_c[K]	Metal	T_c[K]	Alloy	T_c[K]
La	6.05	Ru	0.49	Nb_3Sn	18.05
Ti	0.39	Os	0.66	Nb_3Al	17.5
Zr	0.55	Ir	0.14	V_3Si	17.1
Hf	0.09	Zn	0.85	V_3Ga	16.5
Th	1.37	Cd	0.52	NbN	16.0
V	5.38	Hg	4.15	MoN	12.0
Nb	9.26	Al	1.18	Nb_3Au	11.5
Ta	4.48	Ga	1.09	MoRe	11.5
Mo	0.92	In	3.40	La_3In	10.4
W	0.012	Tl	2.37	TiV	7.3
Tc	7.77	Sn	3.72		
Re	1.7	Pb	7.19		

In order to explain the Meissner effect and persistent current phenomenologically, F. London proposed the following two equations which describe the superconducting current density J_s, electric field E and magnetic field B, i.e.,

$$c \operatorname{rot} \Lambda J_s + B = 0 \tag{5.8.5}$$

$$\frac{\partial}{\partial t}(\Lambda J_s) = E, \tag{5.8.6}$$

where Λ is a constant depending on materials and is defined by $\Lambda = m/n_s e^2$, in which n_s can be interpreted as the superconducting electron density. Equations (5.8.5, 6) combined with the Maxwell equations yield

$$\nabla^2 B = \frac{1}{\lambda^2} B, \quad \lambda^2 = \frac{c^2 \Lambda}{4\pi} \tag{5.8.7}$$

$$\frac{\partial}{\partial t}(B + c\Lambda \operatorname{rot} J_s) = 0. \tag{5.8.8}$$

Equation (5.8.7) shows that the magnetic field cannot penetrate beyond the order of λ from a surface of superconductors, indicating the Meissner effect. The penetration depth λ is given by

$$\lambda = \left(\frac{mc^2}{4\pi n_s e^2}\right)^{1/2} \tag{5.8.9}$$

and estimated as $\lambda \approx 10^{-6}$ cm when $m = 9 \times 10^{-28}$ gm, $e = 5 \times 10^{-10}$ esu and $n_s = 10^{22}$ are used. Integrating (5.8.8) over a closed curve C of area S, and using Stokes theorem, we obtain

$$\int_s B \cdot dA + c\Lambda \oint_c J_s \cdot ds \equiv \Phi(C) = \text{const.} \tag{5.8.10}$$

This gives an explanation of the persistent current phenomena as follows. Let us consider a ring-shaped metal under a uniform magnetic field. On lowering the temperature below T_c, $\Phi(C) = \int B \cdot dA$ is equal to the magnetic flux which penetrates the ring because $J_s = 0$ at the beginning. Now if we turn off the magnetic field $B \to 0$, the supercurrent J_s is induced so as to keep $\Phi(C)$ a constant. This is a persistent current.

In order to establish an electronic theory of superconductivity, we must clarify the microscopic mechanism to produce the superconducting phenomena. The isotope effect mentioned in (4) will give a clue to solving it, that is, it suggests that ionic motions, or in other words the phonons, play an important role in this phenomena. H. Fröhlich pointed out the importance of the electron-phonon interaction for superconductivity. The BCS theory shows that an attractive interaction between electrons is induced via phonon exchange, which causes electron pairing (so-called Cooper pairs) and these pairs behave like Bose particles and can condense into a macroscopic state. Omitting details of the theory, we briefly discuss the conditions for superconductivity to appear and the magnitude of T_c.

We define the matrix element $\alpha_{k',k}$ for the transition in which an electron with a wave vector k is scattered by a phonon into another state with k'. In simple metals, this can be obtained by expanding ionic coordinates of the structure factor $S(q)$ in $\langle k | V_{ps} | k' \rangle = S(k-k')\, v(k-k')$ of (5.4.4) in terms of the normal coordinates of phonons up to the first order. Therefore this is proportional to $v(k - k')$. The electron-electron interaction via phonon exchange is the second-order process of the perturbation. That is, if we consider a process in which a k electron changes into another state with $k - q$ by emitting a q phonon (which is reabsorbed by another k' electron to change into a state with $k + q$), this is equivalent to the electronic process caused by an effective potential V_{ph} in which two electrons with k and k' exchange the momentum q to change into other states with $k - q$ and $k' + q$ respectively. The matrix element of this process is given by

$$\langle k - q, k' + q | V_{ph} | k, k' \rangle = \frac{|\alpha_{k,k-q}|^2}{\varepsilon(k) - \varepsilon(k - q) - \hbar\omega_q}, \tag{5.8.11}$$

where $\varepsilon(k)$ is the electronic energy and $\hbar\omega_q$ is the phonon energy with momentum q. Contrary to (5.8.11), the process in which a k' electron emits a q phonon which is reabsorbed by another k electron is described by the following matrix element:

$$\langle k + q, k' - q | V_{ph} | k, k' \rangle = \frac{|\alpha_{k,k+q}|^2}{\varepsilon(k) - \varepsilon(k + q) + \hbar\omega_q}. \tag{5.8.12}$$

Because the matrix element (5.8.12) with q replaced by $-q$ should give the same matrix element as (5.8.11), one has, noting $\omega_q = \omega_{-q}$,

$$\langle k - q, k' + q | V_{ph} | k, k' \rangle = \frac{|\alpha_{k,k-q}|^2 2\hbar\omega_q}{[\varepsilon(k) - \varepsilon(k - q)]^2 - (\hbar\omega_q)^2}. \tag{5.8.13}$$

This implies that if electrons near the Fermi surface change their states according to the energy conservation law, or if

$$|\varepsilon(k) - \varepsilon(k - q)| \ll \hbar\omega_q \tag{5.8.14}$$

is fulfilled, then

$$\langle k - q, k' + q | V_{ph} | k, k' \rangle \approx - \frac{2|\alpha_{k,k-q}|^2}{\hbar\omega_q} \tag{5.8.15}$$

is valid. This indicates an effective attractive interaction between electrons. Especially if we consider an electron pair with k and $k' = -k$ (when spin is included, this pair should be a spin-singlet with antiparallel spins), it can be seen that through the effective interaction V_{ph}, the pair $(k - k)$ can attract another pair $[k - q, -(k - q)]$. Now let us show that via this effective attractive interaction, a stable bound state of an electron pair may be formed. For simplicity, we neglect the q-dependence of the right-hand side of (5.8.15) and take the attractive interaction as a constant $-V$. The Fermi sphere state filled up to k_F is denoted by $|F\rangle$ and the wave function of an electron pair $(k, -k)$ added to the Fermi sphere by $|k, -k, F\rangle$. Now we look for an eigenstate with an energy lower than the Fermi sphere state $|F\rangle$ by constructing a proper linear combination of $|k, -k, F\rangle$

$$\Phi = \sum_{|k| < k_F} C_k |k, -k, F\rangle . \tag{5.8.16}$$

The Hamiltonian consists of two parts:

$$\mathcal{H} = \mathcal{H}_0 + V_{ph} ,$$

where $\mathcal{H}_0$ is the kinetic energy of electrons, and V_{ph} is the electron-electron interaction via phonons. Its matrix element is given by (5.8.15). If we choose the energy of the state $|F\rangle$ as zero, it is evident that

$$\mathcal{H}_0 |k, -k, F\rangle = 2\varepsilon(k) |k, -k, F\rangle . \tag{5.8.17}$$

Application of V_{ph} to $|k, -k, F\rangle$ leads to another electron pair state $|k', -k', F\rangle$ multiplied by $-V$, which is one of the states described by the linear combination of (5.8.16). The secular equation to determine C_k turns out to be

$$[2\varepsilon(k) - \varepsilon] C_k - V \sum_{k'} C_{k'} = 0 . \tag{5.8.18}$$

Defining $\sum_{|k| > k_F} C_k = A$, we obtain from (5.8.18)

$$C_k = \frac{VA}{2\varepsilon(k) - \varepsilon} . \tag{5.8.19}$$

The above definition leads to

$$A = A \sum_{|k| > k_F} \frac{V}{2\varepsilon(k) - \varepsilon}$$

or

$$\frac{1}{V} = \sum_{|k| > k_F} \frac{1}{2\varepsilon(k) - \varepsilon} , \tag{5.8.20}$$

which determine eigenvalues. We can replace the sum in the above equation by integration over energy $\varepsilon(\boldsymbol{k}) = \zeta$, which is restricted within the average phonon energy $\hbar\omega_0$ near the Fermi energy because of the condition of (5.8.14). Thus we obtain

$$\frac{1}{V} \approx 2N(0) \int_0^{\hbar\omega} \frac{d\zeta}{2\zeta - \varepsilon} = N(0) \ln \left(\frac{2\hbar\omega_0}{-\varepsilon}\right)$$

or

$$\varepsilon = -2\hbar\omega_0 \exp\left(-\frac{1}{N(0)\,V}\right), \qquad (5.8.21)$$

where $N(0)$ is the electronic density of state at the Fermi surface. The negative ε of (5.8.16) means that this coherent state, which is the linear combination of electron pairs of (5.8.16), has lower energy than $|F\rangle$ does.

The BCS theory skillfully treats this condensation phenomenon of electron pairs (Cooper pairs) by a kind of mean field approximation, and theorically verifies the presence of the second-order phase transition to the superconducting state. According to this theory, the transition temperature T_c is given by [Ref. 5.3 Eq. (6.8.5)]

$$k_{\mathrm{B}} T_c = 1.14 \hbar\omega_0 \exp\left(-\frac{1}{N(0)\,V}\right) \qquad (5.8.22)$$

The minimum energy (it is called an energy gap) required to excite the superconductor at $T = 0$ amounts to

$$\Delta_0 = 2\hbar\omega_0 \exp\left(-\frac{1}{N(0)\,V}\right). \qquad (5.8.23)$$

By comparing it with (5.8.21), we readily see a relationship between the superconducting state and Cooper pairs. If we use the Debye temperature Θ in place of ω_0, we get $T_c \propto 1/\sqrt{M}$, or the isotope effect of (5.8.4). Moreover, we can explain the experimental formula (5.8.1) by noting that the specific heat is proportional to $\exp(-\Delta_0/k_{\mathrm{B}}T)$ at low temperatures.

So far, we have only considered the attractive interaction via phonons. However, we must also take into account the direct Coulomb repulsion between electrons. According to the Thomas-Fermi theory, we take the screened Coulomb potential

$$\langle \boldsymbol{k} - \boldsymbol{q}, \boldsymbol{k}' + \boldsymbol{q} | V_{\mathrm{sc}} | \boldsymbol{k}, \boldsymbol{k}' \rangle = \frac{4\pi e^2}{q^2 + k_s^2}, \qquad (5.8.24)$$

which has the negative sign to the matrix element of V_{ph} and hence cancel each other. Therefore, as a first approximation, we can use the sum of V_{ph} and V_{sc} averaged over the Fermi surface:

$$- V = \langle V_{\mathrm{ph}} \rangle + \langle V_{\mathrm{sc}} \rangle \, .$$

However, more accurate treatment gives, instead of this,

$$- V = \langle V_{\mathrm{ph}} \rangle + \frac{\langle V_{\mathrm{sc}} \rangle}{1 + N(0) \langle V_{\mathrm{sc}} \rangle \ln (\varepsilon_{\mathrm{F}}/\hbar \omega_0)} \, . \tag{5.8.25}$$

Introducing

$$\lambda = - N(0) \langle V_{\mathrm{ph}} \rangle \tag{5.8.26}$$

$$\mu^* = \frac{N(0) \langle V_{\mathrm{sc}} \rangle}{1 + N(0) \langle V_{\mathrm{sc}} \rangle \ln (\varepsilon_{\mathrm{F}}/\hbar \omega_0)} \, , \tag{5.8.27}$$

we deduce the transition temperature in the form

$$k_{\mathrm{B}} T_{\mathrm{c}} = 1.14 \hbar \omega_0 \exp \left[- \left(\frac{1}{\lambda - \mu^*} \right) \right] \, . \tag{5.8.28}$$

A necessary condition for superconductivity is therefore

$$\lambda - \mu^* > 0 \, . \tag{5.8.29}$$

As mentioned before, λ could be evaluated in principle once the pseudopotential is given. A detailed expression for λ is written as

$$\lambda \equiv 2 \int_0^{\omega_0} \frac{\alpha^2(\omega)}{\omega} F(\omega) \, d\omega \tag{5.8.30}$$

with

$$\alpha^2(\omega) F(\omega) = \frac{\int_{S_{\mathrm{F}}} d^2k \int_{S_{\mathrm{F}}} \frac{d^2k}{(2\pi)^3 v_{\mathrm{F}}'} |\alpha_{kk'}|^2 \delta(\omega - \omega_{kk'})}{\int_{S_{\mathrm{F}}} d^2k} \, , \tag{5.8.31}$$

where $F(\omega)$ is the frequency distribution function of phonon modes, the integrations in (5.8.31) are evaluated on the Fermi surface, and v_{F}' is the velocity of an electron with the wave vector k' on the Fermi surface. It is evident from these forms that λ is equal to the average value of (5.8.15). λ is also related to the renormalization of an electron mass due to electron-phonon interaction. For instance, the effective mass m^* of an electron is given by

$$m^* = m(1 + \lambda) \, . \tag{5.8.32}$$

It is known that metals which have strong electron-phonon coupling (so-called strong-coupling superconductors) markedly deviate from the BCS theory. We must be careful in taking the above averages near the Fermi surface in such a case. To be more precise, we have to solve integral equations to determine the energy gap and transition temperature. As a result, for a strong-coupling super-conductor, T_c is expressed as [5.14]

$$T_c = \frac{\Theta}{1.45} \exp\left[-\frac{1.04(1 + \lambda)}{\lambda - \mu^*(1 + 0.62\,\lambda)}\right], \tag{5.8.33}$$

where Θ is the Debye temperature. The physical meaning of this equation is as follows. The electron-electron interaction is renormalized by $(1 + \lambda)^{-1}$ due to strong-coupling effects, and λ itself is reduced by the factor $(1 - 0.62\mu^*)$ as a result of Coulomb interactions. Thus a necessary condition for superconducti-vity determined by (5.8.33) now becomes

$$\frac{\lambda}{1 + 0.62\lambda} > \mu^* . \tag{5.8.34}$$

At present, there is no systematic study to examine the condition of (5.8.34) by evaluating λ and μ^* for each atomic element. However, the consistency of λ, which is empirically evaluated using T_c and Θ, with that determined by other experimental data has been successfully checked.

First let us evaluate μ^* empirically. Since μ^* depends on ionic mass through $\omega_0 \propto M^{-1/2}$ in the right-hand side of (5.8.27), the power index α in the isotope effect $T_c \propto M^{-\alpha}$ might deviate from 1/2. In fact some experiments show this deviation. Therefore we can estimate μ^* from this deviation in α as follows. Not-ing (5.8.27), we differentiate the logarithm of T_c with respect to M:

$$\frac{dT_c}{T_c} = -\alpha\frac{dM}{M}$$

from which we have

$$\alpha = \frac{1}{2}\left\{1 - \frac{(1 + \lambda)(1 + 0.62\lambda)\,\mu^{*2}}{[\lambda - \mu^*(1 + 0.62\lambda)]^2}\right\}$$

$$= \frac{1}{2}\left[1 - \left(\mu^* \ln\frac{\Theta}{1.45T_c}\right)^2\frac{1 + 0.62\lambda}{1 + \lambda}\right].$$

If we approximate $(1 + 0.62\lambda)/(1 + \lambda) \simeq 1$, then it turns out that μ^* is related with the observed values of α, Θ and T_c by a formula

$$\mu^* = \frac{(1 - 2\alpha)^{1/2}}{\ln(\Theta/1.45T_c)} . \tag{5.8.35}$$

Table 5.11 shows some examples of μ^* for selected materials calculated from (5.8.35). From this table we may regard that μ^* is approximately the same for all

Table 5.11. Values of several parameters for selected transition metals [5.14]

	α	$T_c[K]$	$\Theta[K]$	μ^*
Zr	0.00 ± 0.05	0.55	290	0.17
Mo	0.37 ± 0.04	0.92	460	0.09
Re	0.38	1.69	415	0.10
Ru	0.00 ± 0.15	0.49	550	0.15
Os	0.21	0.65	500	0.12
Zn	0.30 ± 0.01	0.85	309	0.12

transition metals with an average value $\mu^* = 0.13$. Then we can estimate λ from observed values of T_c and Θ. According to the Fermi gas theory of metals, the electronic specific heat of a normal metal takes on a form

$$C_e = \gamma T, \tag{5.8.36}$$

where γ is proportional to the electronic density of states $N(0)$ at the Fermi surface and the enhancement factor $(1 + \lambda)$ due to the electron-phonon interaction:

$$\gamma = \tfrac{2}{3} \pi^2 k_B^2 (1 + \lambda) N(0). \tag{5.8.37}$$

Therefore if we reduce γ from experiments, we can evaluate $N(0)$ using λ which is determined above. Interesting enough, if we know $N(0)$ for a series of transition metal alloys having a common lattice structure, we can infer the ε-dependence of $N(\varepsilon)$ by adopting the common band model. Figure 5.29 plots the density of

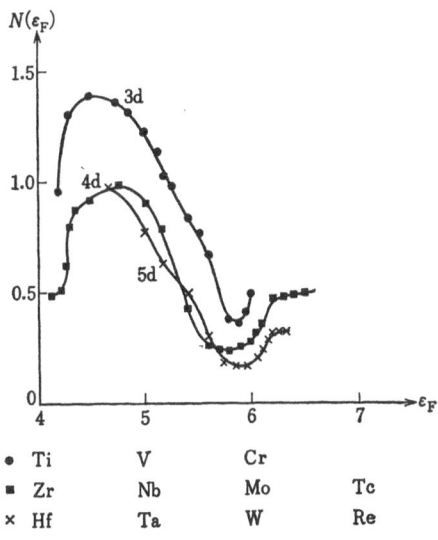

•	Ti	V	Cr	
▪	Zr	Nb	Mo	Tc
×	Hf	Ta	W	Re

Fig. 5.29. Density of states for bcc transition metals [5.14]

states $N(\varepsilon)$ for a bcc structure which is determined from $N(0)$ of $3d$, $4d$, and $5d$ transition metal alloys having the same bcc structure. It is commonly observed that at $n \simeq 4.5$, the curves have a maximum, a deep minimum at $n \simeq 5.8$ and a small wiggle at $n \simeq 6.2$, where n is the valence electron number per atom. These results fit remarkably with the results of band calculations.

6. Nonmetals

Insulators are materials which show a strong contrast in their electric properties to those of metals, which were discussed in Chap. 5. Normally, insulators do not conduct electric currents but only induce electric polarization through the action of applied electric fields, and hence are sometimes called dielectrics. Generally speaking from the band theoretical point of view, insulators are characterized by filled valence bands and empty conduction bands separated from each other by a finite energy gap.

Semiconductors may be regarded as kinds of insulators, but they have relatively small energy gaps, allowing the existence of temperature-dependent conduction electrons which are thermally excited from valence bands or from donor levels inside a forbidden band produced by doping of impurity atoms. Both insulators and semiconductors are sharply distinct from metals in that the former do not have a Fermi surface, which plays an important role in metals. On the other hand, there are other kinds of substances in which conduction bands overlap slightly with valence bands, and a small number of conduction electrons as well as an equal number of positive holes coexist irrespective of temperatures. They are called semimetals and definitely have a Fermi surface. However, the electronic structure of semimetals is rather similar to that of semiconductors, and to understand why and how semimetals can be brought into existence, it is easier to consider them from the insulator side. This is the main reason why we include semimetals in this chapter rather than treating them in connection with metals. Since the theory of the general response of insulators to applied external disturbances has been given in detail in another volume of this series [6.1], the object of this chapter is to review what materials become insulators and why.

Usually insulators are divided into molecular crystals, ionic crystals, covalent crystals, hydrogen-bonded crystals and so on. However, such a classification cannot be made unambiguously, and indeed there are many cases which should be put inbetween. Therefore, in this chapter we shall first follow this classification and give brief descriptions of typical examples in each class, while later sections will be devoted to a unified approach to nonmetals from the electron band theoretical point of view.

6.1 Molecular Crystals

When a crystal is composed of molecules which are saturated in the sense of chemical bond theory, we call it a molecular crystal. A molecular crystal is an insulator because all the electrons are localized in molecules taking part in intramolecular bonding. The cohesive forces forming a crystal come from weak secondary inter-molecular attractions called dispersive forces. Therefore, at room temperatures, molecular crystals are usually in vapour or liquid phases, and even if they were solid, their melting points would be generally very low.

All the elements in Table 5.1 which are classified as nonmetals in the previous section belong to molecular crystals. Among them, inactive elements (inert gases) Ne, Ar, Kr, Xe have a stable closed-shell electronic structure, and naturally they form stable monoatomic molecular crystals. The reason why the other non-metallic elements also become molecular crystals is essentially the same as in the case of hydrogen. For instance, halogens form stable diatomic molecules F_2, Cl_2, Br_2 and I_2, then these diatomic molecules act as building blocks to form molecular crystals. Their crystalline structures belong to a monoclinic system, as shown in Fig. 6.1. As noted in the previous section, the elements which have more than three valence electrons are likely to form covalent bonds, and thence tend to form crystal structures such that the covalent bond condition is satisfied. To be more precise, in order to utilize all unoccupied orbitals, an atom having n outer shell electrons tends to be surrounded by $8-n$ neighboring atoms. Thus for halogens with $n = 7$, each atom combines with $8 - n = 1$ other atoms to form a diatomic molecule. The atoms of the VIth group (S, Se, Te) have $n = 6$ electrons in the outer shell, and form chain-like or ring-like molecules so that they have $8 - n = 2$ neighboring atoms. The molecules thus produced form molecular crystals via weak secondary attractive forces. For example, Se and Te have a crystal structure, as shown in Fig.6.2, and become semiconductors. Sulphur with $Z = 16$ takes many kinds of structures, but a basic form is the S_8 molecule which is composed of eight atoms on an eight-membered ring. The atoms of the Vth group (P, As, Sb, Bi), although they are classified as semiconductors or semimetals, may be treated quite similarly. Since the number of valence electrons is $n = 5$, it is expected that they should have a structure with $8 - n = 3$ neighboring atoms. This is indeed the case, and they all crystallize in As-type

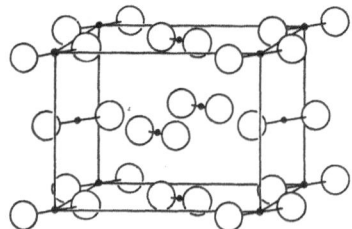

Fig. 6.1. Crystal structure of I_2

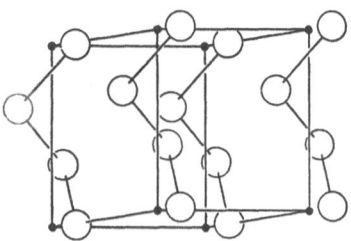

Fig. 6.2. Crystal structure of Se

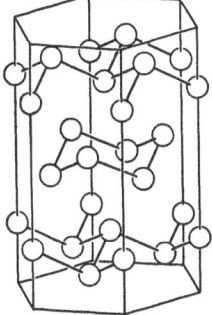

Fig. 6.3. Crystal structure of As

structures depicted in Fig.6.3. We will come back to these substances in the section on semimetals. The IVth group elements which have $n = 4$ valence electrons, form a diamond structure with $8 - n = 4$ neighboring atoms which will be discussed in Sect.6.3.

The exceptions which deviate from the above simple rule are N with $Z = 7$ and O with $Z = 8$. These atoms form stable diatomic molecules via triple bond $N \equiv N$ or double bond $O = O$, and then construct molecular crystals of the diatomic molecules. For such lighter elements, the closed shell $(1s)^2$ is relatively small and hence the intramolecule multiple bonds of valence electrons are rather stabilized. On the contrary, for heavier elements such as P and S, it is energetically more favourable to form covalent bonds with many neighboring atoms rather than forming diatomic molecules via multiple bonds.

The simplest model for the molecular crystal composed of inert element atoms will be a crystal lattice condensed through the Lennard-Jones type central potential treated in Chap. 2

$$U(R) = 4\varepsilon \left[\left(\frac{\sigma}{R} \right)^{12} - \left(\frac{\sigma}{R} \right)^{6} \right]. \tag{6.1.1}$$

Since atoms have a spherical symmetric distribution of electron clouds, we may suppose that these atoms are a sphere with a radius $\sigma/2$ and attract each other with a weak van der Waals force at sufficiently large atomic distances and exert strong repulsion due to the overlap of electron clouds at short distances. Thus the interatomic interaction may be approximated by a potential of the form (6.1.1). This potential contains two parameters σ and ε, which can be estimated by various methods. For instance, the second virial coefficient measured in the gaseous phase can be related to the parameters in the interatomic potential with the help of statistical mechanics of an imperfect gas. The values of σ and ε determined in this way are tabulated in Table 6.1. If one neglects the effect of zero-point vibrations (which seems small for these molecular crystals as seen in Table 3.1) the cohesive energy per molecule can be calculated using (2.3.6) as

$$U_0 = 2\varepsilon \left[\left(\frac{\sigma}{r_{nn}} \right)^{12} C_{12} - \left(\frac{\sigma}{r_{nn}} \right)^{6} C_{6} \right], \tag{6.1.2}$$

Table 6.1. Potential parameters and crystal energy for inert gases

	ε [kcal·mol^{-1}]	σ [Å]	r_{nn} [Å]	$-U_0$ [kcal·mol^{-1}]	$-U_0/\varepsilon$	r_{nn}/σ	$(-U_0)_{cal}$
Ne	0.0719	2.74	3.13	0.450	5.84	1.14	0.431
Ar	0.240	3.40	3.76	1.850	7.71	1.11	1.859
Kr	0.324	3.65	4.01	2.59	8.12	1.09	2.63
Xe	0.461	3.98	4.35	3.83	8.30	1.09	3.83

from which one can guess the most stable crystal structure theoretically. The values of the structure sums C_{12} and C_6 are given in Table 6.2 for main cubic structures. From this table it is concluded that a fcc lattice is most energetically stable. The expression (6.1.2) for a fcc lattice is minimized with respect to r_{nn} to give

$$U_0 = -8.60\varepsilon, \quad \frac{r_{nn}}{\sigma} = 1.09 . \tag{6.1.3}$$

Table 6.2. Structure sums for several crystal structures

	C_{12}	C_6
fcc	12.132	14.454
bcc	9.114	12.253
sc	6.202	8.402
diamond	4.039	5.113

It is known experimentally that all inert gases condense into a fcc crystal at low temperatures. The observed values of U_0 and r_{nn} are also shown in the Table 6.1. The fifth and sixth columns in this table are presented to compare with the theoretical values expected from (6.1.3), revealing fairly good agreement. The fact that the agreement is improved for heavier elements suggests that the small discrepancy is due to the effect of zero-point vibration.

Now let us consider the effect of molecular vibrations in a crude approximation. To look for the ground state of the Hamiltonian given by (2.3.5)

$$\mathscr{H} = \sum_{i=1}^{N} \frac{P_i^2}{2M} + \frac{1}{2} \sum_{i \neq j} U(|x_i - x_j|)$$

following the spirit of the Hartree approximation, we assume a wave function of the form

$$\left. \begin{aligned} \Psi &= \prod_{i=1}^{N} \varphi(R_i + r_i) \\ \varphi(R + r) &= \frac{1}{\sqrt{4\pi}} \frac{u(r)}{r} \end{aligned} \right\} \tag{6.1.4}$$

and adopt the variational principle. Here R_l represent the molecular equilibrium positions in the crystal and are chosen to form a fcc lattice. When the variational calculation is performed, it turns out that the problem reduces to finding the solution $u(r)$ of a differential equation

$$-\frac{\hbar^2}{2M}\frac{d^2u(r)}{dr^2} + w(r)\,u(r) = \varepsilon_0 u(r) \tag{6.1.5}$$

with the boundary conditions

$$u(\infty) = 0, \quad \lim_{r\to 0}\frac{u(r)}{r} = \text{(finite)}, \tag{6.1.6}$$

where $w(r)$ is an averaged potential given by

$$w(r) = \sum_{j}\int_0^\infty v(R_j, \rho, r)|u(\rho)|^2 d\rho$$

with $v(R, \rho, r)$, which is the average of the intermolecular potential $U(|R + \rho - r|)$ between two atoms separated by an equilibrium distance R with respect to the directions of ρ and r as illustrated in Fig. 6.4a

$$v(R, \rho, r) = \langle\!\langle U(|R + \rho - r|)\rangle\!\rangle. \tag{6.1.7}$$

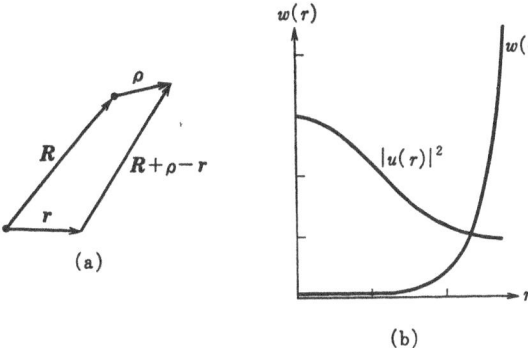

(a)

(b)

Fig. 6.4. (a) Illustration for averaging of interatomic potential. (b) Averaged potential and function $|u(r)|^2$

Figure 6.4b shows an averaged potential determined in this way.

If one could solve the eigenvalue problem (6.1.5) and obtain the lowest eigenvalue ε_0, then the ground state energy per molecule is calculated as

$$U_0 = \varepsilon_0 - \tfrac{1}{2}\int |u(r)|^2 w(r)\,dr. \tag{6.1.8}$$

The results of numerical calculations for inert gases are tabulated in the last column of Table 6.1. It is observed that agreement between the theory and experiments is very much improved.

The Lennard-Jones type central force model can be applied only for the monoatomic molecular crystals such as inert gases, and for other cases the shape of the molecule is far from those spherical in general, so situations become more complicated. For example, consider the next simple case of diatomic molecules such as halogens which crystallize into less symmetric monoclinic structures where, at very low temperatures, molecular axes take definite orientations. For the case of N_2, it has a hexagonal structure at high temperatures and changes to a cubic lattice in which a regular alignment of molecular axes appears at low temperatures, as depicted in Fig. 6.5. This molecular pattern is a typical form often occuring in diatomic molecular crystals at very low temperatures. With increasing temperature, starting with the low temperature phase of a N_2 crystal, the alignment of molecular axes gradually becomes random due to the thermal motions, and eventually the low temperature phase becomes unstable and makes a transition to a hexagonal lattice at a certain temperature T_c. In the high temperature hexagonal structure, all the molecules perform nearly free rotations and hence may be regarded as if they were apparent spherical molecules which are, on the average, closely packed in a hexagonal close-packing (hcp) structure. Such a transition in which molecular axes undergo a change between ordered and random configurations is called rotational phase transition which is a characteristic property seen in molecular crystals in general. In Table 6.3, melting temperatures, rotational phase transition temperatures and changes in crystal structures are summarized for simple diatomic molecular crystals.

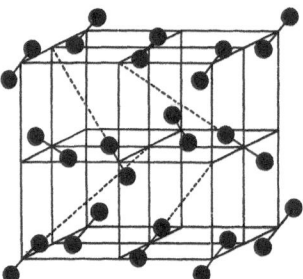

Fig. 6.5. Low temperature phase of N_2 crystal

Among the same class of diatomic molecular crystals, when two atoms are of different species, the molecule can have a permanent electric dipole moment which gives rise to an excess contribution to the cohesive energy due to dipole-dipole interactions, and sometimes causes a large anomaly in the dielectric property at the rotation transition point. Some crystals exist, for instance, such as hydrogen halogenides, which show ferroelectricity in which the permanent dipoles are arranged in one direction. Figure 6.6. shows the crystal structure of DCl at low temperatures where a spontaneous polarization along the b-axis is

Table 6.3. Phase transitions in diatomic molecular crystals. (rho.) rhombohedral; (mon.) monoclinic; (tri) trigonal

	T_m [K]	T_c [K]	Structure change
N_2	63.13	35.61	hcp-fcc
CO	68.09	61.57	hcp-fcc
O_2	54.39	43.76	fcc-rho.
		23.66	rho. -mon.
F_2	53.54	45.55	fcc-mon.
ortho H_2	13.96	2.9	hcp-fcc
para D_2	23.5	4.0	hcp-fcc
HCl	158.94	98.38	
HBr	186.28	116.9	fcc-tri.
HI	222.36	125.68	

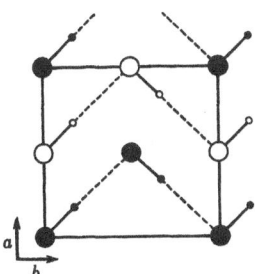

Fig. 6.6. Low temperature structure of DCl. Large circle represents Cl atom and small circle D atom. Filled circle is at the level $z = 0$ and open circle at $z = 1/2$

seen. It is also very probable that in this crystal, hydrogen bonds, which will be discussed in a later section, play some role.

Since one can imagine various molecular crystals composed of many complex molecules, one should say the number of kinds of molecular crystals is infinite. Almost all organic crystals are molecular crystals. They exhibit a lot of varieties in their structures, ranging from a high symmetric lattice composed of an apparently spherical molecule (such as methane) to a very complex crystal in which organic high polymers are cleverly folded in space. Roughly speaking, the principle of determining the crystal structures of organic substances may be stated as:

(I) avoiding the increase in energy due to the short-range intermolecular repulsions and

(II) utilizing all weak attractive forces, van der Waals interaction, dipole-dipole interaction, multipole interaction and so on as much as possible.

Thus, molecules are packed as closely as possible on the one hand, and at the same time molecular axes are arranged in suitable ways so that the above conditions are satisfied. The more complex the molecular structure is, the more possible phases can exist where the differences in free energies are quite slight.

In such situations it sometimes happens that successive phase transitions take places at different temperatures.

Finally, in connection with the molecular crystal, we have to mention a special form of existence, the liquid crystal. As stated above, a molecular crystal composed of nonspherical molecules can generally take an ordered arrangement of molecular axes at low temperatures so that the free energy of the system becomes minimum. With increasing temperatures, the molecular arrangement becomes more and more random to assume a state with large entropy and finally melts into a liquid where a regular periodic arrangement of the centers of gravity of molecule is completely destroyed. However, for a molecular crystal composed of molecules with a very prolonged shape, because a great deal of energy is necessary to change molecular orientation inside the crystal, the random distribution of the molecular axes is not easily realized even at elevated temperatures, and it may happen that the crystal melts, losing the periodic lattice of the centers of gravity of molecules, but still keeping the order of molecular orientation. We call such cases liquid-crystal states because the substance becomes liquid, keeping an internal structure near to a crystal. In Table 6.4. we indicate some typical examples of liquid crystals together with their existing temperature ranges. As indicated in the table, liquid crystals are further classified in three groups, nematic, smectic and cholesteric phases according to the differences in manner of molecular orientations. The characteristics of these three phases are schematically illustrated in Fig.6.7.

Table 6.4. Examples of liquid crystals

	Compound	Range [°C]
nematic-phase	anisylidene-*p*-aminophenylacetate (APAPA)	83 ~ 100
	p-ethoxybenzylidene-*p'*-aminobenzonitrile (PEBAB)	114 ~ 124
	p-azoxyanisole (PAA)	116 ~ 133
smectic-phase	ethyl-*p*-azoxybenzoate	114 ~ 120
	cholesteryl nonanoate	78 ~ 92
cholesteric-phase	cholesteryl chloride	94 ~ 96
	cholesteryl benzoate	143 ~ 177
	cholesteryl acetate	~94.5
	cholesteryl octanonate	69.5 ~96.5

6.2 Ionic Crystals

There are many substances among the compounds belonging to insulators which are classified by the name ionic crystal. An ionic crystal is characterized in that cohesion is mainly due to the Coulomb attraction between positive and negative ions. Positive ions come from electropositive atoms (mainly metallic atoms) which easily lose their electrons to become positively charged, while negative

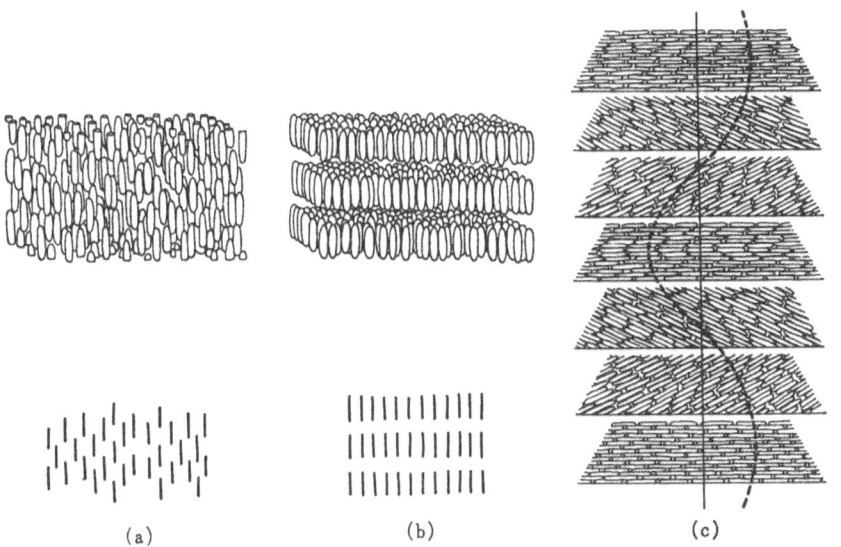

Fig. 6.7a-c. Three types of liquid crystals; **(a)** nematic **(b)** smectic **(c)** cholesteric

ions come from electronegative atoms which have a tendency to become negatively charged. In a typical ionic crystal, electrons are transferred from electopositive metallic atoms to electronegative atoms, yielding $\pm$ ions with stable closed-shell electronic structures, the same as those for inert gas atoms. For instance, alkali halides belong to their group and, e.g., LiF may be expressed as Li^+F^- with

$$Li^+ : (1s)^2, \quad F^- : (1s)^2 \, (2s)^2 \, (2p)^6 \, .$$

The insulating property of an ionic crystal is believed to originate from the fact that all the electrons are localized on positive or negative ions. Indeed, certain experiments by means of x-ray scattering exist which confirm that the measured electron density map corresponds almost completely to the configuration of $\pm$ ions for alkali halides. Let a be the nearest distance between ions and $\pm Ze$ be the ionic charge, then the cohesive energy of an ionic crystal per ion pair can be estimated roughly as $-(Ze)^2/a$ which, on taking $a = 2 \times 10^{-8}$ cm, amounts to 10^{-12} erg $= 1.44$ kcal·mol^{-1}. This value is much larger than the cohesive energy of a molecular crystal, explaining the high melting temperatures of ionic crystals.

Ionic crystals of MX-type generally consist of a $+Z$ ion M^{+z} of the Zth group element and a $-Z$ ion X^{-z} of the $(Z - 8)$th group element. Most of them crystallize in a NaCl structure, as shown in Fig. 6.8a and some others in CsCl (Fig. 6.8b), zincblende (Fig. 6.8c) and wurtzite (Fig. 6.8d) structures. As seen in these figures, a characteristic point of these structures is that each positive ion is surrounded by negative ions quite symmetrically, and visa versa for nega-

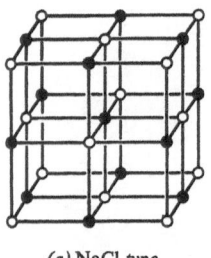

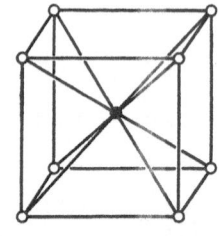

(a) NaCl *type*

(b) CsCl *type*

Fig. 6.8a-d. Crystal structures of MX-type ionic crystals;

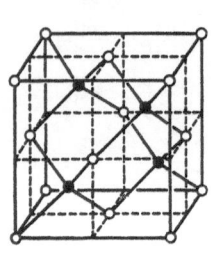

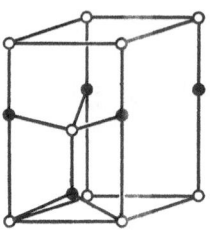

(c) Zincblende *type*

(d) Wurtzite *type*

tive ions. The number of ions surrounding an ion is called the coordination number of that ion. The four structures shown in Fig. 6.8a–6.8d have the coordination numbers 6, 8, 4, 4, respectively, for both ions.

For the ionic crystals of MX_2 or M_2X-type, high symmetric structures, as shown in Fig. 6.9a,b are observed; (a) is known as the CaF_2-type and (b) as the Cu_2O-type. In contrast to Fig. 6.8, there are two kinds of lattice points which are not equivalent, and hence two kinds of coordination numbers. In this case (8.4) and (4.2), respectively, for CaF_2-type and Cu_2O-type. There is another familiar structure of less symmetry, and that is the TiO_2-type lattice as shown in Fig. 6.9c. To describe this structure, one more parameter other than the lattice constant is needed, but we may assign the coordination number as approximately (6.3).

The calculation of the cohesive energy of ionic crystals with these structures and discussion of the stabilities of different structures have been carried out for

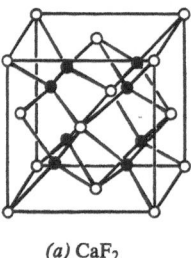

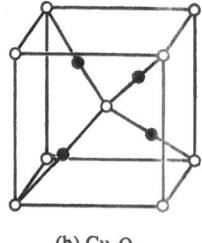

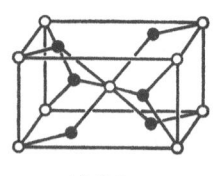

Fig. 6.9a-c. Crystal structures of MX_2-type ionic crystals

(a) CaF_2

(b) Cu_2O

(c) TiO_2

a long time, and indeed they occupied the main part of the classical theory of ionic crystals. It would be unnecessary to describe them here. However, there are some concepts useful even at present, and the classical results might be of interest in comparison with modern theory based on quantum mechnics. We shall give a very brief review of classical theory. The basic idea is that an ionic crystal is assumed to be composed of spherical ions with definite radii and electric charges. The main sources of cohesive force are interionic Coulomb interactions and strong repulsions which act at short distances. In terms of these forces, we can evaluate the binding energy of a crystal. Let r_{ij} be the distance between the ith and jth ions, and let their charges be e_i and e_j respectively, then electrostatic energy can be given by

$$E_c = \frac{1}{2} \sum_{i \neq j} \sum \frac{e_i e_j}{r_{ij}}.$$

If we denote the shortest interionic distance by r, ionic charges by $\pm Ze$ (when there are several kinds of ions, choose the smallest Z) and the total number of ion pairs by N, then the above sum can be eventually put in the form

$$E_c = - N\alpha \frac{(Ze)^2}{r}. \tag{6.2.1}$$

Here α is the Madelung constant, a number dependent only on the crystal structure. The values of α for the crystal structures shown in Figs. 6.8,9 are tabulated in Table 6.5.

Table 6.5. Madelung constants for main cubic structures

Crystal structure	α
CsCl	1,76267
NaCl	1,74756
Zincblende	1.63805
CaF$_2$	5.03878
Cu$_2$O	4.44248

We assume for the repulsive force a potential which depends on the inverse nth power of interionic distance, employed in the previous section. Then the repulsive energy takes on

$$E_r = \frac{1}{2} \sum_{i \neq j} \sum \frac{b_{ij}}{(r_{ij})^n}.$$

Although there are still many issues to be discussed for this energy, they are basic quantum-mechanical effects and it is not meaningful to give details in a classical theory. Therefore, expressing all ionic distances in units of the shortest distance r, we put E_r in the form

$$E_r = N \frac{B}{r^n}. \qquad (6.2.2)$$

B is a parameter depending on b_{ij}, but its detailed form is unnecessary at the moment because B can be eliminated by using the condition that the sum of (6.2.1) and (6.2.2) should be minimum at the shortest ionic distance r_0 in equilibrium. If we neglect all forces other than Coulomb interactions and short-range repulsions, the total crystal energy becomes

$$E = - N \left[\alpha \frac{(Ze)^2}{r} - \frac{B}{r^n} \right]. \qquad (6.2.3)$$

Expressing B in terms of α, r_0 and n form the equation $(\partial E/\partial r)_{r=r_0} = 0$ and inserting it in (6.2.3), we finally obtain

$$E = - N\alpha \frac{(Ze)^2}{r_0} \left(1 - \frac{1}{n} \right). \qquad (6.2.4)$$

It is seen from this result that the repulsive force contributes only about $1/n$ of the Coulomb energy to the crystal energy. In the previous section, we have chosen $n = 12$ for the index of the repulsive potential between inert gas atoms. Because n is thought to be a quantity representing some rigidity of the ion sphere, it would be better to determine n in connection with the compressibility k of crystals, empirically. The compressibility is derived from the crystal energy as follows:

$$\frac{1}{k} = V \frac{d^2 E}{dV^2}.$$

For the cubic crystal of MX-type, it holds that $V = 2Nr_0^3$, and hence from (6.2.4), it immediately follows

$$\frac{1}{k} = (n - 1) \alpha \frac{(Ze)^2}{18r_0^4}. \qquad (6.2.5)$$

If one estimates the value of n by using observed values of k (extrapolated to absolute zero), it turns out that for alkali-halides, n is ranged in between 6–10. However, as mentioned already, since the contribution from the repulsive force to the crystal energy is only 10% of the total energy, it is enough to choose n as an integer or half an integer number within an error of less than a few

Table 6.6. Calculated lattice energy of alkali-halides (in units of $kcal \cdot mol^{-1}$). n is taken to be the average of n's for constituent elements, for instance, in the case of NaBr: $n = (8 + 9)/2 = 8.5$. The values inside ( ) are experimental values

Halogen alkali	F ($n = 6$)	Cl ($n = 8$)	Br ($n = 9$)	I ($n = 11$)
Li ($n = 6$)	240.1	193.3 (198.1)	183.1 (189.3)	170.7 (181.1)
Na ($n = 8$)	215.0	180.4 (182.8)	171.7 (173.3)	160.8 (166.4)
K ($n = 10$)	190.4	164.4 (164.4)	157.8 (156.2)	149.0 (151.5)
Rb ($n = 11$)	181.8	158.9 (160.5)	152.5 (153.3)	144.2 (149.0)
Cs ($n = 13$)	172.8	148.9 (155.1)	143.5 (148.6)	136.1 (145.3)

percent. From this viewpoint, L.Pauling has determined a set of n for the purpose of theoretical calculation, and using the values of n thus determined, *Sherman* [6.2] has evaluated E for many ionic crystals. For later reference, part of his results are shown in Table 6.6.

Besides Coulomb interaction and short-range repulsion, the van der Waals attraction (dispersive force), the zero-point energy of ionic lattice vibrations and others may contribute to the crystal energy. Indeed, elaborate calculation including these effects has also been carried out. However, in ionic crystals, the interionic distance r_0 is grossly determined by the balance between Coulomb attraction and short range repulsion. Thus the contribution from the dispersive forces to the crystal energy is small and amounts to, at most, a few percent of the total energy. Much smaller is the zero-point vibration energy. As seen from Table 6.6, the observed magnitude of the lattice energy is nearly accounted for by the sum of the Coulomb energy and the repulsion energy. For typical ionic crystals such as alkali-halides, detailed theoretical calculations (including all the effects), have been repeated with apparent success, pursuing fine agreement between theory and experiments. However, for other ionic crystals (for instance, oxides of alkaline-earth metals) similar calculations could not afford a good agreement with experiments. One reason for the lack of success is believed to be that, as the constituent ions come nearer to the central part of the periodic table from both sides, electropositivity of cations and electronegativity of anions are gradually weakened, and the crystals partly lose the characteristics of ionic crystals and partly bear covalent characteristics which will be discussed in the next section.

In order to answer the question as to whether a crystal structure is stable for a given material, it is necessary to compare the lattice energies of different structures. Furthermore the detailed form for the coefficient b_{ij} in the repulsive potential becomes important. In this respect, the concept of ionic radii plays an important role in the classical theory. The ground on which this concept stands is the empirical observation that lattice spacings of ionic crystals may be approximately determined by assigning a definite radius to each ion. With slightly different reasoning, several different sets of ionic radii have been tabulated and published by different researchers for main ions. We shall not present such tables here. However, it should be pointed out that the concept of ionic radii is very useful for discussing the manner of packing of ions inside crystals. In conjunction with ionic radii, there is another useful rule from *Pauling* [6.3]. This rule says: the ion arrangement in an ionic crystal is such that all the electric field lines starting from a cation are absorbed by neighboring anions with running distances as short as possible. This rule is indeed fulfilled by all the crystal structures so far given. Even for complex ionic crystals with many kinds of ions, close examination reveals that Pauling's rule is well satisfied without exception. This means that an ion is always surrounded by the ions of opposite charge. Now suppose an ion with a definite radius and make a number of ions with opposite charge contact with the central ion to utilize the Coulomb attraction as much as possible. Then the number of surrounding ions obviously depends on the ratio of ionic radii of the central and surrounding ions. Noting this fact, we can sometimes discuss the stability of different structures of ionic crystals in terms of ratios of ionic radii. Let us mention a few examples. The stability between a NaCl structure with coordination number 6 and a CsCl structure with coordination number 8 is crossed over at $R_+/R_- = \sqrt{2} - 1$, where $R_+(R_-)$ is the ionic radius of a positive (negative) ion. For CaF_2-type and TiO_2-type structures, the former is stable when $R_+/R_- > 0.7$ and the latter when $R_+/R_- < 0.7$. Although such a criterion cannot always discriminate all cases, it seems to explain the gross behaviour of many ionic crystals. As a further elaboration, Pauling proposed a form of the coefficient in the repulsive potential

$$b_{ij} = C_{ij}(R_i + R_j)^{n-1},$$
(6.2.6)

where R_i and R_j are ionic radii of ith and jth ions, respectively, and C_{ij} is a constant which is expressed in terms of the ionic charge numbers Z_i, Z_j, electron numbers in outer shells n_i, n_j and a universal constant B_0 as

$$C_{ij} = B_0 e^2 \left(1 + \frac{Z_i}{n_j} + \frac{Z_i}{n_j}\right).$$
(6.2.7)

This particular form for C_{ij} has been obtained from a quantum mechanical model calculation for a simple ion pair.

Now we are in a position to describe the modern theory of ionic crystals. The quantum mechanical calculations of lattice energy of ionic crystals have been performed by *Landshoff* and *Löwdin* [6.4.6] as an extension of the Heitler-London method. Howland [6.5] also carried out the calculations of band structures and lattice energy of ionic crystals on the basis of the LCAO method. Löwdin evaluated an expectation value of the total Hamiltonian for a whole crystal

$$E = \frac{(\Psi, \mathscr{H} \Psi)}{(\Psi \Psi)} .$$

(6.2.8)

Ψ is chosen to be a Slater determinant constructed from all the atomic orbital functions of electrons localized at ionic sites, and the total energy is calculated by taking into account correctly the nonorthogonality between different atomic orbitals. The cohesive energy E_{coh} is derived from the calculated total crystal energy after subtracting a sum of energies of isolated ions. This is a laborious task, but the final result may be summarized in the form

$$E_{\text{coh}} = E_{\text{elstat}} + E_{\text{exch}} + E_s ,$$

(6.2.9)

where E_{elstat} is electrostatic energy divided further into two parts:

$$E_{\text{elstat}} = E_{\text{Mad}} + E_{\text{corr}} .$$

(6.2.10)

E_{Mad} is the same as the Madelung energy which appears in the classical theory and represents the electrostatic energy of ions in a crystal lattice. E_{corr} is a correction term taking care of the fact that the charge distribution is not point-charge-like but extended as electron clouds. E_{exch} is the exchange energy, E_s (called s-energy) which is due to the overlapping of atomic orbitals, and the sum of $E_{\text{exch}} + E_{\text{corr}}$ corresponds to the contribution from repulsion in the classical theory. In Table 6.7, the numerical values of each term calculated by *Löwdin* [6.6] along with the experimental values of cohesive energy are tabulated. Glancing at this table, one might be surprised by the excellent agreement between theory and experiments. It is also interesting to make a comparison between the classical theory and quantum-mechanical calculation for the corresponding terms. For instance, the classical repulsive energy corresponds to a sum of E_s, E_{exch} and E_{corr} which indicates that the repulsive force introduced phenomenologically cannot be explained simply by quantum mechanics and should be interpreted as a superposition of various effects. Indeed, if one examines the results more carefully, it turns out that, among s–energy terms, particular terms appear which

Table 6.7. Quantum-mechanical calculation of crystal lattice energy

Crystal	E_{Mad}	E_{corr}	E_{exch}	E_s	E_{coh}	E_{exp}
LiCl	− 216.2	+ 5.8	−12.7	+35.4	− 187.7	− 198.1
NaCl	− 210.7	− 7.8	−34.4	+69.7	− 183.2	− 182.8
KCl	− 188.0	− 18.3	−53.7	+93.1	− 166.9	− 164.4

cannot be expressed in the form of ion pair potentials but depend on the nuclear coordinates of more than two ions. Such particular terms have no appreciable size for alkali-halides, but are known to give significant contributions to the evaluation of the elastic constants.

Howland's calculation also gives a result for lattice energy almost identical to that obtained by *Löwdin*. In addition to the lattice energy, *Howland* [6.5] studied the structures of filled valence bands for KCl in detail. To see a general feature of band structures of ionic crystals, in Fig. 6.10, several filled bands are depicted as functions of a lattice constant schematically. The arrows on the right side of this figure show the energy levels of isolated ions, and the reason why the 3*s* and 3*p* levels of Cl⁻ are above the 3*s* and 3*p* levels of K⁺ ion is that electrons are bounded more strongly in a K⁺ ion than a Cl⁻ ion. With decreasing lattice constants, interionic distances become shorter and the Madelung energy becomes more effective, giving rise to shifts and broadening of each levels, as shown in Fig. 6.10. However, it is seen that even at the equilibrium lattice constant, these band widths are still narrow, producing a large forbidden band between filled valence bands and empty conduction bands (4*s* level of K⁺).

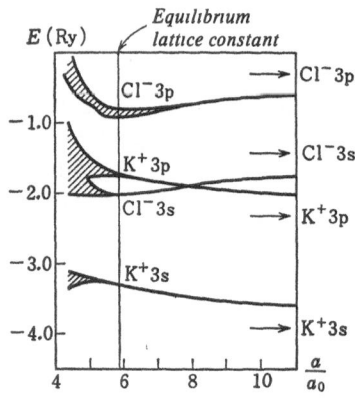

Fig. 6.10. Band structures of KCl as functions of a lattice constant [6.5]

6.3 Covalent Crystals

The elements of the IVth group are in a special situation, since the valencies of their electropositivity and electronegativity are both 4. This comes from the configuration of their outer shell electrons $(ns)(np)^3$. All elements except for Pb crystallize into the diamond structure (see Fig. 6.11), in which four electrons per atom take part in the electron pair bonds, the same as in the case of a H_2 molecule, and each atom extends over four chemical bonds in the tetrahedral directions forming a three-dimensional network. We call such a crystal a covalent crystal (homopolar-bonded crystal or valence-bond crystal). Among them, C (diamond) is an insulator, and Si and Ge are semiconductors. Sn has two kinds

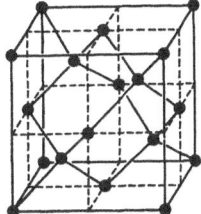

Fig. 6.11. Diamond structure

of crystal structures; grey tin with diamond structure is a semimetal and white tin with another structure is a metal.

The existence of such an electron pair bond with significant directional dependence is explained as follows. There are four valence electron orbitals, one s orbital (abbreviated as s) and three p orbitals (abbreviated as p_x, p_y and p_z). Since the energy difference between these orbitals is small, we can construct, without much energy loss, the following linear combinations of s and p orbitals:

$$
\left.
\begin{aligned}
\psi_{111} &= \tfrac{1}{2}(s + p_x + p_y + p_z) \\
\psi_{1\bar{1}\bar{1}} &= \tfrac{1}{2}(s + p_x - p_y - p_z) \\
\psi_{\bar{1}1\bar{1}} &= \tfrac{1}{2}(s - p_x + p_y - p_z) \\
\psi_{\bar{1}\bar{1}1} &= \tfrac{1}{2}(s - p_x - p_y + p_z)
\end{aligned}
\right\}.
\tag{6.3.1}
$$

These functions have a large amplitude in the directions denoted by the subscripts, respectively. They are called tetrahedral hybrid orbitals. Based on these orbitals, one can further construct bonding orbitals, and antibonding orbitals. Thereby one has to include similar hybrid orbitals of the neighboring atoms. When a pair of electrons with opposite spins is accommodated in a bonding orbital, a covalent bond is completed. It is evident that, in this way, each atom uses up all its four valence electrons to form the bonds in the diamond structure.

The same feature as above may also be described in terms of band theory. Suppose that, keeping the diamond structure, we decrease the lattice constant from a hypothetically large value to the actual value, and look at the changes in the s and p levels during this contraction process. As shown in Fig. 6.12, when the lattice constant is large enough, the s and p orbitals of the valence electrons in all atoms are degenerate at the levels of an isolated atom, respectively. With decreasing lattice constant, the s and p level each form an energy band; the s band can accommodate 2 electrons per atoms and the p–band 6 electrons per atom. For a smaller lattice constant, the s–band and p–bands cross each other and they mix into two bands of completely different character. The lower band corresponds to the band of bonding orbitals of sp^3 hybridization and the higher to that of the antibonding orbitals. Each of them contains four electrons per atom.

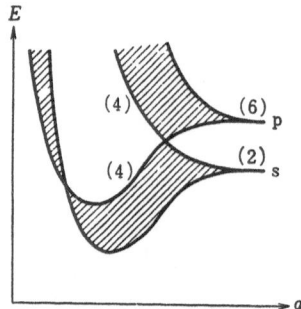

Fig. 6.12. Energy band of diamond structure as a function of a lattice constant

Figure 6.12 shows this situation schematically. The level difference of an isolated atom, the spreading of bands, and the lattice constant at which the crossing of the s and p-bands takes place depend on the value of the principal quantum number n of the (ns) (np) orbitals. In the case of diamond, an equilibrium is attained at a lattice constant for which there is a large forbidden band (of about 7 eV) between the completely filled bonding band and empty antibonding band. On the other hand, Si and Ge have a lattice constant for which the forbidden band is narrower. For grey tin the energy gap is almost zero.

So far we have given only qualitative arguments. However, since Si and Ge are very important and useful semiconductors, experimental studies have been performed on the optical properties, transport coefficients and other physical properties and at the same time detailed theoretical calculations of energy band structures have been carried out in order to gain theoretical information about the electronic states in the conduction bands and the positive hole states in the valence bands. Figure. 6.13 shows a part (i.e., in the vicinity of the energy gap) of their energy bands of Si and Ge as obtained by *Herman* [6.7]. Both energy bands look similar, but close observation reveals some characteristic differences. The top of the filled band is found to be at $k = 0$ for both Si and Ge. However, the minima of the conduction band for Si lie at a point between the origin and the Brillouin zone boundary along the [100] direction and points equivalent to

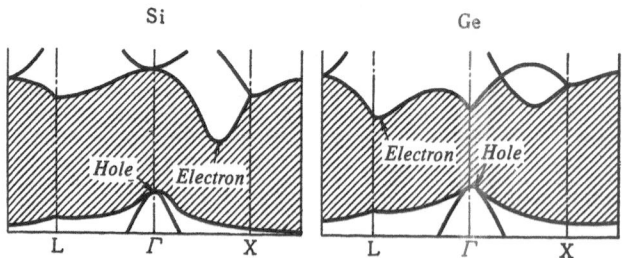

Fig. 6.13. Band structures of Si and Ge calculated by *Herman* [6.7]

it, whereas for Ge, they are at the reciprocal point $(2\pi/a)$ $(1/2, 1/2, 1/2)$ and its equivalent points. Therefore low energy electrons have different effective masses and behave differently for Si and Ge. The existence of these peculiar band structures have been experimentally verified by means of the cyclotron resonance of conduction electrons and positive holes for these substances. However, in contrast to metals, no Fermi surface exists in semiconductors, and various experimental methods to check the band structures are not available in that case. Even the cyclotron resonance just mentioned only gives an information on band structure which is limited to a narrow energy range.

Although we have no direct experimental method for determining band structure of semiconductors and insulators, an alternative method of indirect determination is proposed and successfully carried out with the cooperation of optical absorption experiments and theoretical band calculations. This method is based on the following two important points. The first is a possibility that fine structures found in the light absorption spectra may reflect some characteristics of the band structures. It is well known that absorption of light by dielectrics is proportional to the imaginary part of the complex dielectric constant $\epsilon_2(\omega)$. If we denote the filled valence band energy as a function of k by $\varepsilon_v(k)$ and conduction band energy by $\varepsilon_c(k)$, then $\epsilon_2(\omega)$ can be given as

$$\epsilon_2(\omega) = \frac{4\pi^2 e^2 h}{3m^2\omega^2} \sum_{vc} \int \frac{2}{(2\pi)^3} \delta[\varepsilon_c(k) - \varepsilon_v(k) - \hbar\omega] |M_{vc}(k)|^2 \, dk \,, \tag{6.3.2}$$

where $|M_{vc}(k)|^2$ is the transition probability. If we discard the k dependence of the transition probability, $\epsilon_2(\omega)$ is proportional to the joint density of states defined by

$$J_{vc}(\omega) = \frac{2}{(2\pi)^3} \int \delta[\varepsilon_c(k) - \varepsilon_v(k) - \hbar\omega] \, dk$$

$$= \frac{2}{(2\pi)^3} \int_{\varepsilon_c(k) - \varepsilon_v(k) = \hbar\omega} ds/|\operatorname{grad}_k[\varepsilon_c(k) - \varepsilon_v(k)]| \,. \tag{6.3.3}$$

In the second line of the right-hand side of (6.3.3), the integration should be taken over a surface in k–space defined by

$$\varepsilon_c(k) - \varepsilon_v(k) = \hbar\omega$$

and ds is the surface element. Since this integral becomes singular at the points where the denominator of the integrand vanishes, the same singularities are to appear in $J_{vc}(\omega)$ too. Let $k = k_0$ be one of such points in the k-space where the denominator of the integrand vanishes. Expanding $\omega_{vc}(k) = [\varepsilon_c(k) - \varepsilon_v(k)]/\hbar$ around $k = k_0$, and putting $k - k_0 = \Delta k$, we can write

$$\omega_{vc}(k) = \omega_c + \sum_{\alpha=1}^{3} \alpha_\alpha \Delta k_\alpha^2 \,. \tag{6.3.4}$$

Then $J_{vc}(\omega)$ has a singularity at $\omega = \omega_c$ and exhibits four types of behavior around this point according to the sign of a_α in (6.3.4) as shown in Fig. 6.14. This is called the van Hove singularity. If a singularity such as that given in Fig. 6.14 appears in the absorption spectrum, it indicates that the function $\omega_{vc}(\boldsymbol{k})$ shows a specific $\boldsymbol{k}$ dependence around the point in the $\boldsymbol{k}$-space where $|\operatorname{grad} \omega_{vc}(\boldsymbol{k})| = 0$. Thus, from the shape of the absorption spectrum, we can infer a part of the characteristics in the band structures.

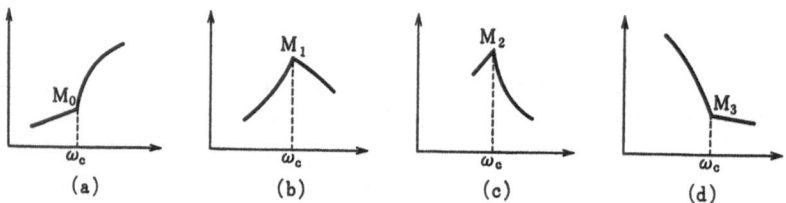

Fig. 6.14a-d. Various types of van Hove singularities: (a) when all a_α are positive; (b) when one of a_1, a_2 and a_3 is negative; (c) when two of a_1, a_2 and a_3 are negative; (d) when all a_α are negative

The second point to be mentioned is the following fact. Although the calculated band structures of Si and Ge seem very complicated, it becomes evident that they are closely related to the energy band of free electrons in the diamond lattice. Because this is similar to the feature which makes the pseudopotential theory of simple metals successful, it is hopefully suggested that an idea similar to the pseudopotential theory would be effective, even to Si and Ge, for reproducing all band structures in terms of a few components of the pseudopotential suitably chosen as empirical parameters. Thus, the following *empirical pseudopotential model* (EPM) comes out. First we analyze the structure of $\epsilon_2(\omega)$ obtained experimentally and estimate the transition energy at the critical points in the

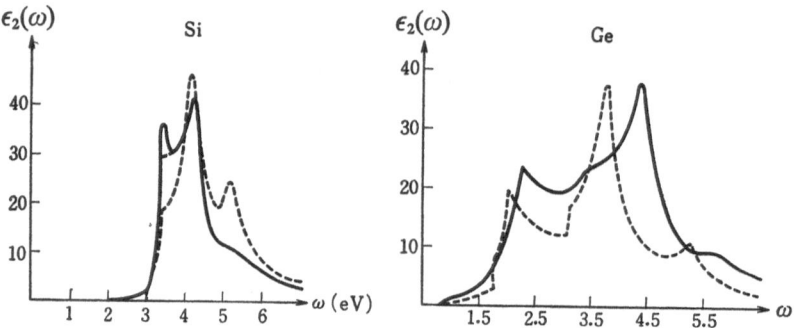

Fig. 6.15. $\epsilon_2(\omega)$ for Si and Ge. Solid lines represent the experimental curves and dotted lines theoretical curves [6.8]

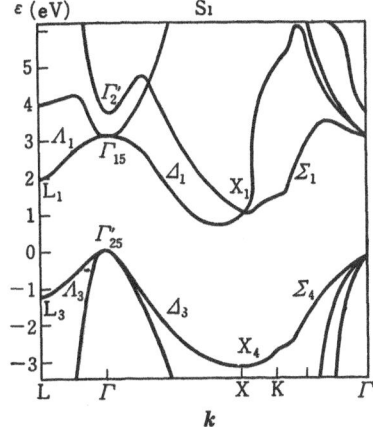

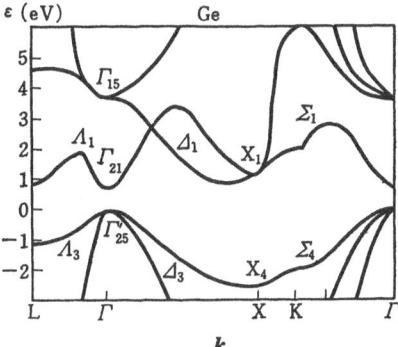

Fig. 6.16. Band structure of Si and Ge determined by means of EPM

k-space. Then we guess suitable components (form factors) of the pseudopotential. By using the pseudopotential thus assumed, the secular equation is solved determining the energy band $\varepsilon(\mathbf{k})$, which is used to calculated $\epsilon_2(\omega)$ to be compared with the absorption spectrum. If the agreement between the calculated imaginary part of $\epsilon_2(\omega)$ and the absorption spectrum is not satisfactory, the origin of the discrepancies is examined and a suitable pseudopotential is reassumed. This process is repeated until the calculations and experiments are in good accord. One example of the results obtained in this method is in Fig. 6.15 which shows comparisons of $\epsilon_2(\omega)$ for Si and Ge between the observed and calculated values. Figure. 6.16 shows the final band structures determined by EPM. The component $v(G)$ of the pseudopotential adopted in this computation has only three values for $G = (2\pi/a)$ (111), $(2\pi/a)$ (220) and $(2\pi/a)$ (113), and a secular equation of dimension 20×20 is solved. For reference, we tabulate below the values of the pseudopotential for Si, Ge and Sn determined by EPM:

	$v(111)$ [Ry]	$v(220)$ [Ry]	$v(112)$ [Ry]
Si	-0.21	$+0.04$	$+0.08$
Ge	-0.23	$+0.01$	$+0.06$
Sn	-0.20	$+0$	$+0.04$

The pseudopotential theory is, as in the case of simple metals, very powerful in providing a basis for calculating various physical properties in a unified way. However, it is not so clear how the theory is related to the concept of covalent bonds mentioned at the beginning of this section. Upon knowing the presence of the pseudopotential which gives the results in good agreement with experiments, is it possible to construct a more simple band model of a covalent crystal?

And can one interpret the characteristics of a covalent bond in the languages of the **k**–space? A model which seems to fulfill this requirement is proposed by *Heine* and *Jones*. [6.9]

Consider the Brillouin zones of a diamond structure and examine an extended Brillouin zone (called Jones zone hereafter) in which four electrons per atom can be accommodated. It is a polyhedron, as shown in Fig. 6.17b, which is of nearly spherical shape constructed from twelve (220) and its equivalent planes. Thus the energy gap between the bonding state and antibonding state takes place just on the surface of this Jones zone and if the free electron approximation is adequate, the energy gap is expected to be nearly constant along this surface. That this is indeed the case is seen in the following way. Suppose the extended Brillouin zone (Jones zone) is mapped back into the reduced Brillouin zone by choosing suitable reciprocal lattice vectors. Then let us examine the changes of energy along the (220) planes in the corresponding reduced zone scheme and compare the results with Fig. 6.16. The X-point (110) corresponds, by subtracting a reciprocal lattice vector $g = (11\bar{1})$, to the X-point (001) in the first Brillouin zone. Similarly, both Γ-points (111) and (020) correspond in the reduced zone scheme to the Γ-point (000) in the first Brillouin zone. When we look at these correspondences on the energy bands diagram given in Fig. 6.16, it is observed that, corresponding to the X-point on the (200) plane of the Jones zone, there are two points X_1 and X_4 separated by an energy gap

$$\varepsilon_G = \varepsilon_{X_1} - \varepsilon_{X_4},$$

corresponding to a movement $X(110) \to \Gamma(020)$ along the (220) plane of the Jones zone. The energy varies along $XK \, \Sigma \, \Gamma$ in Fig. 6.16, and the change X(110)

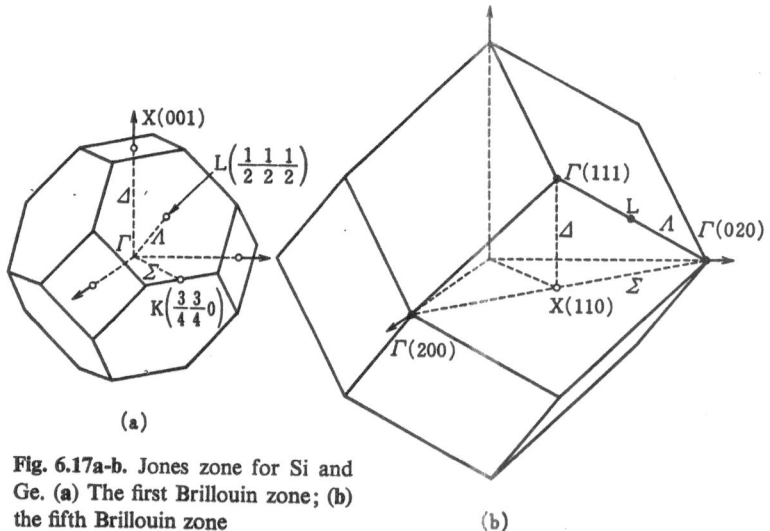

(a)

(b)

Fig. 6.17a-b. Jones zone for Si and Ge. (a) The first Brillouin zone; (b) the fifth Brillouin zone

$\rightarrow \Gamma(111)$ corresponds to the change in energy along $X\Gamma\Delta$ in Fig. 6.16. As is evident from this figure, the energy gap ε_G is kept almost constant for these changes. Therefore if we approximate the Jones zone by a sphere, we are led to a model in which a uniform energy gap ε_G appears on a spherical surface in the k-space (this is a NFE model in the extended zone scheme and called a Penn model).

Let us expand the crystal potential in a diamond lattice as

$$V(r) = \sum_g V(g)\, e^{ig\cdot r}, \tag{6.3.5}$$

then the coefficient $V(g)$ is expressed in terms of the pseudopotential form factor $v(g)$ and structure factor $S(g)$ as

$$V(g) = S(g)\, v(g)$$

and $S(g)$ is given by

$$S(g) = \cos \frac{\pi}{4}\, (g_1 + g_2 + g_3) \tag{6.3.6}$$

$$g = \frac{2\pi}{a}\, (g_1, g_2, g_3). \tag{6.3.7}$$

For $v(g)$ we may use the three values $v(111)$, $v(220)$ and $v(113)$ which were given before. In the NFE approximation, the energy gap is calculated perturbationally to the second order in potential to have

$$\varepsilon_g^0 = 2V(220).$$

However, as already known, for both Si and Ge, $v(111)$ is the largest among $v(g)$ and hence the higher-order contribution from the component $V(111)$ cannot be ignored. Therefore, we try to proceed to a higher-order approximation. Choosing a unit such as $\hbar^2/2m = 1$ for simplicity, we write the matrix element of the Hamiltonian as

$$\mathcal{H}_{gg'} = (k + g)^2 \delta_{gg'} + V(g - g'). \tag{6.3.8}$$

The diagonalization of this matrix gives the energy band, but what we are interested in is the energy gap at the X-point [i.e. $k = (110)$ and $(\bar{1}\bar{1}0)$], so we assume that we could diagonalize it systematically until the two levels we are interested in are isolated from all others, obtaining a 2×2 submatrix of the form

$$\begin{bmatrix} (110)^2 + \text{const} & V_{\text{eff}}(220) \\ V_{\text{eff}}(220) & (\bar{1}\bar{1}0)^2 + \text{const} \end{bmatrix} \tag{6.3.9}$$

with an effective interaction $V_{\text{eff}}(220)$. In principle, this is an exact procedure and we can determine $V_{\text{eff}}(220)$ by a perturbation theory as

$$V_{\text{eff}}(220) = V(220) + \sum_{g}' \frac{\langle 110|V|110-g\rangle\,\langle 110-g|V|\bar{1}\bar{1}0\rangle}{(110)^2 - (110-g)^2} + \cdots ,$$

$$(6.3.10)$$

where the summation over g should be taken over all the reciprocal lattice vectors except for $g = (220)$. In practice, however, it is enough to retain only the terms which give $V(111)$ in the numerator, that is $(110 - g)$ may be limited to (001) and $(00\bar{1})$. Thus, using $V_{\text{eff}}(220)$ determined in this way in (6.3.9), we obtain the energy gap at the X-point as

$$\varepsilon_{\text{G}} = 2V_{\text{eff}}(220) = 2\left[v(220) + \frac{\{v(111)\}^2}{(110)^2 - (001)^2}\right]. \qquad (6.3.11)$$

In Table 6.8 the calculated values of ε_{G} from (6.3.11) are shown along with the experimental values which are deduced from the absorption spectrum given in Fig. 6.15. As seen from Fig. 6.15, a sharp peak in the spectrum around 4 eV = 0.3 Ry is observed for both Si and Ge. This may be attributed to the characteristics of the energy band structures of these substances theoretically deduced above such that there is a nearly constant energy gap of about 0.3 Ry along the (220) plane of the Jones zone in the k–space. It is interesting that a simple band theory such as this can explain the experimental facts in an elementary way.

Table 6.8. Main structure factors of pseudopotentials for Si, Ge and Sn in units of Ry

	$v(220)$	$[v(111)]^2$	$V_{\text{eff}}(220)$	ε_{G}	ε_{exp}
Si	0.04	0.12	0.16	0.31	0.30
Ge	0.01	0.15	0.16	0.32	0.32
Sn	0.00	0.15	0.15	0.30	0.26

Finally we shall examine whether the charge density distribution exhibits any characteristics of covalent bonds in these crystals. In the pseudopotential theory of simple metals, $v(g)$ may be taken as quantities so small that it is enough to calculate the band energy up to the second order and the wave function up to the first order in $v(g)$ perturbationally (Chap. 5). Within such an approximation, the total charge density is expressed as a superposition of that of spherical pseudoatoms. The lattice energy is then a sum over pairs in central potentials. Even in this approximation, some part of the charge density can describe a trend for interatomic bonding, but many-body interactions which are required, for instance, to maintain a constant bond angle do not come out. In the pseudopotential theory for Si and Ge, on the other hand, we have seen that a correct value of the energy gap ε_{G} can be obtained only by including a higher-order perturbation in $v(111)$. This fact is closely related to the characteristics of covalent bonds as will be shown presently. The wave function of the bonding state at the X–point on the Jones zone is given in perturbation theory by

$$\left.\begin{aligned}
\psi &= \frac{1}{\sqrt{2}} [|110\rangle + |\bar{1}\bar{1}0\rangle + \sum_g C_g(|110 + g\rangle + |\bar{1}\bar{1}0 - g\rangle)] \\
C_g &= \frac{V(g)}{(110)^2 - (110 + g)^2}
\end{aligned}\right\} \tag{6.3.12}$$

from which the charge density $\Psi^*\Psi$ is evaluated as

$$\psi^*\psi \approx [1 + \cos(220)\cdot r + \sum_g C_g\{\exp(ig\cdot r) + \exp[i(220 + g)\cdot r]$$
$$+ \text{c.c.}\} + O(C_g^2) \ldots]. \tag{6.3.13}$$

The wave function of a k-state which is sufficiently away from the energy gap is a plane wave, contributing only uniform charge density and hence has nothing to do with the covalent bonding. Important states for covalent bonding are those which have their energies largely reduced by the appearance of the energy gap as in the case of (6.3.12). The number of such states can be roughly estimated as $[4\Omega/(2\varepsilon_F/3)] \times (\varepsilon_G/2)$ where Ω is the atomic volume and the first factor represents the density of states for free electrons. We shall approximate the wave functions of all those states by the function (6.3.12) and its equivalents including twelve equivalent X–points. Then the charge density is evaluated by summing (6.3.13) over the equivalent set and multiplying the total number of similar states $(3\varepsilon_G/\varepsilon_F)$ (Ω is cancelled by the normalization factor). The first term in (6.3.12) yields only a uniform distribution. However, if we take only $g = (111)$ in the terms proportional to C_g, we find as a part of the contribution to the charge density

$$v(111) [\cos(111)\cdot r - \tfrac{2}{9} \cos(311)\cdot r + \cdots]$$

which indicates the presence of bonds extending in the [111] direction. A more detailed calculation shows that this charge distribution is equivalent to that of electrons localized at the midpoint between two bonding atoms with a density

$$Q \approx -\frac{3\varepsilon_G}{2\sqrt{2}\,\varepsilon_F} v(111) \frac{2m}{\hbar^2} \left(\frac{a}{2\pi}\right)^2. \tag{6.3.14}$$

Therefore Q is called the bond charge which, after inserting the values of ε_G, ε_F and $v(111)$, turns out to be

$$Q = 0.19$$

for Si. We show in Fig. 6.18 the spatial variation of the charge density along [111]

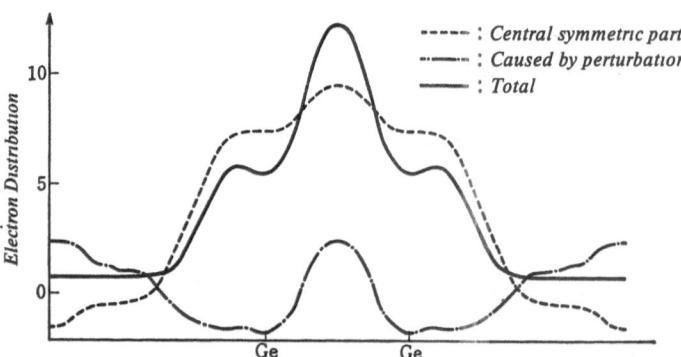

Fig. 6.18. Electron charge distribution along [111] for Ge

in the diamond lattice calculated as from (6.3.12). This clearly shows the electron localization inbetween atoms.

6.4 Hydrogen-Bonded Crystals

A hydrogen bond is a special chemical bond in which a hydrogen atom H, located between two electronegative atoms (or ions) makes them bonded partly in a covalent and partly in an ionic way. Intuitively, its bonding mechanism can be understood in the following way. Consider two electronegative atoms X and Y along with a hydrogen atom H as depicted in Fig.6.19. Let us assume that atom (or ion) X makes a covalent bond with H by sharing two electrons. Since H has only one electron, the screening of the positive electric charge of the hydrogen nucleus, or proton, becomes incomplete and some effect of the positive charge influences the nearby electronegative atom Y. If Y is negatively charged, Coulomb attraction occurs. Even if this is not the case, an electrostatic attraction will be exerted through its polarization effect. Thus, an asymmetrical bond of

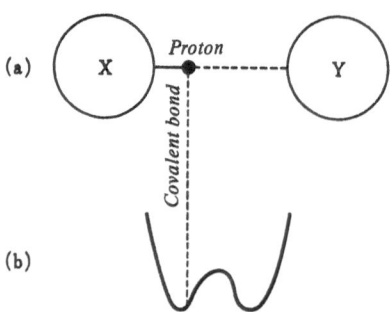

Fig. 6.19a-b. Model of hydrogen bond

the form X — H...Y arises. Naturally the bond of the form X ... H — Y should also exist. Usually the equilibrium positions of the proton are different for these two cases. If two states of different bonding resonate with each other, it is necessary for the proton to go back and forth between two equilibrium positions. When the particular case X = Y, the proton has two equivalent equilibrium positions and moves in a double minimum potential, as shown in Fig. 6.19. The height of the potential barrier is estimated as being of the order of some kilocalories. It is conceived that resonance may take place in some cases by tunneling through the potential barrier in a quantum-mechanical way.

Now the hydrogen bond draws much attention as taking part in the bonds inside an organic molecule as well as between organic molecules. It is known that the hydrogen bond plays an essential role in the secondary structure of the protein molecules and DNA molecules. Other examples in which the hydrogen bond plays an important role are ice and water. Since the positions of protons cannot be determined by x-ray analysis of an ice crystal, various structure models have been proposed for a long time. At present, however, we can get valuable information about the hydrogen configurations in an ice crystal by means of neutron diffraction methods. The Pauling model for the proton configurations which indicates the best agreement with experiments is the following. As shown in Fig.6.20, oxygens of a water molecule H_2O are arranged in a hexagonal lattice with four coordination number. There is one H on every line which connect between two nearest neighboring oxygens. Thus we may say that oxygens in an ice crystal are binded with hydrogen bonds. Each hydrogen has two equilibrium positions separated by a distance of about 0.95 Å along a line connecting two oxygens and is located at either one of two positions. In an ice crystal as a whole, hydrogens are statistically distributed among the assembly of the two positions in such a way that on the average, each oxygen has two nearer and two more distant hydrogens (see Fig.6.20). Sometimes, this restriction imposed on the hydrogen configurations is called the "ice condition". One experimental support for this statistical model is provided by the measured value of the zero-point entropy of ice. If hydrogens are randomly distributed over two respective equilibrium positions fulfilling the ice condition, the excess configurational entropy is shown to be approximately

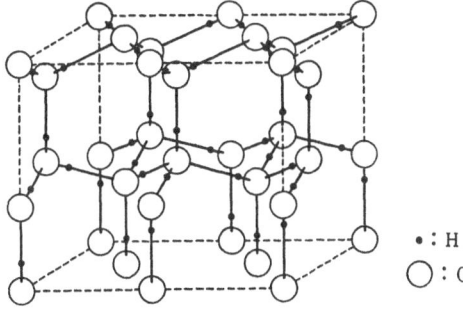

• : H
○ : 0

Fig. 6.20. Crystal structure of ice

$$k_B \log \left(\tfrac{3}{2}\right)^N = R \log \tfrac{3}{2} = 0.806 \quad [\text{cal(mol} \cdot \text{K)}^{-1}]. \tag{6.4.1}$$

Zero-point entropy calculated from the specific heat data of ice above 10 K is $\Delta S = 0.82 \pm 0.05$ cal(mol $\cdot$ K)$^{-1}$ which shows a good agreement with (6.4.1). The energy required to break a hydrogen bond is about 5.7 kcal since the sublimation heat of ice is 11.4 kcal and there are two hydrogen bonds per ice molecule. When ice melts into water, the regular arrangement of molecules depicted in Fig.6.20 will become disordered, but it is supposed that there still remains short-range order similar to that in an ice crystal. In fact, one of the origins of high conductivity in water is believed to stem from a hydrogen ion H$^+$ (proton), which takes part in a hydrogen bond, and is migrating out through some bond defects.

A hydrogen bonded crystal is a crystal which owes its main cohesion to hydrogen bonds and exhibits various interesting properties because of the peculiar nature of hydrogen bonds. It should be noticed, in particular that there is a group of hydrogen bonded crystals which become ferroelectrics. An important role played by the hydrogen bond in crystals is revealed in the existence of so-called crystalline water. We often have materials represented by a molecular formula such as (...) nH$_2$O. This indicates that a water molecule H$_2$O connects two constituent units of the crystal in the form

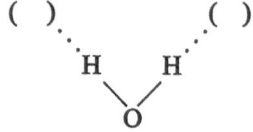

and thus the crystal is reinforced by the additional binding force of hydrogen bonds.

As another interesting example, we shall mention here a substance with a formula Cu(COOH)$_2 \cdot$4H$_2$O. The crystal structure of this substance is shown in Fig. 6.21. As seen from this figure, Cu^{++}ions, surrounded by (COOH)$^-$ ions, form a two-dimensional network. The crystalline waters are sandwiched between successive layers containing Cu^{++} ions and act as binder of the layers to form a three-dimensional lattice. Since a Cu^{++} ion has a certain magnetic moment (Chap. 8), Cu^{++} ions interact with each other only within a two-dimensional network via superexchange through (COOH)$^-$ ions, and give rise to antiferromagnetism at sufficiently low temperatures. The main interest in this substance lies in that it gives a typical example of two-dimensional antiferromagnets on one hand and also exhibits another anomaly in dielectric behaviour at a different temperature. As shown in Fig. 6.22, this substance undergoes a phase transition at $-36°$C accompanied with an anomaly in the dielectric constant. This phase transition is associated with the appearance of a long-range order in the proton configurations below T_c within the layers consisting of water molecules where

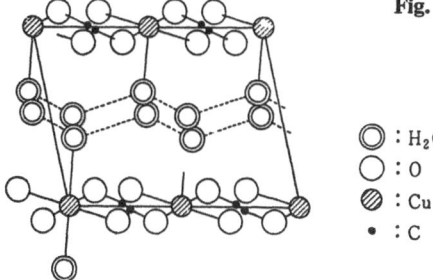

Fig. 6.21. Crystal structure of $Cu(COOH)_2 \cdot 4H_2O$

⊙ : H_2O
○ : O
⊘ : Cu
• : C

oxygen atoms form a hydrogen-bonded honeycomb lattice. The sharp drop in the dielectric constant below T_c is explained by the onset of anti-ferroelectricity producing spontanous polarization within layers which are arranged alternatively to the $+b$ or $-b$ directions along the perpendicular direction, resulting in the cancellation of bulk polarization.

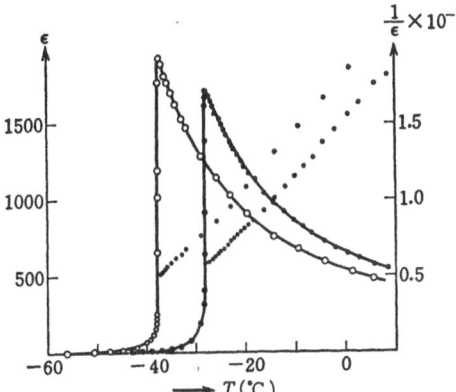

$\frac{1}{\epsilon} \times 10^{-3}$

○ *Crystal containing* $4H_2O$
● *Crystal containing* $4D_2O$

Fig. 6.22. Anomaly of the dielectric constant in $Cu(COOH)_2 \cdot 4H_2O$ [6.10]

As seen in the above example, the networks of hydrogen bonds bring interesting problems to the dielectric properties. It is related to the fact that a kind of cooperative phenomenon takes place by the interactions between hydrogens and some ordered arrangement can appear in the distribution of the hydrogens over two possible equilibrium positions in hydrogen bonds. In most cases the interactions arise indirectly through the coupling between hydrogens and other degrees of freedom. The ordered arrangement of hydrogens affects the other dielectric freedom in the crystal and induces spontanous electric polarization, resulting in ferro- or antiferroelectricity. Some examples of ferroelectrics (or antiferroelectrics) with hydrogen bonds are listed in Table 6.9. In this table, F means a ferroelectric, A an antiferroelectrics, T_{cH} the transition temperature, and T_{cD} the transition temperature when hydrogen is replaced by deuteron. From this table it is observed that the changes in the transition temperature when deuterated are

Table 6.9. Examples of ferroelectric (antiferro) substances with hydrogen bonds. T_{cD} is the transition temperature for deuterized crystal. (F) ferroelectrics, (A) antiferroelectrics

		T_{cH}[K]	T_{cD}[K]
KH_2PO_4	F	123	213
RbH_2PO_4	F	147	218
KH_2AsO_4	F	97	161
RbH_2AsO_4	F	110	173
CsH_2AsO_4	F	143	212
$NH_4H_2PO_4$	A	148	242
NH_4H_2AsO	A	216	304
$NH_4Fe(SO_4)_2 \cdot 12H_2O$	F	88	88
$CH_3NH_3Al(SO_4)_2 \cdot 12H_2O$	F	177	177
$K_4Fe(CN)_6 \cdot 3H_2O$	F	248.5	255
$K_4Rn(CN)_6 \cdot 3H_2O$	F	258.5	265.7
$K_4OS(CN)_6 \cdot 3H_2O$	F	270.6	271.2
$Cu(COOH)_2 \cdot 4H_2O$	A	235.3	245.5
$(NH_2CH_2COOH)_3 \cdot H_2SO_4$	F	322.4	332
$(NH_2CH_2COOH)_3 \cdot H_2BeF_4$	F	348	350
$NaKC_4H_4O_6' 4H_2O$	F	255 ~ 297	251 ~ 308
$(NH_4)_2H_3IO_6$	A	253	260
HCl	F	98.36	105.03
HBr	F	89.75	93.5

sometimes negligible, sometimes rather small, but very large (generally $T_{cD} > T_{cH}$) in a few special cases represented by potassium dihydrogen phosphate (KDP). This remarkable isotope effect in T_c indicates that the mechanism of phase transition is related to proton or deuteron mass. This suggests that proton motion in the hydrogen bond, for instance, the tunneling motion through the barrier of a double minimum potential, may play a crucial role in the phase transition.

The crystalline structure of KH_2PO_4 is shown in Fig.6.23, where two oxygens of two neighboring PO_4 radicals are joined though a hydrogen bond. Each hydrogen is statistically distributed over two equilibrium positions under the constraint of the "ice condition" as in the ice crystal. In a low temperature phase which exhibits ferroelectricity, the positions of all hydrogens are fixed to PO_4 radicals, as shown in Fig. 6.23, and a spontaneous polarization appears along the c-axis. This spontaneous polarization is not attributed directly to H, but to the displacement of K^+ ions and the deformation of the tetrahedra of PO_4 radicals which are induced by the ordered arrangement of H. Above T_c', the configuration of H is random and the spontaneous polarization disappears. Therefore, the ferroelectric phase transition in this substance is an order-disorder transition of H in a network of hydrogen bonds. This view has been justified by neutron diffraction experiments which reveal experimentally the ordered arrangement of H.

The statistical theory of phase transition in KH_2PO_4 was first proposed by *Slater* [6.11], and since then many discussions have followed. It is, however, only

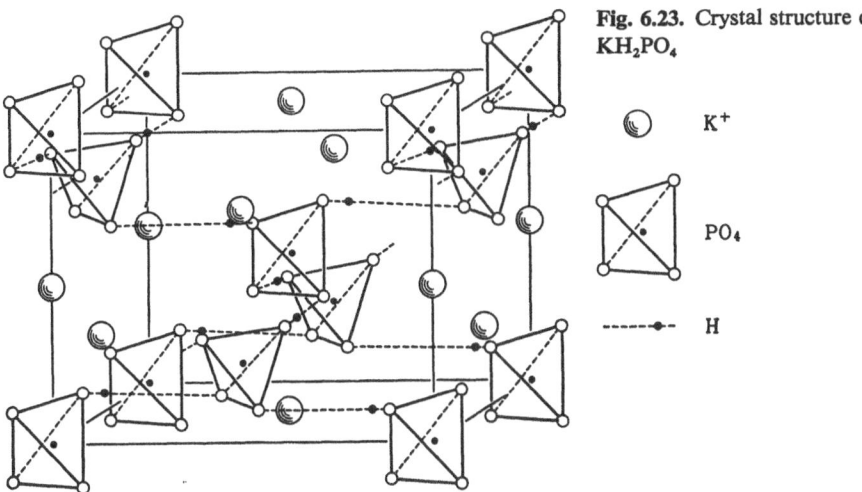

Fig. 6.23. Crystal structure of KH$_2$PO$_4$

K$^+$

PO$_4$

H

recently that the tunneling motion of hydrogen has been included in order to interpret the above-mentioned large isotope effect. We shall present here a pseudo-spin model which is adequate to treat the statistical theory of hydrogen bond networks. This model provides us with a particularly useful means to study the dynamical behavior of a proton system when the tunneling motion of hydrogen along a hydrogen bond becomes important.

In order to represent two possible equilibrium positions of H in a hydrogen bond, we shall associate a pseudo-vector spin σ_j with each hydrogen bond. j is the number attached to each bond. σ is the same as a Pauli spin operator whose z-component may be represented by a 2×2 matrix

$$\sigma^z = \begin{bmatrix} 1 & 0 \\ 0 & -1 \end{bmatrix}. \tag{6.4.2}$$

We assume that two states of σ with the eigenvalues ± 1 correspond to two states of H in a hydrogen bond. In this representation, the x-component of σ can be expressed by a matrix

$$\sigma^x = \begin{bmatrix} 0 & 1 \\ 1 & 0 \end{bmatrix} \tag{6.4.3}$$

which indicates that σ_j^x acts to change the position of the jth H from the 1 position to -1 position or vice versa. In other words, σ^x is an operator making a hydrogen jump between two equilibrium positions. Thus we see that the tunneling effect of protons can be described in terms of σ^x. Now let us consider the Hamiltonian of a whole hydrogen bond system which will consist of two parts, the kinetic energy describing the tunneling motion and the potential energy due to proton-proton interactions. The simplest form is

$$\mathcal{H} = \Gamma \sum_j \sigma_j^x + \sum_{i>j} J_{ij} \sigma_i^z \sigma_j^z , \tag{6.4.4}$$

where Γ represents a frequency of tunneling motion as suggested from the nature of the operator σ^x. The second term indicates that a pair of protons may change energy depending on their positions. More generally, we may expect the terms of the forms $\sigma_i^x \sigma_j^x$ or $\sigma_i^y \sigma_j^y$ in the potential energy, but we shall omit them because they are not essential in the case of KDP. Three-body or high-order interactions should also exist which are neglected for simplicity. It is known that in the case of KDP, a suitable choice of the interaction parameters J_{ij} can reproduce most results of all the previous models as far as the static properties are concerned. On the other hand, since Γ is related to the probability of tunneling motion, it depends on the overlapping between proton wave functions at two equilibrium positions and hence on mass through the wave functions (for instance, the wave function of harmonic oscillators localized at equilibrium positions). Thus Γ should be different for proton and deuteron, and generally

$$\Gamma_H > \Gamma_D \tag{6.4.5}$$

since hydrogen with lighter mass is expected to have large tunneling frequency. Now we can easily imagine the general statistical behaviour of a hydrogen bond system described by the Hamiltonian (6.4.4.). This system is equivalent to a spin system which has a tendency to arrange the spin moments parallel to the z-direction by the exchange interaction under the action of an external magnetic field applied to the x-direction. If the interaction J_{ij} is strong enough, an order of spin arrangement in the z-direction will appear at a low temperature against the magnetic field Γ in the x-direction which tends to disturb the realization of this order. For larger values of Γ, the transition temperature becomes lower. In terms of the hydrogen bond system, the tunneling effect disturbs the occurrence of the order-disorder transition caused by proton-proton interaction. According to (6.4.4), the effect of this disturbance is weaker for deuterons than for protons and hence the transition temperature is higher for deuterons than for protons. It is interesting to note that the Hamiltonian (6.4.4) suggests a possibility of a collective motion of protons in a hydrogen bonded system.

6.5 Electron Theory of Nonmetals (Semiconductors)

Insulators which have a relatively small band gap ε_G are called semiconductors. Si and Ge described in Sect. 6.3 are typical semiconductors, and as element materials Se and Te also belong to this category. As discussed in the following, there are many compounds known as semiconductors and some of them are already utilized in practice. As seen in Sect. 6.2, in the I–VII compounds which are typical ionic crystals, the Madelung energy originated from interionic attraction is a source to produce large energy gaps between filled valence bands and vacant conduction bands. As mentioned briefly there, as in the II-VI or III-V compounds when the constituent atoms approach the VIth elements from both sides, the trend of ionization is weakened and some element of covalency gradually comes into the cohesion mechanism. That is, such a crystal is inbetween an ionic crystal and a covalent crystal. As a result, the energy gap becomes smaller and the crystal becomes more semiconductive. In Table 6.10 the crystal structures and values of ε_G are listed for the II-VI and III-V compounds known as semiconductors. As seen from the table, these compounds are crystallized in either a zincblende structure (Fig.6.8c) or a wurtzite structure (Fig.6.8d) and the value of ε_G decreases as the atomic number increases.

Table 6.10. Semiconducting compounds: II-VI compounds and III-V compounds. (W) Wurtzite-type; (Z) Zincblende-type

II-VI Compounds

	ε_G[eV]	Lattice		ε_G[eV]	Lattice
ZnO	3.2	W	αZnS	3.54	Z
βZnS	3.8	W	ZnSe	2.67	Z
CdS	2.41	W	ZnTe	2.26	Z
CdSe	1.74	W	CdTe	1.44	Z

III-V Compounds

	ε_G[eV]	Lattice		ε_G[eV]	Lattice
GaN	3.39	W	GaAs	1.35	Z
AlP	2.42	Z	InP	1.26	Z
GaP	2.24	Z	GaSb	0.67	Z
AlAs	2.16	Z	InAs	0.35	Z
BP	2.0	Z	InSb	0.18	Z
AlSb	1.6	Z			

The band structures of semiconductors with zincblende or wurtzite structures have been calculated on the basis of the empirical pseudopotential model (EPM) described in Sect.6.3. The results of band calculations seem fairly successful as in the cases of Si and Ge. A main difference compared with the case of Si or Ge is that there are two kinds of atoms in a crystal which are closely related by ionic cohesion. Let us decompose the crystal potential $V(\mathbf{r})$ into Fourier series as

$$V(r) = \sum_g V(g) e^{ig \cdot r} . \tag{6.5.1}$$

In the case of a zincblende lattice, for instance, since there are two atoms in a unit cell, we can choose for simplicity the origin of the coordinates at the midpoint of a line connecting these two atoms (denoted as 1 and 2, respectively). Let us call the radius vectors of two atoms, r_1 and r_2, and the lattice constant a, then we have

$$r_1 = a \left(\frac{1}{8} \frac{1}{8} \frac{1}{8} \right) \equiv \tau, \quad r_2 = - \tau . \tag{6.5.2}$$

The Fourier component $V(g)$ is given by

$$V(g) = \frac{1}{\Omega} \int V(r) e^{-ig \cdot r} dr = \frac{1}{\Omega} \int [V_1(r - \tau) + V_2(r + \tau)] e^{ig \cdot r} dr \tag{6.5.3}$$

which may be also written as

$$V(g) = V_g^s \cos g \cdot \tau + i V_g^a \sin g \cdot \tau , \tag{6.5.4}$$

where

$$V_g^s = \frac{1}{2} [V_1(g) + V_2(g)]. \quad V_g^a = \frac{1}{2} [V_1(g) - V_2(g)] \tag{6.5.5}$$

$$V_j(g) = \frac{2}{\Omega} \int V_j(r) e^{ig \cdot r} dr \quad (j = 1,2) . \tag{6.5.6}$$

If $V_1(g) = V_2(g)$, then $V_g^a = 0$, and this reduces to the case of Si or Ge. For $V_g^a \neq 0$, $V(g)$ is generally a complex number and hence the elements of a determinant in the secular equation also become complex numbers. We shall truncate the infinite sum over g in (6.5.1) by taking a finite number of terms, and limit g in a few vectors in the vicinity of origin. For instance, we choose the following points:

$\frac{a}{2\pi} g = (000)$	$\left(\frac{a}{2\pi}\right)^2 g^2 = 0$	V_0^s	$V_0^a \equiv 0$	
$\frac{a}{2\pi} g = (111)$	$\left(\frac{a}{2\pi}\right)^2 g^2 = 3$	V_3^s	V_3^a	
$\frac{a}{2\pi} g = (311)$	$\left(\frac{a}{2\pi}\right)^2 g^2 = 11$	V_{11}^s	V_{11}^a	
$\frac{a}{2\pi} g = (200)$	$\left(\frac{a}{2\pi}\right)^2 g^2 = 4$	$V_4^s \equiv 0$	V_4^a	
$\frac{a}{2\pi} g = (220)$	$\left(\frac{a}{2\pi}\right)^2 g^2 = 8 .$	V_8^s	$V_8^a = 0$	

Note here that for $(a/2\pi)^2 g^2 = 4$, $V_g^a \equiv 0$; also for $(a/2\pi)^2 g^2 = 0$ and $(a/2\pi)^2 g^2 = 8$, $V_g^a \equiv 0$. Therefore, if we choose the origin of energy so that $V_0^s = 0$, band calculation can be performed by using 6 components of the pseudopotential $(V_3^s, V_8^s, V_{11}^s, V_3^a, V_4^a, V_{11}^a)$. The band structures determined on the basis of EPM are shown in Fig. 6.24 and the values of 6 chosen parameters are tabulated in Table 6.11. Except for minor differences, these results exhibit a remarkable similarity with the band structures for Si and Ge. This fact suggests that even for these compounds the single gap model which is successful for Si and Ge may be applicable within a good approximation. Adopting this model again, we shall try to describe the main properties of these semiconductors in the following.

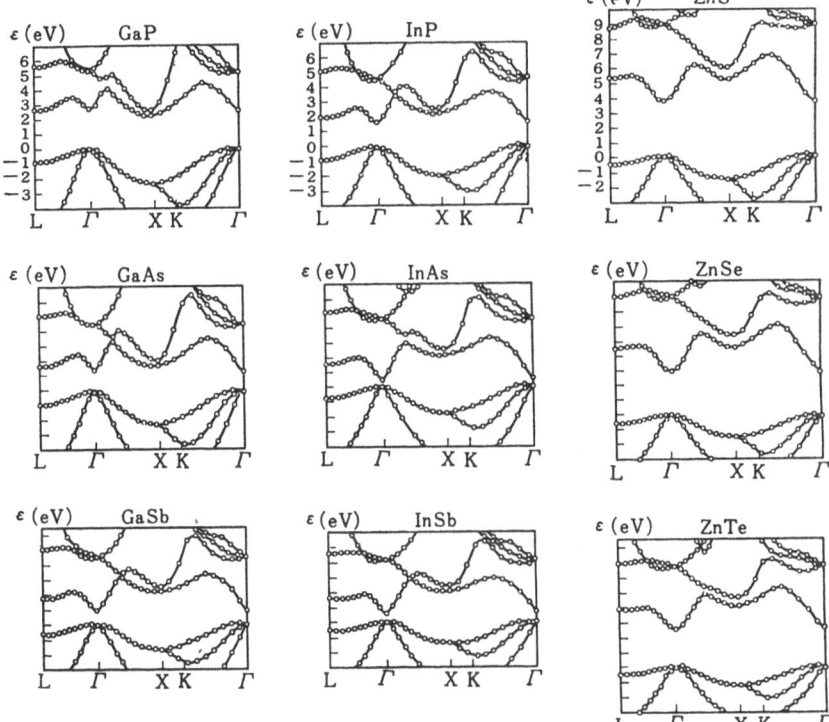

Fig. 6.24. Band structures of compounds determined by means of EPM [6.12]

In the single gap model, we assume that an energy gap exists when the wave vector k is equal to a half of the certain reciprocal lattice vector g, and the energy band can be given as a function of k as

$$\varepsilon(k) = \tfrac{1}{2} \{\varepsilon_0(k) + \varepsilon_0(k - g) \pm \sqrt{[\varepsilon_0(k) - \varepsilon_0(k - g)]^2 + 4|V(g)|^2}\}, \quad (6.5.7)$$

Table 6.11. List of pseudopotentials for compounds

	V_3^s	V_8^s	V_{11}^s	V_3^a	V_4^a	V_{11}^a
GaP	−0.22	+0.03	+0.07	+0.12	+0.07	+0.02
GaAs	−0.23	+0.01	+0.06	+0.07	+0.05	+0.01
Ga Sb	−0.22	0.00	+0.05	+0.06	+0.05	+0.01
InP	−0.23	+0.01	+0.06	+0.07	+0.05	+0.01
InAs	−0.22	0.00	+0.05	+0.08	+0.05	+0.03
InSb	−0.20	0.00	−0.04	+0.06	+0.05	+0.01
ZnS	−0.22	+0.03	+0.07	+0.24	+0.14	+0.04
ZnSe	−0.23	+0.01	+0.06	+0.18	+0.12	+0.04
ZnTe	−0.22	0.00	+0.05	+0.13	+0.10	+0.01

where

$$\varepsilon_0(\boldsymbol{k}) = \frac{\hbar^2}{2m} k^2 \tag{6.5.8}$$

is the energy spectrum of a free electron, and the magnitude of the energy gap is given by

$$\varepsilon_G = 2|V(\boldsymbol{g})| . \tag{6.5.9}$$

If we use (6.5.4) as $V(\boldsymbol{g})$, we may write

$$\varepsilon_G^2 = E_h^2 + C^2$$
$$E_h^2 = 4(V_g^s \cos g\tau)^2, \quad C^2 = 4(V_g^a \sin g\tau)^2 . \tag{6.5.10}$$

Thus E_h is proportional to V_g^s which also exists in covalent crystals. Therefore, E_h may be interpreted as a contribution to the energy gap from a part of the covalent character. On the other hand, C is proportional to V_g^a which is nonzero only when the two atoms in a unit cell are different. Therefore this may be thought to represent a part of the ionic character. Equations (6.5.7–10) are results obtained on the basis of the NFE approximation and invalidated if some components of the pseudopotential were much larger than $V(\boldsymbol{g})$. In such cases $V(\boldsymbol{g})$ should be replaced by a suitable effective value to take into account higher-order correction coming through the large component. For instance, if the energy gap is assumed to occur along the Jones zone constructed from (220)'s planes even for compounds, as in the case of Si and Ge, higher-order perturbation yields

$$\left.\begin{array}{l} \dfrac{E_h}{2} = \dfrac{V^s(220) + \{[V^s(111)]^2 + [V^a(111)]^2\}}{\dfrac{\hbar^2}{2m}\left(\dfrac{2\pi}{a}\right)^2 [(100)^2 - (001)^2]} \\[6mm] \dfrac{C}{2} = \dfrac{2V^s(111)\, V^a(111)}{\dfrac{\hbar^2}{2m}\left(\dfrac{2\pi}{a}\right)^2 [(110)^2 - (001)^2]} \end{array}\right\} \tag{6.5.11}$$

Hereafter, however, we shall forget about (6.5.11), regarding E_h and C as quantities which can be related to other measurable properties. One of the physical quantities closely related to E_h and C is the low frequency dielectric constant $\epsilon(0)$. *Penn*[6.13] has calculated $\epsilon(0)$ on the basis of a single gap model to derive the following results:

$$\epsilon(0) = 1 + \left(\frac{\hbar\omega_p}{\varepsilon_G}\right)^2 \left(1 - \frac{\varepsilon_G}{4\varepsilon_F}\right), \tag{6.5.12}$$

where ε_F is the Fermi energy, or more precisely, the free electron energy for the momentum at which the energy gap appears;

$$\varepsilon_F = \frac{\hbar^2}{2m} k_F^2 \tag{6.5.13}$$

and (6.5.12) is an approximate formula correct up to the first order in $\varepsilon_G/\varepsilon_F$. Since usually $\varepsilon_G/\varepsilon_F = 0.1$, the term of order $\varepsilon_G/\varepsilon_F$ in (6.5.12) may be neglected compared with unity. ω_p represents the plasma frequency which is given in terms of the valence electron density n as

$$\omega_p^2 = \frac{4\pi e^2 n}{m} . \tag{6.5.14}$$

After all, (6.5.12) is expressed approximately as

$$\frac{1}{\epsilon(0) - 1} = \frac{1}{(\hbar\omega_p)^2} (E_h^2 + C^2) \tag{6.5.15}$$

from which C may be evaluated from the observed values of $\epsilon(0)$, ω_p and E_h. First we note that the case $C = 0$ corresponds to the pure covalent crystals (diamond, Si, Ge) for which we have observed values of $E_h = \varepsilon_G$ (Table 6.8). If we assume that E_h is common to elements lying on the same row in the periodic table, then it follows that for the compounds consisting of atoms lying on the same row in the periodic table, C can be calculated from $\epsilon(0)$ and ω_p. Figure 6.25 shows the plots for such compounds of $[\epsilon(0) - 1]^{-1}$ as functions of the square of the difference in the atomic valences of constituent atoms. The fact that the observed points lie almost completely on straight lines suggests that, according to the above assumption,

$$C \propto |Z_A - Z_B| .$$

This indicates that it is reasonable to regard C as a measure of ionicity. In order to extrapolate for the compounds consisting of two atoms which are sitting on different rows in the periodic table, it becomes necessary to know how E_h is changed in the same group as the atomic number is varied. Empirically it is

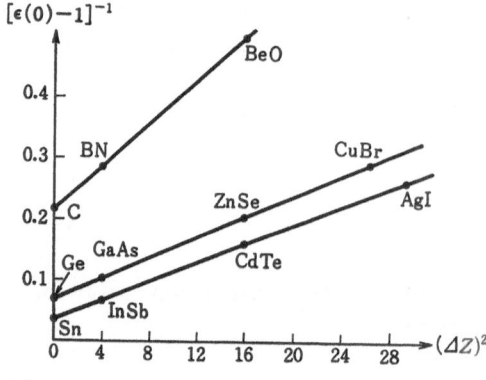

Fig. 6.25. $[\epsilon(0) - 1]^{-1}$ vs $(\Delta Z)^2$

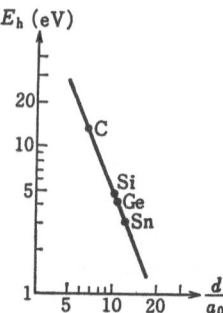

Fig. 6.26. E_h vs d/a_0

is known that E_h depends approximately only on the nearest neighbor distance d of atoms in each material. This is evident, for instance, from Fig. 6.26 which shows the relationship between E_h and d for four elements C, Si, Ge and Sn. Thus we are led to an empirical law

$$E_h = 40d^{-2\,48} \,. \tag{6.5.16}$$

If this relationship holds generally, then E_h is obtained from the observed value of d and hence C is determined from $\epsilon(0)$. The values of E_h and C determined in this way for 32 materials are tabulated in Table 6.12.

It would be very useful if one could find out an empirical formula which correlates the constant $C = C_{AB}$ for a compound consisting of atoms A and B with some physical quantity characterizing A and B atoms. When two atoms A and B lie on the same row in the periodic table, Fig. 6.25 indicates that C is proportional to $|Z_A - Z_B|$. On the other hand, it is seen in Table 6.12 that a remarkable empirical fact exists where, for the compounds consisting of atoms which belong to the same row, the atomic distances are nearly constant. Noting this fact, we can define an atomic radius r for each row in the periodic table by taking half of the atomic distance in the VIth element crystal belonging to each row. Now let us assume the following form for C_{AB}

$$C_{AB} = b \left(\frac{Z_A}{r_A} - \frac{Z_B}{r_B}\right) \exp\left(-k_s \frac{r_A + r_B}{2}\right) \tag{6.5.17}$$

which has a correct dimension, and see to what extent this formula can realize the values of C which have been determined empiricially. In (6.5.17), r_A and r_B are the atomic radii defined above, and k_s is a screening constant for which we use the values evaluated from the Thomas-Fermi approximation. In Table 6.17 the values of b determined on the basis of (6.5.17) are also tabulated. It is seen

Table 6.12. Band parameters for tetrahedrally bonded crystals. [6.14].

Zincblende-type	l.c. (a/a_0)	$\epsilon(0)$	C(eV)	E_h(eV)	b
BN	6.740	4.5	7.8	13.1	1.55
CuF	8.041	2.5	15.8	8.7	1.30
SiC	8.217	6.7	3.9	8.3	1.40
BP	8.576	8.6	0.7	7.4	1.30
BAs	9.027	10.4	0.3	6.6	1.30
BeS	9.165	7.1	4.0	6.3	1.30
BeSe	9.581	7.3	3.4	5.7	1.30
CuCl	10.215	3.7	8.3	4 8	1.50
ZnS	10.222	5.2	6.2	4.8	1.40
GaP	10.300	9.1	3.3	4.7	1.45
AlP	10.301	8.5	3.1	4.7	1.50
BeTe	10.469	11.6	2.1	4.5	1.30
AlAs	10.620	10.2	2.7	4.4	1.50
GaAs	10.684	10.9	2.9	4.3	1.50
ZnSe	10.710	5.9	5.6	4.3	1.45
Inp	11.090	9.6	3.4	3.9	1.40
InAs	11.406	12.3	2.7	3.7	1.30
CuI	11.419	5.5	5.5	3.7	1.55
ZnTe	11.510	7.3	4.5	3.6	1.60
GaSb	11.561	14.4	2.1	3.5	1.75
AlSb	11.593	10.2	3.1	3.5	3.05
Agl	12.232	4.9	5.7	3.1	1.60
InSb	12.242	15.7	2.1	3.1	1.50
CdTe	12.246	7.2	4.4	3.1	1.55
Wurtzite-type					
BeO	7.195	3.0	14.1	11.5	1.55
AlN	8.257	4.8	7.3	8.2	1.20
GaN	8.483	5.0	7.6	7.6	1.30
ZnO	8.628	4.0	9.3	7.3	1.10
InN	9.399	5.5	6.8	5.9	1.30
CuBr	10.865	4.4	6.9	4.1	1.50
CdS	11.047	5.2	5.9	4.0	1.50
CdSe	11.489	5.8	5.5	3.6	1.50

from the table that all b values are similar around the average $b \simeq 1.5$. In other words, the assumed empirical formula (6.5.17) is approximately held.

So far we have examined the compounds having a crystal structure with a coordination number four. It is interesting to generalize the argument for the compounds with rock salt structure such as I-VII compounds. Table 6.13 contains the lists of the values for E_h, C and b determined by exactly the same method as in Table 6.12 for the compounds having NaCl structure. This class of materials is usually regarded as being purely ionic crystals and in fact it is re-

Table 6.13. Band parameters for NaCl-type compounds [6.14]

Crystal	l.c.(a/a_0)	$\epsilon(0)$	C[eV]	E_h[eV]	b
LiF	7.592	1.9	23.0	7.0	1.90
NaF	8.731	1.7	20.9	5.0	1.90
LiCl	9.693	2.7	11.6	3.8	2.30
KF	10.104	1.8	16.1	3.5	1.95
LiBr	10.396	3.2	9.5	3.2	2.30
NaCl	10.639	2.3	11.8	3.1	2.00
RbF	10.658	1.9	13.9	3.0	1.85
NaBr	11.288	2.6	9.8	2.6	2.00
LiI	11.338	3.8	7.4	2.6	2.55
KCl	11.892	2.2	10.4	2.3	2.25
NaI	12.232	3.0	7.8	2.2	2.10
RbCl	12.436	2.2	9.7	2.1	2.15
KBr	12.472	2.4	9.3	2.1	2.30
RbBr	12.952	2.4	8.9	1.9	2.20
KI	13.352	2.7	7.4	1.7	2.35
RbI	13.875	2.7	7.1	1.6	2.25
MgO	7.958	3.0	14.5	6.3	1.45
CdO	8.873	6.2	7.6	4.8	0.9
CaO	9.091	3.3	14.6	4.5	1.90
AgF	9.298	2.9	12.2	4.3	1.30
SrO	9.751	3.2	13.4	3.8	1.90
MgS	9.833	5.1	7.1	3.7	1.50
MgSe	10.301	5.9	6.4	3.3	1.60
AgCl	10.482	4.2	7.8	3.2	1.35
CaS	10.753	4.5	9.1	3.0	2.30
AgBr	10.912	5.0	6.9	2.9	1.35
CaSe	11.168	5.1	8.1	2.7	2.30
SrS	11.376	4.4	8.5	2.6	2.15
SrSe	11.773	4.9	8.0	2.4	2.25
CaTe	11.990	6.3	6.7	2.3	2.65
SrTe	12.227	5.8	6.7	2.2	2.35

flected in the values of C which are clearly larger that those in Table 6.12. Another difference is in the mean value of b which is near 2 for NaCl-structure compounds. However, as a whole, it may be said that the general trend exhibited in the both tables is the same, giving quite reasonable results. To understand the physical implication of such values of E_h and C, let us mention, as an example, a correlation between the strength of ionicity (say magnitude of the ratio C/E_h) and crystal structures. Figure 6.27 is a plot of the material points represented by C (ordinate) and E_h (abscissa) with different symbols for different crystal structures. It is amusing to observe in this figure that a boarder line exists expressed by

$$f = \frac{C}{E_h} = 0.785$$

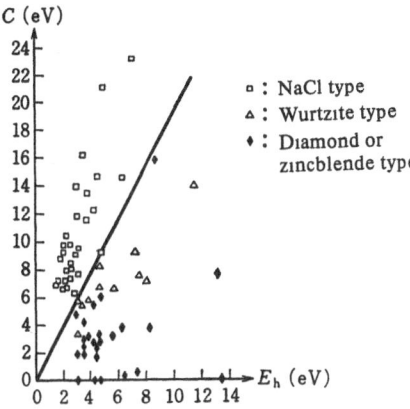

Fig. 6.27. Plots in the C–E_h plane. ($\square$) NaCl type; ($\triangle$) Wurtzite type; (▲) Diamond or zincblende type

□ : NaCl type
△ : Wurtzite type
♦ : Diamond or zincblende type

which separates two regions where NaCl structures and tetrahedrally-bonded structures are realized. Although not so clear, another boarder line seems to exist between the zincblende and wurtzite structures. We can interpret these results qualitatively in the following way. When ionicity is strong enough, the major cohesive force is Coulomb attraction which has no directional dependence, and hence a crystal tends to take a close-packed form with high coordination number 6, namely the NaCl structure. On the other hand, when the covalent character is strong, the major cohesive force is directional and the coordination number, say 4, is first fixed and a crystal takes a form compatible with this coordination number, leading to a rather loosely-packed diamond structure. As to the boundary between zincblende and wurtzite structures, it is not easy to give such a simple interpretation.

6.6 Semimetals

Besides the substances mentioned in Table 6.10 in the previous section, there is another group of semiconductors referred to as IV-VI compounds. In Table 6.14, we list the substances along with their energy gap ε_G and low-frequency dielectric constant $\epsilon(0)$. The characteristics of these substances are that all, except for the

Table 6. 14. Data for IV-VI compounds

	$\varepsilon_G[eV]$	$\epsilon(0)$	$T_m[°C]$
GeTe	~0.1	large	725
SnTe	0.2	T-dep.	780
PbS	0.37	170	1077
PbSe	0.26	250	1062
PbTe	0.30	400	904

low temperature phase of GeTe, crystallize into a NaCl structure and all have very high dielectric constants, the largest among semiconductors.

As to the factors which stabilize this crystal structure, one is inferred from the argument given in the previous section to be the ionicity, but there is another important origin which does not exist in the case of covalent crystals of the IV elements. We have seen that in the IV group, except for the heavier atoms, the bonding states composed of the hybrid sp^3 orbitals play a crucial role in forming a stable diamond structure. For this, however, it is necessary that the difference between the atomic p level and s level is not too large. Now, the IV-VI compounds consist of two atoms lying on both sides of the column of the V group in the periodic table and it can be assumed that both atoms are ionized so as to take the electronic configurations similar to the V group on average. In the V group, the outer s and p orbitals are rather strongly bound to the nucleus, giving rise to a large separation between p and s levels (see for instance, Table 1.1). Because of this, it will be less favourable energetically to form the sp^3 hybrids and more favourable for the s and p orbitals separately to form their bonding and antibonding orbitals and then two s electrons per atom are accommodated in lower s-bands (bonding and antibonding) and three p electrons per atom into the lower bonding p-orbital band, leaving the upper antibonding p-band vacant. This explains qualitatively why these compounds are crystallized in a NaCl structure and become insulators.

If this view were correct, it may be said that, in the IV-VI compounds, the trends of ionization and hybridization are complementary to each other and hence the cohesion mechanism is in between that of an ionic crystal and a covalent crystal. It is expected that the larger the difference in the electronegativity of the constituent atoms, the more the ionic binding stabilizes the crystal structure. Indeed, it is seen in Table 6.14 that GeTe, which contains the relatively lighter atom Ge, assumes the NaCl structure only above 700 K and below this temperature it is crystallized into the so-called As-type structure shown in Fig. 6.28 (cf. Fig. 6.3). This is because GeTe has rather small ionicity so that its bonding

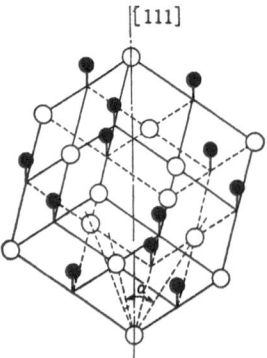

Fig. 6.28. Crystal structure of GeTe

character is akin to that of the Vth elements in which a NaCl structure is forced to deform, as will be discussed shortly.

The Vth elements As, Sb and Bi have the structure shown in Fig.6.3. with structure parameters listed in Table 6.15. This crystal lattice can be thought to be constructed in the following way. First shift the two fcc sublattices of a simple cubic lattice along the [111] direction, so that the nearest distance between the planes perpendicular to [111] becomes u, and then further deform the lattice to have the bond angle α as indicated in Table 6.15. In contrast to the case of IV-VI compounds, there is only one kind of atom in the Vth elements and hence no ionicity, resulting in the destabilization of the NaCl structure. As discussed later, all the Vth elements become semimentals.

Table 6.15. Structural constants for Vth group semi-metals (see also Fig. 6.28). u is the shortest distance between atomic layers perpendicular to the [111] direction divided by the length of the diagonal along [111]

	l.c. a[Å]	α	u
As	4.131	54° 10′	0.226
Sb	4.5066	57° 65′	0.233
Bi	4.7459	57° 14′	0.237
GeTe	4.234	58° 15′	
s.c.		60°	0.25

It is believed that the low temperature phase of GeTe should be ferroelectric, because the different constituent atoms are more or less ionized and relatively shifted, as indicated in Fig.6.28, producing a spontaneous polarization. SnTe, on the other hand, is normally of NaCl structure and has a very large dielectric constant which increases with decreasing temperatures. This behaviour is similar to a typical ferroelectric substance. The origin of this anomaly will be related with the instability of the NaCl structure in this substance tending towards the GeTe structure. The dispersion relations for optical phonons in SnTe have been determined by means of inelastic neutron scattering. The result is shown in Fig. 6.29a where it is clearly observed that the frequency of the transverse optic phonon (TO mode) at Γ-point decreases remarkably with decreasing temperatures. This indicates the fact that this phonon mode becomes soft and the restoring force of the lattice against the deformation associated with this phonon mode is weakened, leading to an instability. Figure 6.29b shows the temperature dependence of the square of the TO mode frequency measured at a long-wave limit. If such a softening of a TO mode proceeds and reaches zero value at a certain temperature, then a permanent deformation of the lattice will take place accompanied by a spontaneous polarization, and thus the substance will become a ferroelectric state. This is a phenomenon taking place in the so-called displacive ferroelectrics and one can expect that this is the case for GeTe below 700 K, although any direct experimental verification does not exist.

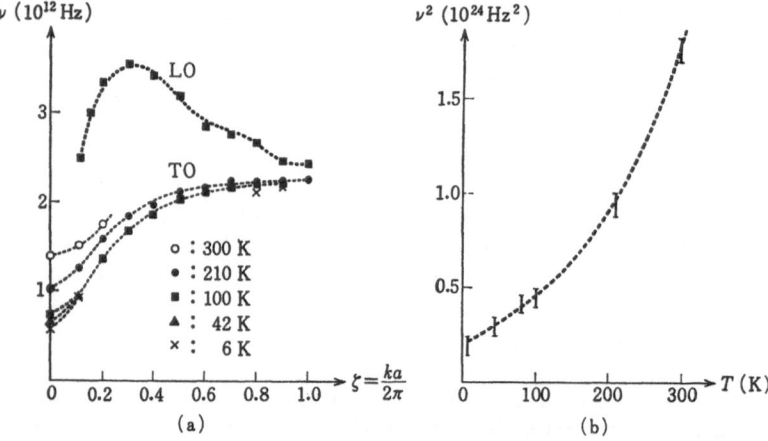

Fig. 6.29a-b. TO soft mode in SnTe. (a) Dispersion curves for LO and TO waves; (b) the square of TO mode frquency at $k = 0$ vs temperature

In order to prepare to understand both the electronic states of the Vth semi-metal elements and the electronic anomaly in SnTe, let us examine the band structures of NaCl-type semiconductors. The band structures of GeTe, SnTe, PbTe and PbSe were calculated on the basis of the relativistic OPW method. Here we demonstrate, as an example, the result of band calculation for SnTe based on EPM [6.15]. As explained in Sect.6.3., in the EPM for SnTe we need to consider both the symmetric part V_g^s and antisymmetric part V_g^a of the pseudopotential. For V_g^a we take the pesudopotential of Sb (which is the atom between Sn and Te) with proper correction for the difference in atomic volume, and choose V_g^a as an adjustable parameter to fit $\epsilon_2(\omega)$ obtained from the optical absorption experiment with that determined theoretically from the results of band calculation. By the symmetry of NaCl structure, there appear in V_g^a only the components of $g = (g_1, g_2, g_3)$ with odd numbers and in V_g with even numbers. When only the components

$$V_{111}^a = 0.56, \quad V_{311}^a = 0.05 \text{ [eV]}$$
$$V_{200}^s = -3.17, \quad V_{220}^s = -0.38, \quad V_{222}^s = 0.22 \text{ [eV]}$$

are retained and all others are discarded, the resulting band structures and ϵ_2 (ω) evaluated thereof are as shown in Fig.6.30, The sharp peak which appeared at about $\hbar\omega = 2$ eV in $\epsilon_2(\omega)$ makes a good correspondence with observation. However, since it is not so clear how this peak comes out from the band structure, let us try to understand the real mechanism by applying the interpretation in terms of the Jones zone used for Si and Ge in Sect.6.5 for the case of SnTe. Because SnTe is composed of Sn with the electronic configuration $(5s)^2 (5p)^2$ and Te with $(5s)^2 (5p)^4$, there are 5 electrons per atom. Therefore we have to

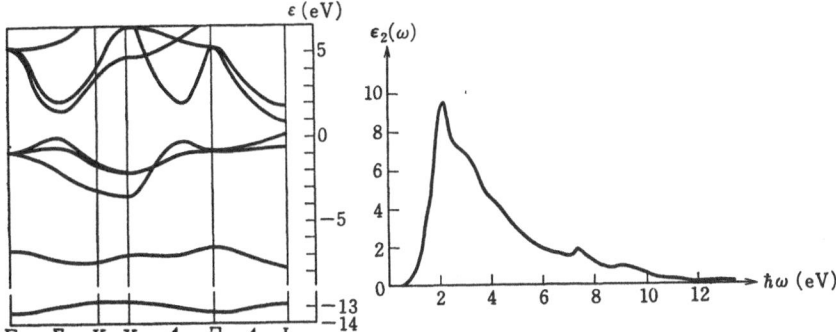

Fig. 6.30. Band structure and $\epsilon_2(\omega)$ of SnTe [6.15]

consider the Jones zones which can accommodate 10 electrons per atom. That is, the Brillouin zones should be taken up to the fifth zones.

In Fig.6.31, the first Brillouin zone and the Jones zones expanded up to the fifth Brillouin zones are depicted for a fcc lattice. The letters indicated in the Jones zone denote the symmetry points which correspond when mapped into the first Brillouin zone. Glancing at this figure, we immediately realize that the Jones zone takes a nearly spherical shape constituted mainly from (311)'s planes. Therefore it will be interesting to calculate the band energies using the same pseudopotential as before along the (311) plane surfaces by varying the wave vector $\mathbf{k}$ as $W \rightarrow Q \rightarrow L \rightarrow \Sigma \rightarrow W$. The result is shown in Fig. 6.32 in which a forbidden band is indicated by the shaded area. It is clearly read off from this figure that the forbidden band has a nearly constant width of 2 eV, explaining quite naturally the reason why a sharp peak exists at around 2 eV in $\epsilon_2(\omega)$. Thus,

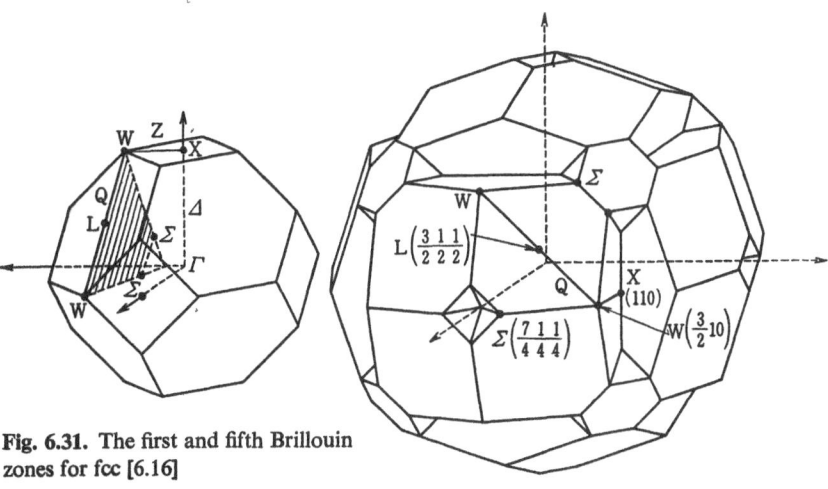

Fig. 6.31. The first and fifth Brillouin zones for fcc [6.16]

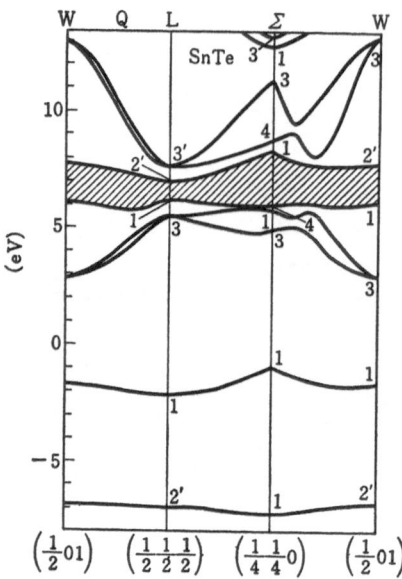

Fig. 6.32. Energy gap for SnTe along the Jones zone [6.16]

we can see that even in SnTe, its characteristics as a semiconductor may be explained in terms of Jones zones. However, further calculations of charge density distribution and other properties are not always successful as in the cases of Si and Ge. For instance, when the magnitude of the energy gap on the (311) plane is evaluated, the previous table for V_g gives value of $V_{311}^s = 0.05$ eV which is too small. Even if we proceed to the next order perturbation theory

$$V_{\text{eff}}(311) = V(311) + \sum_G \frac{\langle k|V|k-G\rangle \langle k-G|V|k-311\rangle}{k^2 - (k-G)^2}, \qquad (6.6.1)$$

this gives at W point

$$V_{\text{eff}}(311) \approx V(311) + \frac{2m}{\hbar^2}\left(\frac{a}{2\pi}\right)^2 V(111)[V(220) + V(200)]$$

$$= 0.05 - 0.54 = -0.49 \quad [\text{eV}] \qquad (6.6.2)$$

and hence

$$e|V_{\text{eff}}(311)| = 1 \quad [\text{eV}] \qquad (6.6.3)$$

which is still too small. This shows that the perturbational calculation is not enough in this case.

Finally, let us look at the problem why the Vth elements become semimetals. With respect to this topic, a model calculation performed by *Cohen* and his coworkers [6.17] is very instructive and given us a deep insight into this problem.

The following discussion is based on their work. As an example, we always consider As and if necessary take the lattice constant and other parameters of the As crystal. The present aim is to explore the relationship between the electronic structure and crystal lattices on the basis of pseudopotential theory. As to the pseudopotential $V(g)$, we adopt a continuous function $V(g)$ of g which is obtained by extrapolating several values of $V(g)$ determined for Ge. We consider four types of crystal lattices: simple cubic (s), rhombohedral (r) derived from a simple cubic by a deforming [111] direction, two displaced fcc lattices (d) with a relative shift along the (111) direction, and finally an As-type lattice (As) which is an addition of (r) and (d). The values of $V(g)$ used are listed in Table 6.16 for each case.

Table 6.16. List of values of $V(g)$ used in the band calculations

g	(s)	(r)	(d)	(As)	I	II
111	0	0	+0.099	+0.108	+0.093	+0.093
11$\bar{1}$	0	0	− 0.034	− 0.032	+0.093	+0.093
200	− 0.177	+0.176	+0.169	+0.168	− 0.177	− 0.177
220	+0.012	− 0.018	+0.010	− 0.015	+0.012	+0.012
2$\bar{2}$0	+0.012	+0.039	+0.012	+0.039	+0.012	+0.012
311	0	0	+0.039	+0.040	+0.025	+0.093
222	− 0.042	− 0.060	− 0.026	− 0.037	+0.042	+0.042
31$\bar{1}$	0	0	− 0.025	− 0.023	+0.025	+0.093
3$\bar{1}\bar{1}$	0	0	+0.009	+0.006	+0.025	+0.093
22$\bar{2}$	− 0.042	− 0.032	− 0.040	− 0.031	+0.042	+0.042

First, the band structure for (s) is shown in Fig. 6.33. The notations of the symmetry points are the same as that of the Brillouin zone for fcc (Fig. 6.31). The Fermi surface appears at the place where 10 electrons can be accommodated. Its energy level is indicated in the figure by a dotted line. The first two lower bands are s-bonding and s-antibonding bands, and the next two (one doubly degenerate, and a single) should be the p-bonding bands. From this figure, it is concluded that the simple cubic lattice is metallic. For the (r) structure, the bands are changed, as shown in Fig. 6.34. In this case the Brillouin zone is shown in Fig. 6.35 (note that the enquivalent L points in fcc are modified to L and T points). Although some degeneracies which have appeared in (s) are lifted in this bands, no essential changes take place, leaving the lattice still metallic. In the case of (d), however, a dramatic change in band structure occurs as shown in Fig. 6.36. On the one hand, the degeneracy of the crossing lines on the X line is completely removed and at the same time new crossing lines $W_1 − T_1$ and $W_2 − T_3$ appear on the $W − T$ line. As seen in this figure, in the case of two displaced fcc lattices, levels of the bonding p orbitals are lowered below the Fermi level and stabilized. This state corresponds to a gapless semiconductor. Finally, the bands

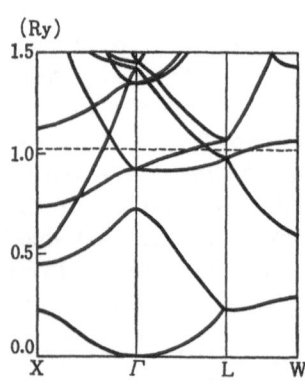

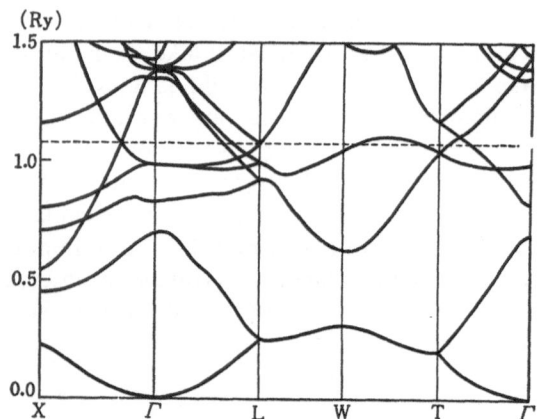

Fig. 6.33. Band structure of As for a simple cubic lattice (s). Dotted line indicates the Fermi surface

Fig. 6.34. Band structure of rhombohedral As (r)

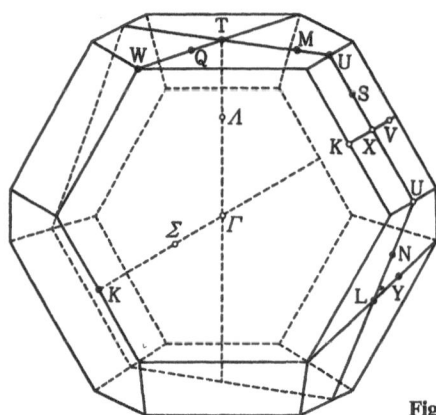

Fig. 6.35. Brillouin zone for an As-type structure

for (As) become as shown in Fig. 6.37. In this case it is observed that the cross point at T is raised above the Fermi level and bands at the L-point are significantly modified, giving rise to positive holes at the T-point and electrons at the L-point. Thus a semimetal is constructed.

Because the above results might depend sensitively on the chosen values of the pseudopotential and the effect of spin-orbit coupling may not be neglected, particularly for heavier atoms, all the calculations given so far should be regarded as giving just some suggestions, but they are still instructive for understanding why some elements exist in a rather special state such as semimetals.

What can we deduce when the above argument is extended to the case of IV-VI compounds by introducing antisymmetric parts of the pseudopotential so that

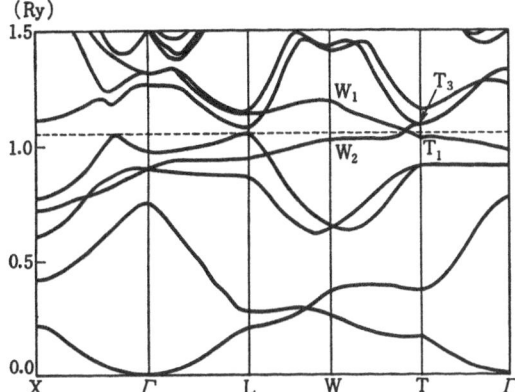

Fig. 6.36. Band structure of As for displaced fcc lattice (d)

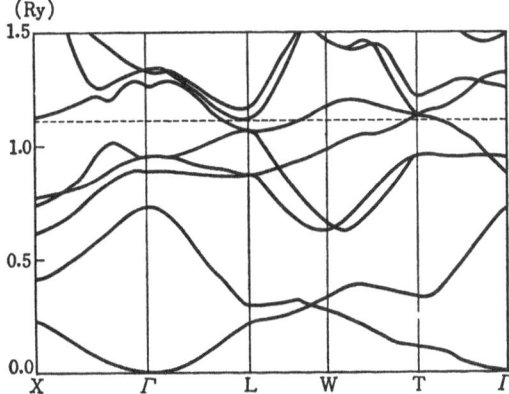

Fig. 6.37. Band structure of As (As)

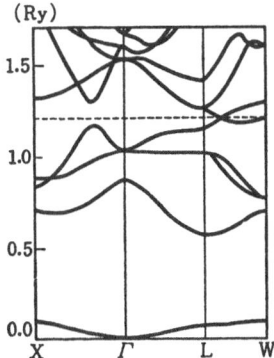

Fig. 6.38. Band structure of As with fcc lattice when I-type V_g^a is included

nonequivalence between IV and VI-atoms may be taken into account? If we consider two kinds of antisymmetric parts V_g^a, type I and type II, as listed in the last two columns of Table 6.16, and perform band calculation for a fcc structure, we obtain a result as shown in Fig. 6.38 for a type I parameter. In this case, it is seen from this figure that s-bands and p-bonding bands are below the Fermi level, making the system a semiconductor. On the other hand, if we adopt a weak antisymmetric potential, the type II parameter, the result is as shown in Fig. 6.39, which reveals no energy gap. It is interesting to note that a small change in the antisymmetric potential may cause a drastic change in the characteristics of a lattice. Thus, we see that the antisymmetric part (which is related with ionicity) can stabilize the semiconducting state. The situation in GeTe may probably be near to type II, and the origin for the lattice deformation which removes the cross degeneracy occurring on the L — W line would be the same as the case of As.

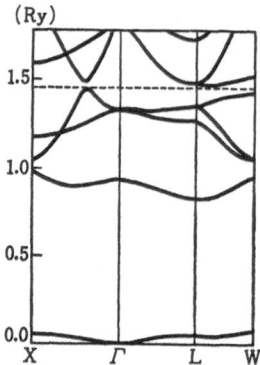

Fig. 6.39. Band structure of As with fcc lattice when II-type V_{g}^{*} is included

6.7 Properties of Matter Under High Pressure

In Chap. 2 we examined the phase diagram of a system composed of protons and electrons, extending the state to very high temperatures and high pressures. We have seen there that with very large changes in temperature and pressure, the normal state realized at low temperature and pressure becomes quite different. This situation seems the same for all materials. What we have discussed so far in this chapter is concerned with the state in a normal condition, that is for instance, the most stable state at $T = 0$ and $P = 0$, and we can expect dramatic changes in these materials when very high pressure is applied. As mentioned repeatedly, the electronic properties of nonmetals depend sensitively on the values of the lattice constants. The energy gap, which is an important quantity to make a discrimination between an insulator, semiconductor and semimetal, is changed with the degree of atomic overlap. Hence it is expected that under sufficiently high pressure, the meaning of the above distinction among nonmetal substances is lost and eventually all the nonmetal substances go to the metallic state. In what follows, referring to the results of experiments done, so far, we shall examine how matters may be changed under very high pressure.

To begin with, let us look at some examples of insulator (molecular crystal)—metal transition under high pressure. The halogen element I_2 is a molecular crystal having a lattice structure, shown in Fig. 6.1. The molecular axes are along the ac-plane and the outlook of the crystal shows a layered structure with a dark violet colour and metallic luster, being rather nearer to a semiconductor. When electric resistivity of this substance is measured with varying pressure, the result is as shown in Fig. 6.40. It is seen that during change of pressure in the range of 100 kbar, the electric resistivity reveals a remarkable decrease within the 5th order of magnitude. Pressure dependence of the energy gap can be also determined by means of optical absorption under pressure. The result is shown in Fig. 6. 41. In this figure, twice the activation energy determined from the temperature dependence of resistivity is also included along with the energy gap obtained

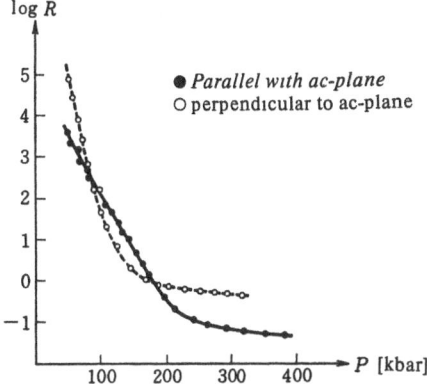

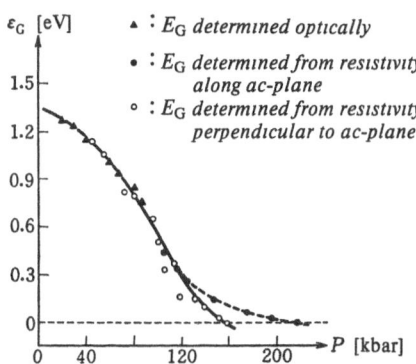

Fig. 6.40. Change in electric resitivity of I_2 under high pressure. (●) Parallel with ac-plane; (○) perpendicular to ac-plane [6.18]

Fig. 6.41. Pressure dependence of energy gap for I_2. (▲) ε_G determined optically; (●) ε_G determined from resistivity along ac-plane; (○) ε_G determined from resistivity perpendicular to ac-plane

optically. This is because the electric conductivity of a semiconductor with an energy gap ε_G depends on ε_G as

$$\sigma \propto \exp\left(-\frac{\varepsilon_G}{2kT}\right) \tag{6.7.1}$$

which gives an experimental check of ε_G independent of the optical measurement. As seen from this figure, the energy gap seems to disappear above the pressure of about 160 kbar, indicating that the substance is going to a metallic state. As a matter of fact, it is observed that above this pressure, resistivity is almost proportional to absolute temperature showing a change of conduction mechanism from a semiconductive to a metallic one (Fig. 6. 42).

Figure 6.43 shows a result of pressure dependence of resistivity measured for semiconducting Se. In contrast to the case of I_2, the resistivity in this case exhibits a discontinous jump at about 130 kbar, indicating that a kind of phase transition of the first order takes place. The mechanism of this phase transition is not known at present. But the high pressure phase is undoubtedly metallic, as verified by an experimental fact that electric resistivity is linear in absolute temperatures in the high pressure phase above 130 kbar.

Now we proceed to the behaviour of semiconductors with diamond or zinc-blende structures under high pressure. From practical as well as theoretical interests, the pressure dependence of band structures of semiconductors such as Si and Ge have been investigated in some detail at least for low pressure. In particular, within a pressure range 0—100 kbar, these semiconductors reveal each characteristic behavior in the pressure dependence and enable us to understand

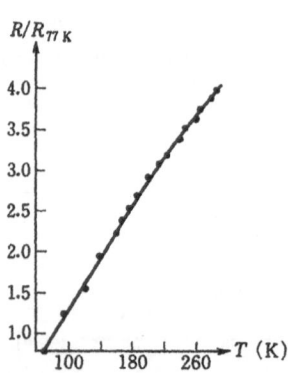

Fig. 6.42. Temperature variation of electron resitivity of I_2 under high pressure (240 kbar)

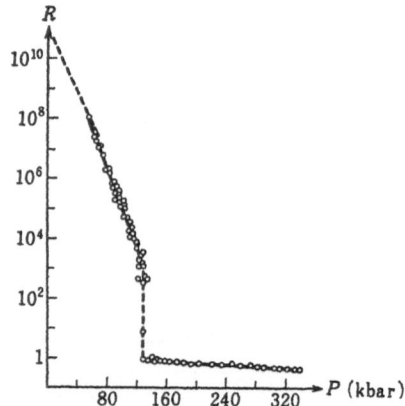

Fig. 6.43. Pressure dependence of the resistivity of Se

the difference in the band structures among them. Figure 6.44 shows the pressure dependence of the characteristic frequency associated with an interband transition corresponding to the optical absorption edge for Si and Ge. It is interesting to note that the energy gap for Si reveals a monotonous red shift with increasing pressure whereas for Ge it first increases (blue shift) up to about 35 kbar and then decreases through a maximum, showing a behavior similar to Si at high pressure. Another example of GaSb in III-V compounds is shown in Fig. 6. 45. In this case, the changes of the energy gap with pressure take place in three steps. It increases linearly with a slope $d\varepsilon_G/dP = 12 \times 10^{-3}$ eV·kbar^{-1}, changes its slope to 7.3×10^{-3} eV·kbar^{-1} at about 18 kbar, takes a maximum at around 50 kbar and then decreases.

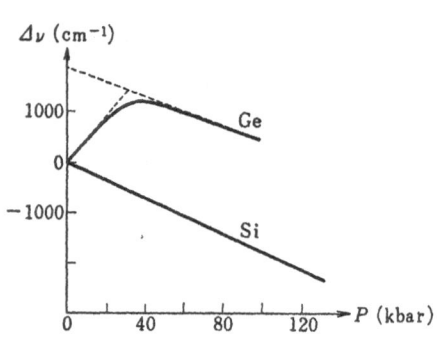

Fig. 6.44. Pressure dependence of the absorption edge for Si and Ge [6.19]

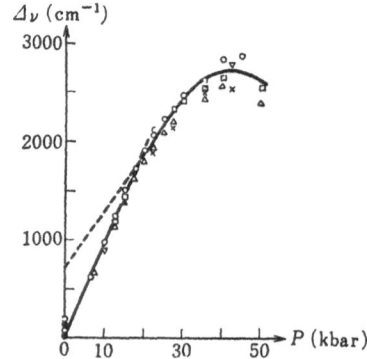

Fig. 6.45. Pressure dependence of absortion edge for GaSb [6.20]

Such different variations of three different substances may be explained by the changes of band structures with pressure. As stated in Sect.5.3, the conduction band of Si has minima along [100] (see Fig. 6.13). If one assumes that the band energy at these minima decreases with increasing pressure, then the variation of the relative distance between the minima and the top of the filled valence band will explain the experimental curve for Si. In contrast to this, the minima of the conduction band of Ge lie along [111]. In order to explain the behavior in Fig. 6.44, one has to assume that the energy at these manima increases with increasing pressure. However, for the case of Ge, above a certain pressure the minima at [100] will go down below the minima at [111] and a further increase in pressure causes the decrease in the energy gap parallel with the case of Si. On the other hand, for the case of GaSb, a minimum of the conduction band is at the point [000], in addition to the extremum points along [111] and [100]. The three-step variation in Fig. 6.45 can be explained by assuming that first, the energy gap is determined by the minimum at [000], then crosses over to [111] and is finally taken over by the minima at [100]. Now, with a further increase in pressure, it happens that at a certain pressure the electrical resistivity decreases to a value corresponding to a metallic state as shown in Fig. 6. 46. That the high pressure phase is metallic is verified experimentally by the observed positive temperature coefficient of resistivity. This discontinous jump in the resistivity should correspond to a certain phase transition of the first order. According to the structure analysis by means of x-ray diffraction, the high pressure phase has a crystal structure the same as that of white tin, indicating a fact that the semiconductor ↔ metal transition at high pressure in Si and Ge is of a similar nature to the phase transition gray tin ↔ white tin.

It is known that the semimetals (As, Sb and Bi) behave in a differently under high pressure. With applied pressure, the As-type structure with the parameters indicated in Table 6.15, is deformed in such manner that the angle α varies toward a value $\alpha = 60$ and the parameter u increases to approach $u = 0.25$, both tending towards the values corresponding to a simple cubic lattice. Thus under high pressure, these semimetals first make a deformation towards the fcc lattice relatively displaced along [111] and then approach a simple cubic struc-

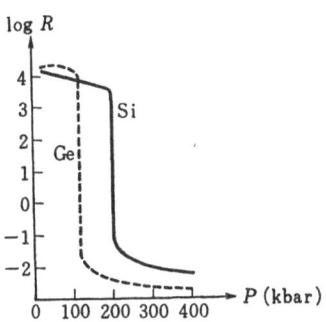

Fig. 6.46. Metallization of Si and Ge under high pressure. Logarithm of resitivity vs pressure. [6.21]

ture. The final structure at very high pressure seems to be hcp. On the other hand, if the electric resistivity is measured, it changes with pressure at first rapidly until it reaches a value corresponding to an insulator (semiconductor), then followed by a decrease to metallic values. This behaviour clearly shows the peculiarity in the electronic structure of these elements and can be understood qualitatively from the band structures discussed in Sect.5.6. As seen from Figs. 6.37, 6.36 and 6.33, the changes in the crystal lattices such as As-type structure → displaced fcc lattice → simple cubic lattice cause changes in the band structures and Fermi surfaces so that the changes semimetal → semiconductor → metal take place.

Under high pressure, metals may undergo phase transitions and change their properties remarkably. Within the hydrostatic pressure of the range 0–500 kbar, the main changes in metals observed so far are summarized as follows:

1) Alkali metals become superconducting under high pressure.
2) Two valent fcc metals become bcc metals through a nonmetallic state.
3) Two valent bcc metals become superconducting.
4) Ferromagnetic Fe becomes paramaganetic under high pressure.

Generally speaking, all the elemental substances would eventually become metals under extremely high pressure. It is believed that even the inert gas elements are not an exception. However, we have not succeeded so far in metallization of the inert gases. According to a theoretical estimation, a super high pressure of the order of 0.7 Mbar would be required to produce a metallic Xe.

What about the possible changes in ionic crystals under high pressure? Many experiments already exist on this point, and it is known that most ionic crystals undergo pressure-induced phase transitions. Since volume is reduced by compression, it is natural that atoms or ions tend to assume a packed configuration as closely as possible. On the other hand, in the case of ionic crystals, there are at least two kinds of ions with different sizes. Generally, we can imagine that anions are more compressible than cations. Therefore, the ratio of ionic radius R_+/R_- is increased with increasing pressure. Because of this, the changes in the crystal structure of ionic crystals take place in such way that the coordination number of an anion surrounding cations may be increased. For instance, the ionic crystals NaCl (295 kbar), KCl (19 kbar) and RbCl (5.3 kbar) undergo a phase transition from a NaCl structure to a CsCl structure at the pressure indicated in the brackets, respectively. That is, this phase transition is accompanied by a change in the coordination number 6 → 8.

7. Localized Electron Approximation

In Sect. 6.5, where the electron theory of nonmetals was described we extended the pseudopotential theory which proved most effective for simple metal (Sect. 5.4). Furthermore we discussed to what extent essentially the ideas of the pseudopotential theory of simple metals may be applied to semimetals and semiconductors for explaining their electronic properties. In essence, the aim of that section was to demonstrate how real properties of nonmetals can be understood in a unified way, using only the information delivered from the pseudopotential common to metals as the sole input source. The fact that many properties of nonmetals can be explained as well from the standpoint of electron band theory may be regarded as demonstrating the success of this unified approach.

However, from the viewpoint of the band structure of nonmetals, it is a characteristic feature that there are the filled valence bands and the vacant conduction bands separated from each other by an energy gap. Thus, no Fermi surface exists on which many nearly free electrons are ready to respond to any disturbances. Therefore, the picture of nearly free electrons will not be applicable as in metals. On the contrary, for nonmetals, the picture of localized electrons is more acceptable in many cases and some times the concept of the bond, familiar to chemists, may be more useful and adequate. This suggests that an electron theory of nonmetals may be constructed from a point of view which is just opposite to that of the pseudopotential theory for metals. If such an approach is established, it will provide useful and instructive information about the electronic behaviour in nonmetals. Furthermore, if one could clarify the interrelationship between two quite different approaches, our insight into a unified view of materials would be very much deepened.

In this chapter, as a supplement to Sect. 6.5, we shall give a brief sketch of recent theories, which start from electron orbitals in atoms constituting nonmetallic substances, and describe the electronic states in terms of linear combination of localized atomic orbitals. This approach will show a success comparable, in some sense, to the pseudopotential theory for simple metals.

7.1 A Simple Model for Tetrahedrally-Bonded Semiconductors

In Sect. 6.3, we explained qualitatively the reasons why the IVth elements, except for heavier ones such as Sn and Pb, are crystallized in the diamond structure and become either very good insulators or typical semiconductors. In this section we shall introduce a very simple but useful model (so-called Weaire-Thorpe model [7.1]) to describe the properties of covalent crystals with tetrahedrally-bonded structure.

Consider an assembly of IVth element atoms, each having four valence electrons. As stated before, each atom is providing one s orbital and three p orbitals as building blocks. We have also learned that from these building blocks we can construct four orbitals (6.3.1) which have the largest amplitudes in the four tetrahedral directions. Let us attach a number i to each atom, and numerate four bonds emerging from each atom as $j = 1, 2, 3, 4$ in a definite order. Thus we discriminate all the orbitals by using the notation ϕ_{ij} which represents the jth bond orbital emerging from the ith atom.

The Weaire-Thorpe model (WT model) contains only two parameters V_1 and V_2, and describes the electronic states of a crystal by a Hamiltonian

$$\mathcal{H} = \sum_i \sum_{j \neq j'} V_1 |\phi_{ij}\rangle \langle \phi_{ij'}| + \sum_j \sum_{i \neq i'} V_2 |\phi_{ij}\rangle \langle \phi_{i'j}| . \tag{7.1.1}$$

Here $|\phi_{ij}\rangle$ is the Dirac ket representing the electron state in ϕ_{ij} orbital, and $\langle \phi_{ij}|$ its Hermite conjugate bra. The first term in the right-hand side of (7.1.1) describes the electron hopping between j and j' orbitals belonging to the same atom i, where V_1 is the overlap integral. The second term describes the electron hopping between two atoms i and i' forming a jth bond. Its overlap integral is given by V_2.

If we put $V_2 = 0$ in (7.1.1), this is equivalent to the Hamiltonian of an assembly of isolated atoms. The remaining first terms give a 4×4 matrix associated with sp^3 orbitals for each atom:

$$\mathcal{H}_1 = \begin{bmatrix} 0 & V_1 & V_1 & V_1 \\ V_1 & 0 & V_1 & V_1 \\ V_1 & V_1 & 0 & V_1 \\ V_1 & V_1 & V_1 & 0 \end{bmatrix} \tag{7.1.2}$$

The eigenvalues of this matrix can be readily obtained as

$$\begin{aligned} E_s &= 3V_1 & \text{(singlet)} \\ E_p &= -V_1 & \text{(triplet)} \end{aligned} \Bigg\} \tag{7.1.3}$$

This is nothing but the s level and p level of the original isolated atom, and $4|V_1|$ (note $V_1 < 0$) gives the level separation between them. On the other hand, if we put $V_1 = 0$ in (7.1.1), for each j we have a 2×2 matrix associated with two orbitals forming the jth bond

$$\mathcal{H}_j = \begin{bmatrix} 0 & V_2 \\ V_2 & 0 \end{bmatrix} . \tag{7.1.4}$$

The eigenvalues are obtained as

$$E = \pm V_2 . \tag{7.1.5}$$

They correspond to the energies of bonding and antibonding orbitals. Thus we see that the ratio of V_1 and V_2 determines the feature of the electronic state of the crystal. In the IVth elements, it is expected that a relation $|V_2| > |V_1|$ is satisfied and hence the level of electrons participating in the bonding is split into bonding and antibonding states by V_2. Each split level is further broadened by V_1 into a bonding band and an antibonding band. Now four electrons per atom may be just accommodated in the lower bonding band, filling the lower band exactly to make the crystal an insulator with a finite energy gap. For this reason we call V_2 the *bonding parameter* and V_1 the *banding parameter*.

In order to know more about the electronic states, it is necessary to diagonalize (7.1.1) to obtain the energy eigenvalues for a whole crystal. As shown in Fig. 7.1, a primitive cell of the diamond lattice contains two atoms and there are eight orbitals $|\phi_{ij}\rangle$ per primitive cell. It is more convenient to denote these orbitals as $|r_i, n\rangle$ instead of $|\phi_{ij}\rangle$. Here r_i is the position vector of the primitive cell and the eight orbitals are labelled by $n = 1, 2 \ldots 8$, as shown in Fig. 7.1.

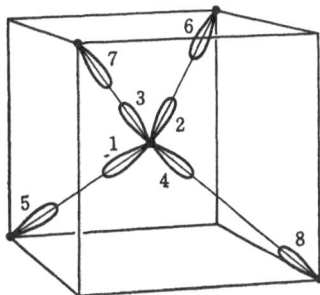

Fig. 7.1. Diamond structure in which 8 hybridized orbitals are shown

Since the Hamiltonian (7.1.1) has the translational symmetry, it should be partially reduced by Bloch functions into diagonal blocks for each wave vector k. That is, let N be the total number of atoms, and define

$$|nk\rangle = \sqrt{\frac{2}{N}} \sum_{r_i} \exp{(ik \cdot r_i)}|r_i n\rangle \qquad (7.1.6)$$

for a new basis, then H is decomposed into an 8×8 matrix for each k. It is shown that each k block has the following form:

$$H(k) = \begin{pmatrix} 0 & V_1 & V_1 & V_1 & V_2 & 0 & 0 & 0 \\ V_1 & 0 & V_1 & V_1 & 0 & V_2 & 0 & 0 \\ V_1 & V_1 & 0 & V_1 & 0 & 0 & V_2 & 0 \\ V_1 & V_1 & V_1 & 0 & 0 & 0 & 0 & V_2 \\ V_2 & 0 & 0 & 0 & 0 & V_1\gamma_y\gamma_z & V_1\gamma_z\gamma_x & V_1\gamma_x\gamma_y \\ 0 & V_2 & 0 & 0 & V_1\gamma_y^*\gamma_z^* & 0 & V_1\gamma_x\gamma_y^* & V_1\gamma_z^*\gamma_x \\ 0 & 0 & V_2 & 0 & V_1\gamma_z^*\gamma_x & V_1\gamma_x^*\gamma_y & 0 & V_1\gamma_y\gamma_z^* \\ 0 & 0 & 0 & V_2 & V_1\gamma_x^*\gamma_y^* & V_1\gamma_z\gamma_x^* & V_1\gamma_z\gamma_x & 0 \end{pmatrix}, \qquad (7.1.7)$$

where

$$\gamma_x = \exp{(ik_x a/2)}, \quad \gamma_y = \exp{(ik_y a/2)}, \quad \gamma_z = \exp{(ik_z a/2)}. \tag{7.1.8}$$

Fortunately the matrix (7.1.7) has a special simple form such that all its eigenvalues may be obtained analytically. Let

$$\alpha_{xyz} = \cos{\frac{k_x a}{2}} \cos{\frac{k_y a}{2}} + \cos{\frac{k_y a}{2}} \cos{\frac{k_z a}{2}} + \cos{\frac{k_z a}{2}} \cos{\frac{k_x a}{2}}, \tag{7.1.9}$$

then it is easily proved that the equation to determine the eigenvalues is factorized into the form

$$[(E_k + V_1)^2 - V_2^2]\{[(E_k + V_1)(E_k - 3V_1) - V_2^2]^2 - 4V_1^2 V_2^2(1 + \alpha_{xyz})\} = 0. \tag{7.1.10}$$

From (7.1.10), we find

$$E_k = -V_1 + V_2 \quad \text{(doublet)} \tag{7.1.11a}$$

$$E_k = -V_1 - V_2 \quad \text{(doublet)} \tag{7.1.11b}$$

$$E_k = V_1 \pm [4V_1^2 + V_2^2 \pm 2V_1 V_2 (1 + \alpha_{xyz})^{1/2}]^{1/2}. \tag{7.1.11c}$$

It should be noted that the k dependence of energy eigenvalues, which is a reflection of the translational symmetry of the crystal, is contained in a single function (7.1.9). If a diamond lattice were formed by atoms having only single s orbitals, then the energy band would, as is easily checked, be given by

$$E_k = E_0 \pm 2V(1 + \alpha_{xyz})^{1/2}, \tag{7.1.12}$$

where E_0 is the s-orbital energy and V the transfer integral between two nearest neighbor s orbitals. Note that the k dependence of the energy band (7.1.11c) is governed by a function $(1 + \alpha_{xyz})^{1/2}$ in the same way as the s-band (7.1.12) is. This similarity is sometimes utilized to derive the density of states of the diamond structure analytically from knowledge of simpler density of states for the energy band (7.1.12).

The density of states calculated in this way is shown in Fig. 7.2 for three different values of V_1/V_2. As seen from this figure and also from (7.1.11a, b), the bonding and antibonding p states have no k dependence, giving rise to the δ-function-like spikes in the density of states. This feature is characteristic of the WT model, and if we include in the model more parameters associated with more distant transfer integrals, the δ-function-like spikes broaden and this singular feature of the WT model disappears.

It is instructive to look upon the band edges as a function of V_1/V_2. From (7.1.11c) we see that the largest band width is obtained for $k = 0$ and $\alpha_{xyz} = 3$. This means that the band edge energy E_0 is given by

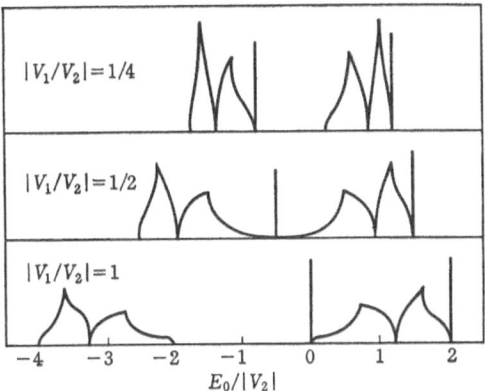

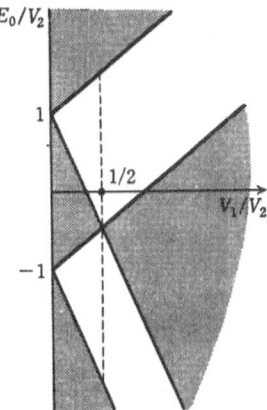

Fig. 7.2. One electron density of states for the WT model

Fig. 7.3. A schematic band structure for the WT model. The shaded parts represent the forbidden bands

$$E_0 = V_1 \pm [(2V_1 \pm V_2)^2]^{1/2} = V_1 \pm (2V_1 \pm V_2)$$
$$= \begin{cases} -V_1 \pm V_2 \\ 3V_1 \pm V_2 . \end{cases} \tag{7.1.13}$$

In Fig. 7.3, E_0/V_2 is plotted as functions of $|V_1/V_2|$, assuming that $V_1 < 0$, $V_2 > 0$. In this figure, the thick solid lines indicate that widthless energy bands contributed from (7.1.11a, b) are added, corresponding to the δ-function-like spikes in Fig. 7.2. Figure 7.3 may be regarded as viewing Fig. 6.12 in Sect. 6.3 from a different side. When atomic distance increases, V_2 should decrease and hence $|V_1/V_2|$ increases, and electron energy band is split into s- and p-bands. Therefore $|V_1/V_2| = 1/2$ is a critical value of the parameter at where a crossover of the bonding scheme takes place. For $|V_1/V_2| < 1/2$, the bands are formed from the bonding and antibonding states of sp^3 hybrid orbitals.

7.2 Bond Orbital Model (BOM)

Harrison and his co-workers [7.2] have generalized the WT model described in the previous section so as to include the zincblende structure crystals composed of two kinds of atoms, and constructed a bond orbital model which may work as a unified model for the tetrahedrally-bonded substances. In this section we shall give a brief account of their approach. It is an attempt to correlate the wide range of properties of four coordinate crystals in terms of a few parameters

Consider a group of the II-VI compounds or III-V compounds (which were a main subject of Sect. 6.5) of two constituent atoms, a more electronegative atom will be called an A atom and a more electropositive one a C atom. If necessary, we will discriminate them by suffices a and c, respectively. In order to realize the regular local tetrahedral arrangement of chemical bonds, we construct four sp^3 hybrid orbitals for each atom. For instance

$$|aj\rangle = \tfrac{1}{2}(|s^a\rangle + \sqrt{3}|p^aj\rangle) \tag{7.2.1}$$

is one of such orbitals constructed for an A atom. $|p^aj\rangle$ represents a proper linear combination of the p orbitals of the A atom having the maximum amplitude in the jth direction. Similarly we prepare $|cj\rangle$ orbitals for the C atom.

Let us define the orbital energies by

$$\left.\begin{aligned}\varepsilon_a &= \langle aj|H|aj\rangle\\ \varepsilon_c &= \langle cj|H|cj\rangle\end{aligned}\right\}. \tag{7.2.2}$$

The difference $\varepsilon_c - \varepsilon_a$ will play an important role in the following. If we put

$$V_3 = \tfrac{1}{2}(\varepsilon_c - \varepsilon_a) \tag{7.2.3}$$

then, when $V_3 > 0$, there is a tendency that an electron moves from the C atom to the A atom in order to decrease the energy of the electron system. Just as in the previous section, we define the matrix elements of energy between different orbitals in the same atom by

$$\left.\begin{aligned}V_1^a &= -\langle aj|H|aj'\rangle\\ V_1^c &= -\langle cj|H|cj'\rangle\end{aligned}\right\} \tag{7.2.4}$$

and the matrix element between two orbitals on the same bond by

$$V_2 = -\langle cj|H|aj\rangle. \tag{7.2.5}$$

Note that we have two V_1 parameters V_1^a and V_1^c instead of single V_1 as in the WT model. The sign of the matrix elements are chosen so that each V parameter may be a positive quantity. The four hybrid orbitals belonging to the same atom can be made orthogonal to each other, but the orbitals belonging to different atoms are generally not orthogonal. Thus we need some corrections due to this nonorthogonality. However, as is proved in a later section, this correction may be taken into account by renormalizing all the parameters with their suitable effective values. For simplicity, therefore, we shall neglect this effect in this section.

In contrast to the case of the IV elements, since the two atoms forming a bond are different, the electron density distribution along a bond is not uniform but polarized in one direction, making the two atoms more or less a positive or a negative ion. To examine this ionization effect, we first determine the electron distribution along a bond by minimuming the energy of the electron in the bond. In effect, this is achieved by constructing bonding and antibonding orbitals from $|aj\rangle$ and $|cj\rangle$ and accommodating two electrons with up and down spins into a lower energy bonding state. Let the bonding orbital be

$$|b\rangle = u_a|aj\rangle + u_c|cj\rangle . \tag{7.2.6}$$

Then the coefficients u_a and u_c may be determined by making $\langle b|H|b\rangle$ minimum under the condition $u_a^2 + u_c^2 = 1$. The results can be expressed in terms of

$$\alpha_p = \frac{V_3}{\sqrt{V_2^2 + V_3^2}} \tag{7.2.7}$$

as

$$u_a = [\tfrac{1}{2}(1 + \alpha_p)]^{1/2}, \quad u_c = [\tfrac{1}{2}(1 - \alpha_p)]^{1/2}, \tag{7.2.8}$$

and the minimum value of the bond energy becomes

$$\langle b|H|b\rangle = \tfrac{1}{2}(\varepsilon_a + \varepsilon_c) - (V_2^2 + V_3^2)^{1/2} . \tag{7.2.9}$$

The quantity α_p defined by (7.2.7) gives $u_a^2 - u_c^2$, which vanishes for the homo-polar case like the IV elements since $u_a = u_c$. Generally, when $\alpha_p \neq 0$, $u_a^2 \neq u_c^2$ and hence the electron distribution along the bond is not symmetric. Therefore, we shall call α_p the polarity parameter. It is consistent that the polarity parameter α_p is proportional to V_3 defined in (7.2.3).

Besides the bonding orbital $|b\rangle$ obtained in (7.2.6), there is another state, the antibonding orbital, which describes a higher energy state associated with the conduction bands. However, it is known that the WT model or its simple extension do not provide good approximations for such high energy states. The bond orbital model consists in keeping only the bonding orbitals and their interactions and in discarding all antibonding orbitals. The merit and disadvantage of this approximate model will become clear in the following discussion.

Since many matrix elements remain which connect different bonding orbitals, an electron accommodated in a bonding orbital may hop to another bonding orbital through these remaining matrix elements. Thus the bond energy (7.2.9) would be broadened into a band. The matrix element between different bonding orbitals are expected to be proportional to V_1^a or V_1^c introduced in (7.2.4). That this is indeed the case can be easily verified by using (7.2.6–8), and we obtain for the A and C atom, respectively,

$$\left. \begin{array}{l} -\langle b|H|b'\rangle_a \equiv A = \tfrac{1}{2}(1 + \alpha_p) V_1^a \\ -\langle b|H|b'\rangle_c \equiv C = \tfrac{1}{2}(1 - \alpha_p) V_1^c \end{array} \right\} . \tag{7.2.10}$$

To calculate the energy band produced by (7.2.10), we look for an appropriate linear combination of $|b\rangle$ in the Bloch wave form. Let us specify each bond orbital $|b\rangle$ by its location vector (for instance, mid-point of the bond) r_j and its orientation in space $\alpha(\alpha = 1, 2, 3, 4)$ as $|b\rangle = |\alpha, r_j\rangle$. Then the Bloch function may be expressed as

$$|\alpha k\rangle = \frac{1}{\sqrt{N}} \sum_j \exp(i k \cdot r_j)|\alpha, r_j\rangle,\qquad(7.2.11)$$

where N is the total number of atoms. The energy matrix with respect to the base functions (7.2.11) can be reduced into a diagonal block for each k vector, and each k block is a 4×4 matrix corresponding to $\alpha = 1, \ldots 4$. Then the problem becomes to determine the eigenvalues of a 4×4 matrix for each k. The reason why we have a 4×4 matrix instead of 8×8 as in the WT model is obvious: we have discarded 4 antibonding orbitals.

It is not difficult to write down the matrix elements as in the case of WT model, but we shall omit the details here and show only the form of the energy band for a special direction of k, i.e., [110],

$$\left.\begin{aligned}
E_1 &= -(A + C) - 2[4AC\cos^4\theta + (A - C)^2]^{1/2}\\
E_2 &= -(A + C) + 2[4AC\cos^4\theta + (A - C)^2]^{1/2}\\
E_3 &= A + C\\
E_4 &= A + C
\end{aligned}\right\},\qquad(7.2.12)$$

where $\theta = (\sqrt{2}/8)ka$, and $\theta = 0, 3\pi/8, \pi/2$ correspond, respectively, to the Γ, K and X points in the Brillouin zone. E_3 and E_4 are degenerate and independent of k (dispersionless). This situation is the same as for the WT model. We see from (7.2.12) that the total band width is given by

$$A + C \equiv V_1 = \tfrac{1}{2}(V_1^a + V_1^c) + \tfrac{1}{2}(V_1^a - V_1^c)\alpha_p.\qquad(7.2.13)$$

For illustration, the energy bands calculated from (7.2.12) are shown in Fig. 7.4. as a function of θ along with the case $A = C$ corresponding to the IV elements. This figure should be compared with Fig. 6.16 or Fig. 6.24. The main concern in this section is not the study of fine details of the band structure, but a qualitative understanding of the general feature of valence bands and average properties derived from the band electrons. The most important feature in this sense will be the splitting of bands at X point; $|X_3 - X_1|$ which is a quantity observed experimentally.

So far four parameters are involved in the present model: V_1^a, V_1^c, V_2 and V_3 or their combinations in α_p. These parameters may be regarded as material constants specific to each compounds. In the following, let us examine how various properties can be correlated in terms of these parameters within the present model.

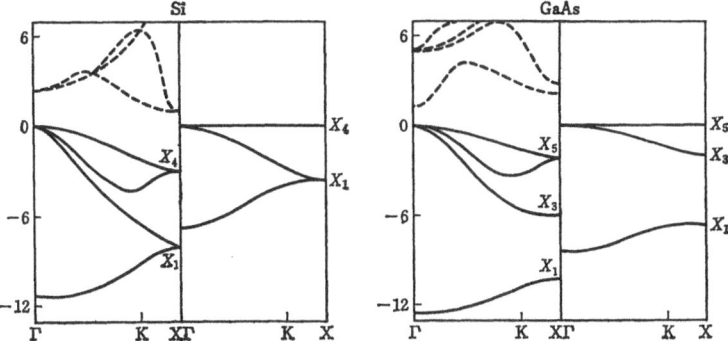

Fig. 7.4. Dispersion curves along the [100] direction (in units of eV). The left half of each figure corresponds to the band structure in the EPM, and the right half to that in the BOM

7.2.1 Effective Charge

Since the distribution of electrons participating in the bonding is not symmetric along the bond, we shall assign two electrons per bond in the ratio

$$\frac{u_a^2}{u_a^2 + u_c^2} : \frac{u_c^2}{u_a^2 + u_c^2}$$

to the negative ion side and positive ion side, respectively. Then the number of bonding electrons contributing to the negative ion will be given by

$$-2e\frac{u_a^2}{u_a^2 + u_c^2} = -e(1 + \alpha_p)$$

and that of the positive ion by

$$-2e\frac{u_c^2}{u_a^2 + u_c^2} = -e(1 - \alpha_p).$$

If we denote the positive charge of the atomic nuclei by $4 + \Delta Z$, taking $\Delta Z = 0$, 1, 2, 3 for the IV, V, VI and VII elements, respectively, then the electric charge of the negative ion in the unit of $-e$ can be given by

$$Z^* = 4(1 + \alpha_p) - (4 + \Delta Z) = 4\alpha_p - \Delta Z. \tag{7.2.14}$$

The positive ion has a charge Z^* just opposite that of the negative ion as it should be. Z^* is not, however, a quantity directly observed. If one tries to detect the effective charge through the observation of the macroscopic polarization produced by an external field, it is necessary to take account of a correction (so-called dynamic effect) due to the ionic displacement induced by the external field. In order to estimate this effect, we first have to know the dipole moment of the bond. This moment is to be proportional to a product of the bond length and the asymmetric part of the electron distribution along the bond $(-ea_p)$, and we may put

$$p = - \gamma ea_p d ,$$

(7.2.15)

where d is a vector connecting a positive ion site and a negative ion site, γ a dimensionless constant, which will be chosen as $\gamma^2 = 2$ later in order to have the best fit to the experiments.

The dynamical effect is essentially the contribution to the polarization due to the change in p induced by the ionic displacements; the most significant contribution will come from a change $\partial p/\partial d$ in (7.2.15). By the definition of a_p, a_p depends on the two parameters V_2 and V_3. Since V_3 is a quantity associated with isolated atoms, it is independent of the ionic separation. Thus the variation $\partial a_p/\partial d$ is mainly caused by a change in the parameter V_2. We shall assume an empirical relationship between V_2 and d

$$V_2 \propto d^{-2} .$$

(7.2.16)

Then, by using (7.2.7)

$$\frac{\partial a_p}{\partial d} = - \frac{1}{2} \frac{V_3}{\sqrt{V_2^2 + V_3^2}} \cdot \frac{2V_2}{V_2^2 + V_3^2} \cdot \frac{dV_2}{dd}$$

$$= \frac{2}{d} a_p(1 - a_p^2) .$$

$(4/3)d$ times this quantity is shown to give a contribution to the effective charge from the dynamical effect. Thus the total effective charge, static plus dynamic, is given by

$$e_T^* = 4a_p - \Delta Z + \tfrac{8}{3} \gamma a_p(1 - a_p^2) .$$

(7.2.17)

This is usually called the transversal effective charge and represents the actual effective charge of an ion in response to an external electric field. Since e_T^* is an observable quantity experimentally, if γ is known, we can experimentally determine a_p for each substance from this formula. On the other hand, if one could know a_p from other sources, (7.2.17) would provide us with a means to test the correctness of the model.

7.2.2 Dielectric Constant

Let us proceed to the calculation of the electronic dielectric constant ϵ_∞. Conventionally, ϵ_∞ is evaluated from the knowledge of the oscillator strength $f_{n'nk}$

for the electronic transition between two energy levels $E_{nk} \rightarrow E_{n'k}$ by using the well-known formula for ϵ_∞:

$$\epsilon_\infty = 1 + \frac{4\pi e^2 h^2}{m} \sum_{nn'} \sum_k \frac{f_{n'nk}}{(E_{n'k} - E_{nk})^2} .$$

In the present model, an electron is accommodated in the bond orbitals, localizing along the bond. It is natural, therefore, to assume that the polarization induced by the external field may be expressed as a sum of the polarizations produced in each bond. Namely, if we denote the total number of bonds per unit volume by N_b, and the polarizability of each bond by α_b, then the electronic dielectric constant ϵ_∞ could be calculated from the equation

$$\epsilon_\infty = 1 + 4\pi N_b \alpha_b . \tag{7.2.18}$$

Here, α_b is defined by a coefficient which relates the external field and the polarization P induced on a bond

$$P = \alpha_b E . \tag{7.2.19}$$

By using standard perturbation theory, we can obtain

$$\alpha_b = \frac{2e^2}{m} \sum_{jj'} \frac{|\langle \phi_j | x | \phi_{j'} \rangle|^2}{E_{j'} - E_j} , \tag{7.2.20}$$

where $-e$ is the electron charge, m the electron mass, and x the electron coordinate along the electric field E. In this expression we use as $|\phi_j\rangle$ and E_j the bond orbital state $|b\rangle$ and its energy E_b for the ground state, and the antibond orbital state $|a\rangle$ and its energy E_a for the excited state. Although we have not given the forms for $|a\rangle$ and E_a explicitly, they are easily derived as

$$\left.\begin{array}{l} |a\rangle = [\tfrac{1}{2}(1 - \alpha_p)]^{1/2} |aj\rangle - [\tfrac{1}{2}(1 + \alpha_p)]^{1/2} |cj\rangle \\ E_a = \tfrac{1}{2}(\varepsilon^a + \varepsilon^c) + (V_2^2 + V_3^2)^{1/2} \end{array}\right\} . \tag{7.2.21}$$

Therefore, the energy denominator in (7.2.20) is given by

$$E_a - E_b = 2(V_2^2 + V_3^2)^{1/2}$$

and the transition matrix element in the nominator of (7.2.20) by

$$\langle b | x | a \rangle = \frac{\sqrt{1 - \alpha_p^2}}{2}[\langle aj | x | aj \rangle - \langle cj | x | cj \rangle] - \alpha_p \langle aj | x | cj \rangle .$$

It is convenient to introduce a parameter α_c defined by

$$\alpha_c \equiv \sqrt{1 - \alpha_p^2} . \tag{7.2.22}$$

We shall call this quantity the degree of covalency. In a sufficiently good approximation, it is verified that the second term in the left-hand side of $\langle b|x|a \rangle$ may be neglected compared with the first term, and thus we may put

$$\langle b|x|a \rangle = \tfrac{1}{8} \gamma d \alpha_c , \tag{7.2.23}$$

where $d = |\boldsymbol{d}|$ is the bond length and γ is the same as defined in (7.2.15) but has a clearer physical meaning in the present definition:

$$\tfrac{1}{4} \gamma d = \langle aj|x|aj \rangle - \langle cj|x|cj \rangle . \tag{7.2.24}$$

Combining all these results into (7.2.18, 20), we have, finally

$$\epsilon_\infty = 1 + \frac{\gamma^2 \pi N e^2 d^2}{3(V_2^2 + V_3^2)^{1/2}} a_c^2 . \tag{7.2.25}$$

It is very interesting to compare this result with the same quantity (6.5.15) derived in Sect. 6.5 by a different argument [note that ϵ_∞ was written as $\epsilon(0)$ in Sect. 6.5]:

$$\frac{1}{\epsilon_\infty - 1} = \frac{1}{(\hbar \omega_p)^2} (E_h^2 + C^2) . \tag{6.5.15}$$

Since $2(V_2^2 + V_3^2)^{1/2}$ is the difference between of the bond orbital energy and antibonding orbital energy, it may be regarded as corresponding to $E_g = (E_h^2 + C)^{1/2}$ in the band theory. Then it turns out that (7.2.25) and (6.5.15) have quite a different dependence on E_g. The expression (6.5.15) was derived on the basis of the Penn model, which is not necessarily justified as a realistic model. In the following, therefore, we shall base our discussion on (7.2.25) and try to determine the parameters V_2 and V_3.

In order to determine V_2 and V_3 from (7.2.25) using the experimental value of ϵ_∞, it is necessary to first determine V_2 by a different method. At this point we make a fundamental assumption, that is, we assume that the parameter V_2 is the same for all elements in the same row of the periodic table. It seems natural from the definition of V_2, that V_2 has a common value for the elements which share the principal quantum number of the valence electrons. By this assumption, it is sufficient to know the values of V_2 for C, Si, Ge, α-Sn, for example. To determine V_2 for these elements, we may use the characteristic sharp absorption of light for the IVth elements which has been discussed in detail in Sect. 6.3; the photon energy corresponding to this sharp peak can be regarded as the transition energy $2(V_2^2 + V_3^2)^{1/2}$ between the bonding and antibonding orbital states ($2V_2$ for the IVth elements since $V_3 = 0$). Thus, once V_2 is determined experimentally, we can use the formula (7.2.25) with $a_c = 1$, $V_3 = 0$,

$$\epsilon_\infty = 1 + \frac{\gamma^2 \pi N e^2 d^2}{3 V_2}$$

to calculate γ from the observed data for ϵ_∞, d and V_2. In Table 7.1, the values of γ thus calculated are listed along with the observed values for V_2 and d. For other compounds, V_2 and V_3 are determined in the following way. For the compounds of atoms which belong to the same row in the periodic table, V_3 may be calculated from (7.2.25) by making use of Table 7.1 and the observed values of ϵ_∞. For the cases of compounds which are composed of two atoms belonging to different rows in the periodic table, we evaluate the geometric means of the two values of V_2 and γ taken from Table 7.1, and then they are combined with the observed value of ϵ_∞ to obtain V_3. In this manner the values of the set (V_2, V_3) for all the compounds with zincblende structure have been determined. In Table 7.2, which will be given in Sect. 7.4, we list all such values of the set (V_2, V_3).

Table 7.1.

	V_2[eV]	γ	d[Å]
C	6.10	1.08	1.54
Si	2.20	1.20	2.35
Ge	2.15	1.42	2.44
α-Sn	1.76	1.69	2.80

In the empirical pseudopotential model (EPM), when various material constants are determined, we often utilize the d-dependence of the parameters. Therefore it would be interesting to test such a d-dependence of V_2 which is derived directly from experiments. In Fig. 7.5, we show a log-log test of the d-dependence of V_2. From this figure we can clearly read off a relation

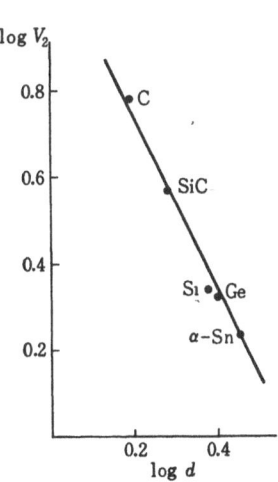

Fig. 7.5. Experimental test of the d-dependence of V_2

$$V_2 \propto d^{-2}.$$

In fact, this relation has already been used for the calculation of the dynamic effective charge.

The BOM has been applied, besides these properties, to calculate the elastic constants [7.3], piezo constant [7.4] and magnetic susceptibility [7.5] and compared with the experimental data for the tests of V_2 and V_3. Since there is no space left to describe them in detail, the readers who are interested in those topics should refer to the references cited at the end of this volume.

7.3 Cohesive Energy and the Stability of Crystal Structures

In the band electron theory of nonmetals developed in Sect. 6.5, the ionicity parameter f_i introduced by *Phillips* [7.6] has played a central role. In the present bond orbital model, there is no such parameter, but instead another α_p appears which bears a meaning similar to f_i. What is then the relationship between f_i and α_p? f_i represents essentially the fraction of the ionic contribution to the magnitude of the energy gap; if the square of the magnitude of the gap is

$$E_g^2 = E_h^2 + C^2,$$

then f_i is defined by

$$f_i = \frac{C}{E_g}.$$

In order to find out the similarity with the BOM, we remind that the relationship between E_g and ϵ is given by (6.5.16) which may be rewritten as

$$E_g^2 = \left[\frac{(\epsilon_\infty - 1)}{N} \frac{m}{4\pi\hbar^2 e^2} \right]^{-1}$$

$$= [(\epsilon_\infty - 1) d^3]^{-1} \times \text{(a universal constant)}. \tag{7.3.1}$$

For the IV elements, since $C = 0$, $E_g = E_h$, $E_h \propto d^{-2\,5}$, we can put

$$1 - f_i^2 = \frac{E_h^2}{E_g^2} \propto \frac{d^{-5}}{[(\epsilon_\infty - 1) d^3]^{-1}} = (\epsilon_\infty - 1) d^{-2}. \tag{7.3.2}$$

On the other hand, from (7.2.25) $\epsilon_\infty - 1$ may be related to α_p as

$$\epsilon_\infty - 1 \propto (1 - \alpha_p^2)^{3/2} d^2. \tag{7.3.3}$$

Thus, combining (7.3.2) with (7.3.3), one has

$$f_i^2 = 1 - (1 - \alpha_p^2)^{3/2}. \tag{7.3.4}$$

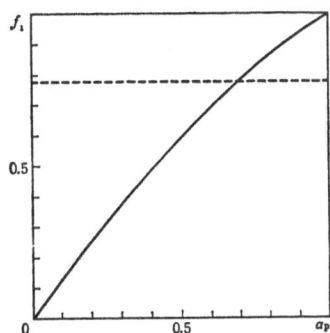

Fig. 7.6. Relationship between f_i and α_p. The dotted line gives the borderline for the destabilization of the 4-coordinated structure

In Fig. 7.6 the relationship between f_i and α_p given by (7.3.4) is shown. It is interesting to observe that f_i is nearly proportional to α_p when α_p is sufficiently small. As has been discussed in Sect. 6.5 and demonstrated in Fig. 6.27, a critical value of f_i exists

$$f_c = 0.785 \qquad (7.3.5)$$

such that, for the I-VII, II-VI and III-V compounds, when $f_i < f_c$, the tetrahedral structure is more stable, and when $f_i > f_c$, the NaCl structure becomes the most stable one. We have not given any explanation for this in Sect. 6.5 and now let us consider this problem, that is, the stability of different crystal structures from the standpoint of the BOM.

In order to discuss the stability of a lattice structure, it is necessary to evaluate the cohesive energy of the crystal lattice. For such a problem, the nonorthogonality between the different atomic hybrid orbitals so far discarded will become important, and therefore it will not be allowed to pass without taking account of the nonorthogonality.

Let us denote the nonorthogonal hybrid orbitals as $|h^a\rangle$, $|h^c\rangle$, etc., and the overlap integrals between them as

$$\langle h^c | h^a \rangle \equiv S . \qquad (7.3.6)$$

Corresponding to V_2 and V_3 in the previous section, we define

$$
\begin{aligned}
\varepsilon^a &= \langle h^a | H | h^a \rangle, \quad \varepsilon^c = \langle h^c | H | h^c \rangle \\
M_3 &= \tfrac{1}{2} (\varepsilon^c - \varepsilon^a) \\
M_2 &= - \langle h^c | H | h^a \rangle
\end{aligned}
\qquad (7.3.7)
$$

Then a simple manipulation reveals that a variation problem to minimize

$$\varepsilon_b = \frac{\langle b | H | b \rangle}{\langle b | b \rangle}$$

with respect to the unknown expansion parameters in the new bond orbital

$$|b\rangle = u_a|h^a\rangle + u_c|h^c\rangle \tag{7.3.8}$$

can be solved with the result that the minimum bond energy is given by

$$\varepsilon_b = \frac{M_2S - [M_2^2S^2 + (1 - S^2)(M_2^2 + M_3^2)]^{1/2}}{1 - S^2}. \tag{7.3.9}$$

This result shows that if we use

$$\left.\begin{array}{l} V_2 = \dfrac{M_2}{1 - S^2} \\[3mm] V_3 = \dfrac{M_3}{\sqrt{1 - S^2}} \end{array}\right\} \tag{7.3.10}$$

in place of M_1 and M_3, respectively, then (7.3.9) can be rewritten as

$$\varepsilon_b = V_2S - (V_2^2 + V_3^2)^{1/2} \tag{7.3.11}$$

and, except for the first term in the left-hand side of (7.3.11), all the results become the same as those of Sect. 7.2, where the effect of S has been neglected. By the way, the antibonding orbital energy may be given in the same notation as

$$\varepsilon_a = V_2S + (V_2^2 + V_3^2)^{1/2} \tag{7.3.12}$$

and the corresponding expansion coefficients are, in contrast to (7.2.8)

$$\left.\begin{array}{l} u_a^2 = \dfrac{1}{2}\left[\dfrac{1 - S\sqrt{1 - \alpha_p^2}}{1 - S^2} + \dfrac{\alpha_p}{(1 - S^2)^{1/2}}\right] \\[4mm] u_c^2 = \dfrac{1}{2}\left[\dfrac{1 - S\sqrt{1 - \alpha_p^2}}{1 - S^2} - \dfrac{\alpha_p}{(1 - S^2)^{1/2}}\right] \end{array}\right\}, \tag{7.3.13}$$

where α_p is formally the same as before

$$\alpha_p = \frac{V_3}{\sqrt{V_2^2 + V_3^2}}.$$

As evident from these results, all the effects of the nonorthogonality can be absorbed into the renormalized V_2 and V_3, except for one correction term in the orbital energy ε_b or α_a, i.e., the first term in (7.3.11) or (7.3.12), which is proportional to S.

To prepare for the discussion on the stability of the different lattice structures, let us first approximately calculate the cohesive energy of the tetrahedrally-bonded crystal, following successive steps.

1) As the first step, we excite the electrons in all the atoms so as to form the sp^3 hybrid orbitals from the s and p orbitals. To this end, it is necessary to give the energy $3V_1^a(3V_1^c)$ to the s electron in the A(C)-atom and the energy $V_1^a(V_1^c)$ to the p electrons in the A(C)-atom. If the number of electrons per atom is 4 $- \Delta Z(4 + \Delta Z)$ for the A(C)-atom, ΔZ being 0, 1,2 for the IV, V, VI element, respectively, then the excitation energy per A-C atom pair is

$$E^* = (4 + \Delta Z) V_1^c + (4 - \Delta Z) V_1^a. \tag{7.3.14}$$

2) Next, we remove ΔZ electrons from a C-atom and add them to an A-atom, so that all the hybrid orbitals may have the same number of electrons. To this end, an extra energy $2\Delta Z V_3$ per pair is required.

3) At this stage, we construct the bonding orbitals, in each of which two electrons with opposite spins are accommodated. The energy gain per pair due to this bond formation is $8\sqrt{V_2^2 + V_3^2} - 8SV_2$.

4) When we take account of the matrix elements between the different bonding orbitals, the bonding orbital energies broaden into bands. However, since the total number of the electrons occupying the bonding orbitals is just fills up the bands, the mean value of the band energy is the same as that obtained from Steps 1–3 because of the invariance property of the trace. This is one of the merits of the BOM; one need not worry about the diagonalization of the Hamiltonian energy matrix.

Combining all these Steps 1–4, we can obtain the released energy per pair when a crystal is formed from an assembly of neutral atoms, i.e., the cohesive energy of the crystal. It turns out to be

$$E = 8(V_2^2 + V_3^2)^{1/2} - 8SV_2 - 2\Delta Z V_3 - E^*. \tag{7.3.15}$$

For the case of the IV elements in particular, since $\Delta Z = 0$ and $V_3 = 0$, the cohesive energy per atom is derived from (7.3.14, 15) as

$$\frac{E}{2} = 4[V_2(1 - S) - V_1]. \tag{7.3.16}$$

This result indicates that the *bonding parameter* V_2 nearly corresponds to the covalent bond energy per electron, which is to be reduced by an amount of the *banding parameter* V_1 per electron when the metallic nature of electrons is included. Hence we define

$$\alpha_m = \frac{V_1}{\sqrt{V_2^2 + V_3^2}} \tag{7.3.17}$$

and call it "metallicity parameter". Then (7.3.16) can be written in terms of α_m as

$$\frac{E}{2} = 4V_2(1 - S - \alpha_m) \tag{7.3.18}$$

which shows that the covalent cohesive energy is decreased as the metallicity is increased. In fact, it is known that in the IV elements, the value of α_m is increased as the mass of the atom is increased. Thus, the heavier element Sn takes either the diamond structure (white tin) or another structure with a coordination number of almost 6 (grey tin), and the heaviest element Pb has no diamond phase but only the fcc structure.

Returning again to (7.3.15), let us consider the case where α_p is very large, i.e., $V_2 \ll V_3$. In this case, (7.3.15) may be approximated as

$$E \approx 2V_3(4 - \Delta Z) - E^* . \tag{7.3.19}$$

The first term of the right-hand side of this equation takes on a form which is a product of the two factors $2V_3$ and $(4 - \Delta Z)$. Since $(4 - \Delta Z)$ is the number of electrons transferred from a C to an A atom, V_3 may be regarded as representing something related to the ionic energy. On the other hand, as suggested by Fig. 7.6, when α_p is too large (when $\alpha_p > 0.7$ from Fig. 7.6), the tetrahedrally-bonded structure is more unstable than the structure with a higher coordination number (6 or possibly 8). It would then be an interesting problem to explain why such a structure change is expected and how to determine where the border line is between the different crystal structures in some relevant variable space characterizing the material.

To begin considering this interesting problem, let us first recall the physical origin of V_3. When we produce the positive and negative ions by transferring the electrons among the neutral atoms, the ionic structure thus created may be stabilized by the so-called Madelung energy due to the Coulomb interaction between ions. At first sight, one might think that such an effect of the Madelung energy has been ignored in the argument so far given. This is not the case, however. Since V_3 is the energy needed to transfer an electron from a C atom to an A atom, it should include the difference in the electrostatic potential energy of the electron before and after the transfer, and hence the Madelung energy should be regarded as being included in V_3. Therefore, let us denote the part of V_3 which is attributed to the contribution from Madelung's energy when positive and negative ions with the effective charges $\pm Z^* e$ form a tetrahedral structure by

$$V_3^M = \frac{Z^* e^2 \alpha}{\epsilon_\infty d} . \tag{7.3.20}$$

Here, α is the Madelung constant and for the zincblende structure

$$\alpha_{ZnS} = 1.638 ,$$

Z^* is the same as defined in (7.2.14).

How do we express the cohesive energies of the 6-coordinated NaCl structure or the 8-coordinated CsCl structure within the same approximation? We can assume $V_2 = 0$ for the 6 (or 8) coordinated structure because it is impossible to form the covalent bonding in these configurations and hence no reduction in cohesive energy is expected due to the covalent bonding. For V_3 we must take into account the change in Madelung's energy and so we put $V_3 + \delta V_3$. Then the cohesive energy of the NaCl structure may be given approximately as

$$E^{\text{NaCl}} = 2(4 - \Delta Z)(V_3 + \delta V_3). \tag{7.3.21}$$

Thus the main difference in the cohesive energy for a 4 and 6-coordinated structure becomes

$$E^{\text{ZnS}} - E^{\text{NaCl}} = 8(V_2^2 + V_3^2)^{1/2} - 8SV_2 - 8(V_3 + \delta V_3) + 2\Delta Z\delta V_3. \tag{7.3.22}$$

The Madelung constant for a NaCl structure and a CsCl structure are, respectively,

$$\alpha_{\text{NaCl}} = 1.763$$
$$\alpha_{\text{CsCl}} = 1.748$$

and their difference is rather small. If we neglect the differences in α, Z^* and ϵ_∞ for these two structures, then the difference in the cohesive energy between 6 and 8- coordinated structures may be ignored compared with the difference between the tetrahedral structure and 6 or 8- coordinated structures. Therefore, let us take $\Delta\alpha = 1.75 - 1.64 = 0.11$ and we may assume δV_3 in (7.3.22) to be

$$\delta V_3 = \frac{0.11 Z^* e^2}{\epsilon_\infty d}. \tag{7.3.23}$$

This gives from (7.3.22) a criterion to determine the border line along which a crossover of the stability of a lattice takes place between 4 and 6 (8)-coordinated structures:

$$8(V_2^2 + V_3^2)^{1/2} - 8SV_2 - 8V_3 - \frac{0.11 Z^* e^2}{\epsilon_\infty d}(8 - 2\Delta Z) = 0. \tag{7.3.24}$$

By making use of the definition (7.2.7), this condition may be expressed as a criterion for α_p:

$$\alpha_p + S\sqrt{1 - \alpha_p^2} = 1 - \frac{0.11}{4} \frac{Z^*(4 - \Delta Z) e^2}{\epsilon_\infty d(V_2^2 + V_3^2)^{1/2}}. \tag{7.3.25}$$

Since this equation is actually satisfied by a fairly large α_p, it may be safely assumed that $\epsilon_\infty \approx 1$, $4 - \Delta Z \approx Z$ and $Z^* \approx Z$. In effect, the tetrahedral structure becomes unstable against the NaCl (CsCl) structure when

$$\alpha_p + S\sqrt{1 - \alpha_p^2} > 1 - 0.028 \frac{Z^2 e^2}{d\sqrt{V_2^2 + V_3^2}}. \tag{7.3.26}$$

For $Z = 1$, $d = 2\text{Å}$, $\sqrt{V_2^2 + V_3^2} = 4$ eV, the right-hand side of (7.3.26) is estimated as

$$\alpha_p + S\sqrt{1 - \alpha_p^2} > 0.95,$$

or if we take $S = 0.35$

$$\alpha_p > 0.7.$$

This is a condition which is also predicted from Fig. 7.6. Thus, although within a crude approximation, we can show why the tetrahedral structure becomes unstable against other cubic structures for large α_p. In a similar manner, it would be possible to predict that the tetrahedral structure becomes unstable against a certain metallic phase when the metallicity α_m becomes sufficiently large.

7.4 Relationship Between the BOM and EPM

Since our main concern was to give a unified explanation of the electronic properties of the tetrahedrally-bonded semiconductors, we have not paid special attention to the detailed structure of the one electron energy band. However, as stated in Sect. 6.3, the band structure of the tetrahedrally-bonded semiconductors are determined in detail on the basis of the EPM and they provide a good starting point from which various properties of these crystals are well understood. It will be then important and instructive to examine the interrelationship between the BOM and EPM.

It turns out that the BOM can almost perfectly realize the fine detail of the energy bands obtained from the EPM if one increases the number of the model parameters sufficiently. There is, however, one reservation: the EPM can produce good information even for the conduction band, whereas the BOM is good only for the valence bands. This point is certainly a disadvantage of the BOM, but as long as the valence bands are concerned, the BOM has a distinguished merit that it gives us detailed information on the relationship between the band structure and the electronic properties of the constituent atoms.

Let us start with the construction of the valence bands from the knowledge of the matrix elements with respect to various bond orbitals. We assume that a bond orbital $|bj\rangle$ is given for each bond j. We can form the Bloch function $|\alpha k\rangle$ by making the linear combination of the bond orbitals with the same orientation $\alpha(\alpha = 1, 2, 3, 4$ corresponding to four tetrahedral directions). Then the calculation of the band energies is reduced to the diagonalization of a 4×4 matrix

$H_{\alpha\beta}(\boldsymbol{k})$ for each $\boldsymbol{k}$. To construct the matrix element $H_{\alpha\beta}(\boldsymbol{k})$, we exercise symmetry consideration to pick up five different kinds of elements among the several bond orbitals as shown in Fig. 7.7; $B_1^A(B_1^C)$ is the matrix element between two bond orbitals emerging from the A(C)-atom, B_4 and B_5 are those between the second neighbor orbitals and B_6, that between the third neighbor orbitals. All other elements will be discarded, assuming they are small.

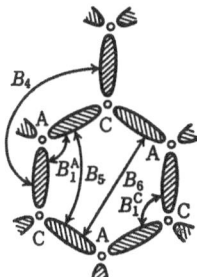

Fig. 7.7. Matrix elements between various bond orbitals

The $\boldsymbol{k}$-dependence of the matrix elements $H_{\alpha\beta}(\boldsymbol{k})$ may be easily determined by referring to the symmetry of the Bloch functions. We shall omit the details of the derivation and give only some of the results. Since $H_{\alpha\beta}(\boldsymbol{k})$'s are generally complex numbers, they are decomposed into two parts, the real symmetric and imaginary asymmetric parts:

$$H_{\alpha\beta}(\boldsymbol{k}) = H_{\alpha\beta}^s(\boldsymbol{k}) + iH_{\alpha\beta}^a(\boldsymbol{k}) . \tag{7.4.1}$$

By making use of the abbreviation such that $x = k_x a/4$, $y = k_y a/4$ and $z = k_z a/4$, we have, for instance,

$$
\left.
\begin{aligned}
H_{11}^s(\boldsymbol{k}) = {} & \varepsilon_b - 2B_4[\cos(2x+2y) + \cos(2y+2z) + \cos(2x+2z)] \\
& - 2B_6[\cos(2x-2y) + \cos(2y-2z) + \cos(2x-2z)] \\
H_{\alpha\alpha}^a(\boldsymbol{k}) = {} & 0 \\
H_{12}^s(\boldsymbol{k}) = {} & -2B_1^s \cos(x+z) - 4B_5 \cos 2y \cos(x-z) \\
H_{12}^a(\boldsymbol{k}) = {} & 2B_1^a \sin(x+z)
\end{aligned}
\right\}, \tag{7.4.2}
$$

where ε_b is the energy of the bond orbital and we have introduced

$$B_1^s = \tfrac{1}{2}(B_1^A + B_1^C), \quad B_1^a = \tfrac{1}{2}(B_1^A - B_1^C) . \tag{7.4.3}$$

All the matrix elements other than (7.4.2) may also be easily derived by certain proper permutations among x, y, and z. The energy eigenvalues of $\|H_{\alpha\beta}(\boldsymbol{k})\|$ can be analytically obtained for special $\boldsymbol{k}$-directions of high symmetry. For instance, along the [100] direction, i.e., on the Δ- line in $\boldsymbol{k}$-space, it holds that

$y = z = 0$ and hence the form of $H_{\alpha\beta}(k)$ becomes very simple. Thus all the eigenvalues are obtained by solving algebraic equations which are at most quadratic:

$$
\left.
\begin{aligned}
\varDelta_5 &= H_{11} - H_{14} = 4(B_4 - B_5 + B_6)(1 - \cos 2x) \\
\varDelta_{3,1} &= H_{11} + H_{14} \pm |H_{13}| \\
&= -4B_1^a - 4B_5(1 + \cos 2x) + 4(B_4 + B_6)(1 - \cos 2x) \\
&\quad \pm 4[(B_1^a)^2 \sin^2 x + (B_1^a + 2B_5)^2 \cos^2 x]^{1/2}
\end{aligned}
\right\}. \tag{7.4.4}
$$

Along the [111] direction, that is, on the $\varLambda$-line in k-space, the eigenvalues are also given in the analytical forms:

$$
\left.
\begin{aligned}
\varLambda_3 &= H_{22} - H_{23} \\
&= 2(B_4 + 2B_6)(1 - \cos 4x) - 4B_5 \sin^2 2x \quad \text{(doublet)} \\
\varLambda_{1,1} &= \tfrac{1}{2}(H_{11} + H_{22} + 2H_{23}) \\
&\quad \pm \{[\tfrac{1}{2}(H_{11} - H_{22} - 2H_{23})]^2 + 3|H_{12}|^2\}^{1/2} \\
&= 2(B_4 + B_6)(1 - \cos 4x) - 4B_5(1 + \cos^2 2x) - 4B_1^a \\
&\quad \pm 2\{[B_1^a + 2B_5 \cos^2 2x + (B_4 - B_6)(1 - \cos 4x)]^2 \\
&\quad + 3[(B_1^a + 2B_5) \cos^2 2x + (B_1^a)^2 \sin^2 2x)]\}^{1/2}
\end{aligned}
\right\}. \tag{7.4.5}
$$

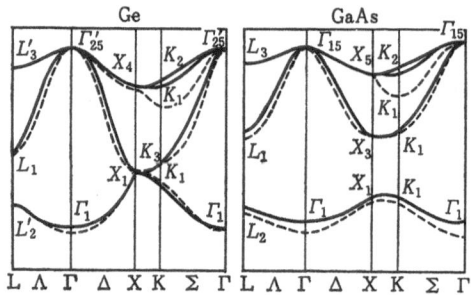

Fig. 7.8. Band structures of Ge and GaAs calculated from the BOM and the EPM (---) [7.7]

Figure 7.8 shows the energy bands for Ge and GaAs calculated from the EPM (dotted lines) and from the BOM (solid lines). In determining the energy dispersion curves in the BOM, the values of the B parameters are chosen in a manner described later. We have already shown the results for the energy-band calculation with two parameters in Fig. 7.4, and it is evident from these figures that the increase in the numbers of the parameters improves the fine detail of the energy bands in great deal.

When the origin of the energy is taken at the $\varGamma_{15}$ (or $\varGamma_{25'}$) point, the main symmetric points in the energy band can be expressed in terms of $\{B\}$ as:

$$\Gamma_1 = -8(B_1^s + 2B_5) \tag{7.4.6a}$$

$$X_s = -8(B_4 - B_5 + B_6) \tag{7.4.6b}$$

$$X_3 = 8(B_4 + B_6) - 4B_1^C \tag{7.4.6c}$$

$$X_1 = 8(B_4 + B_6) - 4B_1^A \tag{7.4.6d}$$

$$L_3 = 4(B_4 - B_5 + 2B_6) \tag{7.4.6e}$$

$$L_{1,2} = 4(2B_4 + B_6 - B_5 - B_1^s) \pm 2[(B_1^s + 2B_4 - 2B_6)^2 + 23(B_1^s)^2]^{1/2} \tag{7.4.6f}$$

Since the values of $\{B\}$ should become smaller as the distances between the involving bonds become larger, one expects that

$$B_4, B_5 \geqslant B_6.$$

That this is indeed the case is seen from the following fact. From (7.4.6) a relation

$$2L_3 - X_s = 8B_6 \tag{7.4.7}$$

may be derived, whereas the result of the EPM, as shown in Fig. 7.8, reveals that $2L_3 \approx X_s$. To be more precise, even for C, (the case of the largest deviation) it holds that $X_s - 2L_3 = 1.5 \sim 2.0$ eV and hence $B_6 \lesssim 0.2$ eV, a much smaller value compared with all other B's. Therefore within a fairly good approximation, one may put

$$B_6 = 0. \tag{7.4.8}$$

The five parameters in (7.4.6) (four parameters if $B_6 = 0$) are involved in the six equations, suggesting that certain interrelationships among the energy values at the high symmetry points must exist. For instance, the energy gap at the X point $\Delta X = X_3 - X_1$, the energy gap at the L point $\Delta L = L_1 - L_2$ and the total band width $\Delta \Gamma = |\Gamma_1|$ are related through an equation

$$4(\Delta L)^2 = (\Delta \Gamma - |X_s|)^2 + 3(\Delta X)^2 \tag{7.4.9}$$

which provides one of the tests for the validity of the BOM.

Since the bond orbitals may be composed of the sp^3 hybrids, the B parameters can also be expressed in terms of the matrix elements between the hybrid orbitals. The relationship between the bond orbitals and the hybrid orbitals may be given either by (7.2.6, 8) or by (7.3.8, 13). In these relationships, the polarity parameter α_p or the covalency parameter α_c which is defined by

$$\alpha_p^2 + \alpha_c^2 = 1 \tag{7.4.10}$$

plays an important role in characterizing the properties of the compounds.

Therefore, it is interesting to examine how the parameters characterizing the band structure depend on the parameter α_p (or α_c). To see this, we label the hybrid orbitals as shown in Fig. 7.9, and assume that the matrix elements between them are nonzero only up to between the second neighbors. Then we have the following five matrix elements:

$$
\begin{aligned}
V_1^A &= - \langle h_a^1 | H | h_a^2 \rangle \\
V_1^C &= - \langle h_c^1 | H | h_c^2 \rangle \\
V_1^X &= - \langle h_a^1 | H | h_c^2 \rangle \\
V_4 &= - \langle h_a^2 | H | h_c^2 \rangle \\
V_5 &= - \langle h_a^3 | H | h_c^2 \rangle
\end{aligned}
\right\} .
\tag{7.4.11}
$$

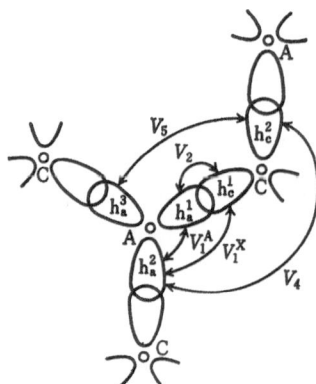

Fig. 7.9. Matrix elements between various hybridized orbitals

Now it is a straightforward task to express the B parameters in terms of these matrix elements. From the assumption made at the beginning, B_6 vanishes automatically; $B_6 = 0$. All other quantities can be written as

$$
\begin{aligned}
B_1^A &= \tfrac{1}{2} (1 + \alpha_p) V_1^A + \alpha_c V_1^X \\
B_1^C &= \tfrac{1}{2} (1 - \alpha_p) V_1^C + \alpha_c V_1^X \\
B_4 &= \alpha_c V_4 / 2 \\
B_5 &+ \alpha_c V_5 / 2
\end{aligned}
\right\} ,
\tag{7.4.12}
$$

or using

$$
\begin{aligned}
V_1^s &= \tfrac{1}{2} (V_1^A + V_1^C) \\
V_1^a &= \tfrac{1}{2} (V_1^A - V_1^C)
\end{aligned}
\right]
\tag{7.4.13}
$$

we have

$$B_1^s = \tfrac{1}{2} V_1^s + \tfrac{1}{2} \alpha_p V_1^a + \alpha_c V_1^x \left.\right\}$$
$$B_1^a + \tfrac{1}{2} \alpha_p V_1^s + \tfrac{1}{2} V_1^a \qquad\right\}. \tag{7.4.14}$$

Here α_p and α_c are, of course, to be given by V_2 and V_3 defined in Sect. 7.2, or by the renormalized V_2 and V_3 introduced in Sect. 7.3. In the BOM, we have assumed that the covalent bond energy V_2 has the same value for the substances with the same electron orbitals; that is, for instance, V_2 is the same for the series of compounds such as Ge–GaAs–ZnSe–CuBr (we shall call this the "iso-orbital" series for brevity). This assumption corresponds to a similar assumption made in the EPM that the symmetric part of the pseudopotential is the same for the "iso-orbital series". On precisely the same ground, we can assume that the present parameters V_1^s, V_1^x, V_4, V_5 are also the same for all the substances in the iso-orbital series. Then, by virtue of this assumption, it becomes possible to correlate the band structures of all tetrahedrally-bonded semiconductors in a systematic way. In what follows, we shall give a brief description of such systematics of the band structures.

When the relations (7.4.12–14) are inserted into (7.4.6), the energy values at the main symmetric points in the Brillouin zone are expressed in terms of the new parameters. The results turn out to reveal several interesting features. First of all, we have from (7.4.6b, 12)

$$X_5 = 4\alpha_c(V_4 - V_5) \tag{7.4.15}$$

which shows that $|X_5|$ varies in proportion to α_c for the iso-orbital series. This is indeed verified experimentally for the Ge series and Sn series, as shown in Fig. 7.10. For the IV elements (where $\alpha_c = 1$), X_5 is solely determined by V_4 and V_5. On the other hand, because V_4 and V_5 are almost the same in nature as V_2 except that they are different in the orientational arrangement of the involved orbitals, V_4 and V_5 are expected to be proportional to d^{-2} as V_2 is. Thus, for instance, one expects

$$|X_5|_{\mathrm{IV}} \propto d^{-2} \tag{7.4.16}$$

which is proved experimentally to hold, as shown in Fig. 7.11.

Next, the total band width Γ_1 is given from (7.4.6a, 12) as

$$\Gamma_1 = -4(V_1^s + \alpha_p V_1^a + 2\alpha_c V_1^x + 2\alpha_c V_5). \tag{7.4.17}$$

For the IV elements, $\alpha_c = 1$, $\alpha_p = 0$ and hence the width $\Delta\Gamma = |\Gamma_1|$ becomes

$$\Delta\Gamma_{\mathrm{IV}} = 4(V_1^s + 2V_1^x + 2V_5) \tag{7.4.18}$$

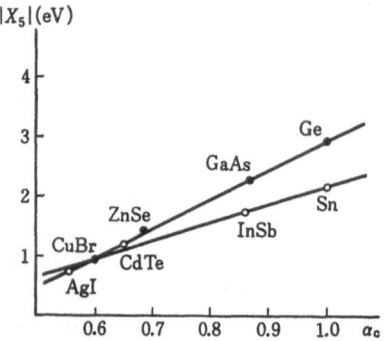

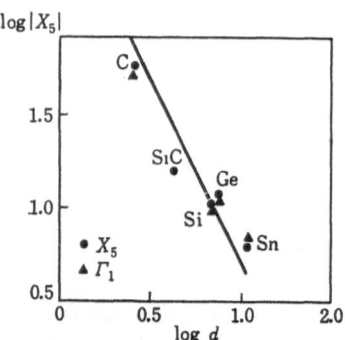

Fig. 7.10. Relationship between the observed values of $|X_s|$ and α_c

Fig. 7.11. Relationship between the observed values of $|X_s|$, $|\Gamma_1|$ for the IV elements and d. The solid line represents the d^{-2} dependence

and in terms of $\Delta\Gamma_{\rm IV}$, it holds generally that

$$\Delta\Gamma = \Delta\Gamma_{\rm IV} + 4[\alpha_p V_1^a - 2(1 - \alpha_c)(V_1^x + V_s)] . \qquad (7.4.19)$$

It is interesting to note that $\Delta\Gamma$ is not much different from $\Delta\Gamma_{\rm IV}$ because the two terms in the square brackets in the right-hand side of (7.4.19) almost cancel with each other. This is indeed the case both in the observed values and the calculated values on the basis of the EPM. As Fig. 7.11 shows, $\Delta\Gamma_{\rm IV}$ is proportional to d^{-2} as $|X_s|$ is.

As to the energy gap at the X point $\Delta X = X_3 - X_1$, (7.4.6c, 6d, 12) give

$$\Delta X = 4V_1^a + 4\alpha_p V_1^s . \qquad (7.4.20)$$

This is an important relation showing ΔX to consist of two parts. That is, a term proportional to the antisymmetric part of the crystal potential and an other term proportional to the polarity parameter α_p. These two terms vanish for the IV elements.

Because the hybridized orbitals (6.3.1) are composed of s and p orbitals, all V_i should be, after all, expressed in terms of the matrix elements between s and p orbitals. Such a transcription may not be of much benefit, but still it would be instructive to recognize that V_i can be eventually correlated to the information associated with the electronic orbitals of the constituent atoms. Here we shall mention only a few important relations.

The relevant matrix elements are the orbital energies

$$\begin{aligned} \varepsilon_p^\lambda &= \langle p^\lambda | H | p^\lambda \rangle \\ \varepsilon_s^\lambda &= \langle s^\lambda | H | s^\lambda \rangle \end{aligned} \quad (\lambda = {\rm A, C}) \qquad (7.4.21)$$

and the matrix elements between the s and p orbitals of neighboring atoms which are shown schematically in Fig. 7.12. We take only these four elements A_{ss}, A_{pp}, $A_{pp\pi}$, and A_{sp}, discarding all other matrix elements between more distant atoms. Then in terms of these quantities, one can derive the following relations:

$$V_1^\lambda = (\varepsilon_p^\lambda - \varepsilon_s^\lambda)/4 \qquad (\lambda = \text{A, C}) \tag{7.4.22a}$$

$$V_1^\mu = \tfrac{1}{8}[(\varepsilon_p^A - \varepsilon_s^A) \pm (\varepsilon_p^C - \varepsilon_s^C)] \quad \left(\mu = \begin{cases} s: + \\ a: - \end{cases}\right) \tag{7.4.22b}$$

$$V_2 = -(A_{ss} + 2\sqrt{3}A_{sp} + 3A_{pp})/4 \tag{7.4.22c}$$

$$V_1^X = -(A_{ss} + \tfrac{2}{3}\sqrt{3}A_{sp} - A_p)_p/4 \tag{7.4.22d}$$

$$V_4 = -(A_{ss} - \tfrac{2}{3}\sqrt{3}A_{sp} + \tfrac{1}{3}A_{pp} - \tfrac{8}{3}B_{pp\pi})/4 \tag{7.4.22e}$$

$$V_5 = -(A_{ss} - \tfrac{2}{3}\sqrt{3}A_{sp} + \tfrac{1}{3}A_{pp} + \tfrac{4}{3}A_{pp\pi})/4 \tag{7.4.22f}$$

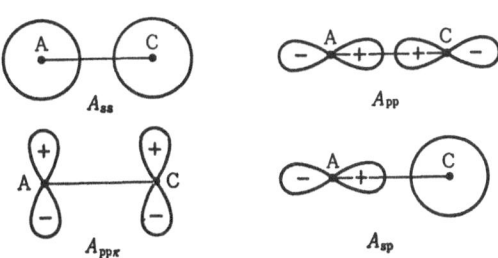

Fig. 7.12. Four different matrix elements between s and p orbitals

An interesting fact deduced from (7.4.22) is

$$X_5 = 2L_3 = 4\alpha_c(V_4 - V_5)$$
$$= 4\alpha_c A_{pp\pi}, \tag{7.4.23}$$

which shows that $X_5 < 0$ because $A_{pp\pi}$ is negative definite.

Finally, we shall close this section by stating a systematic method to determine the parameters which characterize the band structure of the tetrahedrally-bonded semiconductors. A crucial point is the fact that, as already pointed out, the parameters V_1^s, V_1^X, V_4 and V_5 may be regarded as the constants common to the compounds in the iso-orbital series. Thus, for instance, once the values of these parameters are determined for the IV elements, we can estimate the corresponding parameters for any compounds. For a compound composed of atoms lying on the same row in the periodic table, we can use the same value as that of the IV elements belonging to the same row. For a compound composed of two atoms which lie on different rows of the periodic table, we may use geometrical

Table 7.2. From [7.7]

Substance	V_1^a	V_1^x	V_2	V_4	V_5	V_1^a	V_3	α_p	α_c
C						0.00	0.00	0.00	1.00
BN	1.70	1.77	6.10	−1.09	0.41	0.57	2.76	0.41	0.91
BeO						1.10	5.07	0.64	0.77
Si						0.00	0.00	0.00	1.00
AlP	1.40	0.85	2.20	−0.61	0.01	0.34	1.18	0.47	0.88
MgS						0.67	2.12	0.63	0.78
Ge						0.00	0.00	0.00	1.00
GaAs	1.60	0.82	2.15	−0.70	−0.02	0.30	1.21	0.50	0.87
ZnSe						0.58	2.26	0.72	0.69
CuBr						0.70	2.77	0.79	0.61
α-Sn						0.00	0.00	0.00	1.00
InSb	1.30	0.68	1.76	−0.58	−0.03	0.21	1.04	0.51	0.86
CdTe						0.51	2.08	0.76	0.65
AgI						0.51	2.65	0.83	0.56
SiC						0.15	1.54	0.39	0.92
BP	1.54	1.23	3.66	−0.82	0.21	0.29	0.00	0.00	1.00
AlN						0.62	2.68	0.59	0.81
BeS						0.65	1.54	0.21	0.98
BAs						0.36	0.00	0.00	1.00
GaN						0.50	2.59	0.62	0.78
BeSe	1.65	1.21	3.62	−0.87	0.19	0.68	1.23	0.32	0.95
ZnO						1.00	3.56	0.70	0.71
CuF						1.38	5.45	0.83	0.56
InN	1.49	1.10	3.27	−0.80	0.19	0.61	2.75	0.64	0.77
BeTe						0.45	0.00	0.00	1.00
AlAs						0.41	1.06	0.44	0.90
GaP						0.23	1.33	0.52	0.85
MgSe	1.50	0.84	2.18	−0.65	−0.01	0.69	2.06	0.68	0.73
ZnS						0.55	2.32	0.73	0.68
CuCl						0.72	2.47	0.75	0.66
AlSb						0.22	1.26	0.54	0.84
InP	1.35	0.76	1.97	−0.60	−0.01	0.33	1.41	0.58	0.81
MgTe						0.46	1.79	0.67	0.74
CdS						0.62	2.37	0.77	0.64
GaSb						0.11	0.94	0.44	0.90
InAs						0.40	1.22	0.53	0.85
ZnTe	1.44	0.75	1.94	−0.64	−0.03	0.35	1.99	0.72	0.69
CdSe						0.65	2.35	0.77	0.64
CuI						0.44	2.44	0.78	0.63

averages of the two values of the corresponding two IV elements. Thus, after all, the problem is to determine V_1^s, V_1^X, V_4 and V_5 for the IV elements and V_1^a for each compound.

Practically, the three parameters V_4, V_5 and $V_1^s + 2V_1^X$ are determined from the observed values of the band energies at the Γ_1, X_4 and L points for C, Si, Ge and Sn. Then, referring to (7.4.22b), we estimate the values of V_1^s and V_1^a for each element from the spectroscopic data (in doing this estimation, care has to be exercised in making a proper correction so that the determined values are suitable for solid states). In this manner, all the parameters have been determined for all the compounds with a zincblende structure. The results are summarized in Table 7.2. This table should be compared with Table 6.12 given in Sect. 6.5 on the basis of the EPM.

8. Magnetism

In this chapter, discussions will be given on fundamental magnetic properties of solids. Magnetic properties are important in the investigation of condensed matter in general because they reflect the electronic structure and atomic state not only of magnetic substances, but also of magnetically simple substances. The discussions include magnetic responses to an external field, microscopic origin of magnetic interactions and mechanisms and patterns of magnetic orders. Material is spread over insulators, semiconductors and metals. The long-range order takes place among the electronic spins in some substances and this is the type of order not considered in other chapters. The ordered magnetic system exhibits the characteristic collective mode and the phase transition.

8.1 Magnetic Properties of Solids

The electron has a charge and a magnetic moment (related to its spin) and many atoms have nuclear magnetic moment. Thus, all materials show some kind of magnetic response to an applied magnetic field in various manners. Of what nature is this response to an external magnetic field? Furthermore, some solids undergo magnetic phase transitions at certain temperatures below which the solids are in magnetically-ordered states. Why and how can a material be in a magnetically-ordered state? These questions are necessarily related to the electronic structure of the material, so that magnetic properties of materials are quite important in investigating the electronic structures of matter.

Even if only magnetically-ordered materials are considered, they are found in insulators, semiconductors, and metals. In these various materials it can be mentioned that the most dominant (interaction) energy can be different for each material. Therefore, in the usual investigations of magnetism, one takes a model Hamiltonian or an effective interaction which is appropriate for a specific case, and investigates the resulting pattern of magnetic ordering, coilective excitation (or elementary excitation), phase transition, and magnetic response to an external field, and then compares these theoretical results with experiments. However, the spin is essentially a quantum variable and the subject is the many-body problem, so the solution of the above model Hamiltonian is always approximate, except in some special cases like the 1 and 2-dimensional Ising systems (these are not the quantum cases), and one must be careful about the nature of the approximation and the range of validity for the solution.

As can be seen in previous chapters, the electronic states in solids can be classified into two parts:

I)   the electron localized on or near an atom or an ion (localized electron),
II) the electron that moves over a crystal (itinerant electron).

The magnetic properties of these two kinds of electrons are generally different from each other. In real materials, however, the borderline between the above two is not clear; some crystals are typical in either of these, some are intermediate, and some show the phase transition between these two with temperature or pressure. Typical examples of (I) are the electrons in closed orbit in atoms or ions, the $3d$ and the $4f$ electrons in most compounds, electrons trapped at impurities or defects in ionic crystals and in semiconductors. Electrons in the valence and the conduction bands in semiconductors and in the conduction band in metals belong to (II), and they are well described by the Bloch functions.

For all these electrons the fundamental interactions that characterize magnetic properties are the Coulomb interaction, the exchange interaction, the dipolar interaction, and the spin-orbit interaction, in addition to the crystal-field potential, and in the case of metals and semiconductors, the crystalline kinetic energy is also involved. The majority of substances that exhibit magnetically-ordered states, have d or f electrons, regardless of whether they are insulators or conductors. In order to have a general idea of the relative importance of the above fundamental interactions, the orders of magnitude of these interactions are shown in Table 8.1. In ionic crystals, the intra-atomic Coulomb and exchange interactions may be nearly the same as those in free atoms. In metals and semiconductors there are screening effects of conduction electrons and the estimation of these interactions is difficult because the intra-atomic interactions are short-ranged. The width of the $3d$-band in transition metals is also shown in the table for comparison. Although the magnitude of the dipolar interaction between atoms is ~ 1 cm^{-1} at most and therefore quite small, it sometimes determines the easy direction of spins in the ordered state and, if the exchange interaction is quite small in an insulator, it may further provide the ordered-spin structure itself.

Table 8.1. Order of magnitudes of various interaction energies for $3d$ and $4f$ electrons

Energy [cm^{-1}]	10^5–10^4	10^4–10^3	10^3–10^2	$\leq 10^2$
$3d$	Intraatomic Coulomb and exchange interaction		Spin-orbit interaction	
	Crystal-field potential			
		Width of $3d$-band	Interatomic exchange interaction	
$4f$	Intraatomic Coulomb and exchange interaction		Crystal-field potential	
		Spin-orbit int.		
				Interatomic exchange int.

Following Table 8.1, ideas may be obtained concerning the order in which the fundamental interactions should be taken into account. For example, in rare-earth compounds the intra-atomic Coulomb, exchange and spin-orbit interactions are dominant for the $4f$ electrons, and the Russell-Saunders coupling scheme $|LSJM\rangle$ given in Sect. 1.7 is a good starting point. In fact, many compounds of this kind exhibit the magnetic moment related to the g-factor in (1.7.13) at high temperatures ($k_BT >$ crystal-field potential). (The spin-orbit interaction of the $4f$ electron is fairly large and some excited states of rare-earth ions show the intermediate coupling rather than the Russell-Saunders coupling). For the $3d$ electron in transition metal ions, on the other hand, the intra-atomic Coulomb and exchange interactions and the crystal-field potential are dominant, and the spin moment is not tightly bound to the orbitals through the spin-orbit interaction as far as the orbital state is not degenerate. This implies that $g \approx 2$ for the $3d$ electron. The $4f$ orbital is spacially compact so that the $4f$ electron is well-localized even in rare-earth metals, and magnetic properties of the $4f$-electron in rare-earth metals are quite similar to their compounds. In transition metals, the $3d$ orbitals form fairly narrow bands and, even if the screening effect is taken into account, the intra-atomic Coulomb interaction is considered to be still effective. This is in contrast to the simple metals in which conduction bands are wide and the nearly free electron model is valid.

Mechanisms and patterns of magnetic orderings will be discussed in later sections referring to the electronic properties of various materials. Additional importance of magnetism results from the fact that the electronic and nuclear spins play the role of a "test probe" to investigate the nature of materials. For instance, the covalency of substances can be investigated, the electron correlation in metals manifests itself in a characteristic response, and the dynamics of lattice vibration and of chemical reactions are also the object of magnetism. The spins as test probes have a wide range of uses in physics, chemistry, and biology.

8.2 Magnetic Properties of Simple Solids

Let us investigate the magnetic properties of the solids in which an electronic system does not exhibit any magnetic orderings. (a nuclear spin system can be in an ordered state at low temperatures independent of the electronic system in the solid). Such substances are compounds which do not contain the ions with unfilled d or f orbitals, intrinsic semiconductors, and simple metals. In order to investigate the response of such a system against an external field, the following Hamiltonian will be considered:

$$\mathcal{H} = \mathcal{H}_0 - \frac{ie\hbar}{mc}\sum_i A(r_i)\cdot\nabla_i + \frac{e^2}{2mc^2}\sum_i A(r_i)^2 + 2\mu_B\sum_i H\cdot s_i - g_N\mu_N\sum_n H\cdot I_n,$$

$$(8.2.1)$$

where $\mathcal{H}_0$ is the unperturbed Hamiltonian, s_i the electron spin, I_n the nuclear spin, g_N and μ_N are the nuclear g-factor and the nuclear magneton, respectively, and $A(r_i) = (H \times r_i)/2$ (the electronic charge is denoted as $-e$ and $\mu_B = e\hbar/2mc$). $\sum_i$ and $\sum_n$ indicate the sum over all electrons and nuclei, respectively.

Let us consider first the electrons localized in closed orbits on an atom or an ion. It is not necessary to consider the Zeeman energy of spins in this case. The first-order perturbation energy for electrons is obtained from the third term in (8.2.1) to give ($H\|z$)

$$\Delta_e E = \langle 0| \frac{e^2}{2mc^2} \sum_i A(r_i)^2 |0\rangle = \frac{e^2 H^2}{8mc^2} \langle 0| \sum_i (x_i^2 + y_i^2)|0\rangle , \qquad (8.2.2)$$

where $|0\rangle$ means the electronic ground state in the closed orbit. Remembering that the charge distribution of electrons in the closed orbit is isotropic, the corresponding susceptibility is given as

$$\chi_d^{(c)} = - \frac{Ne^2}{6mc^2} \langle 0| \sum_i^{(atom)} r_i^2 |0\rangle , \qquad (8.2.3)$$

where $\sum_i^{(atom)}$ indicates to sum over all electrons in an atom and N is the number of atoms. This is the well-known diamagnetic susceptibility of atoms. The numerical values of the susceptibilities are shown in Table 8.2 for some typical atoms and ions. Generally speaking, the radius of the closed orbit is larger in negative ions and the electron number increases with the atomic number, so that the diamagnetic susceptibility increases in such atoms or ions. This tendency can be clearly seen in the table. For example, H^- and He are iso-electronic but the susceptibility of the former is several times larger, and Cs^+ and Li^+ are both alkaline ions but the former is about ninety times larger. Although the values are not shown in the table, the diamagnetic susceptibity of C^{+4}, C^{-4}, and the benzene are -0.15×10^{-6}, -50×10^{-6}, and -54×10^{-6}, respectively. The diamagnetism is the most important magnetic property of electrons in closed orbits of atoms and molecules.

Nuclear spins play the important role of the test probe in substances which contain only ions with closed orbits. The last term in (8.2.1) is the Zeeman energy of free nuclear spins, and any characteristics of particular materials are not involved in this term. In $\mathcal{H}_0$ in (8.2.1), however, one has the interaction between the electronic and the nuclear spins

$$\mathcal{H}_0^{eN} = 2\mu_B \sum_n \sum_i \mu_n \cdot \left\{ \frac{l_i}{r_{in}^3} - \frac{s_i}{r_{in}^3} + 3\frac{r_{in}(s_i \cdot r_{in})}{r_{in}^5} + \frac{8\pi}{3} s_i \delta(r_i - r_n) \right\}, \qquad (8.2.4)$$

where $\mu_n = g_N \mu_N I_n$ is the nuclear magnetic moment. If electrons are localized at each atom, they interact preferably with the nucleus of the same atom and the sum $\sum_i$ may be replaced by $\sum_i^{(n)}$, which means the sum over electrons localized

Table 8.2. Diamagnetic susceptibilities of some typical atoms and ions. The values shown are
$-\chi_d^{(e)} \times 10^6$ per mole. [8.1]

	H⁻	He	Li⁺	Be⁺²
	H⁻	He	Li⁺	Be⁺²

	H⁻	He	Li⁺	Be⁺²
	8	1.54	0.63	0.34
O⁻²	F⁻	Ne	Na⁺	Mg⁺²
12.6	8.1	5.7	4.2	3.2
S⁻²	Cl⁻	Ar	K⁺	Ca⁺²
40	29	21.5	16.7	13.3
			Cu⁺	Zn⁺²
			13	11
Se⁻²	Br⁻	Kr	Rb⁺	Sr⁺²
70	54	42	35	28
			Ag⁺	Cd⁺²
			44	37
Te⁻²	I⁻	Xe	Cs⁺	Ba⁺²
105	80	66	55	46
			Au⁺	Hg⁺²
			65	55

at the nth atom. The first term in (8.2.4) corresponds to the Biot-Savart law
which denotes the interaction of nuclear spins with electronic orbital motion; the
second and third terms express the dipolar interaction between nuclear and
electron spins, and the last term, the so-called contact term or Fermi interaction
which is present only when the electronic wave function has the amplitude at the
position of nuclear spin. When $I \geq 1$, the nucleus has the electric quadrupole
moment which interacts with the electric field of the same symmetry that is
exerted by surrounding nuclei and electrons. If the symmetry is axial, for instance, the additional term $\sum_n e^2 qQ[3(I_n^z)^2 - I(I+1)]/4I(2I-1)$ is present in
(8.2.4). Here, $eQ = e\int d\mathbf{r}\rho_N(\mathbf{r})(2z^2 - x^2 - y^2)$ is the nuclear quadrupole moment
$[e\rho_N(\mathbf{r})$ is the charge distribution of the nucleus] and q is the electric field that has
the same symmetry as $(2z^2 - x^2 - y^2)$. This field comes not only from neighboring ions, but also from an electron cloud in slightly distorted closed orbitals on
the same atom (Sternheimer effect). When the atom has a partially filled orbital
(whose charge distribution is not spherically symmetric), the electronic cloud can
give rise to a large quadrupole field at the nucleus of the same atom. The contact
term in (8.2.4) also plays an important role in such a case. This is true even if the
unfilled orbital is not the s orbital, because the nonzero spin moment in the unfilled orbital can polarize the inner closed s orbital in an asymmetric way for up
and down-spin states to interact with the nuclear spin.

When a compound contains only ions with closed shells, there is no freedom of electron spins and only the term $2\mu_B \sum_n \overset{(n)}{\sum_i} \mu_n \cdot l_i / r_{in}^3$ should be considered in (8.2.4). Writing the second term in (8.2.1) as $(-ie\hbar/mc) \sum_i A(r_i) \cdot \nabla_i = \mu_B \sum_n \overset{(n)}{\sum_i} H \cdot l_i$, the second-order perturbation energy involving these two term gives

$$\mathscr{H}^{(2)} = 2\mu_B^2 \sum_n \sum_e \frac{\langle n0 | \overset{(n)}{\sum_i} H \cdot l_i | ne \rangle \langle ne | \overset{(n)}{\sum_i} \mu_n \cdot l_i / r_{in}^3 | n0 \rangle}{E_0 - E_e} + \text{c.c.} ,$$

$$\equiv - \sigma_p \sum_n H \cdot \mu_n \tag{8.2.5}$$

where $|n0\rangle$ and $|ne\rangle$ represent the ground and the excited states of the nth ion, respectively, E_0 and E_e their energy eigenvalues and c.c. the complex conjugate. Combining this with the nuclear Zeeman energy, they can be written as

$$-\sum_n (1 + \sigma_p) H \cdot \mu_n.$$

This means that the magnetic field at the nuclear moment is modified by σ_p, which is called the chemical shift. This reflects the nature of materials through E_e and the characteristics of wave functions. In addition to σ_p, the angular momentum operator l is replaced by $l_i + (e/c\hbar) r_i \times A(r_i)$ in a magnetic field and this modification further adds $-\sigma_d$, the diamagnetic shielding effect. The resultant shift is, therefore, $(\sigma_p - \sigma_d)$ and is of the order of $10^{-3} \sim 10^{-5}$. In metals, the electrons in closed orbits also show the diamagnetism given in (8.2.3), conduction electrons have a large effect on nuclear spins and the above chemical shift is rather a minor interaction.

Let us then consider simple metals which do not exhibit any magnetic ordering. As mentioned above, it may be sufficient here to consider only the magnetic properties of conduction electrons. In contrast to the metals that show magnetic ordering, such metals have wide conduction bands and the kinetic energy is large compared with the screened Coulomb and exchange energies, so that the one-electron approximation is generally a good approximation. The metals which have conduction bands of only the s and p orbital belong to this category.

In the one-electron approximation an electron has the energies of $\pm \mu_B H$ according to whether the electron spin is parallel or antiparallel to the external field H (Fig. 8.1). The magnetization of conduction electrons is given by

$$M = \mu_B \int_{-\infty}^{\infty} d\varepsilon N(\varepsilon) [f(\varepsilon - \mu_B H) - f(\varepsilon + \mu_B H)] = - 2H\mu_B^2 \int_{-\infty}^{\infty} d\varepsilon N(\varepsilon) \frac{\partial f}{\partial \varepsilon} ,$$

$$\tag{8.2.6}$$

where $N(\varepsilon)$ is the density of states of the conduction band per spin and $f(\varepsilon) = \{\exp[\beta(\varepsilon - \zeta)] + 1\}^{-1}$ is the Fermi distribution function. Using the well-known formula

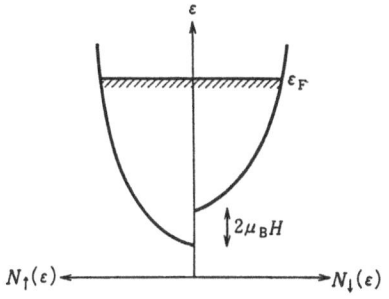

Fig. 8.1. Density of states of the conduction band and the Fermi level of conduction electrons under an external magnetic field

$$-\int_{-\infty}^{\infty} d\varepsilon \phi(\varepsilon)\frac{\partial f}{\partial \varepsilon} = \phi(\zeta) + \frac{\pi^2}{6}(k_BT)^2\left(\frac{\partial^2 \phi(\varepsilon)}{\partial \varepsilon^2}\right)_{\varepsilon = \zeta} + \cdots \tag{8.2.7}$$

and the equation

$$N_e = \int_{-\infty}^{\infty} d\varepsilon N(\varepsilon)\left[f(\varepsilon - \mu_B H) + f(\varepsilon + \mu_B H)\right] = 2\int_{-\infty}^{\infty} d\varepsilon N(\varepsilon)f(\varepsilon) + O(H^2),$$
$$\tag{8.2.8}$$

which provides the chemical potential ζ, the magnetic susceptibility $\chi = (M/H)_{H=0}$ is obtained as

$$\chi_s = 2\mu_B^2 N(\zeta_0)\left\{1 + \frac{\pi^2}{6}(k_BT)^2\left[\frac{1}{N(\varepsilon)}\frac{d^2N(\varepsilon)}{d\varepsilon^2} - \left(\frac{1}{N(\varepsilon)}\frac{dN(\varepsilon)}{d\varepsilon}\right)^2\right]_{\varepsilon = \zeta_0}\right\}. \tag{8.2.9}$$

Here, ζ_0 is the value of ζ at $T = 0$ K and is equal to the Fermi energy ε_F. Equation (8.2.9) is the positive quantity and is called the spin paramagnetism. If the free electron model is valid, (8.2.9) reduces to

$$\chi_s = \frac{3N_e\mu_B^2}{2\varepsilon_F}\left[1 - \frac{\pi^2}{12}\left(\frac{k_BT}{\varepsilon_F}\right)^2\right], \tag{8.2.10}$$

where $\varepsilon_F = \hbar^2k_F^2/2m^*$.

Let us now investigate the response of electron orbitals to an external field. For simplicity, consider the free electron model at $T = 0$ K. Dropping the terms related to the electron spin and nuclear spin, (8.2.1) is written as

$$\mathcal{H} = \frac{1}{2m^*}\sum_i\left[p_i + \frac{e}{c}A(r_i)\right]^2 \tag{8.2.11}$$

and the motion of an electron described by this Hamiltonian will be considered. If one expresses the vector potential to denote the static magnetic field H as $A(r) = (0, Hx, 0)$, (8.2.11) is written as

$$\mathcal{H} = \frac{1}{2m^*}[p_x^2 + (p_y + m\omega_c x)^2 + p_z^2],$$ (8.2.12)

where $\omega_c = eH/m^*c$. The equations of motion for the operator p_y and p_z are, therefore,

$$i\hbar\dot{p}_y = [p_y, \mathcal{H}] = 0, \quad i\hbar\dot{p}_z = [p_z, \mathcal{H}] = 0$$ (8.2.13)

which indicate that p_y and p_z are constants of motion, and they will be denoted as $\hbar k_y$ and $\hbar k_z$, respectively. Writing $x_0 = -\hbar k_y/m^*\omega_c$, (8.2.12) can be rewritten as

$$\mathcal{H} = \frac{1}{2m^*}[p_x^2 + m^{*2}\omega_c^2(x - x_0)^2] + \frac{\hbar^2}{2m^*}k_z^2.$$ (8.2.14)

This means that the motion of an electron in the x-direction is described simply by the one-dimensional harmonic oscillator. The energy eigenvalue of an electron is, therefore, given by

$$\varepsilon(n, k_z) = \hbar\omega_c(n + \tfrac{1}{2}) + \frac{\hbar^2 k_z^2}{2m^*}$$ (8.2.15)

and its eigenfunction by

$$\psi(n, k_y, k_z) = u_n(x - x_0)\exp[i(k_y y + k_z z)].$$ (8.2.16)

Here, $n = 0, 1, 2, \ldots$ and $u_n(x)$ is the eigenfunction of the harmonic oscillator. Such eigenstates quantized in units of $\hbar\omega_c$ under the static magnetic field are called the Landau levels. The motion in the z-direction is not affected by H, however. In the Landau levels there are $m^*\omega_c L_x L_y/2\pi\hbar$ degenerate states with a fixed n and k_z, and this is nothing but the number of states on the $k_x - k_y$ plane in the energy interval $d\varepsilon = \hbar\omega_c$ when $H = 0$. (L_x and L_y are dimensions of the crystal.) The density of states of the nth Landau level is given by

$$N(n, \varepsilon) = \frac{m^*\omega_c L_x L_y}{2\pi\hbar} \frac{L_z}{2\pi} \frac{dk_z}{d\varepsilon} = \frac{Vm^*\omega_c(2m)^{1/2}}{8\pi^2\hbar^2} \frac{1}{[\varepsilon - \hbar\omega_c(n + 1/2)]^{1/2}}.$$ (8.2.17)

This diverges at $\varepsilon = \hbar\omega_c(n + 1/2)$ and it is the consequence of the one-dimensional spectrum in the k_z-direction. $\varepsilon(n, k_z)$ and $N(n, \varepsilon)$ are sketched in Fig. 8.2.

According to the Fermi statistics, electrons occupy the states necessary to accommodate all electrons from the bottom, and let us denote the highest occupied Landau level by n_t. When the magnetic field H increases, the quantized energy interval $\hbar\omega_c = \hbar eH/m^*c$, and hence the degeneracy of states with a given (n, k_z), $m^*\omega_c/2\pi\hbar = eH/2\pi\hbar c$, increases ($L_x = L_y = L_z = 1$ are taken). Therefore, in order to accommodate a given number of electrons, n_t will take a smaller value as H is increased. Each time the Fermi surface moves from a given n_t to another, oscillatory behavior in various physical quantities appears. Let us cut

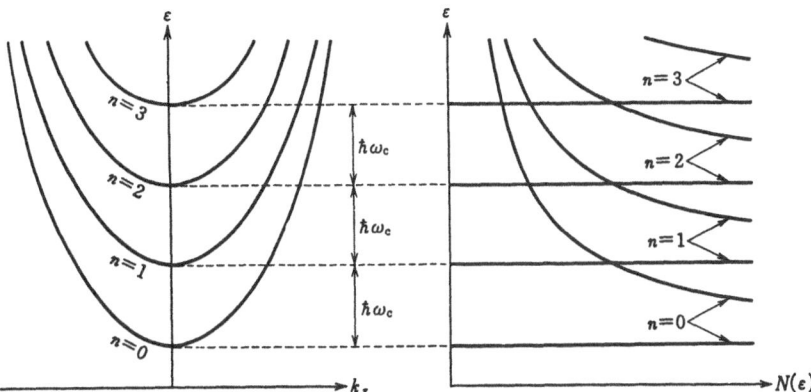

Fig. 8.2. Energy spectrum and the density of states of the Landau levels

the k-space with planes perpendicular to H and with thickness δk_z, and consider one slice of them. As the electrons occupy up to n_t, the electron number in this slice is

$$\delta N_e = \frac{eHn_t\delta k_z}{2\pi^2\hbar c} . \qquad (8.2.18)$$

If n_t is fixed, this δN_e increases linearly with H. But each time n_t changes, δN_e will change quite abruptly. This oscillatory change in δN_e takes place around the mean value $\delta N_e^{(0)}$ which is equal to the electron number contained in the same slice under $H = 0$. It can be shown that the amplitude of $|\delta N_e - \delta N_e^{(0)}|$ is $eH\delta k_z/4\pi^2\hbar c$. The change $(\delta N_e - \delta N_e^{(0)})$ is, of course, the number of electrons that move from (or to) other slices, and this corresponds to the movement of $(\delta N_e - \delta N_e^{(0)})$ electrons to the interested slice near the Fermi surface under $H = 0$. Denoting the Fermi radius on this slice by k_F'(i.e., $k_F^2 = k_F'^2 + k_{F_z}^2$), it will change by an amount $\delta k_F'$ given through $\delta N_e - \delta N_e^{(0)} = 4\pi k_F'\delta k_F'\delta k_z/(2\pi)^3$, and the transferred electrons have an average excess energy of $\hbar^2 k_F'\delta k_F'/m$. Thus, the contribution of the energy change from this slice is

$$\delta E = \frac{2\pi^2\hbar^2}{m\delta k_z}(\delta N_e - \delta N_e^{(0)})^2 . \qquad (8.2.19)$$

As is shown above, δN_e exhibits oscillatory behavior with H so that the same is true for δE. The contribution to the magnetization from this slice is given as $\delta M = -d(\delta E)/dH = -(\varepsilon_F'/H)(\delta N_e - \delta N_e^{(0)})$, where $\varepsilon_F' = \hbar^2 k_F'^2/2m^*$ and the equality $n_t = \varepsilon_F'/\hbar\omega_c$ has been used, and this shows that δM is also oscillatory. As H is increased, the oscillatory behavior takes place each time when n_t changes by one, so that the period is given by $e\hbar/m^*c\varepsilon_F'$ as a function of $1/H$. This period can alternatively be written as $2\pi e/c\hbar S_\perp$ with $S_\perp = \pi k_\perp^2$ so that it is determined by the cross-sectional area of the Fermi surface. One may then expand δM in the Fourier series; $\delta M = \sum_l A_l \sin lx$ with $x = 2\pi m^* c \varepsilon_F' / e\hbar H$. The whole magnetization of the electron system can be obtained by summing up the contributions from all slices, which leads to the Fresnel integral. The main result of this is that each contribution has a different phase and they cancel each other except for the region where the phase becomes stationary. For an arbitrary Fermi surface, the stationary region can be obtained as the maximum or the minimum cross section of the Fermi surface when it is cut by a plane perpendicular to the magnetic field H, and these cross-sectional areas (which replace the above $S_\perp$) determines the period of the oscillation in M. The peak of the oscillation is equidistant when plotted against $1/H$. This phenomenon that the magnetization is oscillatory with H is called the de Haas–van Alphen effect. This effect is very useful in determing the shape of the Fermi surface.

The energy change δE given in (8.2.19) is positive and oscillates between zero and the maximum value which is given as $\delta E_{\max} = (2\pi^2 \hbar^2 / m \delta k_z)(eH\delta k_z / 4\pi^2 \hbar c)^2$ by using (8.2.18). Averaging δE over a period, one gets $\langle \delta E \rangle = e^2 H^2 \delta k_z / 24\pi^2 mc^2$, which provides $\delta \chi = \delta M / H = -e^2 \delta k_z / 12\pi^2 mc^2$. After integration with respect to k_z, this gives

$$\chi_d = -\frac{e^2 k_F}{12\pi^2 mc^2} = -\tfrac{2}{3} N(\varepsilon_F) \mu_B^2 . \tag{8.2.20}$$

This is called the Landau diamagnetism, and its absolute value is equal to 1/3 of the susceptibility of the spin paramagnetism at $T = 0$ K given in (8.2.9).

Thus far, it has been shown that the basic magnetic properties of conduction electrons are the spin paramagnetism and the Landau diamagnetism. In the above investigation, however, any detailed band structures have not been pursued and only a single conduction band is considered. If there are more than two bands lying close to each other, one has the orbital paramagnetism that arises from the orbital freedom of electrons. The paramagnetic contribution arises from the second term of (8.2.1), which can be rewritten as $(-ie\hbar/mc) \sum_i A$ $(r_i) \cdot \nabla_i = \mu_B \sum_i H \cdot l_i$ $(l = -i\hbar r \times \nabla$ is the orbital angular momentum operator). The nondiagonal matrix elements of l_i between different bands can give rise to the average magnetization $\mu_B \langle \sum_i l_i \rangle$ and hence $\chi_0^\alpha = \mu_B \langle \sum_i l_{i\alpha} \rangle / H$ as the susceptibility of the orbital paramagnetism. According to the linear response theory [8.2] this χ_0^α is given as

$$\chi_0^\alpha = \mu_B^2 \sum_{nn'}{}' \sum_k \frac{f(\varepsilon_{nk}) - f(\varepsilon_{n'k})}{\varepsilon_{n'k} - \varepsilon_{nk}} \, |\langle n'k|l_\alpha|nk\rangle|^2 , \tag{8.2.21}$$

where $f(\varepsilon_{nk})$ is the electron distribution function for the k state at the nth band and the prime on the sum symbol indicates the exclusion of the case $n = n'$ in the sum $(\alpha = x, y, \text{ or } z)$. (For the Bloch states, $\langle nk|l|nk\rangle = 0$ holds, in general, by the quenching of the orbital angular momentum except for some special points in k-space where degeneracy exists.) This paramagnetic susceptibility becomes appreciable when $\varepsilon_{n'k}$ is close to ε_{nk}. In metallic vanadium, for instance, the $3d$ bands are close to each other and this type of orbital paramagnetism reaches to some tens of percent of the spin paramagnetism.

Finally, let us consider the interaction between conduction electrons and nuclear spins. As mentioned already, conduction bands in simple metals which do not show any magnetic ordering have partly the character of an s-function, so that the contact term in (8.2.4) may be the dominant interaction. Dealing with this interaction together with the nuclear Zeeman term in (8.2.1), they can be written as $- \sum_{n} \mu_n \cdot [H - (16\pi/3)\mu_B \sum_{i} \langle s_i \rangle \delta(r_i - r_n)]$ which indicates that, as the spin polarization $\langle s_i \rangle$ is antiparallel to H, the contact term has the role of adding an extra field ΔH to H. The quantity $-2\mu_B \sum_{i} \langle s_i \rangle / H$ is nothing but the spin paramagnetism so that the shift due to ΔH is given by

$$\sigma_K \equiv \frac{\Delta H}{H} = \frac{8\pi}{3} \chi_s \langle |\psi_F(0)|^2 \rangle_{AV}, \tag{8.2.22}$$

where χ_s is the susceptibility per atom of the spin paramagnetism, and $\langle |\psi_F(0)|^2 \rangle_{AV}$ is the square of the absolute value of the conduction electron wave functions at a nucleus averaged over the Fermi surface. The more s-character the wave function has, the larger the value of $\langle |\psi_F(0)|^2 \rangle_{AV}$. This shift can be detected in the nuclear magnetic resonance (NMR) and it is called the Knight shift. The quantity σ_K is usually of the order of $10^{-2} \sim 10^{-3}$. When the contact term in (8.2.4) is the dominant interaction between electron and nuclear spins, it can have the role of flipping the direction of the nuclear spins. This means that the relaxation time of $\langle I^z \rangle$, or equivalently the spin-lattice relaxation time T_1, is determined by this interaction. When $\langle I^z \rangle$ changes, $\langle s^z \rangle$ also changes in the opposite direction and the excess angular momentum in $\langle s^z \rangle$ thus transferred disappears into the lattice freedom by the spin-orbit interaction. In the first process above, the change of the nuclear Zeeman energy by the flip of a nuclear spin is quite small so that only electrons close to the Fermi surface are involved in this event by the energy conservation law. This fact suggests that T_1 may have a relation with the above σ_K, and in fact the following relation can be derived:

$$T_1 T = \frac{\mu_B^2}{\pi k_B (g_N \mu_N)^2 \sigma_K^2}, \tag{8.2.23}$$

where k_B is the Boltzmann constant, T the absolute temperature. The relation $T_1 T = \text{const.}$ holds in many metals and in aluminum, for instance, it holds at $1 \sim 1000$ K. Equation (8.2.23) is derived by assuming the presence of the Fermi surface and its validity is important to ascertaining the existence of the Fermi surface just as by the electronic specific heat.

8.3 Mechanism of Spin Polarization and the Electronic State in Solids

As stated in Sect. 2.2, there are two methods of constructing wave functions of many electron systems; one is the Heitler–London method in which each atom always takes a given number of electrons to reflect the electron correlation; another is the molecular-orbital method which takes care of the spatial spread of the one-electron orbital but permits several kinds of valency for each atom. In discussing magnetism, the Heitler–London method may be adequate for insulators where the electron correlation is thought to be large, while in metals the Bloch functions which correspond to molecular orbitals in solids may be the good starting point. In reality, however, various kinds of substances are in between, and it can be only mentioned that magnetic electrons in insulators may be started from the Heitler–London picture and those in metals from the opposite picture. In some extreme cases, electrons change from the localized state to the itinerant state exhibiting the so-called metal-insulator transition. If metallic ions in a compound have an odd number of electrons or a nonintegral number of electrons on the average, this compound should be a conductor by the band theory. However, many compounds of this kind involving the transition metal ions are either insulators or semiconductors (having large electrical resistance and/or negative coefficient in its temperature dependence) under certain temperatures or pressures beyond which they become conductors (with large electrical conductivity and/or negative coefficient in its temperature dependence). In Fig. 8.3 the temperature-dependence of electrical resistance is shown for several kinds of vanadium oxide. In the case of V_2O_3 there is an even number of electrons ($3d^2$) per vanadium ion and it does not correspond to the present example. All other oxides have either an odd number or nonintegral number of $3d$ electrons. They all show clear phase transition (most of them accompany structural distortion of the order of 1% in undergoing the phase transition) and also magnetic orderings at lower temperatures.

The fundamental spin structures of magnetically-ordered states are the ferromagnetism and the antiferromagnetism. In the former, all spins are aligned in one direction and in the latter, the neighboring spins are aligned in the opposite direction without net magnetization as a whole. In order to have such ordered states there must be the interection that is dependent on the spin direction.

First, let us consider the interatomic exchange interaction in the case where the Heitler–London picture is valid. In this case, the free ion wave function may be the good approximate function for electrons on each atom (or ion). Here, a simple case will be considered where only one magnetic electron is present on each atom. As in Sect. 2.2, orbital functions which are centered at the atom a and b are denoted as $\varphi_a(r)$ and $\varphi_b(r)$, respectively (these functions are not necessarily the $1s$-function). There is generally an overlap integral between these functions, but it will be neglected for the moment. Each atom can take two possible spin states, i.e., up-spin state α and down-spin state β, and the two-electron

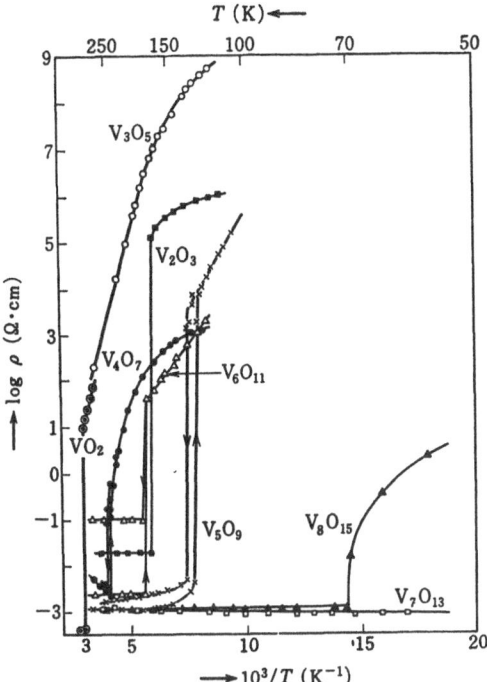

system can take four possible states $\Psi_{\alpha\alpha}$, $\Psi_{\alpha\beta}$, $\Psi_{\beta\alpha}$, and $\Psi_{\beta\beta}$, where $\Psi_{\chi\chi'}$ denotes

$$\Psi_{\chi\chi'} = \frac{1}{\sqrt{2}} \begin{vmatrix} \varphi_a(r_1)\,\chi(s_1) & \varphi_b(r_1)\,\chi'(s_1) \\ \varphi_a(r_2)\,\chi(s_2) & \varphi_b(r_2)\,\chi'(s_2) \end{vmatrix}. \tag{8.3.1}$$

The determinantal equation for the Hamiltonian $\mathscr{H}$ of the electron system can be constructed to yield the energy eigenvalues and the eigenfunctions. Writing

$$\left. \begin{aligned} \mathscr{H}_{ab} &= \iint dr_1 dr_2 \varphi_a^*(r_1)\,\varphi_b^*(r_2)\,\mathscr{H}\varphi_a(r_1)\,\varphi_b(r_2) \\ J_{ab} &= \iint dr_1 dr_2 \varphi_a^*(r_1)\,\phi_b^*(r_2)\,\mathscr{H}\varphi_b(r_1)\,\varphi_a(r_2) \end{aligned} \right\} \tag{8.3.2}$$

one obtains the eigenfunctions

$$\left. \begin{aligned} \Psi_{\alpha\alpha} &= \frac{1}{\sqrt{2}}\,[\varphi_a(r_1)\,\varphi_b(r_2) - \varphi_b(r_1)\,\varphi_a(r_2)]\,\alpha_1\alpha_2 \\ \frac{1}{\sqrt{2}}\,(\Psi_{\alpha\beta} + \Psi_{\beta\alpha}) &= \tfrac{1}{2}\,[\varphi_a(r_1)\,\varphi_b(r_2) - \varphi_b(r_1)\,\varphi_a(r_2)]\,[\alpha_1\beta_2 + \beta_1\alpha_2] \\ \Psi_{\beta\beta} &= \frac{1}{\sqrt{2}}\,[\varphi_a(r_1)\,\varphi_b(r_2) - \varphi_b(r_1)\,\varphi_a(r_2)]\,\beta_1\beta_2 \end{aligned} \right\} \tag{8.3.3}$$

with the eigenvalue $(\mathcal{H}_{ab} - J_{ab})$ and the eigenfunction

$$\frac{1}{\sqrt{2}} (\Psi_{\alpha\beta} - \Psi_{\beta\alpha}) = \tfrac{1}{2} [\varphi_a(r_1)\, \varphi_b(r_2) + \varphi_b(r_1)\, \varphi_a(r_2)] \, [\alpha_1\beta_2 - \beta_1\alpha_2] \quad (8.3.4)$$

with the eigenvalue $(\mathcal{H}_{ab} + J_{ab})$, where α_i and β_i are abbreviations of $\alpha(s_i)$ and $\beta(s_i)$. Three eigenfunctions in (8.3.3) are eigenfunctions of the total spin angular momentum operator $S = s_1 + s_2$ with the eigenvalue $S = 1$ (triplet) and those in (8.3.4) are eigenfunctions with $S = 0$ (singlet). Except the common factor $\mathcal{H}_{ab}$, the energy eigenvalues of the triplet and of the singlet states coincide with the eigenvalues of the following operator:

$$-\tfrac{1}{2} (1 + 4s_a \cdot s_b) \, J_{ab} . \quad (8.3.5)$$

This form of interaction is quite convenient in discussing ordered states of spins in solids. As can be seen from above, the parallel spin state at a and b atoms has lower energy if $J_{ab} > 0$ and the antiparallel one has lower energy if $J_{ab} < 0$.

The overlap integral

$$L = \int dr \varphi_a^*(r)\, \varphi_b(r) \quad (8.3.6)$$

has been neglected in the above consideration. In order to take this into account, one may define a new set of orthogonal functions

$$\left.\begin{aligned}
\psi_a(r) &= (1 - 2\mu L + \mu^2)^{-1/2}[\varphi_a(r) - \mu\varphi_b(r)] \\
\psi_b(r) &= (1 - 2\mu L + \mu^2)^{-1/2}[\varphi_b(r) - \mu\varphi_a(r)]
\end{aligned}\right\}, \quad (8.3.7)$$

where $\mu = L^{-1}[1 - (1 - L^2)^{1/2}]$, and then one can repeat the same discussion as above. Here, $\psi_a(r)$ and $\psi_b(r)$ are no longer the simple functions centered at a and b sèe (Fig. 8.4). The one-electron and the two-electron interactions are involved in the Hamiltonian $\mathcal{H}$. Using the orthogonalized wave functions, only the two-electron (Coulomb) interaction contributes to J_{ab} so that J_{ab} should be positive. If the original functions which are not orthogonalized are used, J_{ab} can be either positive or negative.

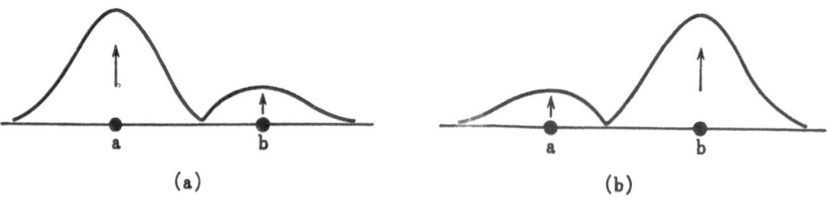

(a)   (b)

Fig. 8.4a, b. Electronic charge distribution in the orthogonalized wave functions. (a) $|\psi_a(r)|^2$, (b) $|\psi_b(r)|^2$

Next, let us take into account the different valencies for each atom. For simplicity, let us consider again the two atoms each having one magnetic electron and presume that the wave functions are orthogonal to each other. In the second quantized form, the Hamiltonian of the two-electron system may be written as

$$\mathcal{H} = \sum_{i=a,b} (\varepsilon_0 n_{i\sigma} + U n_{i\uparrow} n_{i\downarrow}) + \sum_{\sigma} (t_{ab} c_{a\sigma}^\dagger c_{b\sigma} + \text{c.c.}), \tag{8.3.8}$$

where $c_{i\sigma}^\dagger$ ($c_{i\sigma}$) is the creation (annihilation) operator of an electron with spin σ at the ith atom and $n_{i\sigma} = c_{i\sigma}^\dagger c_{i\sigma}$. The term with ε_0 denotes the energy level of the magnetic electron in a free atom, the term with t_{ab} plays the role of transferring an electron from the b-atom to the a-atom (or vise versa) to yield a mixed valence state and the term with U is the intraatomic Coulomb interaction to give the effect that electrons repel each other on the same atom (the electrons with the same spin state cannot occupy the same orbital by the Pauli principle). When U is sufficiently large, the Heitler–London picture may be appropriate. There is also the direct exchange interaction J_{ab} between the two orbitals, but this is suppressed for clarity. Taking the first bracket term as the unperturbed part, the four states $\Psi_{a\alpha,b\alpha}$, $\Psi_{a\alpha,b\beta}$, $\Psi_{a\beta,b\alpha}$, and $\Psi_{a\beta,b\beta}$ are at the energy $2\varepsilon_0$, and these are the states considered in the Heitler–London picture. In addition, there are the other two states $\Psi_{a\alpha,a\beta}$ and $\Psi_{b\alpha,b\beta}$ at $2\varepsilon_0 + U$. For the second bracket term in (8.3.8), the second-order perturbation calculation is carried out to give as matrix elements between the four ground states

$$- \sum_{\sigma\sigma'} \frac{|t_{ab}|^2}{U} \langle G' | c_{a\sigma'}^\dagger c_{b\sigma'} c_{b\sigma}^\dagger c_{a\sigma} + c_{b\sigma'}^\dagger c_{a\sigma'} c_{a\sigma}^\dagger c_{b\sigma} | G \rangle. \tag{8.3.9}$$

Here, $|G\rangle$ and $|G'\rangle$ may be any of the four ground states. Noting the equalities

$$\left. \begin{array}{l} c_{i\uparrow}^\dagger c_{i\uparrow} - c_{i\downarrow}^\dagger c_{i\downarrow} = 2s_i^z \\ c_{i\uparrow}^\dagger c_{i\downarrow} = s_i^x + i s_i^y = s_i^+ \\ c_{i\downarrow}^\dagger c_{i\uparrow} = s_i^x - i s_i^y = s_i^- \end{array} \right\} \tag{8.3.10}$$

and using $\sum_{\sigma} (n_{a\sigma} + n_{b\sigma}) = 2$, (8.3.9) can be transformed to

$$- \tfrac{1}{2} (1 - 4 \mathbf{s}_a \cdot \mathbf{s}_b) \frac{2|t_{ab}|^2}{U}. \tag{8.3.11}$$

Remembering the exchange interaction J_{ab}, (6.3.5,11) give as the effective Hamiltonian

$$\mathcal{H}_{ex} = - \tfrac{1}{2} (J_{ab} + K_{ab}) - 2(J_{ab} - K_{ab}) \mathbf{s}_a \cdot \mathbf{s}_b, \tag{8.3.12}$$

where $K_{ab} = 2|t_{ab}|^2/U$. K_{ab} is clearly positive, so that even if J_{ab} is positive, the effective exchange interaction $(J_{ab} - K_{ab})$ can take either sign. K_{ab} is called the kinetic exchange because this is provided by t_{ab} which gives rise to the kinetic energy of electrons in a crystal, while J_{ab} is called the potential exchange. In most magnetic insulators, the interaction (8.3.12) creates magnetic orderings.

Let us consider the case where each atom accommodates more than one electron. As explained in Sect. 8.1, magnetic substances involve d and f electrons and these orbitals can accommodate up to 10 and 14 electrons, respectively. The electrons in such orbitals usually satisfy the Hund coupling and the ground multiplet has the maximum total spin angular momentum[1] as explained in Sect. 1.6. The intra-atomic interactions which are relevant to the Hund coupling are so large, as is indicated in Table 8.1, that the total spin angular momentum $S_n = \sum_l^{(n)} s_l$ for the nth atom can be regarded as having a given (maximum) magnitude S, and s_a and s_b are replaced by S_a and S_b. If the original interaction is isotropic just like the Coulomb interaction, the resultant effective interaction takes the form of a scalar product between S_n's. The interaction of the form (8.3.12) with s replaced by S is called the Heisenberg interaction.

There is another type of interaction between atoms which makes use of the intra-atomic Hund coupling. Suppose that there can be two possible valence states M^{+n} and $M^{+(n-1)}$ (M:. transition metal atom) in a compound. For definiteness, the electron number n_e at the d–orbital of M^{+n} will be confined to $n_e < 5$. If one presumes the electron system in the M^{+n} state as being an ionic core, the above two-atom system can be considered as $M^{+n} + M^{+(n-1)} = 2M^{+n} + e^-$ and the total energy is the same, regardless of whether the extra electron is in the d orbital of either atom. Let us consider the motion of this extra electron in terms of the Hamiltonian (8.3.8) by extending it to the case with more than one d orbital per atom. Hence, one has to add the intra-atomic exchange interaction between the d orbital to the Hamiltonian. By the Hund rule, this interaction strongly favors the parallel-spin state between the spins of the extra electron and of the ionic core, so that the extra electron can take only the spin state parallel to the spin of the ionic core in transferring from one atom to another. Referring to the total spin S_a on atom a, S_b is now assumed to be titled by θ. This means that the spin function of the extra electron being on atom a can be expressed, when referring to the S_b direction, as

$$\chi(s) = \alpha(s) \cos \frac{\theta}{2} + \beta(s) \sin \frac{\theta}{2} . \tag{8.3.13}$$

This expression is the same if one exchanges the role of S_a and S_b. As the extra

---

[1]As can be seen from Table 8.1, there can be the case where the crystal field exceeds the intra-atomic electron-electron interaction. In such a case, the Hund coupling does not hold and a rearrangement of electron states takes place so as to gain the crystal-field potential. This case is called the strong-field configuration.

electron is either on atom a or b, the Hamiltonian for it can be written in a matrix form as

$$
[\mathscr{H}] = \begin{vmatrix} \left(U - \frac{J_{aa}}{2}\right) n_e & t_{ab} \cos \frac{\theta}{2} \\ t_{ba} \cos \frac{\theta}{2} & \left(U - \frac{J_{bb}}{2}\right) n_e \end{vmatrix},
$$
(8.3.14)

where $J_{aa} = J_{bb}$ is the intra-atomic exchange constant when it is expressed in the form of (8.3.5). One obtains two energy eigenvalues from (8.3.14):

$$
E = \left(U - \frac{J_{aa}}{2}\right) n_e \pm |t_{ab}| \cos \frac{\theta}{2} .
$$
(8.3.15)

The lower eigenvalue of these, depending on the angle between S_a and S_b, favors the ferromagnetic state. This type of interaction is called the double-exchange interaction (Fig. 8.5).

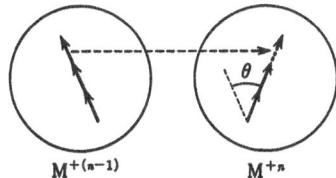

$M^{+(n-1)}$ M^{+n}

Fig. 8.5.
Electron transfer in the double exchange interaction

The tendency to line up spins in an atom is present not only between the same kind orbitals (like $3d$), but also between different orbitals (say, between $3d$ and $4s$). In transition metals and rare-earth metals, the outermost orbitals form wide conduction bands and such interaction is present between the conduction electrons and magnetic electrons in the d and f orbitals, respectively. In transition metals, however, the width of the $3d$-band can be comparable in order of magnitude to the intra-atomic interactions (Table 8.1) and it may not be appropriate to consider that the $3d$ electronic state still obeys the pure Hund rule. However, when transition atoms are dissolved in simple metals, they often show the behaviour of a valid Hund coupling. The $4f$ orbitals in most rare-earth metals also keep the character in free atoms. Let us consider here the $3d$ electrons that still hold the Hund coupling in each atom. Then, the interaction between the $3d$ electron and the conduction electron may be expressed as

$$
\mathscr{H}_{sd} = - \sum_n \sum_i J(r_i - r_n) S_n \cdot \sigma_i ,
$$
(8.3.16)

where $\sigma_i = 2s_i$ is the Pauli spin matrix for the conduction electron and S_n is the total spin operator of the nth magnetic atom. Here, the exchange interaction J is dependent on the coordinate r_t of the conduction electron because conduction electrons are continuously distributed in space and in energy. The Hamiltonian (8.3.16) is written in the second quantized form as

$$\mathcal{H}_{sd} = -\frac{1}{N}\sum_{n}\sum_{kk'} J_{k'k} e^{i(k-k')\cdot r_n} [S_n^z(c_{k'\uparrow}{}^\dagger c_{k\uparrow} - c_{k'\downarrow}{}^\dagger c_{k\downarrow})$$
$$+ S_n^+ c_{k'\downarrow}{}^\dagger c_{k\uparrow} + S_n^- c_{k'\uparrow}^\dagger c_{k\downarrow}], \qquad (8.3.17)$$

where

$$J_{k'k} = N \iint dr_1 dr_2 e^{i(k'-k)\cdot r_n} \psi_{k'}^*(r_1)\, \phi_d(r_1 - r_n)\left(\frac{e^2}{r_{12}}\right)\phi_d^*(r_2 - r_n)\, \psi_k(r_2). \qquad (8.3.18)$$

It is understood that the band index for the Bloch function $\psi_k(r) = \exp(ik\cdot r)u_k(r)$ is included in k, if necessary. The Hamiltonian (8.3.16) or (8.3.17) is called the s-d exchange Hamiltonian. Carrying out the perturbation calculation with respect to conduction electrons up to the second order, one gets the effective Hamiltonian between S_n's:

$$\mathcal{H}_{ss} = \sum_{mn}\sum_{kk'} |J_{k'k}|^2 G(k, k', r_{nm})\, S_m \cdot S_n, \qquad (8.3.19)$$

where

$$G(k, k', r_{nm}) = \frac{2}{N^2}\frac{f_k(1 - f_{k'})}{\varepsilon_k - \varepsilon_{k'}} \exp[i(k - k')\cdot r_{nm}] \qquad (8.3.20)$$

with the Fermi distribution function $f_k = f_{k\uparrow} = f_{k\downarrow}$ and the energy of conduction electrons $\varepsilon_k = \varepsilon_{k\uparrow} = \varepsilon_{k\downarrow}$. Thus, there can be an ordered state of the localized spins by this interaction even if there is no direct interaction between them.

When the localized electron is the f electron instead of the d electron, the above interaction is called the s-f interaction. In this case, the spin-orbit interaction is large (Table 8.1) and $J = L + S$ is the good quantum number even in crystals, so that S_n is replaced by $(g-1)J_n$ in (8.3.19). In rare-earth metals and their alloys, the $4f$ electron is localized on each atom so well that the s-f interaction may be the main interaction for magnetic orderings.

What could be the main interaction giving rise to magnetic orderings in transition metals? As is shown in Table 8.1, it has been experimentally confirmed that the $3d$-band has an appreciable band width. This implies that the $3d$ electrons are not localized on atoms and it is hard to think that the s-d interaction is the main magnetic interaction. Then, there are questions as to whether the $3d$ electrons are well described by the one-electron picture with the Bloch functions and whether it is sufficient to consider the exchange interaction between electrons on this picture. The complete answer for such questions is not available even at present. However, it can be mentioned that the electron correlation effect is larger and more effective for the 3d electrons than for conduction electrons in simple metals. The reason is due to the fact that the electron-electron interaction is primarily stronger between electrons in the 3d orbitals than in the s or p orbitals and the 3d-band width is narrower than the width of conduction bands in simple metals, making the interaction effect important compared to the kinetic energy. This fact is explained alternatively by the nature of the 3d-function whose maximum amplitude is fairly large and located fairly close to the centre of atoms, where the 3d electron–electron interaction is operative. The s or p-function extends more widely than the 3d-function. In transition metals, however, there is a wide conduction band of a mainly 4s-nature and its characteristics can be very similar to the conduction band in simple metals. These 4s electrons may have the role of screening the intra-atomic electron-electron interaction for the 3d electron, but to what extent they screen is not clear quantitatively because this screening effect is short-ranged. There is a more complex situation in real transition metals; the 3d and the 4s-band are degenerate and there is the hybridization effect between these. This hybridization effect is important in determining the 3d-band width and it introduces the mixed nature between the 3d and the 4s-function. In this way, one has to consider various kinds of factors in investigating the magnetism of transition metals.

Now, suppose that energy bands are obtained in an appropriate single particle picture (say, in the Hartree–Fock approximation). Conduction electrons are accommodated in the five d-bands and the s-band. Transforming the Bloch functions thus obtained into the Wannier functions, the Hamiltonian for the electron system may be expressed as

$$
\mathscr{H} = \sum_{p\sigma} \sum_{mn} t_{pn,\,pm} c_{pn\sigma}{}^\dagger c_{pm\sigma}
$$
$$
+ \tfrac{1}{2} \sum_{\substack{mn \\ m'n'}} \sum_{\substack{pqp'q' \\ \sigma\sigma'}} V(p'm', q'n', pm, qn)\, c_{p'm'\sigma}{}^\dagger c_{q'n'\sigma'}{}^\dagger c_{qn\sigma'} c_{pm\sigma} \tag{8.3.21}
$$

where p and q specify bands, m and n lattice sites, σ indicates spin states, and

$$
V(p'm', q'n', pm, qn) = \iint dr\, dr' \varphi_{p'm'}{}^*(r)\, \varphi_{q'n'}{}^*(r') \frac{e^2}{|r - r'|}\, \varphi_{pm}(r)\, \varphi_{qn}(r'). \tag{8.3.22}
$$

The first term in (8.3.21) describes the single particle picture and its Fourier-transform $\sum_{pk} \varepsilon_{pk} n_{pk\sigma}$ is the band energy ($n_{pk\sigma} = c_{pk\sigma}^\dagger c_{pk\sigma}$ and $\varepsilon_{pk} = \sum_n t_{pn,\,pm}$ exp $[ik \cdot (r_n - r_m)]$). Let us suppose that the s-band and the five d-bands can be discriminated with the band index p. The electron screening effect may be taken into account by a proper way of changing V into $\tilde{V}$. The largest interaction is, as explained already, the terms in which $m = n = m' = n'$ and all band indices are assigned as the d-band;

$$\tfrac{1}{2} \sum_{m} \sum_{dd'\sigma\sigma'} [\check{V}_{c}(dm, d'm, dm, d'm) \, (n_{dm\sigma}n_{d'm\sigma'} - \delta_{dd'}\delta_{\sigma\sigma'}n_{dm\sigma})$$

$$- (1 - \delta_{dd'}) \, \check{V}_{ex}(dm, d'm, d'm, dm) \, c_{dm\sigma}^{\dagger}c_{dm\sigma}c_{d'm\sigma'}^{\dagger}c_{d'm\sigma}] , \qquad (8.3.23)$$

where d and d' stand for the five d-bands, $n_{dm\sigma} = c_{dm\sigma}^{\dagger} c_{dm\sigma}$, and the indices on $\check{V}$ indicate the Coulomb and the exchange integrals. When $d = d'$, only the Coulomb term survives and this is nothing but U in (8.3.8) (note that the term involving $\delta_{dd'}\delta_{\sigma\sigma'}$ has the role of excluding the case where two electrons with the same spin occupy the same orbital). The interaction terms given in (8.3.23) may be the most important interaction in investigating the magnetism of transition metals. In a similar way, if one takes $m = m' \neq n = n'$ or $m' = n \neq m = n'$ and assumes m and n to be nearest neighbors, the interactions represent the Coulomb or the exchange interactions between d electrons at nearest neighbor atoms. The latter corresponds to J_{ab} in (8.3.5). If one takes $m = n = m' = n'$ and $(p = p' = d, q = q' = s)$ or $(p' = q = d, p = q' = s)$ (d and s may be interchanged), they represent the Coulomb or the exchange interaction between the d and the s electrons. When the Fourier-transformed operators are used for the s electron in this exchange interaction, it gives the s-d exchange interaction in (8.3.17). Besides (8.3.23), it is possible that these other interactions also contribute to the magnetic properties of transition metals. In fact, the opinion as to which is the most important interaction has changed with time in the history of magnetism.

If the hybridization effect with the $4s$-band is decisive in determining the $3d$-band width rather than the direct transfer integral between the $3d$-orbitals, this effect becomes quite significant for the interaction between magnetic atoms. The hybridization provides the extent to which the d-character remains in an interested band and the degree of localization which is present in that band. In particular, the latter property is intimately related to the problem of localized spins on transition metal atoms as impurities. The interaction between magnetic atoms by this mechanism will be discussed in Sect. 8.6.

Quantitative discussions on such various interactions are still in the future and one needs a detailed knowledge of band structures, wave functions, the screening effect, and other many-body effects.

8.4 Configurations of Spin Polarization in Magnetic-Ordered States

In the preceding section, the fundamental interactions that bring about ordered states of spins were investigated both for insulators and metals. The spin struc-

ture of the ordered state in a crystal depends not only on the sign of the dominant interaction and its functional form with distance, but also on the
magnetic anisotropy energy. They play the role altogether and various kinds of
spin structures are found in real crystals. The spin polarization takes place under
a certain temperature in each crystal and this temperature is called the Curie
temperature (T_C) if the spin structure below this is ferromagnetic, and the Néel
temperature (T_N) if antiferromagnetic.

8.4.1 Magnetic Anisotropy Energy and Crystal Fields

When the exchange interaction is the scalar product between spins as given in
(8.3.12), the spin polarization brought about by this interaction can take any
direction independently of the crystal structure. The stable direction is, however,
determined by the magnetic anisotropy energy. There are two kinds of anisotropy energy; one is obtained with the spin-orbit interaction which provides the
directional energy of spins relative to the orbitals that reflect the crystal structure; another is due to the magnetic dipole–dipole interaction and this provides
the stable spin direction according to a given lattice structure.

First, let us investigate the former which is due to the spin-orbit interaction.
This requires knowing how the electron orbitals reflect the crystal structure. For
smplicity, consider an insulator containing the transition metal ions each of
which accommodates one d electron and assume that the d electron is localized
on the transition metal ion and is well described by the free ion d-functions.
The difference from the free ion is that the electron potential $U(r)$ includes the
effect from neighboring ions, in addition to the free-ion potential $U_0(r)$. The
d electron must satisfy the following Schrödinger equation:

$$\left[-\frac{\hbar^2}{2m} \Delta + U(r) \right] \psi_n(r) = \varepsilon_n \psi_n(r) \tag{8.4.1}$$

In a free transition metal ion the five d orbitals are degenerate. If the energy
separations between the d-orbitals and other orbitals are large compared with
$[U(r) - U_0(r)]$, it is a good approximation for $\psi_n(r)$ to take a linear combination
of the five d orbitals. This situation is satisfied in many compounds and will be
taken here. $V_c(r) = U(r) - U_0(r)$ is called the crystal field. This reflects the
symmetry of the crystal structure and, expanding the functional form of $V_c(r)$
about a particular magnetic ion in terms of the spherical harmonics, it is
expressed as a linear combination of them which is consistent with the symmetry.
When the symmetry is cubic it has the form of

$$V_c(r) = A (x^4 + y^4 + z^4 - \tfrac{3}{5} r^4), \tag{8.4.2}$$

where x, y, and z are coordinates of the d electron [$V_c(r)$ can have a spherical symmetric term but this is irrelevant to the present problem and has been dropped]. Diagonalizing the matrix of $V_c(r)$ with respect to the five d orbitals, one obtains the following energy eigenvalues and eigenfunctions:

$$\varepsilon = \tfrac{3}{5}\Delta : \begin{cases} \psi_1 = \psi_{20}^0 \propto 3z^2 - r^2 \\ \psi_2 = \dfrac{1}{\sqrt{2}}\,(\psi_{22}^0 + \psi_{2-2}^0) \propto x^2 - y^2 \end{cases}$$

$$\varepsilon = -\tfrac{2}{5}\Delta : \begin{cases} \psi_3 = \dfrac{1}{\sqrt{2}}\,(\psi_{22}^0 - \psi_{2-2}^0) \propto xy \\ \psi_4 = \dfrac{1}{i\sqrt{2}}\,(\psi_{21}^0 + \psi_{2-1}^0) \propto xz \\ \psi_5 = \dfrac{1}{i\sqrt{2}}\,(\psi_{21}^0 - \psi_{2-1}^0) \propto yz \end{cases} \qquad (8.4.3)$$

Here, ψ_{2m}^0 is the d-function ($l = 2$, m) of free ions and $\Delta = (4/35)\,A\,\langle r^4\rangle$ [expressing the radial part of ψ_{2m}^0 as $R_2^0(r)$, $\langle r^4\rangle = \int r^2 dr R_2^0(r) r^4 R_2^0(r)$]. The doubly degenerate functions (ψ_1, ψ_2) and the triply degenerate ones (ψ_3, ψ_4, ψ_5) in the cubic field are denoted by the irreducible representation E (or Γ_3) and T_2 (or Γ_5), respectively (alternatively, they are also denoted as $d\gamma$ and $d\varepsilon$, respectively). The orbitals $\psi_1 \sim \psi_3$ are sketched in Fig. 8.6. If the nearest anions are situated on the x, y, and z axes to make a octahedron, the T_2 states have lower energy because their electronic clouds avoid the anions. This means Δ(and A) is positive.

When there is more than one electron per atom, the problem becomes more complicated. If the intra-atomic Coulomb and exchange interactions are larger than the crystal field, the atom still maintains the Hund rule to keep the maximum spin S, while the Hund rule may break and the atom takes another electronic state if opposite. (Table 8.1) These two cases are called the weak-field configuration and the strong-field configuration, respectively, and these cases are shown in Fig. 8.7 for d^4. The weak field case is found in many ionic crystals and the strong field case in some complex salts like those with the cyanic radical. When a crystal distorts from the cubic symmetry, the degeneracy in E and in T_2 states is removed and they split into two or more levels. An example is shown at the extreme right of each figure in Fig. 8.7 for the case where the distortion takes place along the z-axis to change into the tetragonal symmetry. Taking account of the spin states, there are ten states for the d state and many excited states can be provided by considering the crystal field and the Coulomb and the exchange interactions between the d electrons.

Let us now introduce the spin-orbit interaction $V_{so} = \lambda L \cdot S$ (which should be written as $V_{so} = \sum_t \zeta l_t \cdot s_t$ more generally) and the Zeeman energy due to an external magnetic field H. The Hamiltonian for the d electrons of a transition metal ion is written as

$$\mathcal{H} = \mathcal{H}_0' + V_c + \lambda L \cdot S + \mu_B H \cdot (L + 2S), \qquad (8.4.4)$$

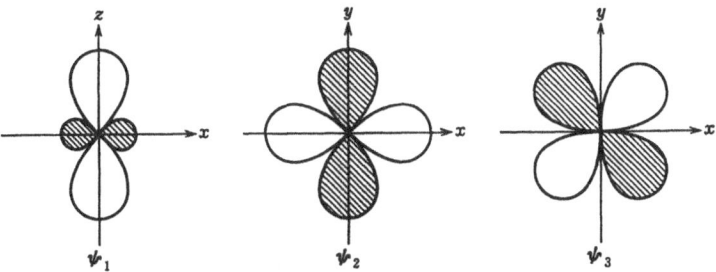

ψ_1 ψ_2 ψ_3

Fig. 8.6. The eigenfunctions ψ_1, ψ_2, ψ_3 in the cubic crystal field

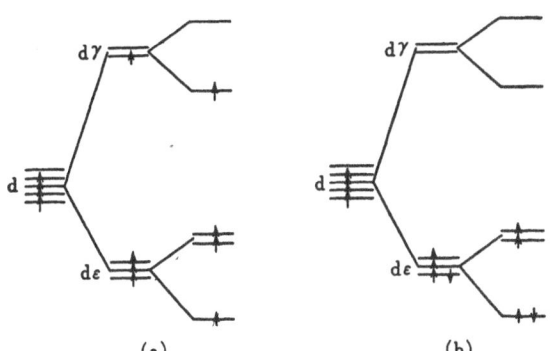

(a)   (b)

Fig. 8.7a, b. Level splitting of the d orbital in the cubic and the tetragonal crystal field. The electron configurations of d^4 are shown in (a) weak field and (b) strong field

where $\mathscr{H}_0'$ is the free ion Hamiltonian minus the spin-orbit interaction. ($\mathscr{H}_0' + V_c$) is the part that has been considered above, and this will be taken as the unperturbed part. The ground orbital state will be assumed to be nondegenerate. As there is still degeneracy with respect to M_S, the magnetic quantum number of S, the ground states are written as $|GM_S\rangle$. It can be proved that the equality $\langle GM_S|L|GM_S\rangle = 0$ holds. Therefore, the first-order perturbation energy comes only from $2\mu_B H \cdot S$. Writing the ground and the excited states as $|GM_S\rangle = |G\rangle|M_S\rangle$ and $|nM_S'\rangle = |n\rangle|M_S'\rangle$, respectively, the second-order perturbation energy for the ground states can be expressed as

$$-\sum_{\substack{nM_S'' \\ (n\neq G)}} \frac{\langle GM_S'|\lambda L\cdot S+\mu_B H\cdot(L+2S)|nM_S''\rangle\langle nM_S''|\lambda L\cdot S+\mu_B H\cdot(L+2S)|GM_S\rangle}{E(n)-E(G)}$$

$$\rightarrow -\lambda^2 \sum_{\alpha\beta} \Lambda_{\alpha\beta} S^\alpha S^\beta - \mu_B\lambda \sum_{\alpha\beta} (\Lambda_{\alpha\beta}+\Lambda_{\alpha\beta}^*) H^\alpha S^\beta - \mu_B^2 \sum_{\alpha\beta} \Lambda_{\alpha\beta} H^\alpha H^\beta .$$

$$(8.4.5)$$

Here, the arrow indicates that the left-side quantity is equivalent to the right, and

$$\Lambda_{\alpha\beta} = \sum_{n(\neq G)} \frac{\langle G|L^\alpha|n\rangle\langle n|L^\beta|G\rangle}{E(n)-E(G)} . \qquad (8.4.6)$$

In deriving (8.4.5), use is made of

$$\sum_{M_S''} \langle M_S' | S^\alpha | M_S'' \rangle \langle M_S'' | S^\beta | M_S \rangle = \langle M_S' | S^\alpha S^\beta | M_S \rangle . \tag{8.4.7}$$

The first term in (8.4.5) is the magnetic anisotropy energy and, as it is a quadratic form of S^α, it can be diagonalized by a transformation of the principal axes to obtain Λ_{xx}, Λ_{yy}, and Λ_{zz} and the transformed form can be expressed in the form

$$D[3(S^z)^2 - S(S+1)] + E[(S^x)^2 - (S^y)^2] , \tag{8.4.8}$$

where $D = -(\lambda^2/3)[\Lambda_{zz} - (\Lambda_{xx} + \Lambda_{yy})/2]$ and $E = -\lambda^2(\Lambda_{xx} - \Lambda_{yy})/2$. In the case of the cubic symmetry, $D = E = 0$ and the anisotropy energy is brought about from a high-order perturbation [to have a higher power in S^α]. It may be of the form $K[(S^x)^4 + (S^y)^4 + (S^z)^4]$ (this quartic form is meaningful only when $S \geq 2$). The second term in (8.4.5) can be combined with $2\mu_B H \cdot S$ to yield the g-shift. Expressing the sum of these two terms as $\mu_B H \cdot g \cdot S$, the g tensor is given by

$$g_{\alpha\beta} = 2\delta_{\alpha\beta} - \lambda(\Lambda_{\alpha\beta} + \Lambda_{\alpha\beta}^*) . \tag{8.4.9}$$

From $\chi_\alpha = -\partial^2 E/\partial(H^\alpha)^2$, the last term of (8.4.5) can provide the orbital paramagnetism which has the same form as (8.2.21). For localized d electrons the first excited state is usually higher than $k_B T$, and this type of paramagnetism is temperature independent and is called the Van Vleck paramagnetism.

Rare-earth ions have the spin-orbit interaction which is usually much larger than the crystal-field as indicated in Table 8.1, so that, in investigating their magnetic properties, it is enough to consider only the ground multiplet in the Russell–Saunders coupling scheme (LSJ). For the ground (LSJ) state there is still the degeneracy with respect to the quantum number M_J. Thus, the crystal-field eigenstates are now linear combinations of M_J states instead of M_L states as for the d electrons. As the spin freedom is already included in M_J states, this means that the crystal-field splitting itself provides the magnetic anisotropy so that the anisotropy energy is quite large.

Finally, the magnetic anisotropy due to the magnetic dipole interaction will be explained briefly. Denoting the magnetic moment at the nth lattice site as μ_n°, the interaction energy between the moments is given by

$$\mathcal{H}_d = \sum_{\langle mn \rangle} \left[\frac{\mu_m^\circ \cdot \mu_n^\circ}{r_{mn}^3} - 3 \frac{(\mu_m^\circ \cdot r_{mn})(\mu_n^\circ \cdot r_{mn})}{r_{mn}^5} \right] , \tag{8.4.10}$$

where r_{mn} indicates the distance vector between μ_m° and μ_n°. Assuming the g-factor to be isotropic and regarding μ_n° to be the magnetic moment associated with the spin polarization, i.e., $\mu_n^\circ = g\mu_B \langle S_n \rangle$, (8.4.10) is equivalent to the spin at r_m being in an effective magnetic field

$$H(r_m) = \sum_n \left[\frac{\mu_n^\circ}{r_{mn}^3} - 3 \frac{r_{mn}(\mu_n^\circ \cdot r_{mn})}{r_{mn}^5} \right]. \tag{8.4.11}$$

Therefore, the energy of (8.4.10) depends on the arrangement of spins and the pattern of the polarization of spins. The sum $\sum_n$ in (8.4.11) is not trivial because the quantity in $\{\cdots\}$ converges with distance slowly as r_{mn}^{-3}. In a ferromagnet, for instance, the contributions to (8.4.11) are classified into three: the contribution from near neighbor lattice sites, the one from distant sites and the surface contribution. The direct lattice sum may be carried out for the first, the Lorentz field $(4\pi/3)M$ can be taken for the second ($M = g\mu_B \langle S \rangle / \Omega$ with the volume Ω per spin), and the last contribution gives the demagnetization field. This anisotropy due to the magnetic dipole interaction plays an important role for magnetic ions with $L = 0$ like Mn^{+2}, Fe^{+3}, and Gd^{+3}.

8.4.2 Magnetic Order in Insulators

Magnetic ions are usually surrounded by anions which have large radii so that the direct interaction between magnetic ions is quite small. Let us consider two transition metal ions on the z-axis intervened by an anion having the p orbital (Fig. 8.8). From symmetry consideration, the $d_{3z^2-r^2}$ orbital can hybridize with the p_z orbital and the d_{xy} orbital with the p_x orbital. The bonding orbitals are mainly p_z or p_x characters and are filled with electrons, while the antibonding orbitals are mainly $d_{3z^2-r^2}$ or d_{xy} characters and are expressed as

$$\left. \begin{aligned} \psi_{3z^2-r^2} &= N_\sigma(d_{3z^2-r^2} - \varepsilon_\sigma p_z) \\ \psi_{xy} &= N_\pi(d_{xy} - \varepsilon_\pi p_x) \end{aligned} \right\}, \tag{8.4.12}$$

where N's are the normalization constants. The coefficients ε are determined by

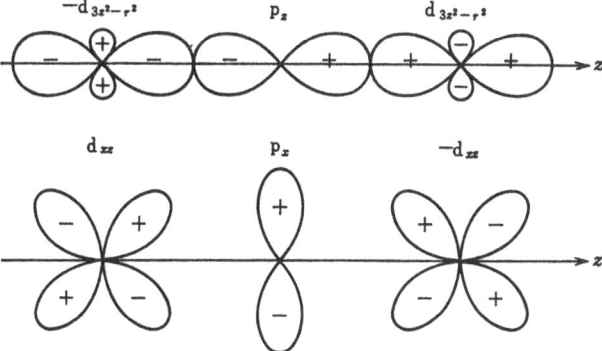

Fig. 8.8. Symmetries of the d and the p orbital

the overlap integral and the covalency between the d and the p-orbitals. With these wave functions, the interaction (8.3.12) between the two magnetic ions is written as

$$\mathcal{H}_{ex} = -2J_{eff}\boldsymbol{S}_a \cdot \boldsymbol{S}_b , \tag{8.4.13}$$

where

$$J_{eff} = J_{dd'} - K_{dd'} . \tag{8.4.14}$$

Here, $J_{dd'}$, and K_{dd}, denote the potential and the kinetic exchange with respect to the wave functions in (8.4.12) and they are of the order of ε^4. Such interaction between magnetic ions which is provided via the intervening anion is called the superexchange interaction.

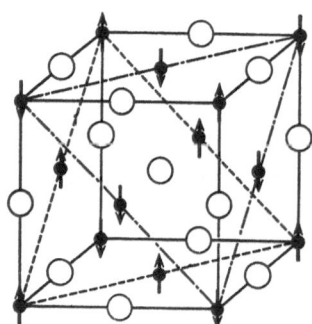

Fig. 8.9. Spin configuration in MnO. The filled circle represents Mn^{+2} and the open circle O^{-2}

If the three ions are on a line as in Fig. 8.8, and if both magnetic ions are the same ion, the observed interaction is antiferromagnetic ($J_{eff} < 0$). In Fig. 8.9 the spin structure observed by the neutron diffraction experiment is shown for MnO ($T_N = 120$ K) which has the NaCl–crystal structure. It shows that the magnetic ions consist of the two sublattices with up-and down-spins and such a spin structure is called the Néel state. For several transition metal oxides with the same crystal structure, the molecular field theory is applied to deduce the exchange constants J_1 and J_2, the values of J_{eff} for the nearest and the next-nearest neighbor magnetic ions, and the results are given in Table 8.3 together with their Néel temperature. If two magnetic ions are on a straight line but the electrons in the two ions occupy different orbitals, say $d_{3z^2-r^2}$ in the left ion and d_{xy} in the right ion, the $d_{3z^2-r^2}$ orbital can hybridize with p_z as given in (8.4.12), while the d_{xy} orbital can hybridize only with p_x. The kinetic exchange does not work between two orbitals with different symmetry and only the potential exchange is effective to provide a ferromagnetic exchange ($J_{eff} > 0$). In fact, in most cases where one magnetic ion has the d orbital occupied by more than

Table 8.3. Values of the effective exchange constants evaluated by the molecular field theory. J_1 and J_2 are for magnetic ions between the nearest and the next-nearest neighbors, respectively. [8.4]

	S	$T_N[K]$	$J_1[K]$	$J_2[K]$
MnO	5/2	116	− 7.2	− 3.5
FeO	2	198	− 7.8	− 8.2
CoO	3/2	292	− 6.9	− 21.6
NiO	1	523	− 50	− 85

half and another by less than half, the interaction between the two ions is fer-romagnetic. On the other hand, when the two magnetic ions make 90 degrees with respect to the anion, the tendency of the exchange constant becomes the opposite; ferromagnetic between the same ions and antiferromagnetic between different kinds of ions. Ferromagnetic insulators that involve a single kind of magnetic ion are rather rare. $CrBr_3$ ($T_C = 37$ K) and EuO ($T_C = 69$ K) are ferromagnetic.

There are magnetic compounds which have antiferromagnetic interaction but have net magnetizations because spin moments for ↑ and ↓ or their numbers are not equal each other. Such compounds are called the ferrimagnets. A typical example is the ferrite which has the chemical formula MFe_2O_4 (M^{+2}: metal ion) and involves eight chemical formulae in the unit cell of the spinel structure. In the unit cell there are 16 positions (called B or 16d site) of metallic ions surrounded by O^{-2} octahedrally and 8 positions (called A or 8a site) surrounded tetrahedrally. There is a strong antiferromagnetic interaction between magnetic ions in A and B sites. This oxide is called the normal spinel if Fe^{+3} occupy the B sites and M^{+2} the A sites, while it is called the inverse spinel if M^{+2} and half of Fe^{+3} occupy the B sites and the remaining half of Fe^{+3} the A sites. For instance, it takes the normal spinel when Mn^{+2} is substituted for M^{+2} and the inverse spinel when Ni^{+2} or Cu^{+2} is substituted. In Table 8.4, the observed and the calculated magnetic moments (per chemical formula) of ferrites are shown together with their Néel temperatures.

There are also some magnetic insulators which are essentially antiferroma-gnetic but show weak ferromagnetism as a whole. These kinds of compounds

Table 8.4. Magnetic moments and Néel temperatures of ferrites. [8.5]

Compound	$MnFe_2O_4$	$FeFe_2O_4$	$CoFe_2O_4$	$NiFe_2O_4$	$CuFe_2O_4$
Magnetic moment					
calc. (μ_B)	5	4	3	2	1
exper. (μ_B)	5.0	4.2	3.3	2.3	1.3
Néel temp. [K]	783	848	793	863	728

usually involve only a single kind of magnetic ion with a definite valency and the spin moments of the ions are arranged almost antiferromagnetically but are tilted slightly with respect to each other (Fig. 8.10). There are two kinds of mechanisms to produce this small tilt of spins. One of these is due to the combined action of the off-diagonal exchange interaction and the spin-orbit interaction. From the second-order perturbation calculation, one obtains the effective interaction between the spins S_a and S_b which is anisotropic and can not be expressed in the form of a scalar product. In particular, it involves an interaction antisymmetric with respect to the interchange of S_a and S_b:

$$\mathcal{H}_{ex}^a = D \cdot [S_a \times S_b] \tag{8.4.15}$$

where D is a vector. Apparently S_a and S_b tend to become perpendicular to each other by this interaction. Assuming that the off-diagonal exchange is the same order of magnitude as the diagonal one J, one gets $|D| \sim |J\lambda/\Delta E| \sim |J(2 - g)|$ by making use of (8.4.9), where ΔE is the energy difference between the ground and the intermediate states. αFe_2O_3 (α-hematite) is an example that shows the weak ferromagnetism by this mechanism and the ferromagnetic component is $\sim 10^{-2} \mu_B$ per Fe^{+3} ion.

Fig. 8.10. Schematic picture to show the weak ferromagnetism. M_1 and M_2 are the magnetizations of two sublattices

There is another mechanism of tilting spins and NiF_2 belongs to this class. NiF_2 has the rutile structures (see Fig. 8.11) and the unit cell involves two Ni^{+2} ions. Ni^{+2} ions have the anisotropy energy of the type given in (8.4.8) with $D > 0$ so that the easy axis is either the x or y-axis according to the sign of E. It can be seen from Fig. 8.11 that the x and y-axis are interchanged for two kinds of Ni^{+2}

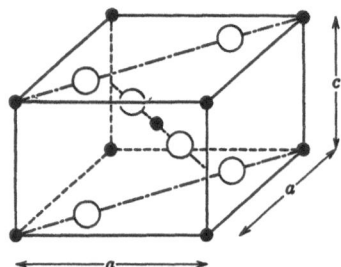

Fig. 8.11. Lattice structure of the rutile type. Filled circle represents metallic ions and open circle anions

Fig. 8.12a, b. Spin structures of (a) the helical (or screw)-type and (b) the sinusoidal-type

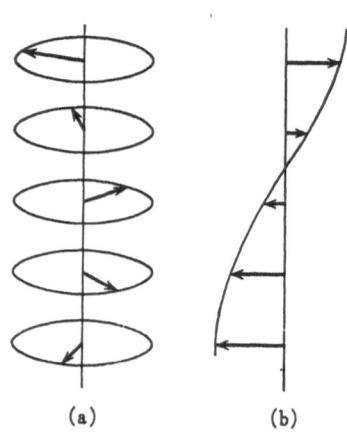

(a)   (b)

positions and this indicates that the easy axes are perpendicular to each other at these positions. On the other hand, there is a strong antiferromagnetic interaction between Ni^{+2} ions at these positions. Combining these two effects, Ni^{+2} ions make an almost antiferromagnetic configuration but with a small tilt to produce a small ferromagnetic component (again $\sim 10^{-2}\mu_B$ per atom).

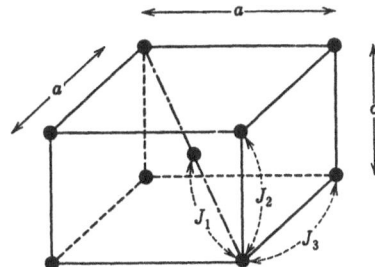

Fig. 8.13. Exchange interactions between magnetic ions in the rutile-type structure. The positions of the anions are suppressed (Fig. 8.11)

So far, one has considered only the spin configurations that are determined by the nearest-neighbor exchange interaction, the interactions beyond the nearest neighbors having been implicitly considered. However, in the helical-(or screw-) type and the sinusoidal-type spin polarizations (Fig. 8.12), the exchange interactions beyond the nearest neighbor are vitally important. The first observation of this kind of spin polarization was for MnO$_2$ which has the rutile type lattice structure (Fig. 8.11), As is shown in Fig. 8. 13, these kinds of exchange interaction, J_1, J_2 and J_3, will be considered and the spin Hamiltonian is written as

$$\mathcal{H}_{ex} = -2J_1 \sum_{\langle ln \rangle} S_l \cdot S_n - 2J_2 \left(\sum_{\langle ll' \rangle_z} S_l \cdot S_{l'} + \sum_{\langle nn' \rangle_z} S_n \cdot S_{n'} \right)$$
$$- 2J_3 \left(\sum_{\langle ll' \rangle_{xy}} S_l \cdot S_{l'} + \sum_{\langle nn' \rangle_{xy}} S_n \cdot S_{n'} \right) \tag{8.4.16}$$

where l and l' indicate the lattice sites of one sublattice consisting of the corner positions of the parallelepiped in Fig. 8.13, n and n' the one consisting of the center position (each sublattice having $N/2$ sites), $\langle ll' \rangle_z$ and $\langle nn' \rangle_z$ the nearest-neighbor pairs of magnetic ions in the z-direction and $\langle ll' \rangle_{xy}$ and $\langle nn' \rangle_{xy}$ in the $x-y$ plane. Taking the molecular field approximation and regarding S as a polarization vector with the Fourier transforms $S_l = \sum_k S(k) \exp(k \cdot r_l)$ and $S_n = \sum_k S'(k) \exp(ik \cdot r_n)$, (8.4.16) can be rewritten as

$$\mathcal{H}_{ex} = -\frac{N}{2} \sum_k \left\{ 16J_1 \cos\left(\frac{k_x a}{2}\right) \cos\left(\frac{k_y a}{2}\right) \cos\left(\frac{k_z c}{2}\right) S(k) \cdot S'(-k) \right.$$
$$+ 2[J_2 \cos k_z c + J_3(\cos k_x a + \cos k_y a)]$$
$$\left. \cdot [S(k) \cdot S(-k) + S'(k) \cdot S'(-k)] \right\}. \tag{8.4.17}$$

Let the magnitudes of the polarizations be $S_l^2 = S_n^2 = S^2$. Here, two spin configurations will be investigated. One is the spin configuration in which each sublattice takes the ferromagnetic structure but two sublattices are arranged antiferromagnetically to each other as is observed in MnF_2. In this case, only the Fourier component of $k_x = k_y = k_z = 0$ contributes and $S(k = 0) = -S'$ $(k = 0) = S$. From (8.4.17) one obtains

$$\frac{E}{(N/2)} = (16J_1 - 4J_2 - 8J_3)S^2 . \tag{8.4.18}$$

Another spin configuration is the helical structure which is expressed as

$$\left.\begin{aligned} S_l &= S[i \cos (k \cdot r_l) + j \sin (k \cdot r_l)] \\ S_n &= S[i \cos (k \cdot r_n) + j \sin (k \cdot r_n)] \end{aligned}\right\} . \tag{8.4.19}$$

This indicates that one is considering only the $\pm k$ Fourier components and $|k|$ provides the pitch of the screw structure. Taking $k_x = k_y = 0$ and $k_z \neq 0$, (8.4.17) gives

$$\frac{E}{(N/2)} = -\left[16J_1 \cos \left(\frac{k_z c}{2}\right) + 4J_2 \cos k_z c + 8J_3\right] S^2 . \tag{8.4.20}$$

Taking the variation of E with respect to k_z, one gets

$$\cos \left(\frac{k_z c}{2}\right) = -\frac{J_1}{J_2} . \tag{8.4.21}$$

Substituting (8.4.21) back into (8.4.20), it provides

$$\frac{E}{(N/2)} = \left[8\left(\frac{J_1^2}{J_2}\right) + 4J_2 - 8J_3\right] S^2 . \tag{8.4.22}$$

From (8.4.21), it can be seen that a meaningful solution can be obtained only when $|J_2| \geq |J_1|$. J_3 plays no important role. Assuming $J_1 < 0$ and $J_2 < 0$, the screw structure is more stable than (8.4.18) as far as $|J_2/J_1| > 1$. In MnO_2, the spin configuration is the screw structure of this type with $k_z \approx 2\pi/7c$.

8.4.3 Magnetic Order due to the Double Exchange Interaction

When the same transition metal ions can take two valence states in a compound, the extra electrons can be transferred from ion to ion to produce the interaction $-|t_{ab}| \cos(\theta/2)$, as is given in (8.3.15). If there are Nx extra electrons, N being the number of magnetic ions in the crystal, the energy gain due to this double exchange is

$$E_{doub} = -Nzx|t|\cos \frac{\theta}{2} , \tag{8.4.23}$$

where z is the number of nearest-neighbor magnetic ions. Supposing that, in addition, the superexchange interaction exists between M^{+n} ions, then the energy gain due to this is given in the molecular field approximation as

$$E_{ex} = -NzJS^2 \cos\theta, \qquad (8.4.24)$$

where S indicates the magnitude of spin polarization of a M^{+n} ion. The energy (8.4.23) prefers the ferromagnent, while the energy (8.4.24) prefers the antiferromagnetism if $J < 0$. Minimizing the sum of these, θ is given by

$$\cos\frac{\theta}{2} = \frac{|t|x}{4|J|S^2}. \qquad (8.4.25)$$

This again shows that spin polarization takes a canted spin configuration as in αFe_2O_3 with the antisymmetric exchange interaction or NiF_2 with the perpendicular easy axes. In the present case, thermally excited spin fluctuations give rise to different temperature dependences for $\langle\cos(\theta/2)\rangle_T$ and $\langle\cos\theta\rangle_T$ and one obtains the phase diagram of spin configurations as shown in Fig. 8.14.

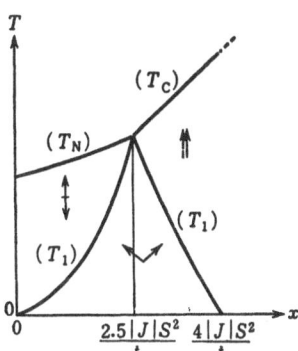

Fig. 8.14. Phase diagram of the system which has both the double exchange and the superexchange interactions. Ferromagnetic, antiferromagnetic, and canted spin configurations are possible [8.6]

An example of such a system is the mixed crystal of $LaMnO_3$ and $CaMnO_3$, $(La_{1-x}Ca_x)(Mn^{+3}_{1-x}Mn^{+4}_x)O_3$. It can be thought that there is an extra electron in Mn^{+3} when $x \sim 1$ and an extra hole in Mn^{+4} when $x \sim 0$ and that these extra electrons and holes are mobile. In fact, this mixed crystal exhibits the properties mentioned above at $0 < x < 0.25$, and it is a good insulator at $x = 0$ and $x = 1$ and a fairly good conductor when $x \neq 0$.

8.4.4 Magnetic Order in Metals

The most familiar ferromagnetic metals are iron, cobalt, and nickel. The $ZrZn_2$ alloy whose elements are not separately magnetic and the gadolinium metal of the rare-earth group are also ferromagnetic.

The magnetism of transition metals is apparently due to the d electrons, and experimentally the d electrons show the characteristics of band-electrons as explanined in Sect. 8.3. An example of such experimental facts is the spontaneous magnetization per atom at $T = 0$ K in transition metal alloys and it is shown in Fig. 8.15. It should be noticed that the spontaneous magnetication is almost determined only by the electron number of alloys, and this curve is called the Slater–Pauling curve.

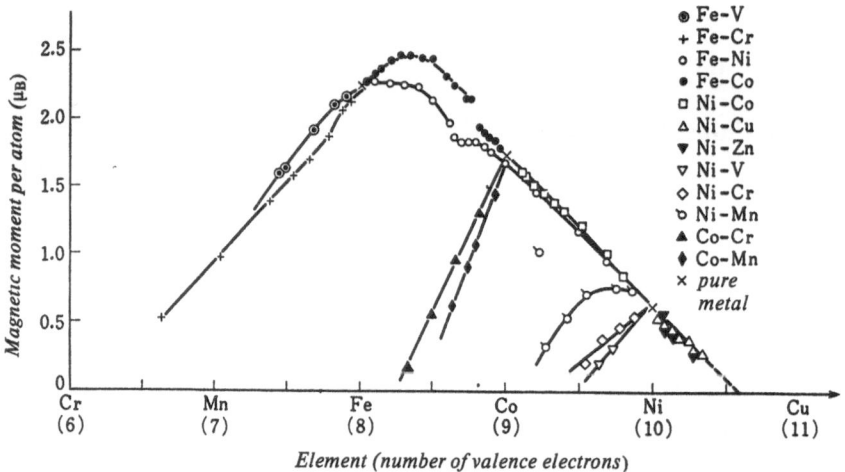

Fig. 8.15. The Slater-Pauling curve which shows the spontaneous magnetization of various 3d-metals and their alloys. The numbers in parentheses indicate the numbers of valence electrons in 3d and 4s-bands [8.7]

For simplicity, let us consider a single band and take it as the free-electron type. Furthermore, take only the exchange term in (8.3.21) and put $m' = n$ and $m = n'$ which are the nearest neighbors to each other [notice that the exchange term in (8.3.23) vanishes in the case of single band]. Transforming into the Bloch representation it becomes

$$\mathcal{H}_{\text{ex}} = - \sum_{kk'q} \sum_{\sigma\sigma'} \frac{1}{2N} \bar{V}_{\text{ex}'} c_{dk+q\sigma}^\dagger c_{dk\sigma'} c_{dk'-q\sigma'}^\dagger c_{dk'\sigma} . \tag{8.4.26}$$

Let us suppose that $\bar{V}_{\text{ex}'}$ is a constant independent of k. In the Hartree approximation this energy is expressed as

$$E_{\text{ex}} = - \frac{\bar{V}_{\text{ex}'}}{N} \sum_{k\sigma} N_\sigma \bar{n}_{k\sigma}$$

$$= - \frac{\bar{V}_{\text{ex}'}}{2N} \sum_{k} [(N_\uparrow + N_\downarrow)(\bar{n}_{k\uparrow} + \bar{n}_{k\downarrow}) + (N_\uparrow - N_\downarrow)(\bar{n}_{k\uparrow} - \bar{n}_{k\downarrow})], \tag{8.4.27}$$

where $\bar{n}_{k\sigma} = \langle c_{dk\sigma}^\dagger c_{dk\sigma}\rangle$ and $N_\sigma = \sum_k \bar{n}_{k\sigma}$; ($\langle\cdots\rangle$ means the average of the Hartree approximation). The first term of (8.4.27) is simply a constant. The second term depends on the magnetization $M = \mu_B(N_\uparrow - N_\downarrow)$ and can be expressed as $-(\bar{V}_{ex'}/2N\mu_B)M \sum_k (\bar{n}_{k\uparrow} - \bar{n}_{k\downarrow})$ $(g = 2$ has been assumed). This can be interpreted as the conduction electrons with $\uparrow$ and $\downarrow$ spins interacting with the effective field

$$H_e = \frac{\bar{V}_{ex'}M}{2N\mu_B^2} \equiv \gamma\sigma, \tag{8.4.28}$$

where $\sigma = M/N\mu_B$ and $\gamma = \bar{V}_{ex'}/2\mu_B$. H_e is called the Weiss field or the molecular field. If H is replaced with $H + H_e$ in (8.2.6,8) they can be combined to give

$$\left. \begin{aligned} \int d\varepsilon N(\varepsilon)f(\varepsilon - \mu_B H - \mu_B\gamma\sigma) &= \frac{N_e}{2}(1 + \sigma) \\ \int d\varepsilon N(\varepsilon)f(\varepsilon + \mu_B H + \mu_B\gamma\sigma) &= \frac{N_e}{2}(1 - \sigma) \end{aligned} \right\} \tag{8.4.29}$$

They are the equations to determine σ. Here, it is not always valid to assume $\mu_B\gamma\sigma \ll \varepsilon_F$ so that one cannot expand these equations in powers of $\mu_B\gamma\sigma$. In order to have a nonzero σ_0, spontaneous magnetization at $H = 0$ and $T = 0$ K, it is required that

$$\frac{\mu_B\gamma}{\varepsilon_F} > \frac{2}{3} \quad \text{or} \quad N(\varepsilon_F)\gamma' > 1, \tag{8.4.30}$$

where $\gamma' = \bar{V}_{ex'}/2N$. Both of these inequalities are equivalent. In particular, if $\mu_B\gamma/\varepsilon_F > 2^{-1/3}$, the solution is $\sigma_0 = 1$ and one obtains a saturated ferromagnet where all conduction electrons occupy the states with the same spin. Ni belongs to this type of ferromagnet with 0.6 holes per atom in the d–band, all of which have the same spin at $T = 0$ K. The present approximation provides a very clear model where the exchange interaction is expressed as the Weiss field for the conduction (d) electrons. This is called the Stoner model. Here, the interatomic exchange interaction $\bar{V}_{ex'}$ is thought to play the most important role, but at present it is considered that the intra-atomic interaction in (8.3.23) is more essential for magnetism.

Nevertheless, the result in (8.4.30) is instructive and it may indicate that ferromagnetism can take place in metals which have a large exchange interaction and/or large density of states at the Fermi surface. Several band calculations have been done for Ni metal and an example for ferromagnetic Ni is shown in Fig. 8.16. The band structure is not very simple because there are five sub-bands for the d-band but one can see that the density of states at the Fermi level is quite large (see also Sect.5.6).

From Fig. 8.16 it can also be observed that widths of the d-bands are narrow, and this implies that the electron correlation given by the Coulomb interaction

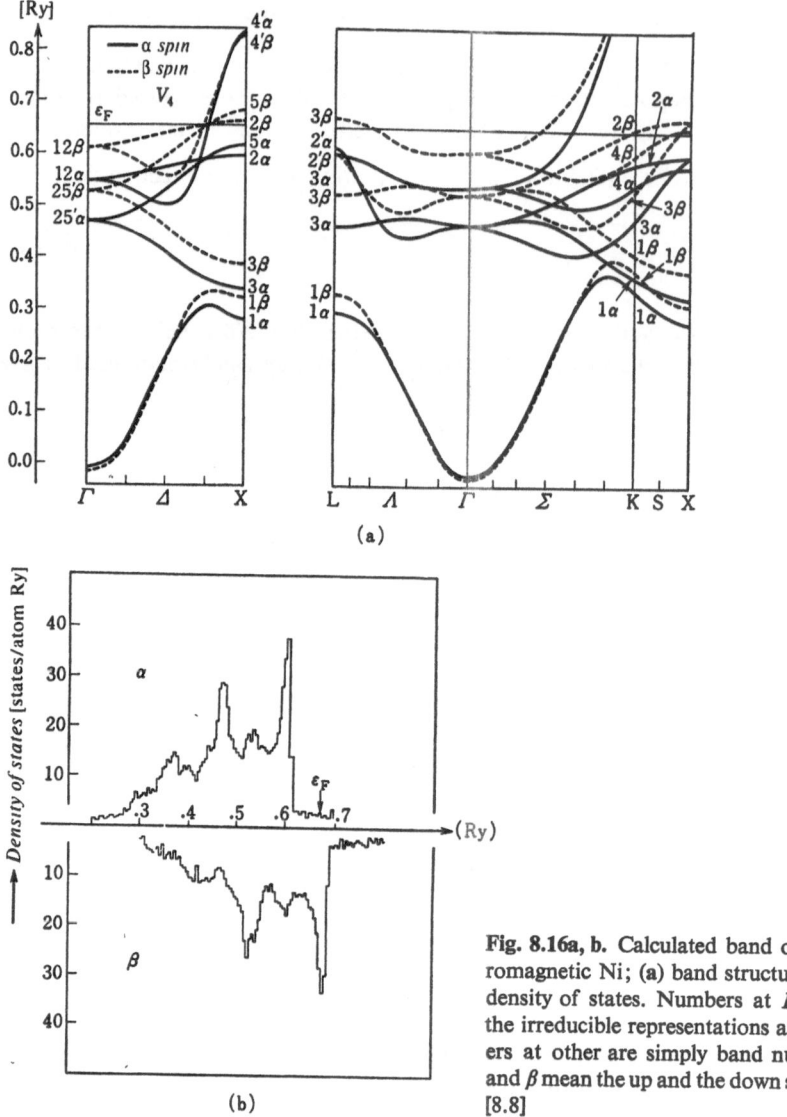

Fig. 8.16a, b. Calculated band of the ferromagnetic Ni; (a) band structure and (b) density of states. Numbers at Γ indicate the irreducible representations and numbers at other are simply band numbers. α and β mean the up and the down spin states [8.8]

of the first term in (8.3.23) may play an important role. In particular, if electrons are in the same sub-d-band ($d = d'$), this correlation energy is effective only when they are in different spin states so that this energy has the following characteristics. Transforming the correlation energy into the Bloch scheme and supposing that $\tilde{V}_C$ is a constant independent of the wave vector k it can be written as

$$\mathscr{H}_{C_-}^{\text{intra}} = \sum_{kk'q} \frac{1}{N} \tilde{V}_C c_{k+q\uparrow}^\dagger c_{k\uparrow} c_{k'-q\downarrow}^\dagger c_{k'\downarrow} , \tag{8.4.31}$$

where the suffix d is abbreviated because only a single sub–band is being consi-dered. When one takes only $q = 0$ in (8.4.31), it becomes $E_C^{\text{intra}} = (\tilde{V}_C/N) \sum_{kk'} n_{k\uparrow}$ $n_{k'\downarrow}$. From the Fermi liquid theory in Chap.3, the energy of a quasiparticle is given by $\varepsilon_{k\sigma} = \delta E/\delta n_{k\sigma}$ and the effective interaction between quasiparticles is provided through $f_{k\sigma, k'\sigma'} = \delta\varepsilon_{k\sigma}/\delta n_{k'\sigma'}$. Thus, one has $f_{k\uparrow k'\downarrow} = f_{k'\downarrow, k\uparrow} = \tilde{V}_C/N$ and all other f's are zero. If the effective interaction is defined by the difference in energies between different spin configurations of an electron-pair as

$$\tilde{V}_{\text{eff}} = f_{k\uparrow, k'\downarrow} - f_{k\uparrow, k'\uparrow} = \frac{\tilde{V}_C}{N}, \tag{8.4.32}$$

this can be considered to correspond to $(\tilde{V}_{\text{ex}}/2N)$ in (8.4.27) or (8.4.28), or this may replace γ' above. Therefore, the criterion for ferromagnetism in (8.4.30) is rewritten as

$$N(\varepsilon_F)\,\tilde{V}_{\text{eff}} > 1. \tag{8.4.33}$$

The effective interaction $\tilde{V}_{\text{eff}} = \tilde{V}_C/N$ is due to the Hartree approximation, and it may overestimate the interaction. Under the condition that the number of elec-trons (or holes) in a band is relatively small, a more advanced calculation has been done to take into account the multiple scattering between electrons to yield as the effective interaction between the electrons at the k- and k'- state

$$\tilde{V}_{\text{eff}}(k, k') = \frac{\tilde{V}_C/N}{1 + \tilde{V}_C g(k, k')}, \tag{8.4.34}$$

where

$$g(k, k') = \frac{1}{N} \sum_{q} \frac{(1 - \bar{n}_{k+q})(1 - \bar{n}_{k'-q})}{\varepsilon_{k+q} + \varepsilon_{k'-q} - \varepsilon_k - \varepsilon_{k'}}. \tag{8.4.35}$$

Here, $\bar{n}_k$ is the occupation number of electrons in the k-state. As the number of electrons (or holes) has been assumed to be small, k and k' may be taken as those in the bottom (or top) of the band and then $g(k,k') \sim 1/W$ (W being the width of the band). Thus, one gets $\tilde{V}_{\text{eff}} \approx (\tilde{V}_C/N)/[1 + (\tilde{V}_C/W)]$ and the electron–electron interaction is reduced by the factor $1/[1 + (\tilde{V}_C/W)]$. This $\tilde{V}_{\text{eff}}$ will enter into (8.4.33). The present $\tilde{V}_{\text{eff}}$ indicates that, however large $\tilde{V}_C$ is, $\tilde{V}_{\text{eff}}$ is at most of the order of W.

As explained in Sect. 8.3, the magnetism of rare-earth metals is due to the s–f exchange interaction. In several rare-earth metals and in Cr metals, the screw or the sinusoidal spin structure has been observed. In Cr metals ($T_N = 310$ K), this spin polarization is called the spin-density wave which is due to the itinerant d electrons in contrast to the magnetism due to the localized f electrons in rare-earth metals. Therefore, the mechanisms to induce these polarizations are different to each other but in both cases the shape of the Fermi surface is essentially important. The case of rare-earth metals will be considered below.

The interaction between localized spins originates from the s–f interaction and it is given by (8.3.19,20) Taking the s–band as the free-electron type and carrying out the k-sum in (8.3.19) in $T = 0$ K, the interaction is expressed as

$$\mathcal{H}_{JJ} = - \sum_{mn} \sum_{q} (g - 1)^2 J(q)^2 f(q) \exp(i q \cdot r_{nm}) \, J_m \cdot J_n \,, \tag{8.4.36}$$

where $q = k - k'$, $J(q) = J_{k'k}$ and

$$f(q) = \frac{3N_e}{8\varepsilon_F N^2} \left(1 + \frac{4k_F^2 - q^2}{4k_F q} \ln \left|\frac{2k_F + q}{2k_F - q}\right|\right). \tag{8.4.37}$$

If the system is one-dimensional, q becomes the one-dimensional q and $f(q)$ is replaced by

$$f(q) = \frac{m}{\pi \hbar^2 N^2} \ln \left|\frac{2k_F + q}{2k_F - q}\right|. \tag{8.4.38}$$

The function $f(q)$ in (8.4.37) is monotonically decreasing with q, but $f(q)$ in (8.4.38) is divergent at $q = 2k_F$. The behaviour of $f(q)$ is shown in Fig. 8.17 for 1, 2, and 3–dimensional cases.

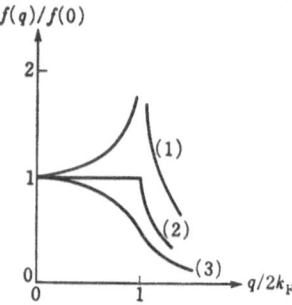

Fig. 8.17. The function $f(q)$ that is involved in the s-d interaction. The numbers in parentheses indicate dimensions [8.9]

Fig. 8.18. (a) The lower half of the Brillouin zone of hcp structure. (b) The Fermi surface of Dy metal calculated by the relativistic APW method. Conduction electrons occupy four bands, and the shaded areas with lines and dots correspond, respectively, to the electron and the hole parts. If one takes $2k_F^* = Q_2$ in the double zone scheme, the observed screw structure can be explained. [8.10]

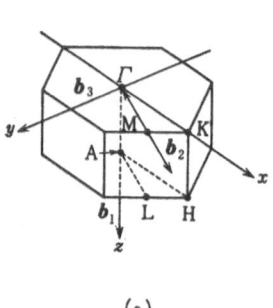

(a)

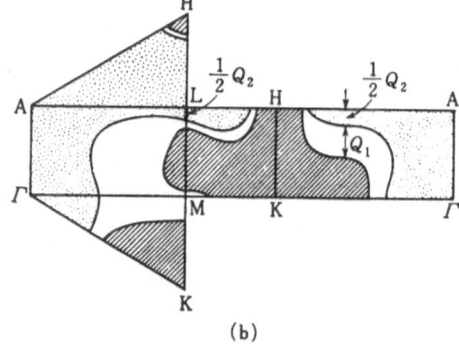

(b)

Most of the actual rare-earth metals have either a hcp (ABAB... type) or dhcp (ABACABAC... type) crystal structure. In Fig. 8.18, the shape of the Fermi surface determined from a band calculation is shown for Dy metal with hcp structure. It is much more complicated than that expected from the free-electron approximation. However, it may be noticed that there is a fairly flat part in the direction perpendicular to k_z. This part may contribute to $f(q)$ as if the system is nearly one-dimensional in the z-direction. In (8.4.36), one can write $N^{-2} \sum_{mn} \exp$

$(i\boldsymbol{q} \cdot \boldsymbol{r}_{nm})\boldsymbol{J}_m \cdot \boldsymbol{J}_n = J(\boldsymbol{q}) \cdot J(-\boldsymbol{q})$ so that $-(g-1)^2 J(\boldsymbol{q})^2 f(\boldsymbol{q})$ corresponds to the coefficient in (8.4.17). Thus, one may expect that the stable-spin structure is the screw structure determined by $\boldsymbol{q} = (0, 0, 2k_F^*)$, where k_F^* is a quantity describing the actual Fermi surface when the system is translated into a one-dimensional system in the k_z-direction. In rare-earth metals with the hcp lattice structure, the spin polarization of the screw type along the c-axis is actually found. However, the anisotropy energy is very large in these metals (cf. Sect. 8.4.1) and one has to discuss the actual spin structure by taking this energy into account. In Tb and Dy metals, the screw spin structure has $\langle \boldsymbol{J} \rangle$ only in the direction perpendicular to the c-axis, while in Ho and Er metals, $\langle \boldsymbol{J} \rangle$ has components both along and perpendicular to the c-axis. In Nd metals a more complicated spin structure has been observed.

The s–d (or s–f) interaction reflects the shape of the Fermi surface as explained above. If one takes the Fermi surface of the three-dimensional free-electron gas and puts $J(\boldsymbol{q}) \approx J(0)$, the sum $\sum_{\boldsymbol{q}}$ can be carried out in (8.4.36) to give

$$\mathcal{H}_{ss} = \left(\frac{3N_e}{2N}\right)^2 \frac{2\pi}{\varepsilon_F} J(0)^2 \sum_{mn} \left[\frac{\cos(2k_F r_{nm})}{(2k_F r_{nm})^3} - \frac{\sin(2k_F r_{nm})}{(2k_F r_{nm})^4}\right](\boldsymbol{S}_m \cdot \boldsymbol{S}_n) \qquad (8.4.39)$$

(here, $(g-1)\boldsymbol{J}$ has been replaced by $\boldsymbol{S}$). This interaction decays with distance rather slowly as r_{nm}^{-3} and osciallates with the period $r_{nm} \approx \pi/k_F$. If the localized spins are in a polarized state, they induce a polarization of conduction electrons and its density is given by

$$\rho_{\uparrow}(\boldsymbol{r}) - \rho_{\downarrow}(\boldsymbol{r}) = -\frac{9\pi N_e^2}{N\varepsilon_F} J(0) \sum_n \left[\frac{\cos(2k_F |\boldsymbol{r} - \boldsymbol{r}_n|)}{(2k_F |\boldsymbol{r} - \boldsymbol{r}_n|)^3} - \frac{\sin(2k_F |\boldsymbol{r} - \boldsymbol{r}_n|)}{(2k_F |\boldsymbol{r} - \boldsymbol{r}_n|)^4}\right] S_n^z .$$
$$(8.4.20)$$

8.5 Collective Motion in Ordered States and Phase Transitions

The lowest excited states are thermally excited at first at the lowest temperature and, in an ordered-spin system, they are the collective mode of spins. The elementary excitation in spin systems will be explained in detail in [8.11], and only a brief explanation for the collective mode of spins will be given here.

Consider a ferromangetic insulator with the exchange interaction of the Heisenberg type. Including an external field, the Hamiltonian is given by $\mathscr{H} = -\sum_{\langle mn \rangle} 2J_{mn} \mathbf{S}_m \cdot \mathbf{S}_n - g\mu_B H \sum_n S_n^z$. The statistical average of an operator A at temperature T will be denoted as $\langle A \rangle_T$ below. The equation of motion of spin at the lattice site l is expressed as

$$\dot{\mathbf{S}}_l = \frac{1}{i\hbar} [\mathbf{S}_l, \mathscr{H}]$$

$$= \frac{1}{\hbar} \sum_n 2J_{ln} (\mathbf{S}_l \times \mathbf{S}_n) + \frac{g\mu_B H}{\hbar} (iS_l^y - jS_l^x). \tag{8.5.1}$$

For the perpendicular components $S^{\pm} = S^x \pm iS^y$, this is written as

$$\dot{S}_l^{\pm} = \pm \frac{i}{\hbar} [\sum_n 2J_{ln}(S_l^z S_n^{\pm} - S_l^{\pm} S_n^z) - g\mu_B H S_l^{\pm}] \tag{8.5.2}$$

(the same sign of $\pm$ should be taken at both sides).
Let us linearize the right-hand side of (8.5.2) by replacing S^z with $\langle S^z \rangle_T$. Introducing the Fourier transform $\mathbf{S}_l = \sum_k \mathbf{S}_k \exp(i\mathbf{k} \cdot \mathbf{r}_l)$, (8.5.2) can be expressed as

$$\dot{S}_k^{\pm} = \mp i \frac{\varepsilon_k}{\hbar} S_k^{\pm}, \tag{8.5.3}$$

where

$$\varepsilon_k = \sum_n 2J_{nl} \langle S^z \rangle_T [1 - \exp(i\mathbf{k} \cdot \mathbf{r}_{nl})] + g\mu_B H. \tag{8.5.4}$$

This ε_k is the excitation energy of the collective mode, i.e., spin wave or magnon. If the exchange interaction is only between the nearest-neighbor spins, it can be written as

$$\varepsilon_k = 2zJ \langle S^z \rangle_T (1 - \gamma_k) + g\mu_B H, \tag{8.5.5}$$

where z is the number of the nearest-neighbor spins and $\gamma_k = z^{-1} \sum_n^{(nn)} \exp(i\mathbf{k} \cdot \mathbf{r}_n)$ ($\sum_n^{(nn)}$ indicates to sum over the nearest-neighbor spin sites). When magnetic ions constitute the simple cubic (sc) lattice, one has $\gamma_k = 3^{-1}(\cos ak_x + \cos ak_y + \cos ak_z)$, and when they constitute the bcc lattice, $\gamma_k = \cos(ak_x/2) \cos(ak_y/2) \cos(ak_z/2)$, etc... [$a$ = lattice constant].
If one set $H = 0$ in (8.5.4), $\varepsilon_k = 0$ at $k = 0$. This results from the interaction of the Heisenberg type being isotropic. Let us denote the ground state of the ferromagnetic state by $|0\rangle$ and and its energy is $E_0 = -zJNS^2$ (with $H = 0$). Operating both sides of (8.5.3) to $|0\rangle$, one obtains $(\mathscr{H} - E_0)S_k^-|0\rangle = \varepsilon_k S_k^-|0\rangle$ and this means that the state with a single magnon is given by $S_k^-|0\rangle$.

In an antiferromagnet, the Néel state may be taken as the ground state and a similar argument can be made to obtain

$$\varepsilon_k = 2z|J\langle S^z\rangle_T|\sqrt{1-\gamma_k^2} \pm g\mu_B H \tag{8.5.6}$$

corresponding to (8.5.5). Due to the presence of two equivalent sublattices, the magnon dispersion is doubly degenerate everywhere, but this degeneracy is removed by an external field and the presence of $\pm$ signs in (8.5.6) indicates this fact. Again, one has $\varepsilon_k = 0$ at $k = 0$ (with $H = 0$) as in ferromagnets. However, the dispersion in antiferromagnets is different from that in ferromagnets. In the sc lattice $\gamma_k \approx 1 - (a^2 k^2/6)$ near $k \sim 0$ so that $\varepsilon_k \propto k^2$ in ferromagnets and $\varepsilon_k \propto k$ in antiferromagnets. In Fig. 8.19, ε_k along the k_x axis is shown for ferro and antiferromagnets. The branch with $-g\mu_B H$ in (8.5.6) becomes negative at $k \to 0$. This indicates that the spin structure with the spontaneous polarization in the $\pm$ z-direction is unstable under a magnetic field in the z-direction, and the polarization tends to be in the direction perpendicular to the external field (refer to the explanation below). Furthermore, in contrast to the ferromagnet, the Néel state itself is not a quantum-mechanical eigenstate at $T = 0$ K. One may, however, consider that the Néel state is stabilized by taking account of the anisotropy energy.

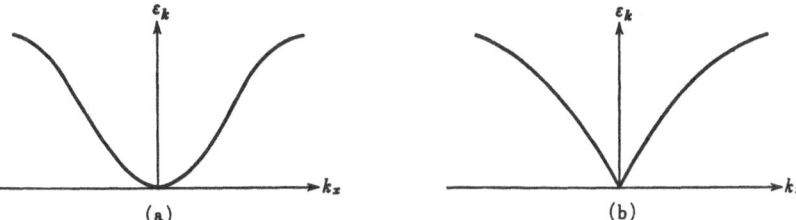

Fig. 8.19. Energy spectra of spin waves for (a) ferromagnets and (b) antiferromagnets

The effect of the thermally excited magnons appears at $\langle S^z\rangle_T$ which decreases with temperature. This simply indicates in (8.5.1) that the average torque due to the exchange interaction decreases. It is not a complete description of the real situation because many magnons will be excited with temperature and they will collide with each other very often. At very low temperatures, however, the number of thermally excited magnons is small and the picture of independent magnons may not be bad. In such a case, one may treat them as the Bose particles and the magnetization of the ferromagnet is given as

$$\frac{M}{Ng\mu_B} = S - \frac{1}{N}\frac{1}{(2\pi)^3}\int dk \frac{1}{\exp(\beta\varepsilon_k)-1}. \tag{8.5.7}$$

For the sc lattice, one may put $\varepsilon_k \approx 2JS(ka)^2$ to obtain

$$\frac{M}{Ng\mu_B} = S - \zeta\left(\frac{3}{2}\right)\left(\frac{kT}{8\pi JS}\right)^{3/2} - \cdots ,\tag{8.5.8}$$

where $\zeta(3/2) = 2.612$ is the Riemann ζ-function. The magnetic energy is given by $E_m = E_0 + (2\pi)^{-3} \int dk\varepsilon_k/[\exp(\beta\varepsilon_k) - 1]$ so that the specific heat related to it is

$$C_m = \frac{\partial E_m}{\partial T} = \frac{15}{4} Nk\zeta\left(\frac{5}{2}\right)\left(\frac{kT}{8\pi JS}\right)^{3/2} + \cdots .\tag{8.5.9}$$

In contrast to the ferromagnet, the sublattice magnetization of the antiferromagnet varies with temperature as T^2 and its magnetic specific heat is proportional to T^3.

In metals, the collective motion of spins can be investigated similarly. For simplicity, consider a ferromagnetic metal with a single conduction band. If the operators $S_q^z = (1/2)\sum_k (c_{k+q\uparrow}{}^\dagger c_{k\uparrow} - c_{k+q\downarrow}{}^\dagger c_{k\downarrow})$, $S_q^+ = \sum_k c_{k+q\uparrow}{}^\dagger c_{k\downarrow}$, and $S_q^- = \sum_k c_{k+q\downarrow}{}^\dagger c_{k\uparrow}$ are introduced, they satisfy the same commutation relations as those of the (Fourier-transformed) spin operators of $S = 1/2$. Thus, investigating the equation of motion of $S_q^\pm$ with the interaction (8.4.26) or (8.4.31), one gets the dispersion of the collective mode without an energy gap. In metals, however, there is the single particle excitation in addition to the collective mode. Let us take the band as the free-electron type and suppose that the majority spin ($\uparrow$) is filled up to $k_{F\uparrow}$ and the minority spin ($\downarrow$) to $k_{F\uparrow}$ at $T = 0$ K. The difference in the Fermi energy is $\Delta\varepsilon_F = (\hbar^2/2m)(k_{F\uparrow}^2 - k_{F\downarrow}^2)$. As in the magnon excitation, there is a single particle excitation with spin-flip; $(k, \uparrow)$ electron is transfered into $(k + q, \downarrow)$ state. If $q = 0$, the excitation energy is $\Delta\varepsilon_F$, and, if $q = k_{F\uparrow} - k_{F\downarrow}$, the excitation energy is zero. The region where such single particle excitation can take place is shown in Fig. 8.20 by the shaded region together with the magnon dispersion.

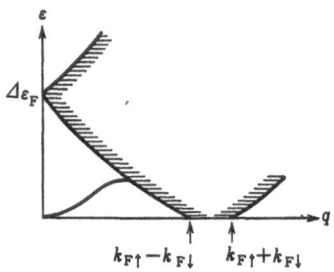

Fig. 8.20. Spectrum for spin wave and single-particle excitations in metals

When the temperature is raised near to T_C or T_N, the picture of the collective mode may be obscured even in insulators. Let us return to the molecular field approximation and write the Hamiltonian for the nth spin as $\mathcal{H}_n = -2Jz\langle S\rangle_T \cdot S_n - g\mu_B H \cdot S_n$ (For convenience, take $\mu_B < 0$. Only the nearest-neightbor exchange interaction has been taken.) This is equivalent to having an effective magnetic field $H_{\text{eff}} = (2Jz\langle S \rangle_T/g\mu_B) + H$ for S_n, and one gets, by taking $\langle S \rangle_T$ and H in the z-direction,

$$\langle S_n^z \rangle_T = SB_S \left(\frac{2JzS\langle S^z \rangle_T + g\mu_B SH}{kT} \right), \tag{8.5.10}$$

where

$$B_S(x) = \frac{2S+1}{2S} \coth \frac{(2S+1)x}{2S} - \frac{1}{2S} \coth \frac{x}{2S} \tag{8.5.11}$$

is the Brillouin function. Now, put $H = 0$ and $\langle S_n^z \rangle_T = \langle S^z \rangle_T$. $\langle S^z \rangle_T$ is small near T_C and one may expand the right-hand side of (8.5.10) in powers of $\langle S^z \rangle_T$ up to the 3rd order [there is no quadratic term in $B_S(x)$] to yield

$$\langle S^z \rangle_T = \sqrt{\frac{10}{3}} \frac{S(S+1)}{[S^2+(S+1)^2]^{1/2}} \left(\frac{T_C - T}{T_C} \right)^{1/2}, \tag{8.5.12}$$

where

$$T_C = \frac{2JzS(S+1)}{3k} \tag{8.5.13}$$

is the Curie temperature. The magnetic energy is $E_m = -NzJ\langle S^z \rangle_T^2$, and the magnetic specific heat $C_m = \partial E_m/\partial T$ increases with temperature up to T_C to reach $C_m(T_C) = 5NkS(S+1)/[S^2+(S+1)^2]$ and then it drops suddenly to zero. There is no spontaneous polarization above T_C, but the susceptibility can be obtained by reserving the external field term

$$\chi = \left(\frac{Ng\mu_B\langle S^z \rangle_T}{H} \right)_{H=0} = \frac{C}{T - T_C}, \tag{8.5.14}$$

where $C = N(g\mu_B)^2 S(S+1)3k$ is called the Curie constant. The spontaneous polarization, susceptibility and the magnetic specific heat obtained in the molecular field approximation are sketched in Fig. 8.21.

In order to investigate the antiferromagnet in the molecular field approximation, the Néel state will be assumed. If the exchange interaction is present only between the nearest-neighbor magnetic ions and if a given magnetic ion in a sublattice is surrounded only by the magnetic ions of another sublattice (say, bcc structure), the above consideration for the ferromagnet can be applied to the antiferromagnet with a slight modification. Let $\langle S^z \rangle_T$ and $\langle S^{z\prime} \rangle_T$ be the polarizations of two sublattices which are parallel and antiparallel to H, respectively. The argument of B_S in (8.5.10) is replaced by $(-2z|J|S\langle S^{z\prime} \rangle_T + g\mu_B SH)/kT$ and $(-2z|J|S\langle S^z \rangle_T + g\mu_B SH)/kT$ for $\langle S^z \rangle_T$ and $\langle S^{z\prime} \rangle_T$, respectively. The Néel point T_N is given by (8.5.13) with $|J|$ instead of J and the susceptibility χ by

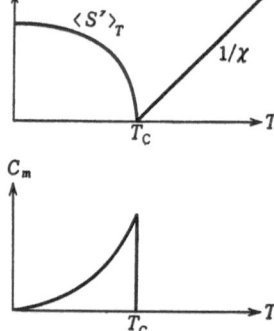

Fig. 8.21. Sketch of the spontaneous polarization $\langle S^z\rangle_T$, the susceptibility χ, and the magnetic specific heat C_m obtained for the ferromagnet in the molecular field approximation

(8.5.14) with T_C replaced by $-T_N$. (In the susceptibility χ, one usually writes θ instead of T_C or T_N. This paramagnetic Curie temperature θ does not necessarily agree with T_C or T_N in real system.) The sponteneous polarizations of two sublattices cancel each other below the Néel point so that the macroscopic net polarization is only the induced polarization by an external field. Assuming that the spontaneous polarizations are in the $\pm z$-direction in the absence of the external field, $\langle S^z\rangle_T (>0)$ will grow slightly and $\langle S^{z\prime}\rangle_T (< 0)$ will shrink slightly if an external field H is applied in the z-direction, and $\langle S\rangle_T$ and $\langle S'\rangle_T$ will tilt slightly toward H if it is applied in a direction perpendicular to the z-axis. The net polarization is thus induced to give the susceptibility as

$$\chi_\parallel = \frac{N(g\mu_B S)^2 B_S'(2z|J|S\overline{S^z}/kT)}{T+(2z|J|S^2/k)B_S'(2z|J|S\overline{S^z}/kT)} \tag{8.5.15}$$

and

$$\chi_\perp = \frac{N(g\mu_B)^2}{4z|J|} \tag{8.5.16}$$

corresponding to the above two cases. Here, $\overline{S^z} = \langle S^z\rangle_T = -\langle S^{z\prime}\rangle_T$ and B_S' means the differentiation of B_S. Schematic pictures of $\chi_\parallel$ and $\chi_\perp$ are shown in Fig. 8.22. As explained below (8.5.6), the anisotropy energy is necessary in order to make the sublattice polarizations in the $\pm z$-direction stable [if this anisotropy energy becomes appreciable and is not negligible compared with $z|J|$, some modification is necessary for (8.5.16)]. However, when the external field applied in the z-direction is large, $\langle S\rangle_T$ and $\langle S'\rangle_T$ tend to become perpendicular to H even if they loose the anisotropy energy because $\chi_\perp > \chi_\parallel$ and the gain of the Zeeman energy is overwhelming. This phenomenon is called the spin-flopping. If one takes $-D(\overline{S^z})^2$ as the anisotropy energy, it can be balanced with the gain of the Zeeman energy $(\chi_\perp - \chi_\parallel)H^2/2$ at $H_c \approx [2D(\overline{S^z})^2/(\chi_\perp - \chi_\parallel)]^{1/2}$, where the spin-flopping can be expected. This transition is of the 1st order. In Fig.8.23, the phase diagram of the antiferromagnet is shown, where the region with the symbol $\nwarrow\nearrow$ indicates the state after the spin-flopping.

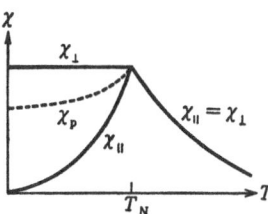

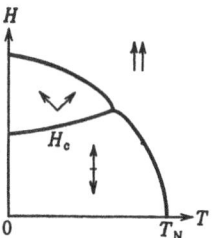

Fig. 8.22. Schematic picture of the susceptibility of the antiferromagnet obtained in the molecular field approximation. $\chi_p = (\chi/3) + (2\chi_\perp/3)$ is the susceptibility for a polycrystalline sample

Fig. 8.23. Phase diagram of an antiferromagnet where an external field is applied in the direction of the easy-axis

Ferrimagnets can be dealt with in the MFA just as for antiferromagnets. Crystal structures of the ferrimagnet are in general not simple so as to produce the net spontaneous mangetization as a bulk, and the intra-sublattice exchange interaction should be taken into account as well as the intersublattice one (the intra-sublattice exchange interaction cannot always be neglected even in the antiferromagnet). Typical examples of the spontaneous magnetization in ferrimagnets obtained in the MFA are shown in Fig. 8.24. In Fig. 8.24c there is the temperature where the sublattice magnetizations cancel each other and the magnetization appears in a reverse direction above that temperature.

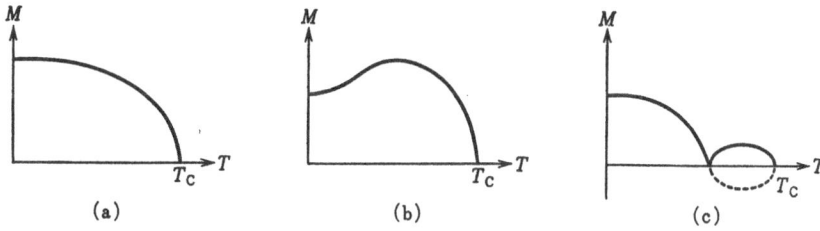

Fig. 8.24a-c. Examples of the spontaneous magnetization in ferrimagnets [8.12]

For ferromagnetic metals, one can give a systematic derivation of the spontaneous polarization, the susceptibility and the specific heat by taking account of only the single particle excitation in the Stoner model. However, one also has to consider the effect of magnons and the effect of the electron correlation. There have recently been many attempts to improve the Stoner model at finite temperatures and with some appreciable advancements. This topic will be deferred to other appropriate books[2] [8.13, 14]

---

[2] For example, S. Nakajima, Y. Toyozawa, R. Abe: *The Physics of Elementary Excitations*, Springer Ser. Solid-State Ser., Vol. 12 (Springer, Berlin, Heiderberg, New York 1980)

Near the Curie (or Néel) temperature it may be expected that the spin fluctuation from its average value becomes appreciable, and the MFA does not provide the correct behavior of the spin correlation because it simply replaces the neighboring spins by $\langle S^z \rangle_T$. The exact correlation cannot be obtained for such many-body problems (except some special cases). For example, in the computer simulation, the quantum magnetic system with the Heisenberg-type interaction has been solved for only up to 11 spins. A general trend can be mentioned, however; there is always a fluctuation that gives a deviation from $\langle S^z \rangle_T$ and, if the nth spin S_n^z takes the value S accidentally, the neighboring spins have higher probability of taking larger values than $\langle S^z \rangle_T$ through the magnetic coupling. When such a fluctuation effect is taken into account, the resulting value of T_C (or T_N) is generally lower than that obtained in MFA. In particular, this discrepancy is exaggerated in systems with lower dimensions. In the 1-dimensional case, it can be proved that there is no finite T_C even though the MFA provides $T_C > 0$. In 3–dimensional spin systems, expected features near T_C are $\langle S^z \rangle \propto (T_C - T)^\beta$ at $T \lesssim T_C$ and $\chi \propto (T - T_C)^{-\gamma}$ at $T \gtrsim T_C$ with $\beta \approx 1/3$ and $\gamma \approx 4/3$ and the magnetic specific heat diverging as $\ln |T - T_C|$. β and γ are called the critical indices. In the MFA, $\beta = 1/2$ and $\gamma = 1$ from (8.5.12,14) The exact solution of the 2–dimensional Ising system with nearest-neighbor interaction provides $\beta = 1/8$, $\gamma = 7/4$, and the logarithmically diverging specific heat. Corresponding to real systems, it may be an interesting many-body problem to investigate the dependence of critical indices on the nature of interaction, the dimensionality and the symmetry of crystals and the dimensionality of order parameters.

8.6 Magnetic Properties of Dilute Alloys

When a small amount of the $3d$-transition-metal atoms are doped in simple metals or in metals which do not show any magnetic orderings by themselves, localized moments are often observed on the doped atoms. For instance, this is

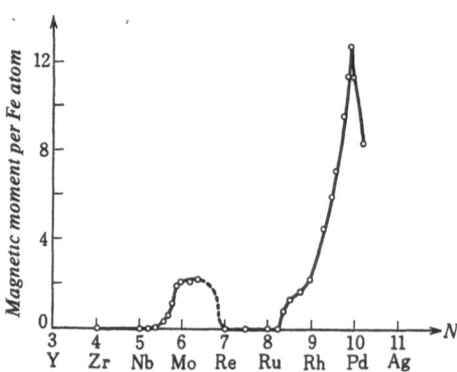

Fig. 8.25. Magnetic moment per Fe atom when Fe is doped by 1 % in metals or alloys between Y and Ag. [8.15]

observed for Mn and Fe atoms doped in noble metals or Fe atoms doped in the 4d–transition metals. The latter case is shown in Fig.8.25. A very large moment is found in Pd metal and this is due to the presence of the appreciable interaction between electrons in Pd. In fact, it is observed experimentally that a magnetic moment is present over about 100 Pd atoms around each Fe atom.

In order to investigate the magnetic moment localized on impurity magnetic ions, the following simple model will be introduced. The conduction band (called the s-band) of the host metal is of the free-electron type and one impurity atom which has a single d orbital with the interaction (8.3.23) is embedded in it ($\bar{V}_{ex}$ has no role here). Furthermore, the mixing effect is assumed to be present between the d orbital and the s-band orbitals. The Hamiltonian of such a system is written in the Hartree approximation as

$$\mathcal{H} = \sum_{k\sigma} \varepsilon_k n_{k\sigma} + \sum_{\sigma} (E_d + U\langle n_{d-\sigma}\rangle) n_{d\sigma} + \sum_{k\sigma} (V_{kd} c_{k\sigma}^\dagger c_{d\sigma} + V_{dk} c_{d\sigma}^\dagger c_{k\sigma}),$$

$$(8.6.1)$$

where $n_{k\sigma} = c_{k\sigma}^\dagger c_{k\sigma}$, $n_{d\sigma} = c_{d\sigma}^\dagger c_{d\sigma}$, E_d is the energy level of the d orbital before mixing and $\langle n_{d-\sigma}\rangle$ is the Hartree average of $n_{d-\sigma}$. The new operator that diagonalizes (8.6.1) will be written as $c_{n\sigma}^\dagger = \sum_k \langle n\sigma | k\sigma \rangle c_{k\sigma}^\dagger + \langle n\sigma | d\sigma \rangle c_{d\sigma}^\dagger$ and its energy eigenvalue as $\varepsilon_{n\sigma}$. The Green's function $G(\varepsilon) = (\varepsilon + is - \mathcal{H})^{-1}$ satisfies the equation

$$(\varepsilon + is - \mathcal{H}) G(\varepsilon) = 1.$$

$$(8.6.2)$$

Solving this equation in terms of the original representation $|k\sigma\rangle$ and $|d\sigma\rangle$, one obtains

$$\langle d\sigma | G(\varepsilon) | d\sigma \rangle = \left[\varepsilon - E_d - U\langle n_{d-\sigma}\rangle - \mathscr{P} \sum_k \frac{|V_{dk}|^2}{\varepsilon - \varepsilon_k} \right.$$
$$\left. + i\pi \sum_k |V_{dk}|^2 \delta(\varepsilon - \varepsilon_k) \right]^{-1},$$

$$(8.6.3)$$

where $\mathscr{P}$ indicates the principal value. The $|d\sigma\rangle$ state which is now mixed into the new eigenstates $|n\sigma\rangle$'s has a width in energy. The energy distribution is given by

$$N_{d\sigma}(\varepsilon) = \sum_n |\langle n\sigma | d\sigma \rangle|^2 \delta(\varepsilon - \varepsilon_{n\sigma})$$

$$= -\frac{1}{\pi} \cdot \sum_n \langle d\sigma | n\sigma \rangle \cdot \text{Im} \left\{ \left\langle n\sigma \left| \frac{1}{\varepsilon - \mathcal{H} + is} \right| n\sigma \right\rangle \right\} \cdot \langle n\sigma | d\sigma \rangle$$

$$= -\frac{1}{\pi} \text{Im} \{ \langle d\sigma | G(\varepsilon) | d\sigma \rangle \},$$

$$(8.6.4)$$

where Im indicates the imaginary part and use has been made of $\lim_{s\to +0} 1/(x + is)$

$= \mathscr{P}(1/x) - i\pi\delta(x)$. Using the density of states of the original s band $N(\varepsilon) = \sum_k \delta(\varepsilon - \varepsilon_k)$, let us introduce $\delta = \mathscr{P} \sum_k |V_{dk}|^2/(\varepsilon - \varepsilon_k)$ and $i\Delta = i\pi \sum_k |V_{dk}|^2 \delta(\varepsilon - \varepsilon_k) = i\langle |V_{dk}|^2\rangle_\varepsilon N(\varepsilon)$ ($\langle\dots\rangle_\varepsilon$ means the average over $|k\sigma\rangle$ states that satisfy $\varepsilon = \varepsilon_k$). Neglecting the ε–dependency of δ and Δ, one gets from (8.6.3,4)

$$N_{d\sigma}(\varepsilon) = \frac{1}{\pi} \frac{\Delta}{(\varepsilon - E_d - U\langle n_{d-\sigma}\rangle - \delta)^2 + \Delta^2}. \tag{8.6.5}$$

When electrons are filled up to ε_F in this $N_{d\sigma}(\varepsilon)$, one obtains

$$\langle n_{d\sigma}\rangle = \int_{-\infty}^{\varepsilon_F} d\varepsilon N_{d\sigma}(\varepsilon) = \frac{1}{\pi} \cot^{-1} \frac{E_d + U\langle n_{d-\sigma}\rangle + \delta - \varepsilon_F}{\Delta}. \tag{8.6.6}$$

Taking ↑ and ↓ for σ, one gets two equations from (8.6.6) and they comprise a simultaneous equation to be solved for $\langle n_{d\uparrow}\rangle$ and $\langle n_{d\downarrow}\rangle$. $\langle n_{d\uparrow}\rangle = \langle n_{d\downarrow}\rangle$ is always one solution which corresponds to a nonmagnetic state of the impurity atom. If U/Δ becomes so large as to satisfy

$$UN_{d\sigma}(\varepsilon_F) > 1, \tag{8.6.7}$$

another solution of the type $\langle n_{d\uparrow}\rangle \neq \langle n_{d\downarrow}\rangle$ appears. In this case, there is a two-fold degeneracy corresponding to $\langle n_{d\uparrow}\rangle > \langle n_{d\downarrow}\rangle$ and $\langle n_{d\uparrow}\rangle < \langle n_{d\downarrow}\rangle$. This implies that the impurity atom is in a magnetic state with a moment which may be either in the upward or the downward direction. A schematic picture of the nonmagnetic and the magnetic states are shown in Fig. 8.26. Such a d state which has a finite width but still has a localized nature is called the virtual bound state.

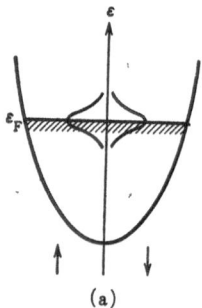
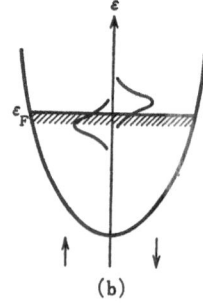

Fig. 8.26a, b. Density of states of the conduction band and the virtual bound state. The virtual bound state is in (a) the nonmagnetic and (b) magnetic states

Let us then consider the system where two such impurity atoms of the transition metal are involved. If these two atoms are far apart they are independent of each other and the above investigation is applicable for each atom. If they are close to each other, interaction exists between them. Suppose that the two atoms are in the magnetic state and that they may not be the identical atoms. Extending (8.6.1), the Hamiltonian for the present two-impurity system may be written as

$$\mathcal{H} = \sum_{k\sigma} \varepsilon_k n_{k\sigma} + \sum_{m=1,2} (E_m + U_m \langle n_{m-\sigma} \rangle) n_{m\sigma} + \sum_{m=1,2} \sum_{k\sigma} (V_{km} c_{k\sigma}^\dagger c_{m\sigma}$$

$$+ V_{mk} c_{m\sigma}^\dagger c_{k\sigma}) + \sum_\sigma (V_{12} c_{1\sigma}^\dagger c_{2\sigma} + V_{21} c_{2\sigma}^\dagger c_{1\sigma}), \tag{8.6.8}$$

where the suffix d has been dropped and the number of the impurity atoms has been attached instead. The last term in (8.6.8) involves the direct transfer integral between the d orbitals. Just as before, write down the equations for all Green's functions, eliminate the Green's function which connects the d and the s orbital, and derive the equation for Green's function that is related solely to the d orbital.

The resulting equation for $\langle m\sigma | G(\varepsilon) | n\sigma \rangle$ [which is now written as $G_{m\sigma, n\sigma}(\varepsilon)$] is

$$(\varepsilon + is - E_{m\sigma} + i\Delta_m) G_{m\sigma, n\sigma}(\varepsilon) = \delta_{mn} + \sum_{l(\neq m)} \left(V_{ml} + \sum_k \frac{V_{mk} V_{kl}}{\varepsilon + is - \varepsilon_k} \right) G_{l\sigma, n\sigma}(\varepsilon),$$
$$\tag{8.6.9}$$

where $E_{m\sigma} = E_m + U_m \langle n_{m-\sigma} \rangle + \delta_m$, and δ_m and Δ_m are the quantities defined below (8.6.4) for the mth impurity atom. The last term indicates that for the two impurity atoms there is the indirect interaction $\sum_k V_{mk} V_{kl}/(\varepsilon + is - \varepsilon_k)$ via the s electron in addition to the direct interaction V_{ml}. The sum of these interactions will be written as V to denote the effective interaction between the two impurity atoms. Again, as before, one can calculate the number of the electrons which is accommodated in the d orbitals that now have finite widths and can then obtain the total energy of the system. As the width of the d orbitals depends naturally on V, the total energy is a function of V. The energy change of the total system referred to the case of $V = 0$ is given, when V is small, as

$$\delta E = - V^2 \sum_\sigma \frac{\langle n_{1\sigma} \rangle^{(0)} - \langle n_{2\sigma} \rangle^{(0)}}{E_{2\sigma} - E_{1\sigma}}, \tag{8.6.10}$$

where $\langle \ldots \rangle^{(0)}$ is the average in the case of $V = 0$. If the two magnetic impurities are identical atoms, one has $\langle n_{1\uparrow} \rangle^{(0)} = \langle n_{2\downarrow} \rangle^{(0)} \neq \langle n_{1\downarrow} \rangle^{(0)} = \langle n_{2\uparrow} \rangle^{(0)}$ and $E_{1\uparrow} = E_{2\downarrow} \neq E_{1\downarrow} = E_{2\uparrow}$ for an antiferromangetic arrangeement and $\langle n_{1\uparrow} \rangle^{(0)} = \langle n_{2\uparrow} \rangle^{(0)} \neq \langle n_{1\downarrow} \rangle^{(0)} = \langle n_{2\downarrow} \rangle^{(0)}$ and $E_{1\uparrow} = E_{2\uparrow} \neq E_{1\downarrow} = E_{2\downarrow}$ for a ferromagnetic arrangement. In the latter case, (8.6.10) becomes the derivative of $\langle n_{1\sigma} \rangle^{(0)}$ so that the expression under the sum reduces to $N_{1\sigma}(\varepsilon_F)$. δE is generally different for the antiferromagnetic and the ferromagnetic arrangements of the impurity atoms and, therefore, (8.6.10) provides an effective interaction between the magnetic moments of the impurity atoms. This interaction may be present in the transition metals and their alloys and must be added to the interaction given in Sect. 8.3.

Now, returning to (8.6.1), let us consider the case where $(E_d - \varepsilon_F) \ll -|V_{dk}|$ and $(U + E_d - \varepsilon_F) \gg |V_{dk}|$ are satisfied. In such a case, only $\uparrow$ spin (or $\downarrow$ spin) state is occupied in the d orbital and one can put $\langle n_{d\uparrow} \rangle = 1$ and $\langle n_{d\downarrow} \rangle = 0$ (or vice versa). Taking the former case for definiteness, let us evaluate the effect of

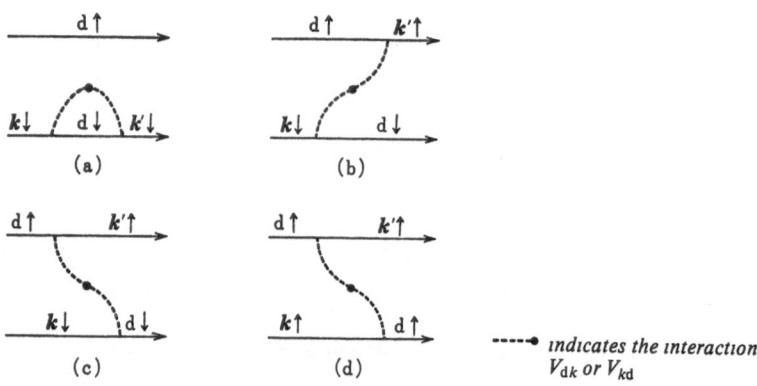

Fig. 8.27a-d. Scattering processes by V_{dk} or V_{kd} in the second order in the case of $\langle n_{d\uparrow}\rangle = 1$ and $\langle n_{d\downarrow}\rangle = 0$

V_{dk} by the perturbation. In the 2nd-order perturbation, one has the scattering processes shown in Fig. 8.27, where the dotted line indicates V_{dk} or V_{kd}. From the process (b), for instance, one gets

$$\frac{V_{k'd}V_{dk}}{\varepsilon_k - U - E_d}\, c_{k'\uparrow}^{\dagger} c_{d\uparrow}\, c_{d\downarrow}^{\dagger} c_{k\downarrow}\,. \tag{8.6.11}$$

Summing up Fig. 8.27 a-c and the corresponding cases with $\langle n_{d\downarrow}\rangle = 1$ and $\langle n_{d\uparrow}\rangle = 0$ and using the equalities in (8.3.10), one obtains contributions of two types. One of them is independent of spin in the d orbital and is given by

$$\mathcal{H}_{\mathrm{imp}}^{\mathrm{eff}} = \sum_k \frac{|V_{kd}|^2}{E_d - \varepsilon_k}$$
$$+ \sum_{kk'\sigma} V_{k'd}V_{dk}\left[\frac{1}{\varepsilon_k - U - E_d} - \frac{1}{2}\left(\frac{1}{\varepsilon_k - U - E_d} + \frac{1}{E_d - \varepsilon_{k'}}\right)\right] c_{k'\sigma}^{\dagger} c_{k\sigma}\,. \tag{8.6.12}$$

This involves the term denoting the self-energy of the d electron at the impurity atom and that denoting the ordinary impurity scattering. Another type of contribution involves the spin operator of the d orbital and is expressed as

$$\mathcal{H}_{sd}^{\mathrm{eff}} = -\sum_{kk'} V_{k'd}V_{dk}\left(\frac{1}{\varepsilon_k - U - E_d} + \frac{1}{E_d - \varepsilon_{k'}}\right)$$
$$\cdot \left[(c_{k'\uparrow}^{\dagger} c_{k\uparrow} - c_{k'\downarrow}^{\dagger} c_{k\downarrow})\, S^z + c_{k'\uparrow}^{\dagger} c_{k\downarrow} S^- + c_{k'\downarrow}^{\dagger} c_{k\uparrow} S^+\right]\,. \tag{8.6.13}$$

This has exactly the same form as the s-d exchange interaction given in (8.3.17) replacing the exchange integral by $J_{k'k}^{\mathrm{eff}}/N = V_{k'd}V_{dk}[(\varepsilon_k - U - E_d)^{-1} + (E_d - \varepsilon_{k'})^{-1}]$. As scatterings must take place effectively at $\varepsilon_k \sim \varepsilon_{k'} \sim \varepsilon_F$, one can set equal those quantities in $J_{k'k}^{\mathrm{eff}}$. Combining the assumption made above (8.6.11), one gets $J^{eff}/N \approx - V^2[(U + E_d - \varepsilon_F)^{-1} + (\varepsilon_F - E_d)^{-1}] < 0$ which indicates that J^{eff} is antiferromagnetic.

When a small amount of the transition metal atoms is doped in the noble metals (Fe in Cu, for instance), the resistance minimum is often observed at low temperature. This phenomenon can be explained by the s-d interaction with $J < 0$. Let us consider a simple metal which involves one magnetic impurity. The Hamiltonian of this system is given by

$$\mathcal{H}' = - \frac{J}{N} \sum_{kk'} [S^z(c_{k'\uparrow}^\dagger c_{k\uparrow} - c_{k'\downarrow}^\dagger c_{k\downarrow}) + S^+ c_{k'\downarrow}^\dagger c_{k\uparrow} + S^- c_{k'\uparrow}^\dagger c_{k\downarrow}] . \tag{8.6.14}$$

For this perturbation the transition probability will be calculated according to the Born approximation:

$$P_{\beta\alpha} = \frac{2\pi}{h} \delta(E_\beta - E_\alpha) \left[\mathcal{H}'_{\alpha\beta} \mathcal{H}'_{\beta\alpha} + \sum_{\gamma(\neq\alpha)} \left(\frac{\mathcal{H}'_{\alpha\gamma} \mathcal{H}'_{\gamma\beta} \mathcal{H}'_{\beta\alpha}}{E_\alpha - E_\gamma} + \text{c.c.} \right) + \cdots \right]. \tag{8.6.15}$$

From (8.6.14), the matrix elements of $\mathcal{H}'$ with respect to $|...k\sigma...; M\rangle$'s(M being the eigenvalue of S^z) are given as

$$\left. \begin{array}{l} \langle k' + ; M - 1 | \mathcal{H}' | k - ; M \rangle = - \dfrac{J}{N} \sqrt{(S + M)(S - M + 1)} \\[2ex] \langle k' - ; M + 1 | \mathcal{H}' | k + ; M \rangle = - \dfrac{J}{N} \sqrt{(S - M)(S + M + 1)} \\[2ex] \langle k'\sigma; M | \mathcal{H}' | k\sigma; M \rangle = - \dfrac{J}{N} M, \end{array} \right\} \tag{8.6.16}$$

where $\uparrow$ and $\downarrow$ spin states have been denoted by $+$ and $-$, respectively, and the states of conduction electrons which are irrelevant to the scattering have been dropped. Substituting (8.6.16) into the first term of (8.6.15) and summing M over $(2S + 1)$ degenerate states, the scattering probabilities in the first Born approximation can be obtained; $P_{k\sigma,k\sigma}^{(1)} = (2\pi J^2 S(S + 1)/3\hbar N^2)\delta(\varepsilon_k - \varepsilon_{k'})$ and $P_{k'-\sigma,k\sigma}^{(1)} = [4\pi J^2 S(S + 1)/3\hbar N^2]\delta(\varepsilon_k - \varepsilon_{k'})$. If there are Nc impurity spins (c being the concentration of impurities) and if they contribute to the scattering of the s electrons independently, $NcP^{(1)}$ provides the electrical resistance which is proportional to $J^2 S(S + 1)$ and is independent of temperature.

In the second Born approximation there is also the spin-conserving transition probability $P_{k\sigma,k\sigma}^{(2)}$ and the spin-flopping one $P_{k'-\sigma,k\sigma}^{(2)}$. Confinting ourselves to the former transition probability, there are two types of contributions as shown in Fig. 6.28. In the figure, (......x) indicates the interaction (8.6.14) and one should put $\sigma' = \sigma$ for S^z and $\sigma' = - \sigma$ for $S^\pm$. In Fig. 6.28b, the first scattering is $|q\sigma'\rangle \rightarrow |k'\sigma\rangle$ and the second is $|k\sigma\rangle \rightarrow |q\sigma'\rangle$, thereby terminating the

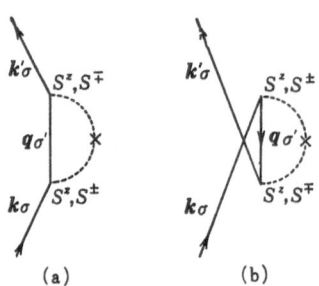

Fig. 8.28a, b. Processes of the second Born approximation in the s–d scattering. Only processes in which spin states are conserved in the initial and the final states are shown

-----×*Indicates the s-d*
interaction

$|q\sigma'\rangle$ line. Therefore, a vacant $|q\sigma'\rangle$ state contributes in process (a) and a filled $|q\sigma'\rangle$ state in (b). The contribution from the case of $\sigma' = -\sigma$ is quite different in nature from that of the first Born approximation and it is given by

$$P^{(2-\text{flip})}_{k'\sigma,\,k\sigma} = \frac{4\pi}{\hbar} \delta(\varepsilon_k - \varepsilon_{k'}) \left(-\frac{J}{N}\right)^3 \sum_M M$$

$$\cdot \sum_q \left[(S-M)(S+M+1)\frac{1-f_q}{\varepsilon_k - \varepsilon_q} - (S+M)(S-M+1)\frac{f_q}{\varepsilon_q - \varepsilon_{k'}}\right]$$

$$= \frac{8\pi J^3 S(S+1)}{3\hbar N^2}\frac{1}{N}\sum_q \frac{f_q}{\varepsilon_q - \varepsilon_k}\delta(\varepsilon_k - \varepsilon_{k'}) +$$

(terms that do not include f_q) $\hfill (8.6.17)$

where f_q is the Fermi distribution function for the s electron. When there are Nc impurity atoms, it must be multiplied by (8.6.17) as before. In the first equality of (8.6.17), the term with $(1-f_q)$ corresponds to the above process (a) and that with f_q to (b). In the second equality, the term without f_q gives a contribution which is similar to the first Born approximation, and the anomalous term with f_q is brought about by the fact that the operators S^+ and S^- appear in opposite order in (a) and (b) and they do not commute with each other $(S^+S^- - S^-S^+ \neq 0)$. A similar result is also obtained for the scattering which does not conserve the spin in the initial and the final states. For simplicity, let us take a simple density of states for the s-band; $N(\varepsilon) = \text{const.} = \rho$ if $-D \leq \varepsilon \leq D$ and zero otherwise. Carrying out the sum $\sum_q$ in (8.6.7), one gets

$$\frac{1}{N}\sum_q \frac{f_q}{\varepsilon_q - \varepsilon_k} = \rho \ln \left|\frac{\varepsilon_F - \varepsilon_k}{D + \varepsilon_k}\right|. \hfill (8.6.18)$$

When the transition probability that involves such $\ln|\varepsilon_F - \varepsilon_k|$ dependence is inserted into the expression of the electrical resistivity, one has the intergral $-\int d\varepsilon_k N(\varepsilon_k)\ln|\varepsilon_F - \varepsilon_k|(\partial f_k/\partial \varepsilon_k)$ and this provides the temperature dependence of $\ln T$. This reflects the fact that the electrons associated with the electrical conductivity are in the region of $|\varepsilon_F - \varepsilon_k| \sim kT$. Summing up all contributions, the electrical resistivity can be expressed up to the 2nd Born approximation as

$$R = R_t + c\left(R_t + J^3\xi \ln \frac{kT}{D}\right), \tag{8.6.19}$$

where R_t is the resistivity due to phonons, R_t the temperature-independent residual resistivity and $\xi (> 0)$ is a constant. Taking $J < 0$ and $R_t = aT^5$, the resistivity in (8.6.19) has the minimum at the temperature

$$T_{\min} = \left(\frac{|J|^3\xi}{5a}\right)^{1/5} c^{1/5}. \tag{8.6.20}$$

If one proceeds to the higher Born approximation, terms with the higher power of $\ln (k_B T/D)$ appear. This situation is similar in other physical quantities. An example is the average spin moment $\langle S \rangle_T$ of the impurity atom. If one picks up only the term with the highest power of $\ln (k_B T/D)$ in each order of perturbation, they constitute a geometric series and can be expressed as[3]

$$\langle S \rangle_T = S\left\{1 + \frac{1}{2}\left(\frac{J\rho}{N}\right)^2 \ln \frac{kT}{D} \cdot \left[1 - \frac{J\rho}{N} \ln \frac{kT}{D}\right]^{-1}\right\}, \tag{8.6.21}$$

where S is the polarization of the localized spin assumed before the s–d interaction is turned on. It can also be shown that the polarization $\langle s \rangle_T = (J\rho/4N) \langle S \rangle_T$ takes place in conduction electrons. As can be seen from (8.6.21), even if one assumes $S > 0$, $\langle S \rangle_T$ becomes small with decreasing temperature and diverges to $-\infty$ at $T_K = (D/k_B) \exp (N/J\rho)$, T_K being called the Kondo temperature. This may imply that the perturbation calculation is not valid there. From the shrinking tendency of $\langle S \rangle_T$ it is conjectured that the ground state at $T = 0$ K is the singlet state.

8.7 Applications of Magnetic Measurements for Condensed Matter

If one has a good knowledge of the nature of magnetic interactions and of the magnetic response to an external field, then electronic properties and the dynamical behavior of matter can be observed through magnetic measurements. This is due to the presence of the electron spins or nuclear spins, both of which play the role of test probe in matter. They are utilized widely in the electron spin resonance (ESR), the nuclear magnetic resonance (NMR), the Mössbauer effect, neutron diffraction experiment, etc.. Typical examples are measurements of the degree of covalency by ESR or NMR, or the determination of the wave functions of shallow donor states in semiconductors by ENDOR (i.e., simultaneous

---

[3] This result has been obtained for the Hamiltonian (8.6.14) with the coefficient $(-J/2N)$ instead of $(-J/N)$.

application of ESR and NMR). Similarly one can investigate the collective modes in magnetic substances by neutron scattering or one can analyze the relationship between the spontaneous spin polarization and the electronic state in magnetic substances through the Mössbauer effect by observing the internal field that constituent atomic nuclei feel. As the explanation of these phenomena is beyond the scope of this book, we refer the reader to other books of the respective field.

The reason why the spins can be utilized as a probe is due to their weak coupling to other modes in crystals (e.g. lattice vibrations). In some cases, however, the spins interfere with other modes or other order parameters in crystals rather intensively and such phenomena also provide interesting problems. These include the Jahn-Teller effect vs magnetism, or other structural changes (ferroelectricity, Martensitic transition, etc.) vs magnetism, and the phenomenon of superconductivity vs magnetism.

9. Magnetic Properties of Dilute Alloys — the Kondo Effect

In this chapter, an important topic which has been omitted in Chap. 8 is supplemented. This is the Kondo effect which has been one of the central problems of magnetism and seems to have been settled in a satisfactory resolution. There are two kinds of models to consider with this problem; the Anderson Hamiltonian and the s-d Hamiltonian. As the Anderson Hamiltonian may cover a wider region in reality, the approaches from this Hamiltonian are the main subject in this chapter.

9.1 Recent Investigations of the Kondo Effect

As explained in Sect. 8.6, there are two models for magnetic impurity atoms in metals. One is the model expressed by (8.6.1) which is the so-called Anderson Hamiltonian

$$\mathcal{H} = \sum_{k\sigma} \varepsilon_k n_{k\sigma} + \sum_{\sigma} E_d n_{d\sigma} + U n_{d\uparrow} n_{d\downarrow} + \sum_{k\sigma} (V_{kd} c_{k\sigma}^\dagger c_{d\sigma} + V_{dk} c_{d\sigma}^\dagger c_{k\sigma}). \quad (9.1.1)$$

The notation is the same as in (8.6.1). In (9.1.1), however, the original Hamiltonian is shown without making the Hartree approximation. The five-fold degeneracy of the d-orbital has been neglected and only a single orbital has been considered. Another model is given by (8.6.14) which is called the s-d Hamiltonian;

$$\mathcal{H} = -\frac{J}{2N} \sum_{kk'} [S^z (c_{k'\uparrow}^\dagger c_{k\uparrow} - c_{k'\downarrow}^\dagger c_{k\downarrow}) + S^+ c_{k'\downarrow}^\dagger c_{k\uparrow} + S^- c_{k'\uparrow}^\dagger c_{k\downarrow}] \quad (9.1.2)$$

In Sect. 8.6, the transition from the nonmagnetic to the magnetic state has been investigated by starting from (9.1.1). It is related to the localization of magnetic electrons. When it is assured that the magnetic electron is well localized, the electron has only the spin degree of freedom and the problem of the resistance minimum has been discussed with the Hamiltonian (9.1.2). However, the Hamiltonian (9.1.1) can cover a wider field than (9.1.2), and (9.1.1) is thought to be a more fundamental model. In fact, as can be derived from the result in Sect. 8.6, the Anderson model reduces to the s-d model when $E_d < \varepsilon_F$, $E_d + U > \varepsilon_F$, and $\Delta \ll U$.

As is shown in Sect. 8.6, the electrical resistivity increases as $|\ln(kT/D)|$ with decreasing temperature in metals involving magnetic impurity atoms and exhibits minimum resistance. Such logarithmic temperature dependence appears in other physical quantities as well, say, in spin polarization of the magnetic impurity. If one sums the contributions from higher-order perturbations, the resistance increase with decreasing temperature becomes more rapid, and it diverges at $T_K = (D/k) \exp(- N/|J|\rho)$. The spin polarization shrinks with decreasing temperature and diverges to the negative direction at T_K. On the other hand, it is apparent that the simple metal involving one magnetic impurity atom (or a very small amount of magnetic impurity atoms so dilute that they do not interact with each other) cannot show any magnetic-phase transition at a finite temperature. This indicates that the divergence of spin polarization has no meaning. Thus, the real properties of such a system has not been fully understood for a long time.

There have been a large number of theoretical and experimental works, and recent investigations elucidated that at low temperatures ($T \ll T_K$) the coupling between the localized spin and the conduction electron spins becomes effective to form a singlet state. The representative works in this direction are the theory of the singlet ground state by *Yosida* and *Yoshimori* [9.1] and the theory based on the renormalization group by *Wilson* [9.2]. These theories are based on the *s–d* model, however, The more recent theory by *Yamada* [9.3] and *Yosida* and *Yamada* [9.4] starts from the Anderson Hamiltonian and this seems to be more important in understanding the nature of the magnetic *d* electron at low temperatures ($T \ll T_K$). In this chapter, the properties that localized spins may exhibit at low temperatures will be explained mainly according to the Yosida-Yamada theory.

9.2 A Survey of Recent Experiments

Before going into an explanation of theoretical investigations, let us look at the recent experimental data and see what takes place at $T \lesssim T_K$ in real systems. In Fig. 9.1, the experimentally estimated Kondo temperatures are plotted for Ti, V, Cr, Mn, Fe, and Co impurities doped in Cu, Ag, Au, and Al metals. All curves have minima around Cr ∼ Mn and T_K spreads over the region 10^{-2}K ∼ 10^3K. That T_K spreads in such a wide region is not unnatural from the expression of T_K given in Sect. 6.6, but it is not effective to make experiments at extremely high or low temperatures. Furthermore, samples must be of a high purity containing only a small amount of magnetic impurity atoms as explained in Sect. 9.1, and one has to record a small response from such magnetic impurities. Therefore, it is only recently that reliable data has been provided. There is also a problem of how to estimate T_K correctly because T_K is not connected with any sharp phase transition.

One of the fundamental physical quantities is the magnetic susceptibility due to the magnetic impurity atoms, and in Fig. 9.2, that for CuFe is shown. It is of the Curie–Weiss type at high temperatures and continuously changes into the form of a–bT^2. The susceptibility of AuV is shown in Fig. 9.3 which indicates not only such temperature-dependence but also the importance of controlling the

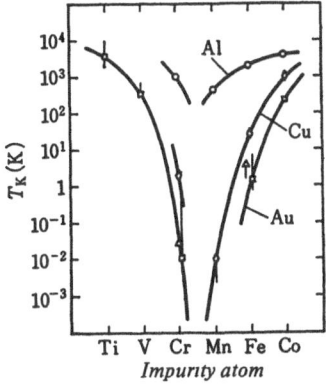

Fig. 9.1. The Kondo temperature T_K of dilute alloys which involve a small amount of the 3d-transition metal atoms. Mother metals are Cu($\diamond$), Ag($\triangle$), Au ($\square$), and Al($\circ$). [9.5]

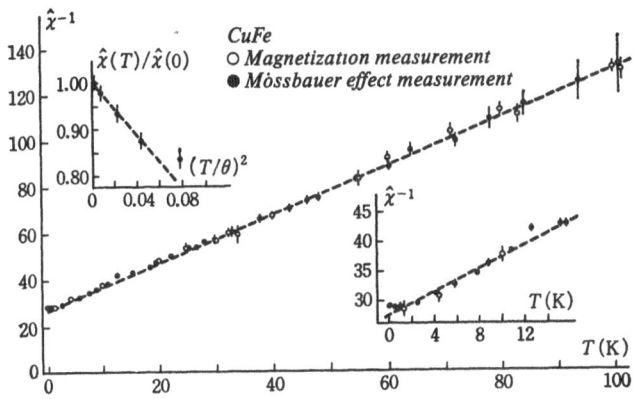

Fig. 9.2. Temperature-dependence of the magnetic susceptibility χ for CuFe. The inset shows a deviation of χ from the Curie-Weiss law at low temperatures and it has the form of $\chi \propto 1 - (T/\theta)^2$. [9.6]

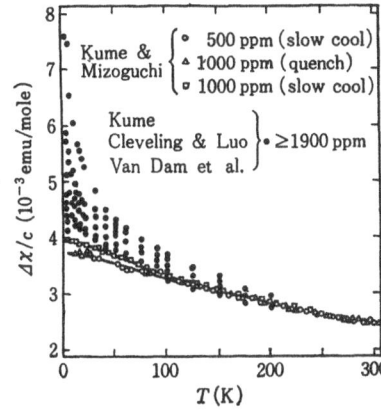

Fig. 9.3. Temperature-dependence of the magnetic susceptibility χ of AuV. χ is the difference between the observed and the pure Au susceptibilities. c denotes the concentration of V atoms. Filled circles are the previous data for samples with higher V atom concentrations. The data under 1000 ppm are from the recent experiments. [9.7]

V atom concentration down to ~500 ppm. One can summarize the magnetic susceptibility due to the d electron on magnetic impurity atoms as follows:

$$\left. \begin{array}{ll} T = 0\text{K}: & \chi_d = \dfrac{g^2 \mu_B^2}{k_B \theta_0} \\[2ex] T < T_K: & \chi_d \propto 1 - \left(\dfrac{T}{\theta_1}\right)^2 \\[2ex] T > T_K: & \chi_d \propto \dfrac{1}{T + \theta_2} \end{array} \right\}. \tag{9.2.1}$$

The temperatures θ_0, θ_1, and θ_2 are nearly equal each other and they provide approximate T_K. The contribution to the susceptibility from impurities is only a small part so that its estimation is not easy. Experiments of NMR or the Mössbauer effect may be profitable for the present purpose. In fact, it has been confirmed in NMR experiments that the Knight shift does not change for AuV in the 1000 ~ 10 ppm region. The internal field at Fe nucleus has also been observed for CuFe by the Mössbauer experiment and it is shown in Fig. 9.4. At high temperatures the internal field follows the Brillouin function but tends to a constant value at low temperatures in agreement with (9.2.1).

The specific heat due to the magnetic impurity atom is plotted in Fig. 9.5 for CuCr. It has a gentle peak at $T \sim T_K$ and continuously decreases at both sides. The dashed line is the extrapolated curve to lower temperatures by assuming a linear dependence on T. The entropy associated with this specific heat is very close to $\Delta S = R \ln(2S + 1)$ with $S = 3/2$. The electrical resistivity passes through a minimum with decreasing temperature, shows the $\ln T$ dependence and finally changes more gradually with temperatures to reach to a certain value. Such a common feature is shown in Fig. 9.6 for several dilute alloys. The observed resistivities have the form

$$R \propto 1 - \left(\frac{T}{\theta_R}\right)^2, \tag{9.2.2}$$

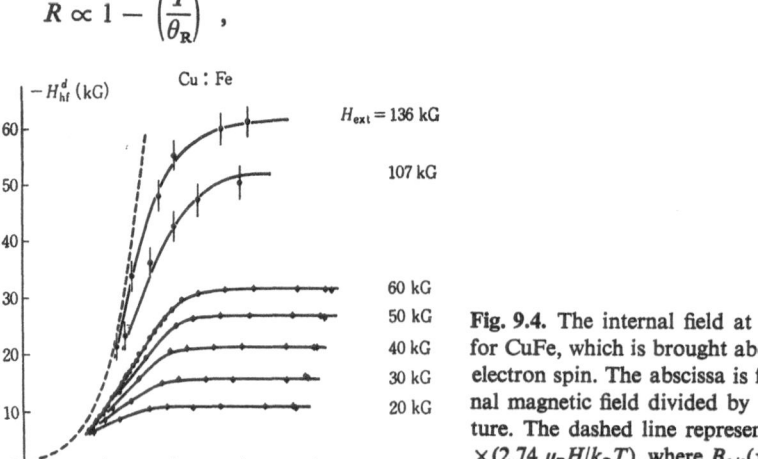

Fig. 9.4. The internal field at ^{57}Fe nucleus for CuFe, which is brought about by the d-electron spin. The abscissa is for the external magnetic field divided by the temperature. The dashed line represents $-151 B_{3/2} \times (2.74 \, \mu_B H/k_B T)$, where $B_{3/2}(x)$ is the Brillouin function with $S = 3/2$. [9.6]

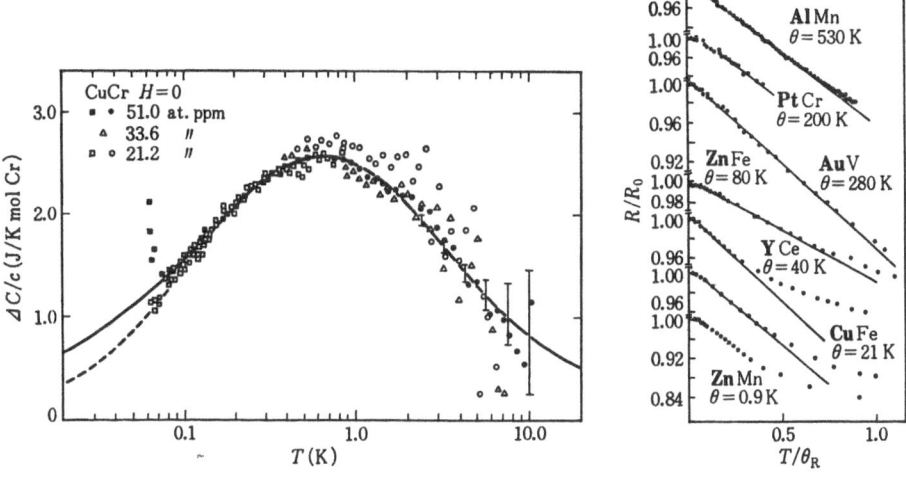

Fig. 9.5

Fig. 9.6

Fig. 9.5. The specific heat of CuCr at low temperatures. C represents the difference of the observed specific heat and the specific heat of Cu. c is the concentration of Cr atoms. [9.8]

Fig. 9.6. Temperature-dependence of electrical resistivity for several dilute alloys. θ_R is the temperature corresponding to T_K which is estimated from the resistivity. Resistivity values are normalized to unity at $T = 0$ K. [9.9]

where θ_R is again approximately equal to T_K. There are, in addition, other kinds of experiments including T_l in NMR, the electron fluctuation time τ_{elec} and the effect of magnetic impurity atoms in superconductors.

From the experimental data given above, it may be seen that, when a small amount of magnetic impurity atoms is doped in noble metals, the systems do not show any phase transition at a finite temperature and the spin associated with impurities loose the free spin state as $T \to 0$ K to reach to the singlet ground state. This agrees with what has been explained in Sect. 8.6. According to the theory based on the $s-d$ model, the magnetic susceptibility of the impurity atom [with $S = 1/2$] in the singlet ground state at $T = 0$ K is given by (9.2.1) with $\theta_0 = 4T_K$;

$$\chi_d = \frac{\mu_B^2}{k_B T_K}. \tag{9.2.3}$$

9.3 Yosida–Yamada Theory

The Yosida–Yamada theory not only provides the knowledge on the temperature-dependence of physical quantities associated with the d electron of the magnetic impurity atoms, but also the information about the relationship between the physical quantities and the localization of the d electron. As explained before, the s–d model is a limiting case of the Anderson model and in reality U may take an arbitrary value. In order to avoid a mathematical complication, let us consider only the case of $E_d = -U/2$ in (9.1.1) and take the origin of energy at the Fermi level ($\varepsilon_F = 0$). In such cases the electron-hole symmetry is present for the d orbital and the impurity atom is always in a definite valency to give $\langle n_{d\uparrow} + n_{d\downarrow}\rangle = 1$. Therefore, (9.1.1) can be rewritten as

$$
\left.
\begin{aligned}
\mathcal{H} &= \mathcal{H}_0 + \mathcal{H}' \\
\mathcal{H}_0 &= \sum_{k\sigma} \varepsilon_k n_{k\sigma} + \sum_{k\sigma} (V_{kd} c_{k\sigma}^\dagger c_{d\sigma} + V_{dk} c_{dk}^\dagger c_{k\sigma}) - \frac{U}{4} \\
\mathcal{H}' &= U\,(n_{d\uparrow} - \tfrac{1}{2})\,(n_{d\downarrow} - \tfrac{1}{2})
\end{aligned}
\right\}.
\tag{9.3.1}
$$

$\mathcal{H}'$ will be taken as the perturbation. (If there is no electron-hole symmetry, some of the following results are still valid but others need some modifications.). The unperturbed thermal Green's function for the d electron is given by

$$
G_{jl} = - \langle T_\tau c_{d\sigma}(\tau_j)\, c_{d\sigma}^\dagger(\tau_l)\rangle ,
\tag{9.3.2}
$$

where T_τ is the (imaginary) time-ordering operator with which the operator with greater τ is put on the left, and for an arbitrary operator O, $O(\tau) = \exp(\mathcal{H}_0 \tau)\, O \exp(-\mathcal{H}_0 \tau)$ and $\langle O \rangle = \mathrm{Tr}\{\exp(-\beta\mathcal{H}_0)\,O\}/\mathrm{Tr}\{\exp(-\beta\mathcal{H}_0)\}$. Using the Fourier-transformed form it becomes

$$
\left.
\begin{aligned}
G_{jl} &= \frac{1}{\beta} \sum_{\omega_n} G(\omega_n) \exp(-i\omega_n \tau_{jl}) \\
G(\omega_n) &= \frac{1}{i(\omega_n + \Delta\,\mathrm{sgn}\{\omega_n\})}
\end{aligned}
\right\},
\tag{9.3.3}
$$

where $\tau_{jl} = \tau_j - \tau_l$, $\omega_n = (2n+1)\pi/\beta$ (note that $\varepsilon_F = 0$), and $\Delta = \pi\rho|V|^2$ (V_{kd} is written simply as V). This form may easily be guessed from (8.6.3).

Let us consider the free energy F which is given by

$$
e^{-\beta F} = \mathrm{Tr}\{e^{-\beta\mathcal{H}}\} \quad (\beta = 1/k_B T),
\tag{9.3.4}
$$

where $\mathcal{H}$ is given in (9.3.1). The well-known formula

$$
\mathrm{Tr}\{e^{-\beta\mathcal{H}}\} = Z_0 \left\{ 1 + \sum_{n=1}^{\infty} (-1)^n \int_0^\beta d\tau_n \int_0^{\tau_n} d\tau_{n-1} \right.
$$
$$
\left. \cdots \int_0^{\tau_2} d\tau_1 \, \langle \mathcal{H}'(\tau_n)\, \mathcal{H}'(\tau_{n-1}) \cdots \mathcal{H}'(\tau_1)\rangle \right\}
\tag{9.3.5}
$$

will be used. Here, $Z_0 = \mathrm{Tr}\{\exp(-\beta\mathcal{H}_0)\}$. As explained before, the present system does not show any phase transition so that the spin ↑ and ↓ states are even and Green's function in (9.3.2) has the expression independent of σ [this is the reason why the index σ has been dropped in (9.3.2)]. Thus, the condition of the electron-hole symmetry can be expressed as

$$\langle c_{d\sigma}^\dagger c_{d\sigma}\rangle = \langle n_{d\sigma}\rangle = \frac{1}{2}. \tag{9.3.6}$$

From the Wick theorem, $\langle \mathscr{H}'(\tau_n) \dots \mathscr{H}'(\tau_1)\rangle$ in (9.3.5) is given as

$$\langle [c_{d\uparrow}^\dagger(\tau_n) c_{d\uparrow}(\tau_n) - \tfrac{1}{2}][c_{d\downarrow}^\dagger(\tau_n) c_{d\downarrow}(\tau_n) - \tfrac{1}{2}] \times \dots$$
$$\times [c_{d\uparrow}^\dagger(\tau_1) c_{d\uparrow}(\tau_1) - \tfrac{1}{2}][c_{d\downarrow}^\dagger(\tau_1) c_{d\downarrow}(\tau_1) - \tfrac{1}{2}]\rangle$$
$$= \left[\sum_P{}' (-1)^P \langle T_\tau c_{d\uparrow}^\dagger(\tau_n) c_{d\uparrow}(\tau_{P[n]})\rangle \langle T_\tau c_{d\uparrow}^\dagger(\tau_{n-1}) c_{d\uparrow}(\tau_{P[n-1]})\rangle \dots \right.$$
$$\left. \times \langle T_\tau c_{d\uparrow}^\dagger(\tau_1) c_{d\uparrow}(\tau_{P[1]})\rangle \right]^2, \tag{9.3.7}$$

where P denotes a permutation among n τ's, $P[m]$ the mth number in a certain permutation P, and $\sum_P{}'$ indicates the sum over all possible P's without including those P which involve $P[m] = m$. It is clear from (9.3.6) that the P's which involve $P[m] = m$ do not contribute in (9.3.7). Thus, one gets from (9.3.4,5)

$$\exp(-\beta F) = \exp(-\beta F_0) \sum_{n=0}^{\infty} \frac{(-1)^n U^n}{n!}$$
$$\cdot \int_0^\beta d\tau_n \dots \int_0^\beta d\tau_1 [D^{(n)}(1,2,\dots,n)]^2, \tag{9.3.8}$$

where $\exp(-\beta F_0) = Z_0$ and $D^{(n)}$ is the $n \times n$ antisymmetric determinant which is given by

$$D^{(n)}(1,2,\dots,n) = \begin{vmatrix} 0 & G_{12} & G_{13} & \dots\dots & G_{1n} \\ G_{21} & 0 & G_{23} & \dots\dots & G_{2n} \\ \vdots & & & & \vdots \\ G_{n1} & G_{n2} & G_{n3} & \dots\dots & 0 \end{vmatrix}. \tag{9.3.9}$$

From the antisymmetry, $D^{(n)}$ is identically zero if n is odd. Taking the logarithm of (9.3.8), one has the contributions only from the connected diagram and the free energy can be expressed as

$$F = F_0 - \sum_{n=1}^{\infty} \frac{(-1)^n U^n}{n!} \frac{1}{\beta} \int_0^\beta d\tau_1 \dots \int_0^\beta d\tau_n [D^{(n)}(1,2,\dots,n)]^2_{\text{conn.}}. \tag{9.3.10}$$

In order to obtain the low temperature specific heat associated with the impurity, expand F in power of T and retain only up to the T^2 term. For this purpose, let us expand the one-particle Green's function G_{jl} involved in $D^{(n)}$ in powers of T. Now, if a function $f(\omega)$ has a discontinuity at $\omega = 0$, its sum with respect to ω_n can be expanded in powers of T^{2n} as follows:

$$\frac{1}{\beta} \sum_{\omega_n} f(\omega_n) = \frac{1}{2\pi} \int_{-\infty}^{\infty} d\omega f(\omega) - \frac{1}{12\pi} \left(\frac{\pi}{\beta}\right)^2 [f'(0_-) - f'(0_+)] + \cdots . \quad (9.3.11)$$

In fact, $G(\omega)$ has a discontinuity at $\omega = 0$ so that (9.3.11) can be applied to G_{jl} to obtain the T^2 term corresponding to the second term of (9.3.11)

$$(\text{the } T^2 \text{ term of } G_{jl}) = -\frac{1}{6}\left(\frac{\pi}{\beta}\right)^2 \frac{1}{\pi\Delta} \tau_{jl} . \quad (9.3.12)$$

In order to obtain the T^2 term associated with the impurity in (9.3.10), let us expand one of the $D^{(n)}$'s into a sum of products of G_{jl} and its cofactor and use (9.3.12) to the factorized G_{jl}. The contribution to the free energy from this part is given as

$$F^{(2)} = -\frac{1}{3}\left(\frac{\pi}{\beta}\right)^2 \frac{1}{\pi\Delta}\left\{1 - \sum_{n=1}^{\infty} \frac{(-1)^n U^n}{n!} \right.$$
$$\left. \cdot \lim_{\beta \to \infty} \frac{1}{\beta} \int_0^\beta d\tau_1 \cdots \int_0^\beta d\tau_n \sum_{ij} [\tau_{jl} D_{ji}^{(n)} D^{(n)}]_{\text{conn}} \right\}, \quad (9.3.13)$$

where $D_{ij}^{(n)}$ is the cofactor of $D^{(n)}$ against G_{jl}. This first term is the contribution from F_0. As only the connected diagrams are taken in the second term of (9.3.13), this factor does not depend on temperature. Therefore, the low temperature specific heat is given by

$$C_V = -\frac{2}{3}\pi^2 k_B^2 \frac{T}{\pi\Delta}\left\{1 - \sum_{n=1}^{\infty} \frac{(-1)^n U^n}{n!} \right.$$
$$\left. \cdot \lim_{\beta \to \infty} \frac{1}{\beta} \int_0^\beta d\tau_1 \cdots \int_0^\beta d\tau_n \sum_{ij} [\tau_{jl} D_{ji}^{(n)} D^{(n)}]_{\text{conn}} \right\}. \quad (9.3.14)$$

This indicates that the specific heat associated with the d electron at the impurity is proportional to T and its coefficient is a function of Δ and U.

Let us next consider the susceptibility χ_d due to the d electron at the magnetic impurity atom. It is expressed as

$$\chi_d = \left(\frac{g\mu_B}{2}\right)^2 \frac{1}{\beta} \int_0^\beta \int_0^\beta d\tau \, d\tau' \langle\!\langle T_\tau [\tilde{n}_{d\uparrow}(\tau) - \tilde{n}_{d\downarrow}(\tau)][\tilde{n}_{d\uparrow}(\tau') - \tilde{n}_{d\downarrow}\tau')]\rangle\!\rangle$$
$$\equiv \chi_{\uparrow\uparrow} + \chi_{\uparrow\downarrow}, \quad (9.3.15)$$

where $\langle\!\langle O \rangle\!\rangle = \text{Tr}\{e^{-\beta\mathcal{H}}O\}/\text{Tr}\{e^{-\beta\mathcal{H}}\}$ and $\tilde{n}_{d\sigma} = n_{d\sigma} - 1/2$. In (9.3.15), $\chi_{\uparrow\uparrow}$ and $\chi_{\uparrow\downarrow}$ are defined as

$$\chi_{\uparrow\uparrow} = \frac{(g\mu_B)^2}{2} \frac{1}{\beta} \int_0^\beta \int_0^\beta d\tau \, d\tau' \langle\!\langle T_\tau \tilde{n}_{d\uparrow}(\tau) \, \tilde{n}_{d\uparrow}(\tau') \rangle\!\rangle$$

$$\chi_{\uparrow\downarrow} = -\frac{(g\mu_B)^2}{2} \frac{1}{\beta} \int_0^\beta \int_0^\beta d\tau \, d\tau' \langle\!\langle T_\tau \tilde{n}_{d\uparrow}(\tau) \tilde{n}_{d\downarrow}(\tau') \rangle\!\rangle$$

(9.3.16)

A similar calculation for the free energy can be made for χ. Here, however, the extra factor $\tilde{n}_{d\sigma}$ is present which can be absorbed into $D^{(n)}$. For $\chi_{\uparrow\uparrow}$, the two $\tilde{n}_{d\uparrow}$'s are incorporated with one $D^{(n)}$ to give

$$\chi_{\uparrow\uparrow} = \frac{(g\mu_B)^2}{2} \sum_{n=0}^\infty \frac{(-1)^n U^n}{n!} \frac{1}{\beta} \int_0^\beta \cdots \int_0^\beta d\tau \, d\tau' d\tau_1 \ldots d\tau_n$$
$$\cdot [D^{(n+2)}(\tau, \tau', 1, \ldots, n) \, D^{(n)}(1, 2, \ldots, n)]_{\text{conn}} .$$

(9.3.17)

and for $\chi_{\uparrow\downarrow}$, each $\tilde{n}_{d\sigma}$ is incorporated with each $D^{(n)}$ to give

$$\chi_{\uparrow\downarrow} = -\frac{(g\mu_B)^2}{2} \sum_{n=1}^\infty \frac{(-1)^n U^n}{n!} \frac{1}{\beta} \int_0^\beta \cdots \int_0^\beta d\tau \, d\tau' d\tau_1 \ldots d\tau_n$$
$$\cdot [D^{(n+1)}(\tau, 1, \ldots, n) \, D^{(n+1)}(\tau', 1, \ldots, n)]_{\text{conn.}} .$$

(9.3.18)

Here, $D^{(n+2)}(\tau, \tau', 1, \ldots, n)$, for instance, is given by

$$D^{(n+2)}(\tau, \tau', 1, \ldots, n) = \begin{vmatrix} 0 & G_{\tau\tau'} & G_{\tau 1} & \cdots\cdots & G_{\tau n} \\ G_{\tau'\tau} & 0 & G_{\tau' 1} & \cdots\cdots & G_{\tau' n} \\ \vdots & & & & \vdots \\ G_{n\tau} & G_{n\tau'} & G_{n1} & \cdots\cdots & 0 \end{vmatrix} .$$

(9.3.19)

Because of the antisymmetry of $D^{(n)}$, $\chi_{\uparrow\uparrow}$ involves only even powers of U and $\chi_{\uparrow\downarrow}$ only odd ones. The first term n (9.3.17) ($n = 0$) is $\chi_{\uparrow\uparrow}^{(0)} = 1/\pi\Delta$. The prescription of the succeeding calculation is the following: take out the Green's function involving τ and τ' from $D^{(n+2)}$ and $D^{(n+1)}$ and reduce them to $D^{(n)}$ or cofactors of the Green's function. For instance, $D^{(n+2)}$ can be expressed as

$$D^{(n+2)} = - G_{\tau\tau'}G_{\tau'\tau}D^{(n)}$$
$$+ \sum_{ij} (G_{\tau\tau'}G_{j\tau}G_{\tau'i}D^{(n)}_{ji} + G_{\tau'i}G_{\tau i}G_{j\tau'}D^{(n)}_{ji} + \sum_{i'j'} G_{\tau i}G_{j\tau}G_{\tau'i'}G_{j'\tau'}D^{(n)}_{jij'i'}) .$$

(9.3.20)

Substituting this into (9.3.17), the terms involving $G(\omega_n)^3$ and $G(\omega_n)^2 \cdot G(\omega_{n'})^2$ appear. When one is interested in the susceptibility at $T \to 0$ K, he can make the replacement $\beta^{-1} \sum_n \to (2\pi)^{-1} \int_{-\infty}^\infty d\omega$. The following equality can be obtained by integrating by parts

$$\frac{1}{\pi} \int_{-\infty}^\infty d\omega \, G(\omega)^3 \exp(-i\omega\tau_{ji})$$

$$= -\frac{1}{2\pi} \tau_{jt} \int_{-\infty}^{\infty} d\omega \, G(\omega)^2 \exp(-i\omega\tau_{jt}) \tag{9.3.21}$$

with which the power of $G(\omega)$ can be reduced. Furthermore, the equalities

$$\left.\begin{aligned}
\tau_{jt} D_{jt}^{(n)} &= -\sum_{i'j'} \tau_{j'i'} G_{j'i'} D_{jij'i'}^{(n)} \\
\tau_{jt} G_{jt} &= -\frac{1}{\pi\Delta} - \frac{1}{2\pi} \int d\omega \, G(\omega)^2 \exp(-i\omega\tau_{jt})
\end{aligned}\right\} \tag{9.3.22}$$

can be utilized to erase $D_{jij'i'}^{(n)}$ in (9.3.20). Thus, $\chi_{\uparrow\uparrow}$ at $T \to 0$ K is given as

$$\chi_{\uparrow\uparrow} \Big/ \frac{(g\mu_B)^2}{2}$$

$$= \frac{1}{\pi\Delta}\left\{1 - \sum_{n=1}^{\infty} \frac{(-1)^n U^n}{n!} \lim_{\beta\to\infty} \frac{1}{\beta} \int_0^\beta \ldots \int_0^\beta d\tau_1 \ldots d\tau_n \sum_{ij} [\tau_{jt} D_{jt}^{(n)} D^{(n)}]_{\mathrm{conn}}\right\}. \tag{9.3.23}$$

In a similar way, $\chi_{\uparrow\downarrow}$ can also be calculated to give

$$\chi_{\uparrow\downarrow} \Big/ \frac{(g\mu_B)^2}{2}$$

$$= -\frac{1}{\pi\Delta}\sum_{n=1}^{\infty} \frac{(-1)^n U^n}{n!} \lim_{\beta\to\infty} \frac{1}{\beta} \int_0^\beta \ldots \int_0^\beta d\tau_1 \ldots d\tau_n \sum_{ijl'j'} [D_{jt}^{(n)} D_{jl'j'}^{(n)}]_{\mathrm{conn}}. \tag{9.3.24}$$

If (9.3.23) is compared with (9.3.14), it turns out that the quantities in $\{\ldots\}$ are exactly the same in both equations so that the following equality holds:

$$\frac{\chi_{\uparrow\uparrow}T}{C_V} = \frac{3}{4}\frac{(g\mu_B)^2}{(\pi k_B)^2}. \tag{9.3.25}$$

Using $\chi_d = \chi_{\uparrow\uparrow} + \chi_{\uparrow\downarrow}$, this can be rewritten as

$$\frac{\chi_d T}{C_V} = \frac{3}{4}\frac{(g\mu_B)^2}{(\pi k_B)^2}\left(1 + \frac{\chi_{\uparrow\downarrow}}{\chi_{\uparrow\uparrow}}\right). \tag{9.3.26}$$

This is an important expression because it shows how the ratio of χ to C_V varies as a function of U. When $U = 0$, $\chi_{\uparrow\downarrow} = 0$ from (9.3.18). On the other hand, $\chi_{\uparrow\uparrow}$ and $\chi_{\uparrow\downarrow}$ are the even and the odd functions with respect to U so that

$$\chi_d(-U) = \chi_{\uparrow\uparrow}(U) - \chi_{\uparrow\downarrow}(U). \tag{9.3.27}$$

In the limiting case of $U \to \infty$ which corresponds to the s–d model, the left-hand side of (9.3.27) must vanish because the state with the filled d orbital is stable and the magnetic state with single d electrons cannot take place. Thus, $\chi_{\uparrow\uparrow} = \chi_{\uparrow\downarrow}$ when $U \to \infty$. To summarize, $(1 + \chi_{\uparrow\downarrow}/\chi_{\uparrow\uparrow})$ in (9.3.26) is equal to 1 when $U = 0$ and to 2 when $U \to \infty$ and it may vary continuously between these two values with U. In order to obtain more definite trends of $\chi_{\uparrow\uparrow}$ and $\chi_{\uparrow\downarrow}$, the perturbation calculation has been carried out up to the fourth order of U and they are given, under the notation of $\chi_d = (g^2\mu_B^2/2\pi\Delta)(\tilde{\chi}_{\uparrow\uparrow} + \tilde{\chi}_{\uparrow\downarrow})$, as

$$\tilde{\chi}_{\uparrow\uparrow} = 1 + \left(3 - \frac{\pi^2}{4}\right)\left(\frac{U}{\pi\Delta}\right)^2 + 0.0553\left(\frac{U}{\pi\Delta}\right)^4 + \cdots$$

$$\tilde{\chi}_{\uparrow\downarrow} = \frac{U}{\pi\Delta} + \left(15 - \frac{3\pi^2}{2}\right)\left(\frac{U}{\pi\Delta}\right)^3 + \cdots$$

Making use of this result, the variation of $\chi_{\uparrow\downarrow}/\chi_{\uparrow\uparrow}$ with respect to U is shown in Fig.9.7. This calculated value is decreasing at the large U region (the portion of the dashed curve) and this decrease may be due to neglect of the higher-order contribution. If these contributions are taken into account, it is expected that the curve tends to tanh $(U/\pi\Delta)$. In the limit of the s–d model, (9.2.3) can be used in (9.3.26) to yield

$$C_V = \frac{\pi^2}{6} k_B \frac{T}{T_K}. \tag{9.3.29}$$

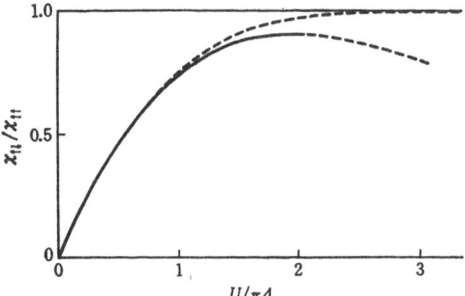

Fig. 9.7. Variation of $\chi_{\uparrow\downarrow}/\chi_{\uparrow\uparrow}$ as a function of U. The solid line is due to the perturbation calculation up to the fourth order of U and the decreasing part of the dashed line beyond this line is also the calculated one which is thought to be spurious. This decrease may be due to the neglect of higher-order contributions. The upper dashed line represents tanh$(U/\pi\Delta)$. [9.3]

Now, the electrical resistivity associated with the magnetic impurity atom will be investigated. The general formula for the electrical conductivity is expressed as

$$\sigma = -\tfrac{2}{3} e^2 \int d\varepsilon_k \tau_k v_k^2 \rho(\varepsilon_k) \frac{\partial f}{\partial \varepsilon_k}. \tag{9.3.30}$$

Our present task is to calculate the mean free time τ_k in the presence of the impurity. Here, $\rho(\varepsilon)$ is the density of states of conduction electrons. Using the t-matrix, τ_k is expressed as

$$\frac{1}{\tau_k(\omega)} = -2\,\mathrm{Im}\{t_{kk}(\omega)\}. \tag{9.3.31}$$

The t-matrix is given in terms of the scattering potential V and the retarded Green's function of the d-electron $G_{d\sigma}^R(\omega)$ as

$$t_{kk',\sigma}(\omega) = V_{kd}G_{d\sigma}^R(\omega) V_{dk'} \tag{9.3.32}$$

so that

$$\text{Im}\{t_{kk,\sigma}(\omega)\} = |V_{kd}|^2 \, \text{Im}\{G_{d\sigma}^R(\omega)\} = -\pi|V_{kd}|^2 N_{d\sigma}(\omega), \tag{9.3.33}$$

where $N_{d\sigma}(\omega) = (-1/\pi)\text{Im}\{G_{d\sigma}^R(\omega)\}$ is the density of states of the d electron. Thus, the problem is reduced to obtain $N_{d\sigma}(\omega)$. Assuming the electron-hole symmetry, the density of states in the unperturbed state is given, as can be seen from (9.3.3) or (8.6.5), as

$$N_{d\sigma}^{(0)}(\omega) = \frac{1}{\pi}\frac{\Delta}{\omega^2 + \Delta^2}, \tag{9.3.34}$$

where ω is the real frequency and $\hbar = 1$ has been taken. The self-energy in $G_{d\sigma}^R(\omega)$ must be calculated to obtain $N_{d\sigma}(\omega)$ in the presence of perturbation. As for the free energy or the susceptibility, the Green's function $\bar{G}_\sigma(\tau - \tau') = -\langle\!\langle T_\tau \tilde{c}_{d\sigma}(\tau) \tilde{c}_{d\sigma}^\dagger(\tau')\rangle\!\rangle$ ($\tilde{}$ denotes the Heisenberg representation) can be expanded to

$$\bar{G}_\sigma(\tau - \tau')$$
$$= \sum_{n=0}^{\infty} \frac{(-1)^n U^n}{n!} \int_0^\beta \cdots \int_0^\beta d\tau_1 \cdots d\tau_n [D_\sigma^{(n+1)}(\tau, \tau', 1, \ldots, n) D_{-\sigma}^{(n)}(1, \ldots, n)]_{\text{conn.}}$$
$$= G_{\tau\tau'} - \sum_{n=1}^{\infty} \frac{(-1)^n U^n}{n!} \int_0^\beta \cdots \int_0^\beta d\tau_1 \cdots d\tau_n \sum_{ij} [G_{\tau i}G_{j\tau'}D_{ji}^{(n)} D^{(n)}]_{\text{conn.}} , \tag{9.3.35}$$

where σ has been suppressed in the last equality because it has no particular σ-dependence. In the Fourier-transformed form it may be expressed as

$$\bar{G}(\omega_n) = G(\omega_n) + G(\omega_n) \Sigma'(\omega_n)G(\omega_n), \tag{9.3.36}$$

where the self-energy $\Sigma'(\omega_n)$ involves the reducible diagrams as well. The self-energy that involves only the irreducible diagrams can be calculated, and up to the fourth order of U and with the real frequency ω, this self-energy for the retarded Green's function is given by

$$\Sigma^R(\omega) = -\alpha_R\omega - i\alpha_I\Delta\left[\left(\frac{\omega}{\Delta}\right)^2 + \left(\frac{\pi k_B T}{\Delta}\right)^2\right] + \cdots, \tag{9.3.37}$$

where

$$\alpha_R = \left(3 - \frac{\pi^2}{4}\right)\left(\frac{U}{\pi\varDelta}\right)^2 + 0.0553\left(\frac{U}{\pi\varDelta}\right)^4 + \cdots \Bigg]$$

$$\alpha_I = \frac{1}{2}\left[\left(\frac{U}{\pi\varDelta}\right)^2 + 6\left(5 - \frac{\pi^2}{2}\right)\left(\frac{U}{\pi\varDelta}\right)^4 + \cdots\right]\Bigg].$$

(9.3.38)

Making a correction of this self-energy in Green's function and using the equality given below (9.3.33), the density of states for the d electron is expressed at small ω and T as

$$N_{d\sigma}(\omega) = \frac{1}{\pi\varDelta}\left\{1 - \alpha_I\left[\left(\frac{\omega}{\varDelta}\right)^2 + \left(\frac{\pi k_B T}{\varDelta}\right)^2\right] - (\alpha_R + 1)\left(\frac{\omega}{\varDelta}\right)^2 + \cdots\right\}. \quad (9.3.39)$$

If this expression is used in (9.3.30–33) the resistivity is given by

$$R = R_0\left\{1 - \frac{\pi^2}{3}\left(\frac{k_B T}{\varDelta}\right)^2 [4\alpha_I + (\alpha_R + 1)^2]\right\}. \quad (9.3.40)$$

Comparing (9.3.38) with (9.3.28), it holds up to the fourth order of U that

$$\left.\begin{array}{l}\alpha_R + 1 = \tilde{\chi}_{\uparrow\uparrow} \\ 2\alpha_I = \tilde{\chi}_{\uparrow\downarrow}^2\end{array}\right\} \quad (9.3.41)$$

so that (9.3.40) may alternatively expressed as

$$R = R_0\left[1 - \frac{\pi^2}{3}\left(\frac{k_B T}{\varDelta}\right)^2 (\tilde{\chi}_{\uparrow\uparrow}^2 + 2\tilde{\chi}_{\uparrow\downarrow}^2)\right]. \quad (9.3.42)$$

A more detailed examination indicates that (9.3.42) holds independently of the order of U. This form of R agrees with that of (9.2.2) and the comparison between them shows that, as $\chi_{\uparrow\uparrow} = \chi_{\uparrow\downarrow} = \chi/2$ holds in the s–d limit, θ_R in (9.2.2) is given by

$$\theta_R = \frac{4}{\pi^2} T_K. \quad (9.3.43)$$

This indicates that θ_R is also of the order of T_K.

The density of states of the d electron

$$N_{d\sigma}(\omega) = \frac{1}{\pi}\frac{\varDelta - \mathrm{Im}\{\Sigma^R(\omega)\}}{[\omega - \mathrm{Re}\{\Sigma^R(\omega)\}]^2 + [\varDelta - \mathrm{Im}\{\Sigma^R(\omega)\}]^2} \quad (9.3.44)$$

is plotted in Fig.9.8 by using (9.3.37,38) for Σ^R. As can be seen from the figure, a broad peak appears at $\omega \approx U/2$ when $U/\pi\varDelta \gtrsim 2$.

The present theory has been extended by *Yoshimori* [9.10] to the case where degeneracy of the d orbital is present. Reserving the electron-hole symmetry and taking the s–d limit ($U \to \infty$), (9.3.26) is replaced by

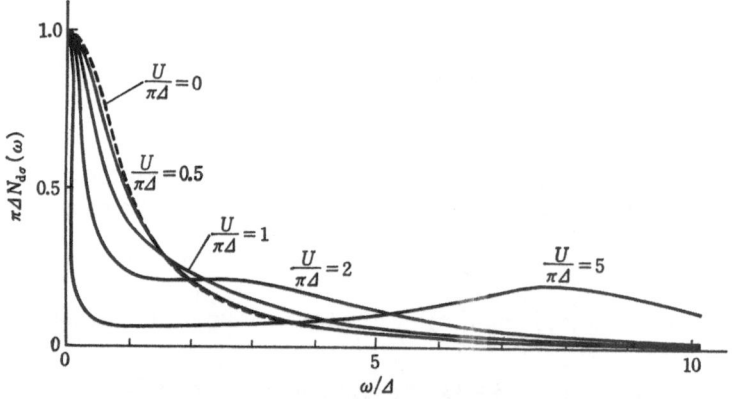

Fig. 9.8. Density of states of the d-electron, $N_{d\sigma}(\omega)$, for several values of $U/\pi\Delta$. When $U/\pi\Delta$ $\gtrsim 2$, a broad peak appears at high energy. [9.3]

$$\frac{\chi_d T}{C_V} = \frac{3}{4} \frac{(g\mu_B)^2}{(\pi k_B)^2} \frac{4}{3} (S+1),$$ (9.3.45)

where S is the magnitude of spin of the magnetic impurity. From this, the following equalities may be inferred in this limit:

$$\left. \begin{array}{l} \chi_d = \dfrac{S(S+1)}{3} \dfrac{(g\mu_B)^2}{k_B T_K} \\[3mm] C_V = \tfrac{1}{3} \pi^2 k_B S \dfrac{T}{T_K} \end{array} \right\}.$$ (9.3.46)

From what has been explained above, one can have the following picture for dilute alloys involving magnetic impurity atoms. Substituting (9.3.41) into (9.3.39), the density of states of the d electron can be expresed as

$$N_{d\sigma}(\omega) \cong \frac{1}{\pi\Delta} \left[1 - \frac{\pi^2}{2} \tilde{\chi}_{\uparrow\uparrow}^2 \left(\frac{k_B T}{\Delta}\right)^2 - (\tfrac{1}{2} \tilde{\chi}_{\uparrow\uparrow}^2 + \tilde{\chi}_{\uparrow\downarrow}^2)\left(\frac{\omega}{\Delta}\right)^2 + \cdots \right]$$ (9.3.47)

at small ω and T. Put $T = 0$ in (9.3.47). When $U = 0$, $\tilde{\chi}_{\uparrow\uparrow} = 1$ and $\tilde{\chi}_{\uparrow\downarrow} = 0$ so that (9.3.47) agrees with $N_{d\sigma}(\omega)$ in (9.3.34) by expanding in powers of ω/Δ. Thus, $N_{d\sigma}(\omega)$ has a peak at $\omega = 0$ and a width of $\sim\Delta$. When $U \to \infty$, using $\tilde{\chi}_{\uparrow\uparrow} = \tilde{\chi}_{\uparrow\downarrow}$ and (9.2.3) for the equation $\chi_d = (g^2\mu_B^2/2\pi\Delta)\,(\tilde{\chi}_{\uparrow\uparrow} + \tilde{\chi}_{\uparrow\downarrow})$, one obtains $\tilde{\chi}_{\uparrow\uparrow} = \tilde{\chi}_{\uparrow\downarrow}$ $= \pi\Delta/4k_B T_K$. Substituting this into (9.3.47), the width of $N_{d\sigma}(\omega)$ turns out to be $\sim k_B T_K$. Such behavior is sketched in Fig.9.9, where the curve, monotonically decreasing with U, represents the width $\delta\omega_{1/2}$ of the peak in $N_{d\sigma}(\omega)$. When U is increased, an additional peak appears at $\omega \approx U/2$ and its position is indicated by the straight line in Fig.9.9. If the ordinate is translated into a temperature, the lowest temperature region is the nonmagnetic singlet state. The left-hand side

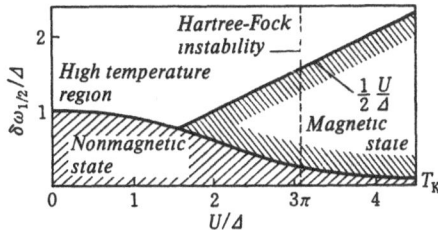

Fig. 9.9. Half width $\delta\omega_{1/2}$ of the peak in $N_{d\sigma}(\omega)$ as a function of U. The region below the curve stands for the non-magnetic singlet state and the straight line represents the position where an additional peak appears. [9.11]

of this region corresponds simply to the state in which electrons partially occupy the extended d state in the degenerate distribution according to the Fermi–Dirac statistics and make the singlet state as a whole, whereas the extreme right-hand side corresponds to the singlet bound state which is provided by the localized spin and the conduction electrons through the s–d Hamiltonian. Both types of the singlet state change continuously with U. The region bounded by the above-mentioned curve and the straight line correspond to the magnetic state obtained in Sect. 6.6 and the threshold of this magnetic state obtained in the Hartree approximation is indicated by the dashed line. In reality, each state has not such sharp boundaries and the solid curve and the solid line give only rough measures for each region. From Fig. 9.9, it may be seen that impurities in metals can be classified into two groups: either there is no magnetic state at any temperature; or a magnetic state is present at $U/2k_B > T > T_K$.

9.4 Renormalization Group Approach

There have been further developments in the theory of the Kondo problem. The new developments are noticeable in the asymmetric case ($E_d \neq -U/2$) of the Anderson Hamiltonian and in the "exact" solution of the s–d Hamiltonian.

The asymmetric case of the Anderson Hamiltonian may be particularly interesting in connection with the valence fluctuations which have actually been observed in rare-earth metals and their compounds, for instance. A number of discussions of this Hamiltonian have been done along compounds. The new developments in the s–d Hamiltonian include the "exact" solution at T = 0 K based on the Bethe ansatz. Since present interest has been mainly in the Anderson Hamiltonian which may cover a wider range of real systems, a brief outline of the work of *Krishna murthy* et al.[9.12] will be given below. It relies on the renormalization group and is consistent with the main results of Haldane's investigation (see [9.13]) It discusses various features of the system described by the Anderson Hamiltonian over all temperatures and the whole range of the parameter space (U, Δ), Δ being defined below (9.3.3).

The Hamiltonian (9.1.1) is rewritten as

$$\mathscr{H} = -\tfrac{1}{2}U + \sum_{k\sigma} \varepsilon_k n_{k\sigma} + \sum_{\sigma}(E_d + \tfrac{1}{2}U)\,n_{d\sigma}$$

$$+ \sum_{k\sigma}(V_{kd}c^\dagger_{k\sigma}c_{d\sigma} + V_{dk}c^\dagger_{d\sigma}c_{k\sigma}) + \tfrac{1}{2}U(\sum_\sigma n_{d\sigma} - 1)^2 \,. \tag{9.4.1}$$

Let us further simplify the model as follows; the conduction band is isotropic and extends in energy from $-D$ to D with a constant density of states; V_{kd} is taken as constant; the basis states for the conduction electrons are expanded in terms of spherical waves about impurity and take only the s-wave that couples to impurity. Now, label the s-wave states by energy ε and write $k = \varepsilon/D$. The creation operator of an s wave is then written as $a^\dagger_{k\sigma}$ which satisfies $\{a_{k\sigma}, a^\dagger_{k'\sigma'}\} = \delta_{\sigma\sigma'}\delta(k - k')$. In order to proceed along the renormalization group, let us introduce a logarithmic discretization of k-space, as shown in Fig.9.10, where $\varLambda > 1$. Let us further define a complete set of orthonormal functions in the k-space by the Fourier series in each interval:

$$\psi^\pm_{np}(k) = \begin{cases} \dfrac{\varLambda^{n/2}}{(1 - \varLambda^{-1})^{1/2}} \exp(\pm i\omega_n pk) & \text{for } \varLambda^{-(n+1)} < \pm k < \varLambda^{-n}, \\ 0 & \text{otherwise,} \end{cases} \tag{9.4.2}$$

where interval index n takes $0,1,2,\ldots$, the Fourier harmonic index p takes on all integral values from $-\infty$ to $+\infty$, and

$$\omega_n = \frac{2\pi}{\varLambda^{-n} - \varLambda^{-(n+1)}} = \frac{2\pi\varLambda^n}{1 - \varLambda^{-1}} \tag{9.4.3}$$

is the fundamental Fourier frequency in the nth interval. The superscript $\pm$ in (9.4.2) stands for positive and negative k. Using $\psi_{np}(k)$, $a_{k\sigma}$ can be expanded to

$$a_{k\sigma} = \sum_{np}[a_{np\sigma}\psi^\dagger_{np}(k) + b_{np\sigma}\psi^-_{np}(k)] \,. \tag{9.4.4}$$

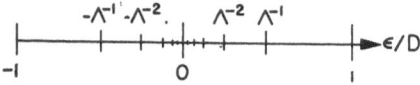

Fig. 9.10. Logarithmic discretization of the conduction band. The Fermi level is at zero and the band spreads from $-D$ to D. [9.12]

It is a good approximation to retain only the terms containing a_{np} and b_{np} with $p = 0$ in the term of $\sum_{k\sigma}\varepsilon_k n_k$ in (9.4.1), and the subscript p will be dropped hereafter. The Hamiltonian (9.4.1) is then written with the new operators as

$$\mathscr{H}/D = \tfrac{1}{2}(1 + \varLambda^{-1})\sum_{n=0}^\infty \sum_\sigma \varLambda^{-n}(a^\dagger_{n\sigma}a_{n\sigma} - b^\dagger_{n\sigma}b_{n\sigma}) + \sum_\sigma \frac{1}{D}(E_d + \tfrac{1}{2}U)c^\dagger_{d\sigma}c_{d\sigma}$$

$$+ \left(\frac{2\varDelta}{\pi D}\right)^{1/2}\sum_\sigma(f^\dagger_{0\sigma}c_{d\sigma} + c^\dagger_{d\sigma}f_{0\sigma}) + \frac{1}{2}\frac{U}{D}(\sum_\sigma c^\dagger_{d\sigma}c_{d\sigma} - 1)^2 \,, \tag{9.4.5}$$

where

$$f_{0\sigma} = [\tfrac{1}{2}(1 - \Lambda^{-1})]^{1/2} \sum_{n=0}^{\infty} \Lambda^{-n/2}(a_{n\sigma} + b_{n\sigma}) . \tag{9.4.6}$$

Let us further make a unitary transformation from $(a_{n\sigma}, b_{n\sigma})$ to a new orthonormal set $(f_{n\sigma})$ such that $f_{0\sigma}$ is still given by (9.4.6) and the first term of (9.4.5) exhibits nearest-neighbor coupling. Such a transformation was explicitly shown by *Wilson* in his previous paper [9.2], and the Hamiltonian (9.4.5) is expressed as

$$\mathcal{H} = \lim_{N\to\infty} \tfrac{1}{2}(1 + \Lambda^{-1}) D\Lambda^{-(N-1)/2} \mathcal{H}_N , \tag{9.4.7}$$

where

$$\mathcal{H}_N = \Lambda^{(N-1)/2}\left[\sum_{n=0}^{N-1}\sum_{\sigma} \Lambda^{-n/2} \xi_n(f_{n\sigma}^{\dagger} f_{n+1\sigma} + f_{n+1\sigma}^{\dagger} f_{n\sigma}) + \sum_{\sigma}\tilde{\delta}_d c_{d\sigma}^{\dagger} c_{d\sigma}\right.$$
$$\left. + \sum_{\sigma}\tilde{\Delta}^{1/2}(f_{0\sigma}^{\dagger} c_{d\sigma} + c_{d\sigma}^{\dagger} f_{0\sigma}) + \tilde{U}(\sum_{\sigma} c_{d\sigma}^{\dagger} c_{d\sigma} - 1)^2\right] \tag{9.4.8}$$

with

$$\xi_n = (1 - \Lambda^{-n-1})(1 - \Lambda^{-2n-1})^{-1/2}(1 - \Lambda^{-2n-3})^{-1/2} ,$$
$$\tilde{\delta}_d = \frac{2}{1 + \Lambda^{-1}}\frac{1}{D}(E_d + \tfrac{1}{2}U) = \tilde{E}_d + \tilde{U} , \tag{9.4.9}$$
$$\tilde{\Delta} = \left(\frac{2}{1 + \Lambda^{-1}}\right)^2 \frac{2\Delta}{\pi D} .$$

It is apparent that $\tilde{\delta}_d = 0$ in the symmetric case. The factor $\Lambda^{(N-1)/2}$ in (9.4.8) makes the lowest energy scale (1st term) of the order of 1. Corresponding to a given Hamiltonian $\mathcal{H}_N$, let us introduce the spin operator S_N and the charge operator Q_N:

$$S_N = \tfrac{1}{2}\sum_{n=0}^{N}\sum_{\sigma\sigma'} f_{n\sigma}^{\dagger}\sigma_{\sigma\sigma'} f_{n\sigma'} + \sum_{\sigma\sigma'}\tfrac{1}{2} c_{d\sigma}^{\dagger}\sigma_{\sigma\sigma'} c_{d\sigma'} ,$$
$$Q_N = \sum_{n=0}^{N}(\sum_{\sigma} f_{n\sigma}^{\dagger} f_{n\sigma} - 1) + (\sum_{\sigma} c_{d\sigma}^{\dagger} c_{d\sigma} - 1) . \tag{9.4.10}$$

As these operators commute with $\mathcal{H}_N$, the eigenstates of $\mathcal{H}_N$ can be chosen to be also the eigenstates of Q_N, $(S_N)^2$, and S_{Nz} and they can be labeled by the corresponding quantum number Q, S, and m_s. The energy eigenvalues are independent of m_s and this may be dropped. The sequence of the Hamiltonian $\mathcal{H}_N$ provides the recursion formula

$$\mathcal{H}_{N+1} = \Lambda^{1/2}\mathcal{H}_N + \xi_N(f_{N\sigma}^{\dagger} f_{N+1\sigma} + f_{N+1\sigma}^{\dagger} f_{N\sigma}) - E_{G,N+1} , \tag{9.4.11}$$

where $E_{G,N+1}$, the ground state energy of $\mathcal{H}_{N+1}$, is substracted for convenience. This recursion formula can be utilized in obtaining many-body eigenstates and energy levels of $\mathcal{H}_{N+1}$ from those of $\mathcal{H}_N$. The diagonalization of the Hamiltonian can be carried out in each (Q, S, m_s) subspace.

In calculating thermodynamic quantities, one needs the operator

$$\exp\left[-\beta \tfrac{1}{2}(1 + \Lambda^{-1}) D \Lambda^{-(N-1)/2} \mathcal{H}_N\right] = \exp(-\bar{\beta}_N \mathcal{H}_N), \tag{9.4.12}$$

where

$$\bar{\beta}_N = \tfrac{1}{2}(1 + \Lambda^{-1}) \Lambda^{-(N-1)/2} \frac{D}{k_B T}. \tag{9.4.13}$$

Now, let us pick a fixed number $\bar{\beta}$ and choose an N such that $\bar{\beta}_N = \bar{\beta}$. From (9.4.13) this gives

$$T_N = \frac{D}{k_B} \frac{1}{2}(1 + \Lambda^{-1}) \Lambda^{-(N-1)/2} \frac{1}{\bar{\beta}}. \tag{9.4.14}$$

The impurity contribution to the susceptibility at the temperature T_N can be expressed as

$$k_B T_N \chi(T_N) = (g\mu_B)^2 \left[\frac{\text{Tr}\{(S_{Nz})^2 \exp(-\bar{\beta}\mathcal{H}_N)\}}{\text{Tr}\{\exp(-\bar{\beta}\mathcal{H}_N)\}} - \frac{\text{Tr}\{(S_{Nz}^0)^2 \exp(-\bar{\beta}\mathcal{H}_N^0)\}}{\text{Tr}\{(\exp(-\bar{\beta}\mathcal{H}_N^0)\}}\right], \tag{9.4.15}$$

where the second term is the quantity without impurity and the g factor is taken to be equal for the d-electron and the conduction electron. Thus, the index N defines a logarithmic temperature scale.

When (9.4.11) is formally expressed as $\mathcal{H}_{N+1} = \mathcal{T}[\mathcal{H}_N]$, the fixed point may be given by $\mathcal{T}[\mathcal{H}^*] = \mathcal{H}^*$. $\mathcal{T}$ does not have any fixed point but $\mathcal{T}^2$ does. However, as the present system does not show any phase transition, fixed points appear in the course of the renormalization-group transformation in the form that $\mathcal{T}^2[\mathcal{H}_N]$ is very close to $\mathcal{H}_N$ itself.

An example of the symmetric case is shown in Fig.9.11 for the case of odd N. The lowest-lying free-electron fixed-point levels are shown for odd $N(\eta_N^*)$ and even $N(\hat{\eta}_N^*)$ at left and right, respectively. The region $5 < N < 15$, $23 < N < 51$, and $61 < N$ correspond, respectively, to the free-orbital regime, local-moment regime and strong-coupling regime. The typical pictures of these three regimes may be described as follows: in the free-orbital regime one may put $\tilde{\Delta} = \tilde{U} = 0$ and $n_d = 0$, 1, and 2 states are all populated thermally; in the local-moment regime $\tilde{U}$ can be regarded as being so large compared with other parameters that states with energy $\tilde{U}$ above the ground state can be neglected; in the strong-coupling regime one may regard $\tilde{\Delta} \rightarrow \infty$ and the impurity is strongly coupled to

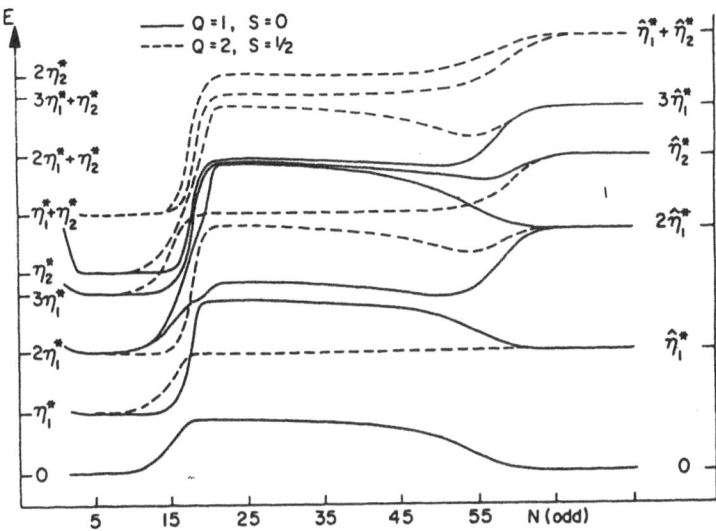

Fig. 9.11. Low-lying energy levels of $\mathscr{H}_N$ as a function of $N(\text{odd})$ for $U/D = 10^{-3}$, $U/\pi\Delta = 12.66$, and $\Delta = 2.5$ in the symmetric case

the conduction-electron states. The crossover between such regimes can also be seen in the suceptibility, as shown in Fig.9.12, where the temperatures are translated from N through (9.4.14). The above three regimes correspond to $T\chi = 1/8$, $T\chi = 1/4$, and $\chi = \text{const.}$, respectively. It can be seen from Fig.9.12 that, when $U \sim \pi\Delta$, there is a direct crossover from the free-orbital to the strong-coupling regime. The schematic diagram of the various regimes is shown in Fig.9.13. If the space in Fig.9.13 is cut by a plane with $\Delta = \text{const.}$, the picture on the plane is quite similar to Fig.9.9. The free energy can be calculated in a similar way and the specific heat associated with the impurity is shown to be linearly proportional to temperature at low temperatures in the strong coupling regime. Furthermore, the quantity corresponding to $(1 + \chi_{11}/\chi_{11})$ in $\chi_d T/C_V$ of (9.3.26) is confirmed by numerical calculations to be equal to 2 in the case of $U \gg \Delta$ and to 1 in the case of $\Delta \gg U$, in agreement with the Yosida-Yamada theory.

In the asymmetric case ($E_d \neq - U/2$), the situation is a little more complicated but calculations can be done similarly. In this case it may be convenient to rewrite the Hamiltonian (9.4.1) slightly to give, instead or (9.4.8), the sequence of Hamiltonians

$$\mathscr{H}_N = \Lambda^{(N-1)/2} \left[\sum_{n=0}^{N-1} \sum_{\sigma} \Lambda^{-n/2} \xi_n (f_{n\sigma}^{\dagger} f_{n+1\sigma} + f_{n+1\sigma}^{\dagger} f_{n\sigma}) + \sum_{\sigma} \tilde{E}_d c_{d\sigma}^{\dagger} c_{d\sigma} \right.$$

$$\left. + \sum_{\sigma} \tilde{\Delta}^{1/2} (f_{0\sigma}^{\dagger} c_{d\sigma} + c_{d\sigma}^{\dagger} f_{0\sigma}) + 2\tilde{U} n_{d\uparrow} n_{d\downarrow} \right],$$

$$(9.4.16)$$

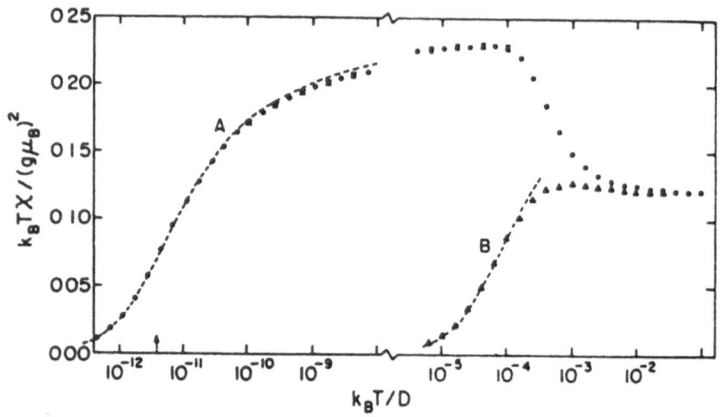

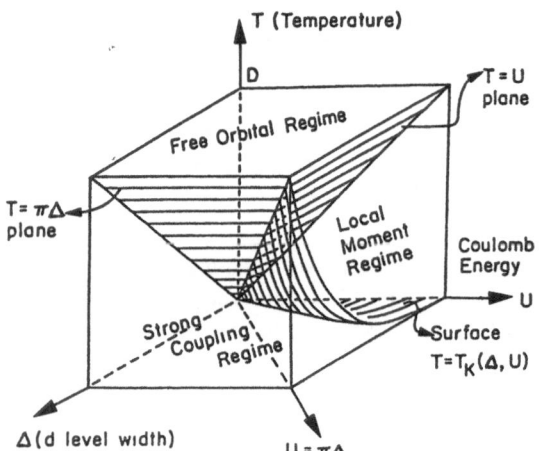

Fig. 9.12. Plots of $K_B T\chi(T)/(g\mu_B)^2$ as a function of $\ln(k_B T/D)$ for $U/D = 10^{-3}$, $U/\pi\Delta = 12.66(A)$ and 1.013(B). The arrows on the abscissa mark the effective Kondo temperature

Fig. 9.13. Schematic sketch of the various regimes for the symmetric Anderson model

where

$$2\tilde{U} = \frac{2}{1 + \Lambda^{-1}} \frac{U}{D},$$

$$\tilde{E}_d = \frac{2}{1 + \Lambda^{-1}} \frac{E_d}{D} = \tilde{\delta}_d - \tilde{U}.$$
(9.4.17)

Other notations are the same as before. In the present case the electron-hole transformation changes $\mathscr{H}$ into itself with E_d replaced by $-(E_d + U)$, so that it is sufficient to discuss the region of the parameter space for which $E_d > -U/2$. Now, suppose that $E_d < 0$ and $|E_d| \ll U$. When the temperature is in the region $-E_d \ll T \ll U$, there can be a new regime in which both $n_d = 0$ and $n_d = 1$ subspaces are thermally populated. This may be called the valence-fluctuation regime and is charaterized by $T\chi = 1/6$. This regime does not appear in the symmetric case. In Fig. 9.14, an example of the lowest-lying energy levels of $\mathscr{H}_N$

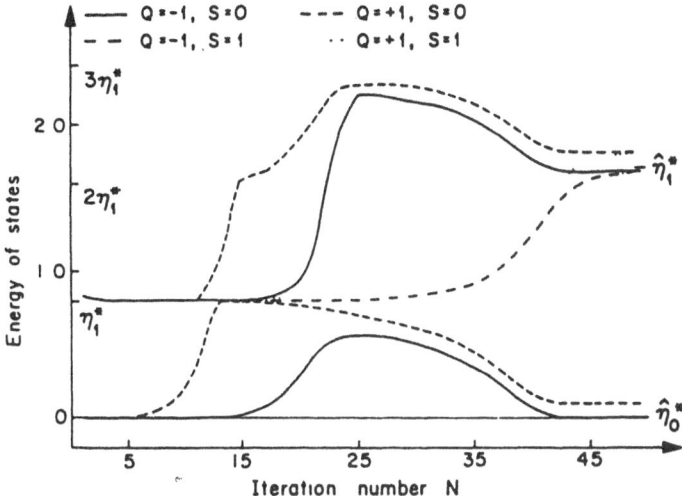

Fig. 9.14. Low-lying energy levels of $\mathcal{H}_N$ as a function of N(odd) for $U/D = 10^{-3}$, $-E_d/D$ $= 10^{-5}$, $2\Delta/\pi D = 10^{-6}$, and $\Lambda = 3$ in the asymmetric case

is shown for the case of odd N. The region $N < 9$, $13 < N < 17$, $23 < N < 31$, and $41 < N$ correspond, respectively, to the free-orbital regime, valence-fluctuation regime, local-moment regime, and frozen-impurity regime. The frozen-impurity regime is described by states in which only the impurity state with $n_d = 0$ is thermally populated and thus the impurity degree of freedom is frozen out. Such behavior arises because of the fact that the impurity orbital energy E_d is effectively replaced by

$$E_d'(T) = E_d + \frac{\Delta}{\pi} \ln \frac{U}{T}. \qquad (9.4.18)$$

The susceptibility is plotted as a function of temperature in Fig.9.15 for several sets of parameters and the crossover can be seen between the free-orbital regime $(T\chi = 1/8)$, the valence-fluctuation regime $(T\chi = 1/6)$, the local-moment regime $(T\chi = 1/4)$ and the frozen-impurity regime $(T\chi = 0)$. The impurity specific heat can be shown to be linearly proportional to temperature at low temperatures in the frozen-impurity regime and this contribution of the specific heat comes from the interaction between renormalized conduction-electron degrees of freedom which include potential scattering.

The asymmetric case can exhibit more variety of states than mentioned above according to the choice of parameters. In Fig.9.15, the crossover between the local-moment and the frozen-impurity regimes takes place at temperatures of order T_K. If $|E_d|$ is very small, say, $|E_d| \sim \Delta$, the direct crossover from the valence-fluctuation regime to the frozen-impurity regime takes place. When $\Delta \gg U$, the free-orbital regime makes the crossover into the frozen-impurity

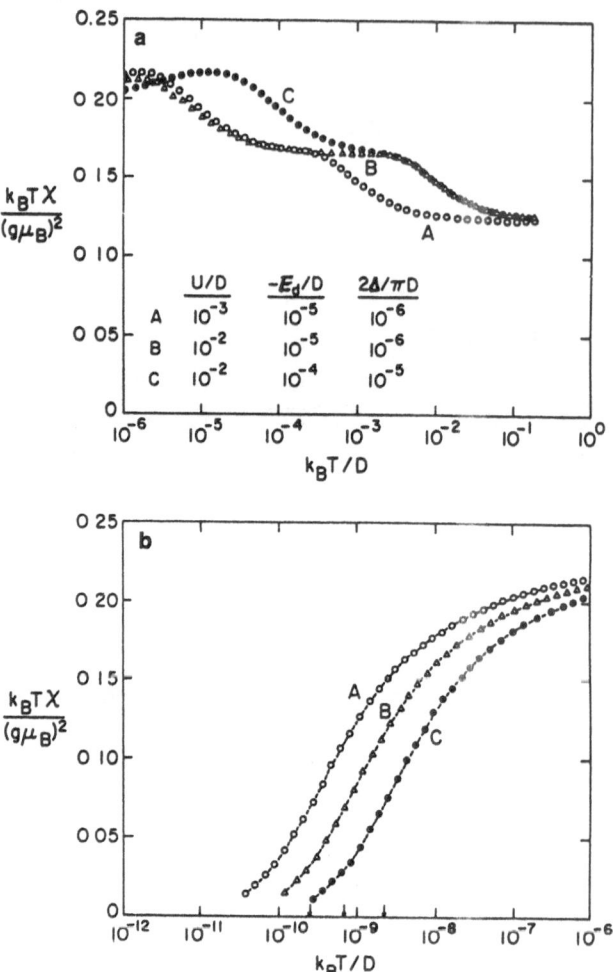

Fig. 9.15a,b. Plots of $k_B T \chi(T)/(g\mu_B)^2$ vs $\ln(k_B T/D)$ for several sets of parameters

regime and there is no intervening valence-fluctuation and local-moment regimes. In the special case of $\Delta = 0$, the final regime at the lowest temperature is
the local-moment regime if $-E_d \ll U$, the valence-fluctuation regime if $E_d = 0$,
and the frozen-impurity regime if $E_d > 0$, while in the symmetric case it is simply
the local-moment regime in this special case.

10. Random Systems

In this chapter various types of disorder in real substances are discussed. After a general discussion of periodic and aperiodic systems brief reviews are given of point defects, line imperfections and dislocations, melting phenomena of crystals and order-disorder phase transitions in alloys. Then some recent experimental facts on noncrystalline solids and liquid metals are sketched and discussion is closed with some general arguments on the electronic and vibrational states of random systems.

10.1 Periodic and Aperiodic Systems

In preceding chapters, we have discussed how the phases, i.e., solid, liquid and gas phases, of various materials appear as the temperature and pressure are changed. In these discussions, atoms in the solid phase have been assumed to form a periodic crystal lattice. What is the periodicity? Let $\rho(r)$ be the probability density for finding an atom (or an ion) at a position r in a given system during a physically feasible observation time. In perfectly periodic crystals in the three-dimensional space, there are three linearly independent vectors a_α such that

$$\rho(r + a_\alpha) = \rho(r) \tag{10.1.1}$$

and vectors a_α have a nonzero minimum length.

It should be remarked here that $\rho(r)$ in (10.1.1) is defined as the probability density in a given period of a physical observation. Therefore, the validity of (10.1.1) is sensitive to the observation time in question. For example, atoms in crystals continually perform slight deviations from their equilibrium position by the thermal vibration and/or quantum effects, and obviously an instantaneous atomic configuration does not satisfy (10.1.1). Nonetheless, since the normal frequency of an atom is the order of 10^{14}s^{-1} in most cases, an observation which takes much longer than 10^{-14}s averages out the instantaneous motion of atoms. The subsequent probability density $\rho(r)$ is a time average of the instantaneous density. It will restore the periodicity and will be described by a canonical distribution function for a sufficiently long observation.

For liquids and gases at room temperature and pressure, the probability density $\rho(r)$ determined by an observation during a period less than 10^{-10}s or so does not have any periodicity and does not satisfy (10.1.1). However, a sufficiently long observation gives a constant $\rho(r)$ and hence (10.1.1) holds for any a_α or the minimum value of $|a_\alpha|$ which satisfies (10.1.1) is always zero. Consider a sufficiently slow observation and take a limit of the infinite system size in order to avoid the difficulty of the surface effects. Then, systems whose $\rho(r)$ is a constant will be called a fluid and those which have a nonzero minimum value of $|a_\alpha|$ satisfying (10.1.1) will be called a crystal.

Now, is there any possibility of finding a system which does not belong to either fluids or crystals defined above? The first plausible case is a system where the length of one or two of three vectors a_α can take any infinitesmal value, in other words, a system whose probability density $\rho(r)$ takes a form of $\rho(x, y)$ or $\rho(x)$. Using the following arguments, *Landau* [10.1] has shown that there are no such systems that have $\rho(r) = \rho(x)$, except when $\rho(x) = $ const.

Suppose a displacement in a small region of a system from the equilibrium position produced by the thermal fluctuations. If the displacement diverges with the system size, the particle position smears. This is inconsistent with the starting assumption that $\rho(r)$ is not a constant but a periodic function. First, we show that this inconsistency does not exist in three-dimensional crystals.

Let a vector $u(x, y, z)$ characterize the displacement of the small volume element at (x, y, z). The displacement vector u can be expanded into a Fourier series

$$u = \sum_f u_f e^{i f \cdot r} . \tag{10.1.2}$$

Here, we have assumed that the system obeys periodic boundary conditions with a sufficiently long period L, and hence

$$f_x = \frac{2\pi}{L} v_x, \quad f_y = \frac{2\pi}{L} v_y, \quad f_z = \frac{2\pi}{L} v_z \tag{10.1.3}$$

$$\left[v_\alpha = -\frac{L}{2d} + 1, \ldots, -2, -1, 0, 1, 2 \ldots, \frac{L}{2d}; \ \alpha = x, y, z \right],$$

where d denotes the edge length of the small volume element.

Now, the increment ΔF of the free energy produced by the displacement (10.1.2) can be obtained in a form of a power series in u. Since the uniform displacement, which is equivalent to a translation of the whole body, does not change the free energy, the power series does not contain u itself but the derivatives of u (x, y, z). Moreover, $u = 0$ corresponds to an equilibrium position and hence a minimum of the free energy. Therefore, the series does not have the first derivative or u either. The second derivatives of u are the lowest-order terms of the series. Neglecting higher-order terms of the infinitesimal displacement, we now have

$$\Delta F = \sum_{\alpha, \beta} C_{\alpha, \beta} \int \frac{\partial u}{\partial x_\alpha} \cdot \frac{\partial u}{\partial x_\beta} \, dr$$

$$= V \sum_f |u_f|^2 \phi(f_x, f_y, f_z) , \tag{10.1.4}$$

where $C_{\alpha,\beta}$ are constants ($\alpha, \beta = 1, 2, 3$), $V = L^3$, $\phi(f_x, f_y, f_z)$ is a quadratic function of f_x, f_y, f_z. Note that $u_f^* = u_{-f}$, because u is real. It is easy to see that for a sufficiently slow observation we can apply the standard statistical mechanics. The mean square of the fluctuation is given by

$$\overline{|u_f|^2} = \frac{\int |u_f|^2 \, e^{-\Delta F/kT} \, du_1 \cdots du_f \cdots}{\int e^{-\Delta F/kT} \, du_1 \cdots du_f \cdots} = \frac{1}{2} \frac{kT}{V\phi(f_x, f_y, f_z)} \tag{10.1.5}$$

and the mean square of the displacement (exclude the uniform displacement) is given by

$$\overline{u^2} = \sum_{f(\neq 0)} \overline{|u_f|^2} \,. \tag{10.1.6}$$

Since the summation over f converges as $L \to \infty$, the displacement is finite regardless of the system size. Thus, the necessary condition for the existence of crystals is certainly satisfied in the three-dimensional space. Incidentally, in the two-dimensional space the summation (10.1.6) diverges logarithmically with L. It follows that the crystal defined above cannot exist in the two-dimensional space (and this is also the case in the one-dimensional space in a stronger sense).

Let us now consider a system with density $\rho = \rho(x)$. Because the density of this sytem is a constant in the y and z directions, a large fluctuation in these directions does not cause any contradictions. Among the small displacements, only u_x, the displacement in the x direction, may change the free energy. Moreover, the derivatives $\partial u_x/\partial y$ and $\partial u_x/\partial z$ are excluded from the expansion of the free energy, because the uniform $\partial u_x/\partial y$ and $\partial u_x/\partial z$ correspond to a rotation of the whole system around the z and y axes, respectively, and these rotations do not change the free energy. Consequently, in the expansion of the free energy, it is sufficient to take account only of the following second-order terms:

$$\left[\frac{\partial u_x}{\partial x}\right]^2, \quad \frac{\partial u_x}{\partial x}\left[\frac{\partial^2 u_x}{\partial y^2} + \frac{\partial^2 u_x}{\partial z^2}\right], \quad \left[\frac{\partial^2 u_x}{\partial y^2} + \frac{\partial^2 u_x}{\partial z^2}\right]^2. \tag{10.1.7}$$

Insertion of (10.1.3) into (10.1.7) yields such terms as

$$|u_{xf}|^2 f_x^2, \quad |u_{xf}|^2(f_y^2 + f_z^2)f_x, \quad |u_{xf}|^2(f_y^2 + f_z^2)^2. \tag{10.1.8}$$

Thus, the change ΔF in the free energy has the form

$$\Delta F = V\sum_f |u_{xf}|^2 \phi(f_x, f_y^2 + f_z^2), \tag{10.1.9}$$

where ϕ is a quadratic function of two variables f_x and $f_y^2 + f_z^2$. We now have, instead of (10.1.5,6)

$$\overline{u_x^2} = \frac{kT}{2V}\sum_f' \frac{1}{\phi(f_x, f_y^2 + f_z^2)}\,. \tag{10.1.10}$$

The summation runs over all f except for $f_x = 0$, since the uniform displacement in the x direction is excluded. The summation diverges logarithmically with L, and this is inconsistent with the assumption that $\rho(x)$ is an intrinsic periodic function.

If $\rho = \rho(x, y)$, it can be shown that $\overline{u_x^2}$ and $\overline{y_y^2}$ do not depend on the system size. However, this type of system have not yet been observed. Thus, the existence of a periodic $\rho(r)$ requires a proper condition, even if the observation is carried out sufficiently slowly. Some organic materials, especially some molecular crystals composed of linear or planar molecules are known to tend to form an ordered array of molecular directions. If one increases the temperature, these materials are first transformed, not into an isotropic liquid, but into an intermediate state as shown in Fig. 10.1. The liquid phase appears at a higher temperature. The system in the intermediate state is known as a liquid crystal, since they show a fluidity as liquids as well as the multiple refraction characteristic to uniaxial crystals. Figure 10.1 shows two examples of liquid crystals. The density function $\rho(r)$ is a constant for the nematic liquid crystal, while it looks like $\rho(r) = \rho(x)$ for the smectic liquid crystal. The latter case contradicts Landau's theory. So far, x-ray experiments have shown only that the probability density of finding a molecule is approximately a periodic function of x in the coordinate system fixed on the molecule. It has not been elucidated whether $\rho(r) = \rho(x)$ holds exactly or not.

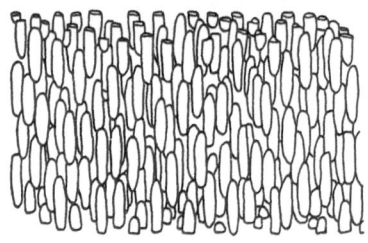

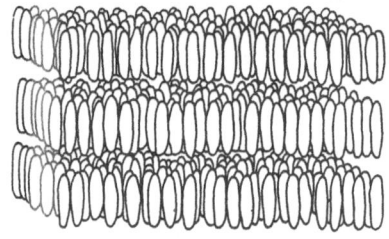

Nematic liquid crystal   *Smectic liquid crystal*

Fig. 10.1. Liquid crystals

We now turn to actual crystals. If the temperature is increased, atoms sometimes escape from the equilibrium lattice points and relocate at an interstitial position, or form a new crystal lattice on the surface. Equation (10.1.1), however will still be valid in a single crystal for a sufficiently long observation time. The characteristic time scale for an atom to jump from an equilibrium position to another is of the order of 1 s at room temperature. Therefore, a longer time than this characteristic time is required for an experiment to preserve the condition (10.1.1). As the temperature is reduced, this time scale is increased drastically, longer than the attainable time scale in laboratories. In addition to the point imperfections described in the above, real solids have line imperfections called dislocations, crystal grains and planar imperfections such as the stacking fault. These imperfections are not in thermal equilibrium, but in a nonequililibrium state in which they are pinned at the position where they were produced during the crystal growth. Within an observation time attainable in laboratories, these defects destroy the periodicity (10.1.1.).

It is an interesting but so far unsolved problem to precisely discuss, on the basis of the particle interaction, if $\rho(r)$ for a sufficiently long observation becomes a periodic function, or a constant, or an aperiodic random function when the system size is increased infinitely. In any event, when they are cooled, liquids are not always transformed into crystals; some change into a glassy amorphous solid. To discuss the nature of these materials, we must take the aperiodic atomic array as a staring point. Another example is liquid metals. Since atoms (ions) move much more slowly than electrons, the density function $\rho(r)$ of liquid atoms may be assumed to be constant for an observation time scale longer than the characteristic time of the structure change of liquids. In order to discuss the electronic state of liquids, however, the randomness of the atomic arrangement must be taken into account properly. The point defects and dislocations are spatially localized and thus each defect can be treated independently in the first approximation. On the other hand, it is very difficult to theoretically discuss the effect of the aperiodic randomness on the physical properties of matter. This is the reason why the theoretical research on the effects of randomness has been far behind that of localized imperfections. In recent years, however, much effort has been made to study random systems. In the first half of this chapter we shall discuss the fundamental properties of point defects and dislocations. Then, we shall mention the mechanism of the generation of randomness. Finally, the properties of amorphous solids and liquid metals, as examples of random systems, and the elementary method of handling these systems will be discussed.

10.2 Point Defects

The simplest form of a crystal defect is the vacant lattice site, where an atom is missing from the lattice site. The proportion of lattice sites vacant at temperatures just below the melting point is about $10^{-3} \sim 10^{-4}$ in metals with close-packed structures. But in some alloys, in particular in the hard transition metal carbides such as TiC, the proportion of vacancies of one component occasionally amounts to 50%. The existence of vacant sites is very important as a possible mechanism of diffusion of atoms through solids.

Now, let us evaluate the concentration of vacant lattice sites in monatomic crystals in thermal equilibrium. Let E_v be the energy required to bring an atom from a lattice site inside the crystal to a lattice site on the surface. Then, the energy necessary to produce n isolated vacancies is given by $E = nE_v$. The entropy of the formation of n vacancies is written as

$$S = nS_v + S_c,$$ (10.2.1)

where S_v is the entropy of formation of a single vacancy and S_c is the configurational entropy. Explicitly, S_c is given by

$$S_c = k \ln \frac{N!}{(N-n)\, n!},$$ (10.2.2)

since the number of arrangements of n vacancies in N sites is $\binom{N}{n}$. Therefore, when n vacancies are formed the free energy is increased by

$$F = E - TS = n(E_v - TS_v) - kT \ln \frac{N!}{(N-n)!n!}.$$ (10.2.3)

In thermal equilibrium, n must minimize the free energy F, and hence

$$\left[\frac{\partial F}{\partial n}\right]_T = E_v - TS_v - kT \ln \frac{N-n}{n} = 0, \quad \text{or}$$ (10.2.4)

$$\ln \frac{n}{N-n} = -\frac{E_v - TS_v}{kT}.$$ (10.2.5)

In particular, if $n \ll N$, we have from (10.2.5)

$$n \approx N \exp\left[-\frac{E_v - TS_v}{kT}\right].$$ (10.2.6)

Experimentally, the concentration of vacancies n/N is measured from the difference between the linear dilation coefficient and the fractional change of the lattice constant determined by x-ray diffraction when the crystal is heated. Table 10.1 lists the parameters S_v and E_v for various materaials estimated by this method. By inserting these data into (10.2.5), the vacancy concentration in thermal equilibrium is evaluated which are shown in Fig. 10.2.

In the foregoing discussion, we have assumed that, when a vacancy is formed, an atom is transferred to the crystal surface. The vacancy produced by this mechanism is called the Schottky defect. Another type of vacancy is the Frenkel defect, in which an atom is brought to an interstitial position inside the crystal

Table 10.1.

	S_v/k	E_v[eV]
Cu	1.5 (assumed values)	1.17
Ag	1.5	1.09
Au	1	0.94
Al	2.4	0.75

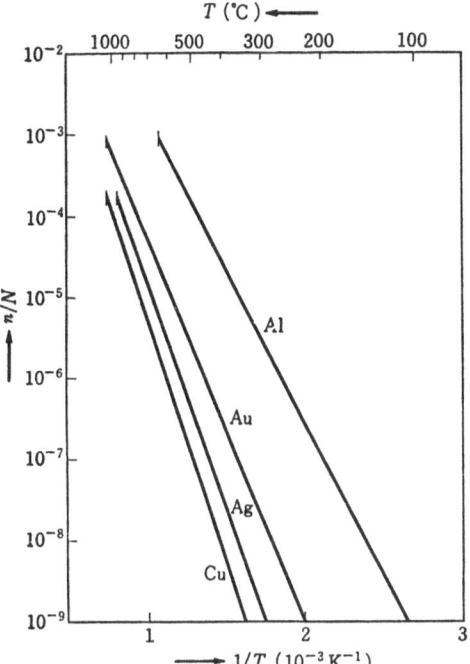

to form a lattice vacancy. Figure 10.3 illustrates schematically both mechanisms. The equilibrium number n of the Frenkel defects is calculated by minimizing the free energy as before. If n is much smaller than the number of lattice sites N and the number of interstitial sites N', we find

$$n \approx (NN')^{1/2} \exp\left[-\frac{E_i - TS_i}{2kT}\right]. \tag{10.2.1}$$

Here E_i is the energy required to remove an atom from a lattice site to an interstitial position and S_i is the entropy change of the process. The production

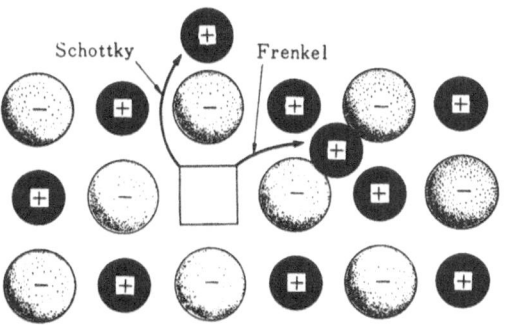

Fig. 10.3. Schottky and Frenkel defects

of the Frenkel defects does not change the crystal density in contrast to the Schottky defects. However, the defects greatly affect the diffussion of atoms through crystals and the ionic conductivity in ionic crystals, from which we can estimate the defect concentration. According to this estimation, the most common vacancies are Schottky defects in alkali halides and Frenkel defects in silver halides.

If a crystal is grown at an elevated temperature and then cooled suddenly, vacant sites are also quenched and the actual concentration of vacancies is higher than the equilibrium value at the observation temperature. When an ionic crystal is exposed to metallic vapors, metal ions (cations) sometimes squeeze into the crystal more preferably than anions. Then, vacancies are created at anion sites and electrons are trapped at these sites to preserve the charge neutrality of the whole crystal. The vacant sites become charged defects. The electron bound to defect can absorb visible light, and hence this type of vacancy is known as an F center (Farbzentrum). Pure alkali halides do not absorb visible light and are transparent throughout the visible region. They are tinged with some color by such point defects. Many types of point defects created by more than two vacancies are known and each vacancy absorbs light at its characteristic frequency range.

Various mechanisms are supposed to be responsible for the diffusion of atoms in crystals. Figure 10.4 shows three examples; the interchange of positions between adjacent atoms, migration through interstitial sites, and the exchange of occupied and vacant sites or equivalently the diffusion of vacancies. If the vacancies trap electrons to dress a negative charge, the negative charge travels with the vacancy. These processes require an activation energy for the moving atom to pass over the potential barrier produced by the surrounding atoms. Let E_b denote the activation energy and ν s^{-1} be a characteristic frequency of the lattice vibration. Then, the probability for a given atom migrating during one second is

$$p \approx \nu \exp\left[-\frac{E_b}{kT}\right]. \qquad (10.2.8)$$

Here, ν is of the order of 10^{14}.

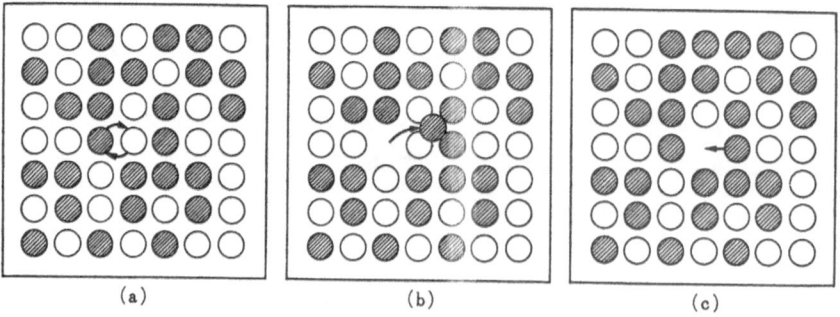

(a)   (b)   (c)

Fig. 10.4a–c. Basic mechanisms of diffusion: (a) Interchange of two atomic positions (b) Diffusion through interstitial sites (c) Exchange of atomic position with vacant lattice sites

Now, we consider two parallel planes normal to the x-axis in the crystal. The planes are separated by lattice constant a and there are S diffusers (impurity atoms or vacant sites) on one plane and $S + a\,dS/dx$ on the other. The net number of diffusers moving in the x direction between two planes per unit time is given by $-pa\,dS/dx$. If the number of diffusers per unit volume is n, then a volume element between two planes with cross sectional area 1 cm² contains $S = an$ diffusers. Therefore, the number of diffusers crossing unit area of the plane normal to the x-axis in unit time is given by

$$J = - pa^2 \frac{dn}{dx} .$$
(10.2.9)

On the other hand, according to the empirical Fick's law for diffusion, the flux J is related to the concentration gradient:

$$J = - D \operatorname{grad} n ,$$
(10.2.10)

where D is called the diffusion constant. Comparing (10.2.10) with (10.2.9), we have

$$D = va^2 \exp\left[-\frac{E_b}{kT}\right] .$$

Here, use has been made of (10.2.5).

If the diffuser carryies a charge q, the ionic mobility μ obeys the Einstein relation

$$kT\mu = qD ,$$
(10.2.11)

and hence μ is writen as

$$\mu = \frac{qva^2}{kT} \exp\left[-\frac{E_b}{kT}\right] .$$
(10.2.12)

Thus the ionic conductivity is given by

$$\sigma = nq\mu = \frac{nq^2va^2}{kT} \exp\left[-\frac{E_b}{kT}\right] .$$
(10.2.13)

The density of vacancies n depends on the temperature, as we have discussed in Sect. 10.1. However, for the monovalent ionic crystals such as alkali metal halides and silver halides, n at low temperatures can be regarded as a constant determined by the concentration of divalent impurities in the crystal. Thus, one can estimate the activation energy E_b in this temperature range from the slope of an experimental plot of $\ln \sigma$ vs $1/kT$. The estimated values of E_b are given in Table 10.2, together with E_v and E_i mentioned before.

Table 10.2 Values of E_b, E_v, E_i for ionic crystals

Crystal	E_b[eV]	E_v[eV] or E_i in parentheses
NaCl	0.86	2.02
LiF	0.65	2.68
LiCl	0.41	2.12
LiBr	0.31	1.80
LiI	0.38	1.34
KCl	0.89	2.1 ~ 2.4
AgCl	0.39	(1.4)
AgBr	0.25	(1.1)

When impurity atoms are brought into crystals, they occupy either the equilibrium atomic positions substitutionally or occupy interstitial sites. These impurities yield a deviation of periodic structure in the solid. They scatter the wave of electron or lattice vibration and affect the mobility of electrons and the thermal conductivity. This effect is prominent at extremely low temperatures where few excitations of the microscopic motion are excited. Impurity atoms occasionally donate free electrons into the crystal to become positive ions or accept electrons from the crystal to become negative ions. These impurities contribute largely to the electric conductivity of a crystal by transforming an insulator into a semiconductor. Furthermore, an impurity can localize the vibration of atoms in its vicinity and this characteristic frequency is sometimes observed in infrared absorption experiments. Point defects thus greatly affect the properties of the crystal. Beside this effect, point defects are a possible probe of characteristics of the pure crystal itself because in lower concentrations of the impurities they can be treated independently, neglecting interactions between defects. The analysis is rather clear in most cases, and hence observable quantities become abundant by the impurity doping. In fact, enormous numbers of theoretical and experimental studies have been carried out. However, we do not describe the present situation in detail but instead we proceed to the discussion of line imperfections.

10.3 Line Imperfections and Dislocations

To understand the plasticity of common solid crystals, it is quite important to take account of the fact that the crystal does not have a perfect periodicity and has line imperfections. In fact, as we shall see below, the observed values of the elastic limit of some perfect crystals are 10^{-4} times smaller than theoretical estimates for defect-free crystals. This discrepancy is attributed to the effect of line imperfections or so-called dislocations in crystals.

Suppose a defect-free single crystal depicted in Fig. 10.5 We consider a shear displacement of the two planes separated by a distance d. For the small elastic deformation, the stress σ is related to the displacement x by

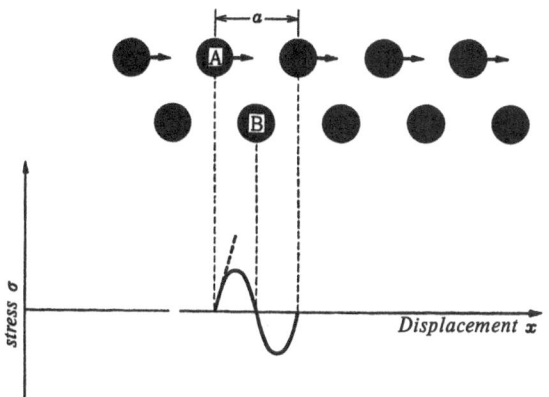

Fig. 10.5. Elastic deformation of perfect crystals

$$\sigma = \frac{Gx}{d}, \qquad (10.3.1)$$

where G is the shear modulus of the crystal. Let a be the interatomic spacing in the direction of shear. Then, σ for a general deformation must be a periodic function of x with a period a. Thus, invoking (10.3.1), we might write the first approximation for the stress-displacement relation as

$$\sigma = \frac{Ga}{2\pi d} \sin \frac{2\pi x}{a}. \qquad (10.3.2)$$

Consequently, the crystal lattice becomes unstable at the critical shear stress σ_c given by the maximum of (10.3.2)

$$\sigma_c = \frac{Ga}{2\pi d}. \qquad (10.3.3)$$

Hence, putting $a \approx d$, we find $\sigma_c \approx G/2\pi$. Table 10.3 shows the observed values of shear modulus G and the critical stress σ_c at the elastic limit. The observed ratio G/σ_c is far above the foregoing prediction 2π. Even if we employ a higher order approximation by considering realistic intermolecular forces and by taking account of the fact that other atoms can occupy a stable position as

Table 10.3 Experimental data of shear modulus G and the critical stress σ_c at the elastic limit.

	$G[\text{dyn} \cdot \text{cm}^{-2}]$	$\sigma_c[\text{dyn} \cdot \text{cm}^{-2}]$	G/σ_c
Sn (Single crystal)	1.9×10^{11}	1.3×10^7	15000
Ag (Single crystal)	2.8×10^{11}	6×10^6	45000
Al (Single crystal)	2.5×10^{11}	4×10^6	60000
Al (Poly crystal)	2.5×10^{11}	2.6×10^8	900
Fe (Poly crystal)	7.7×10^{11}	1.5×10^9	500

it is sheared, we will still have $G/\sigma_c \approx 30$, which does not differ much from the previous theoretical estimation 2π. To explain the difference between theoretical values and experimental data in Table 10.3, we must take account of the existence of line imperfections (dislocations), which we shall discuss below.

When we dealt with the shear displacement of atomic planes in Fig. 10.7, we assumed that the crystal on one side of the plane slides as a unit against another side. This assumption is reasonable if the crystal has a perfect periodicity. However, even if the crystal deviates slightly from the perfect periodic structure, we may rather expect that the crystal on one side of the slip plane is separated into the slipped and unslipped regions by a boundary plane.

In Fig. 10.6, for instance, let the slip plane be perpendicular to the page face and pass through the broken line. Suppose that a plane α through A perpendicular to both the page face and the slip plane divide the slipped side of the crystal into two parts, namely the right-hand side of the plane α slides a distance a, the atomic spacing, in the slip direction while the left-hand side of the plane α does not. The atomic configuration of Fig. 10.6a stores a high potential energy of the strain on the plane α and spontaneously changes to a more stable configuration depicted in Fig. 10.6.b. This atomic configuration differs from the perfect crystal only in the vicinity of the line l_A passing through A and perpendicular to the page face. Therefore, the stress energy is stored near the line l_A. This line imperfection is called the dislocation. The configuration of Fig. 10.6b is also regarded as through an extra atomic plane passing through A and perpendicular to the page face is inserted beneath the slip plane. Since the line l_A looks like an edge, the dislocation of Fig. 10.6b is called the edge dislocation.

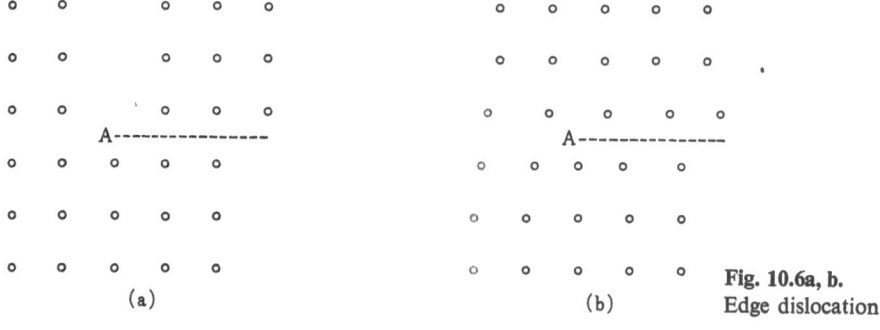

(a)   (b)   Fig. 10.6a, b.
                                                                 Edge dislocation

Now, in the foregoing consideration we have introduced three fundamental quantities, that is, the slip plane, a vector defining the slip and the boundary line between the slipped and unslipped regions. If the direction and magnitude of the vector is selected to be the slip direction and the atomic spacing in that direction, respectively, then this vector is called the Burgers vector. The boundary line stated above is known as the dislocation (line). In the edge dislocation, the Burgers vector is perpendicular to the dislocation line. As a different case, we can take the Burgers vector parallel to the dislocation line as shown in Fig. 10.7a. Here, the slip plane is on the page face, the boundary line of the slip is denoted by the dotted line, the Burgers vector is parallel to the boundary line, and solid and broken lines represent the atomic array above and below the page face, respectively. The atomic configuration of Fig. 10.7a will be spontaneously transformed into that of Fig. 10.7b, since in the former configuration, the potential energy of the strain is accumulated in the vicinity of the line A − A. Figure 10.8 is a sketch of the latter. The broken line denotes the boundary of slip, i.e., the dislocation. This line imperfection is called the screw dislocation, since as we can see from Fig. 10.8, a spiral is formed when we go around the dislocation on an atomic plane.

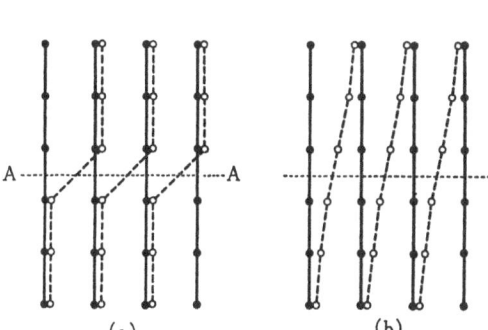

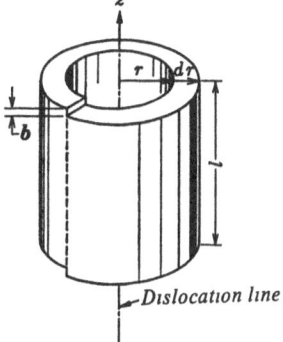

Fig. 10.7a, b. Screw dislocation

Fig. 10.8. Sketch of the screw dislocation

As we have seen in these examples, the line defect can be generally understood as the boundary line between slipped and unslipped parts of the crystal, provided a slip plane is in the crystal. The dislocation line of the edge and screw dislocations are, respectively, perpendicular and parallel to the Burgers vector. The general dislocation line can intersect the Burgers vector at any angle. However, the dislocation is defined as a boundary on the slip plane, and thus it must be either closed in itself or terminated at the crystal surface.

To estimate the energy of dislocation, let us now suppose a cylindrical crystal shown in Fig. 10.8. Here, the length of the crystal is l and it has an axial screw dislocation whose Burgers vector b is along the axis of the crystal. If one moves a distance $2\pi r$ in the circumferential direction, a displacement b is accompanied in the z direction. Therefore, the shear strain is given by

$$e_{z\theta} = \frac{b}{2\pi r} \qquad (10.3.4)$$

and the corresponding shear stress is

$$\sigma_{z\theta} = Ge_{z\theta} = \frac{Gb}{2\pi r},$$ (10.3.5)

where G is the shear modulus. It should be remarked that this expression does not hold in the region immediately around the dislocation, for the continuum approximation cannot be applied in this region. The elastic energy of the cylindrical shell of length l between radii r and $r + dr$ is written as

$$dE = \tfrac{1}{2} Ge_{z\theta}^2 l(2\pi r dr) = \frac{Gb^2}{4\pi} l \frac{dr}{r}.$$ (10.3.6)

Integration of this from r_0 to R gives

$$E = l \frac{Gb^2}{4\pi} \ln \frac{R}{r_0}$$ (10.3.7)

for the total elastic energy of the cylidrical shell between radii r_0 and R. The outer radius R cannot exceed the dimension of the crystal. We put $r_0 = b$ in (10.3.7) and neglect the actual strain energy in the very vicinity of the dislocation, which is not included in (10.3.7) since it is quite small compared with the total energy. The total energy is in proportion to the natural logarithm of R/r_0, and thus a slowly varying function of R/r_0. Therefore, for the sake of simplicity, we may assume $\ln R/r_0 = 4\pi$. Within these approximations, the energy of the screw dislocation of length l is expressed as

$$E \approx lGb^2.$$ (10.3.8)

On the other hand, the energy of the edge dislocation of length l is given by

$$E = \frac{1}{1-\nu} \frac{lgb^2}{4\pi} \ln \frac{R}{r_0} \approx \frac{lGb^2}{1-\nu},$$ (10.3.9)

where ν denotes the Poisson ratio. The Poisson ratio of most materials is about 0.3. Inserting $\nu = 1/3$ into (10.3.9), we find that the energy of the edge dislocation is roughly 3/2 times as large as that of the screw dislocation. Since the energy of both edge and screw dislocations is in proportion to b^2, the most stable dislocation is in the direction that minimizes the Burgers vector, or the direction in which the atomic spacing is the smallest. The energy of dislocations (10.3.8) or (10.3.9) is proportional to its length l. Therefore, the dislocation has a tension

$$T = \frac{\partial E}{\partial l} \approx Gb^2.$$ (10.3.10)

It follows that a closed dislocation tends to disappear if no external force acts on it. The dislocation which ends on the crystal surface is pinned there and can exist stably.

If impurity particles are deposited in the crystal, they are occasionally not so vulnerable to a shear because host atoms and dislocations cannot move across those impurity particles. Thus, the dislocation created in the process of the crystal growth or by applying an external strain on the body can exist in the crystal as a quasi-equilibrium state. In fact, taking $b \approx 2 \times 10^{-8}$cm and $G \approx 10^{12}$ dyn·cm^{-2} from Table 10.3, we have $E/l \approx 4 \times 10^{-4}$erg·cm^{-1}, that is, the energy of dislocation per atom along the dislocation line is about 8×10^{12} erg or 6×10^{4} K in temperature scale. Consequently, the dislocation at room temperature can exist only in quasi-equilibrium states.

If dislocations exist in crystals in such a manner, the dislocations move when a stress is applied in the crystal and a shear strain accompanies the motion of dislocations. The stress required to yield a dislocation-medicated plastic strain is much less than that estimated in (10.3.6), in the latter strain, all atoms on one side of the slip plane must be moved coherently, while in the former case, a displacement of a small part of atoms can produce the motion of dislocation line as shown in Fig.10.9. Actually, this mechanism of the strain accounts for the observed values of the stress σ_c at the elastic limit.

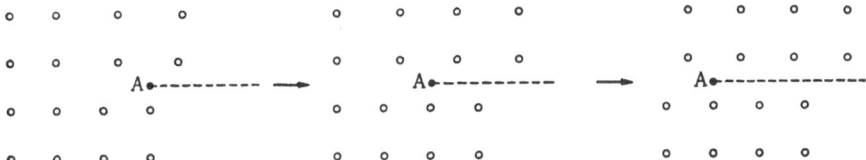

Fig. 10.9. Motion of the dislocation line A and the plastic deformation of crystals

Since the crystal near dislocation is severely deformed, a chemical reaction takes place easily in their vicinity. If the crystal surface is exposed to a corrosive, the vicinity of dislocations is affected and corroded holes appear near the terminal of dislocation on the crystal surface. Observing the corroded holes with a microscope, we can measure the density of dislocation up to 4×10^{8} dislocations/cm^{2}. For a thin film of crystal, the electron microscopy enables us to measure the density up to $10^{11} \sim 10^{12}$ dislocations/cm^{2}. According to these measurements, the dislocation density is normally $10^{2} \sim 10^{3}$ dislocations/cm^{2} in the best Si and Ge crystals, and $10^{11} \sim 10^{12}$ dislocations/cm^{2} in heavily deformed metal crystals.

Under conditions of high supersaturation, we can obtain thin hairlike crystals, called whiskers, of radius about 10^{-4}cm. The whisker is not supposed to contain any dislocations except for a uniaxial screw dislocation that aids their one-dimensional crystal growth. The stress of the whisker at the elastic limit σ_c is about $G/30$, as estimated earlier in this section, and about 1000 times greater than that of usual three-dimensional perfect crystals.

10.4 Cradle of Random Systems—Melting of Crystals and Order-Disorder Phase Transition in Alloys

A solid transformed into a liquid by heating is a common phenomenon in our daily life; for example, ice melts to become water, a solder melts, and so on. In 1910, *Lindemann* [10.1] put forward the following argument: If the amplitude of thermal vibration of atoms in a solid is enhanced by heating, the action sphere of atoms finally touches each other. Then the atoms directly interchange their energy by collision as in the gas phase, and the shift of the equilibrium position of atoms and the distribution of crystal, i.e., the melting, takes place. Lindemann assumed that the ratio of the radius of the action sphere and the interatomic spacing R_0 in the equilibrium state of the crystal is a constant independent of materials. Therefore, the above argument is identical to assuming that the ratio of R_0 and the atomic displacement from its equilibrium position is a universal constant, independent of materials, at the melting point. Let $\langle u^2 \rangle$ be the mean square of one component u of the displacement of atoms. Then, Lindemann's assumption is expressed as

$$\frac{\langle u^2 \rangle}{R_0^2} = \delta^2 , \tag{10.4.1}$$

where δ is a constant.

If we employ the Einstein model for the atomic motion in the crystal, in which each atom is assumed to oscillate independently as being constrained to a fixed point by a harmonic spring with the spring constant K, the equation of motion for an atom of mass m reads as

$$m \frac{d^2 u}{dt^2} = - Ku . \tag{10.4.2}$$

A stationary solution of (10.4.2) is written as

$$u = A \sin(\omega t + \phi), \quad \omega = \sqrt{\frac{K}{m}}, \tag{10.4.3}$$

where A and ϕ are constants, and the total energy E of the stationary solution is given by

$$E = \frac{m}{2} \dot{u}^2 + \frac{K}{2} u^2 = \frac{m\omega^2}{2} A^2 = m\omega^2 \langle u^2 \rangle . \tag{10.4.4}$$

For temperatures which satisfy $kT \gg \hbar\omega$, E is roughly equal to kT and hence $\langle u^2 \rangle$ is given by

$$\langle u^2 \rangle = \frac{kT}{m\omega^2} . \tag{10.4.5}$$

Inserting (10.4.5) into Lindemann's assumption (10.4.1), we have

$$\frac{\langle u^2 \rangle}{R_0^2} = \frac{kT_m}{m\omega^2 R_0^2} = \delta^2 \tag{10.4.6}$$

at the melting temperature T_m. Therefore, T_m is expressed in terms of mole number M and molar volume V as

$$T_m = CN\omega^2 V^{2/3}, \tag{10.4.7}$$

where C is a constant.

Lindemann compared the characteristic frequency ω of various materials obtained by the specific heat measurements with that estimated from (10.4.6) with experimental values of T_m and R_0. He pointed out from this comparison that δ. in (10.4.6) can be assumed to be a constant. If we use the Debye temperature Θ instead of ω in (10.4.7), the melting temperature is given by

$$T_m = CM\Theta^2 V^{2/3}, \tag{10.4.8}$$

where C is a corresponding new constant. This expression is known as Lindemann's formula of melting.

Lindemann's assumption (10.4.1) is not guaranteed by the statistical mechanical point of view. Actually, the melting occurs when the chemical potential of solid and liquid phases become equal at a pressure P and a temperature T. Therefore, it is hard to believe that the melting temperature is generally determined only by the nature of the solid phase. Nonetheless, Lindemann's assumption (10.4.1) has been empirically confirmed to be quantitatively correct if a realistic model is employed for the atomic vibration. Suppose a monatomic crystal where atoms perform harmonic oscillations. The mean square of the amplitude of oscillation (in one direction) is generally given by

$$\langle u^2 \rangle = \frac{1}{3mN} \sum_{k,j} \frac{\langle E_{k,j} \rangle}{\omega_j^2(k)}. \tag{10.4.9}$$

Here, m is the atomic mass, N is the number of atoms, and $\langle E_{k,j} \rangle$ and $\omega_j(k)$ are the average energy and the angular frequency, respectively, of a normal mode of wave vector k and polarization j. As shown in Fig.10.4, the melting temperature T_m for most materials is higher than the Debye temperature Θ. Therefore, we can put $\langle E_{k,j} \rangle \approx kT_m$ at the melting point, and from (10.4.1,9), Lindemann's assumption is rewritten as

$$\delta^2 = \frac{\langle u^2 \rangle}{R_0^2} \approx \frac{kT_m}{3mNR_0^2} \sum_{k,j} \omega_j(k)^{-2}, \tag{10.4.10}$$

where δ is a constant. Recently, to estimate δ from (10.4.10), *Singh* and *Sharma* [10.2] employed Krebs' model which is expected to give a good description of the atomic vibration in metals, while *Shapiro* [10.3] employed the model of spherically symmetric interaction including first and second-neighbor interactions. They used experimental values of elastic constant to determine parameters in the model, and evaluated $\omega_j(\mathbf{k})$ and in turn estimated δ from (10.4.10) using observed T_m. Table 10.4 shows their estimation of δ. It can be concluded from this table that (I) materials with the same crystal lattice have nearly equal δ if they belong to a common family, but if they belong to different families δ varies slightly, and (II) δ for the bcc lattice is larger than that for the fcc lattice.

Table 10.4 Melting temperature T_m and Debye temperature Θ and other parameters of metals

Metal	T_m[K]	Θ[K]	Group in period system	Lattice structure	δ [10.3]	δ [10.2]
Li	459	335	I a	bcc	0.116	0.089
Na	370.7	172	I a	bcc	0.111	0.083
K	335.2	91.1	I a	bcc	0.112	0.074
Rb	311.7	55.5	I a	bcc	0.115	0.080
Cs	301.7	39.5	I a	bcc	0.111	0.082
Cu	1356	315	I b	fcc	0.069	0.061
Ag	1233.8	215	I b	fcc	0.071	0.061
Au	1336	162.4	I b	fcc	0.074	0.061
Al	932.7	398	III b	fcc	0.072	0.061
Pb	600.4	88	IV b	fcc	0.065	0.059
V	2070	374	V a	bcc	—	0.070
Nb	2220	—	V a	bcc	—	0.080
Ta	3300	—	V a	bcc	—	0.067
Cr	2050	485	VI a	bcc	—	0.050
Mo	2900	379	VI a	bcc	—	0.051
W	3700	310 ∼ 384	VI a	bcc	—	0.052
Fe	3510	420 ∼ 519	VIII	bcc	—	0.057
Ni	3350	375 ∼ 413	VIII	fcc	0.077	0.055
Pd	1810	270	VIII	fcc	—	0.062

In recent years, high speed computers have made great progress and investigations based on computer simulations are flourishing in two methods, the molecular dynamics and the Monte Carlo method. In the fomer, the time evolution of a set of classical particles moving under a given interparticle potential is examined by solving the initial value problem for the classical equation of motion, while in the latter, the canonical average of physical quantities are evaluated by the computer. At present, only a system of less than 1000 atoms can be examined because of the limitation of core memory and the operating speed of the computer. Nonetheless, the equations of state which are obtained by these methods coincide not only with one another, but also are in good agreement with that of asufficiently large system given by a theory based on the low density-high temperature expansion. Therefore, those methods serve as a powerful tool for observing the behavior of a group of classical particles with mutual interactions.

Using the computer simulation, *Alder* and *Wainwright*, [10.4] first predicted that a system of hard spheres would show a phase transition similar to the liquid-solid transition and the ratio of the volume V per atom in the solid phase and V_0 in the close-packed state took $V/V_0 \approx 1.5$ at the melting point. The observed equation of state for hard sphere molecules is shown in Fig. 10.10. Recently, *Alder* and *Hoover* [10.5] carried out a computer simulation for a system of molecules with pair potential $\phi(r) = Cr^{-n}$ ($C > 0, n = 4, 6, 9, 12$) and estimated the parameter δ at the melting point. According to their result, the critical value of δ is almost independent of n and takes a value in the range $0.07 < \delta < 0.09$. Thus, Lindemann's assumption seems to hold as a good approximation, and hence, except in solids such as helium and hydrogen in which a large quantum effect is expected, atoms in the crystal remain in the very vicinity of their equilibrium position, even just below the melting point. However, if we deal with much more general systems, it is not clear whether the validity of Lindemann's assumption is limited or not. In fact, while the melting temperature is increased monotonically by applying a pressure for almost all materials except for several cases such as Ga, Ge, Bi, Sb, ice and so on, the melting temperature of Cs, Rb, K and others has extrema as a function of pressure, followed by a decrease. There are many unsettled arguments about the reason responsible for this phenomenon. It will be a further problem if Lindemann's assumption is still valid in the region near the extremum or the negative-slope region of the melting temperature.

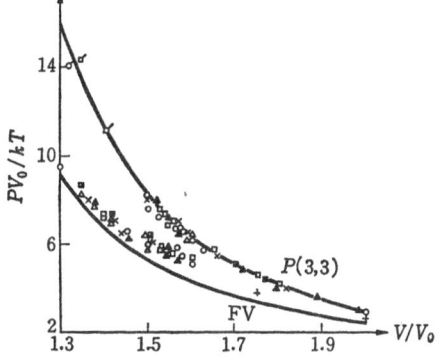

Fig. 10.10. Equation of state for hard sphere molecules. V_0 denotes the volume in the close-packed state. Several results of various computer simulations for different number of molecules are plotted. FV and P(3, 3) are the result obtained by the free volume approximation and P(3, 3) Padé approximation based on the calculation of the virial coefficients

There has been a completely different approach to the melting, namely the lattice or cell model, where the melting is treated by a statistical mechanical method based on a rather abstract model, though still taking account of both liquid and solid phases. To derive the thermal equilibrium state of liquid from a simple molecular theory, *Lennard-Jones* and *Devonshire* [10.6] assumed that each molecule in the liquid state moves independently in a cell assigned to it, and they extended this idea to discuss the melting on the basis of the cell model. Atoms in a solid phase form an array of the regular crystal lattice at low temperatures, and when the temperature is raised, they deviate from the lattice points with increasing probability of coming close to other atoms. Although atoms originally locate in a continuous space, the position of atoms is specified, for the sake of simplicity, by either the cell centered on a lattice point of the perfect crystal or the cell corresponding to an interstitial site of the perfect crystal. Each cell is assumed to admit not more than one atom. For instance, if we divide the simple cubic lattice into two sublattices A and B as shown in Fig. 10.12, each sublatttice forms an fcc lattice and we can regard the sublattice A as the lattice point of the perfect lattice and the sublattice B as the interstitial sites.

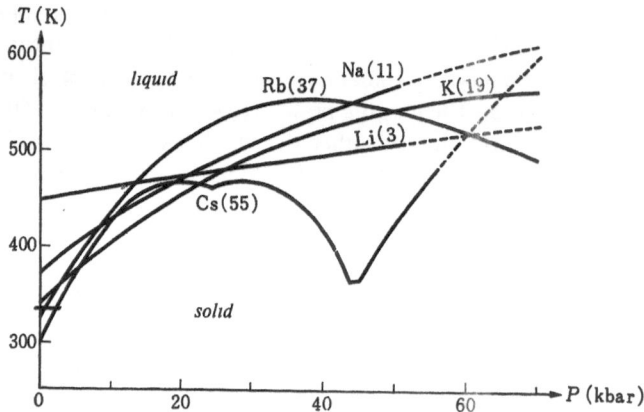

Fig. 10.11. Melting curves of the alkali metals. The number in parentheses denotes the atomic number

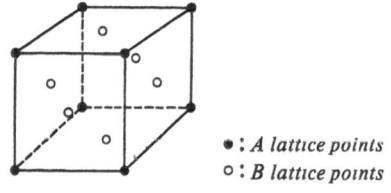

● : *A lattice points*
○ : *B lattice points*

Fig. 10.12. Cell model: A lattice points and B lattice points

At the absolute zero temperature, all of the cells centered at A-sublattice points (A cells) are occupied by atoms, while those centered at B-sublattice points (B cells) are empty. As the temperature is increased, some atoms move to occupy B cells, and at temperatures higher than the melting point, the probabilities of atoms staying in A cells or B cells are equal, which is understood to represent the liquid state. The reason that less atoms are distributed in B cells at lower temperatures is that transferring an atom from an A cell to a B cell increases the total energy due to the interatomic repulsion, since almost all atoms are in A cells. The melting temperature T_m is roughly estimated from the interatomic potential as follows.

Suppose that N atoms are in a box of volume V which makes contact with a heat bath of temperature T. The box is divided into $2N$ fictitious cells, half of which is the A cell and remaining half is the B cell. The number of atoms occupying A and B are denoted by N_A and N_B, respectively, and their fractions are denoted by

$$x_A = \frac{N_A}{N}, \quad x_B = \frac{N_B}{N}, \quad x_A + x_B = 1 . \tag{10.4.11}$$

Let us find the Helmholtz free energy under given N, T, V, x_A and x_B. The potential energy Φ for an atomic configuration where all atoms locate at the center of each cell is assumed to be given by an average over atomic configurations:

$$\begin{aligned}
\Phi &= Ne(v, x_A, x_B) \\
&= \tfrac{1}{2} N(E_{AA} x_A^2 + E_{BB} x_B^2 + 2E_{AB} x_A x_B) .
\end{aligned} \tag{10.4.12}$$

Here, E_{AA}, E_{BB} and E_{AB} are constants which depend upon the cell size, or in this case on the specific volume $v = V/N$. Each atom can deviate from the center within its own cell. The contribution of this deviation to the free energy is assumed to be $Nf_0(v, T)$, independent of the various distributions of atoms into the cells. Here $f_0(v, T)$ is a function only of v and T. The number of arrangements of atoms in the cells is given by $\binom{N}{N_A}\binom{N}{N_B}$. Therefore, the corresponding entropy is written as

$$S = k \ln\left[\binom{N}{N_A}\binom{N}{N_B} \right] \approx -2kN(x_A \ln x_A + x_B \ln x_B) . \tag{10.4.13}$$

Thus, the Helmholtz free energy is written as

$$F(N, T, V, x_A, x_B) = N[e(v, x_A, x_B) + f_0(v, T) - Ts(x_A, x_B)] , \tag{10.4.14}$$

where

$$s(x_A, x_B) = -2k(x_A \ln x_A + x_B \ln x_B) . \tag{10.4.15}$$

We define an order parameter $\sigma (0 \leq \sigma \leq 1)$ by

$$x_A = \tfrac{1}{2}(1 + \sigma), \quad x_B = \tfrac{1}{2}(1 - \sigma) . \tag{10.4.16}$$

Then, $\sigma = 0$ corresponds to the liquid state and $\sigma > 0$ to the crystalline state. In particular, $\sigma = 1$ represents the crystal at 0 K. The Helmholtz free energy is now expressed in terms of σ as

$$\begin{aligned}
F(N, T, V, x_A x_B) &= F(N, T, V, \sigma) \\
&= N\Big\{ E_0 + \frac{1}{2}(E_{AA} - E_{BB})\sigma - W\sigma^2 + f_0(v, T) \\
&\quad + kT[(1 + \sigma)\ln(1 + \sigma) + (1 - \sigma)\ln(1 - \sigma) - 2\ln 2] \Big\},
\end{aligned} \tag{10.4.17}$$

where

$$E_0 = \tfrac{1}{8}(E_{AA} + E_{BB} + 2E_{AB}),\tag{10.4.18}$$

$$W = -\tfrac{1}{8}(E_{AA} + E_{BB} - 2E_{AB}).\tag{10.4.19}$$

From the symmetry between A cell and B cell, we may assume $E_{AA} = E_{BB}$. The actual order parameter σ in equilibrium minimizes F. To guarantee that $\sigma = 1$ at 0 K, W must be positive. Now, from the condition $\partial F/\partial\sigma = 0$, we easily find

$$\sigma = \tanh\left[\frac{W\sigma}{kT}\right].\tag{10.4.20}$$

and the second derivative reads as

$$\frac{\partial^2 F}{\partial\sigma^2} = N\left[-2W + kT\left(\frac{1}{1+\sigma} + \frac{1}{1-\sigma}\right)\right].\tag{10.4.21}$$

If follows that if $kT > W$, $\sigma = 0$ minimizes F and the liquid state is realized, while if $kT < W$, a nonzero $\sigma > 0$ minimizes F and a crystalline state is realized. Therefore, the melting temperature T_m is given by $T_m = W/k$. Inserting σ determined by (10.4.20) into (10.4.17), we obtain the pressure through $P = -(\partial F/\partial V)_{N,T}$ and in turn we find the equation of state for a given temperature the relationship between volume and pressure. It is well known that the region where $(\partial P/\partial v)_T > 0$ in Fig.10.13 is unstable as a single phase, and the equation of state corrected by the Maxwell construction yields a phase separation.

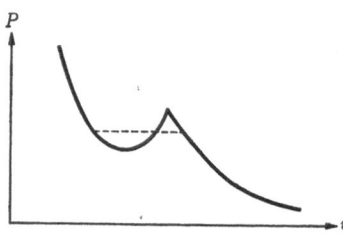

Fig. 10.13. Equation of state given by the cell model

Assuming $W = W_0(V_0/V)^4$, Lennard-Jones and Devonshire [10.6] obtained the equation of state for inert gases such as Ar near the melting temperature. The functional form $W = W_0(V_0/V)^4$ corresponds to an assumption that the interatomic repulsive potential is proportional to r^{-12} (r is the interatomic distance). The constants V_0 and W_0 are determined from a comparison with experimental data. According to their results for Ar, the melting is a first-order phase transition for temperatures below 132 K and pressures below 2400 atm, while it is a second-order phase transition for temperatures above 132 K or for pressures above 2400 atm. This behavior is attributed to the fact that W is constant for the fixed volume, and the transition determined by (10.4.20) always becomes second order because of the symmetry between A and B cells. Therefore, if we break the symmetry between A and B lattice points, the transition always becomes first order. However, it is still an unsettled question that to what extent these cell models can represent the actual melting phenomenon. In fact, according to the model of Lennard-Jones et al., atoms deviate largely from their equilibrium position before the melting takes place when the temperature is increased at a fixed volume, which is in distinct contrast with the prediction by Lindemann's theory that the melting takes place at a small deviation, about 10% of the nearest-neighbor distance. As mentioned above, computer simulations and other evidence support Lindemann's prediction. If the melting is understood by a larger displacement of atoms as in the cell model, the melting temperature is estimated to be much higher than the actual one, since a larger displacement requires a higher energy.

The cell model of melting was originally introduced by analogy to the order-disorder phase transition in alloys. For instance, 50%–50% CuZn alloy takes a bcc structure at elevated temperatures and is a so-called disordered alloy, where all lattice points are equivalent for the distribution of Cu and Zn atoms. The bcc lattice is composed of two simple cubic sublattices, whose lattice points are direct neighbors. If the temperature is reduced lower than about 460°C, the system turns into the ordered alloy in which Cu atoms are distributed mostly in one sublattice and Zn atoms mostly in another sublattice. This transformation between ordered and disordered phases is known as the order-disorder phase transition. Let us assume that tne interatomic interaction is restricted within the nearest-neighbor lattice points. Let e_{aa}, e_{bb}, and e_{ab} be the potential energy between atomic pairs Cu-Cu, Zn-Zn and Cu-Zn, respectively, and N_A and N_B be the number of Cu atoms occupying A sublattice and B sublattice, respectively. Then, the total potential energy analogous to (10.4.12) is written approximately as

$$\Phi \approx NZ\{e_{aa}x_A x_B + e_{bb}(1-x_A)(1-x_B) + e_{ab}[x_A(1-x_B) + x_B(1-x_A)]\}$$
$$= NZ[e_{ab} + (e_{aa} + e_{bb} - 2e_{ab})x_A x_B], \tag{10.4.22}$$

where the definition (10.4.11) has been used, and Z denotes the number of nearest-neighbor sites; for the bcc lattice, $Z = 8$. The potential energy (10.4.22) is equivalent to that of (10.4.12) if one sets

$$E_{AA} = E_{BB} = 0, \quad E_{AB} = Z(e_{aa} + e_{bb} - 2e_{ab}). \tag{10.4.23}$$

Therefore, the parameter W reads from (10.4.19) as

$$W = \frac{Z}{4}(e_{aa} + e_{bb} - 2e_{ab}), \tag{10.4.24}$$

and hence the transition temperature is given by

$$T_c = \frac{W}{k} = \frac{Z}{4k} (e_{aa} + e_{bb} - 2e_{ab}) . \tag{10.4.25}$$

Here, we have assumed $W > 0$. If $W < 0$, a phase separation takes place at lower temperatures and the transition temperature is given by $T_c = |W|/k$.

10.5. Noncrystalline Solids

In Sect. 10.4, we discussed the melting of solids from the thermodynamic point of view, that is, we have seen of which states, crystal or liquid, materials take in thermal equilibrium. According to a computer simulation for a system of about 100 atoms confined in a container with periodic boundary conditions, the system remains in the crystal state for a while as the temperature is increased higher than the melting temperature and eventually melts at a definite temperature. In actual materials, though, such a super-heating of crystals has been barely attained, partly because the heating is carried out under a fixed pressure. On the other hand, the super-cooling state of liquids is obtained quite commonly and, unless there are crystal nuclei in the liquid, crystals do not grow immediately in the liquid even if its temperature is lowered to the melting point.

Turnbull and Fisher [10.7] proposed the following mechanism of crystal growth. Let A be the free energy for the nucleus formation in the crystal growth and N be the number of sites where nuclei will be formed. Then, the number of nuclei at a temperature T is roughly given by

$$n \approx N \exp\left[-\frac{A}{kT}\right] . \tag{10.5.1}$$

Now, if a crystal nucleus of volume a^3 and surface area $\alpha a^2 (\alpha \sim 6)$ is formed in a liquid, this process requires a free energy W:

$$W = a^3 \Delta G_v + \alpha a^2 \gamma , \tag{10.5.2}$$

where ΔG_v is the difference in the free energy per unit volume between solid and liquid and γ is the surface energy per unit area. The free energy W is schematically plotted in Fig. 10.14 as a function of a. The actual free energy A of the nucleus formation will be given by the maximum value of W:

$$A = \frac{4\alpha^3 \gamma^3}{27(\Delta G_v)^2} . \tag{10.5.3}$$

The rate of nucleus formation will read as

$$J = BN \exp\left[-\frac{A}{kT}\right] , \tag{10.5.4}$$

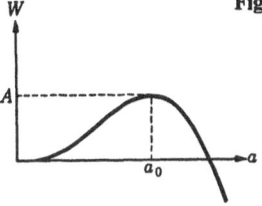

Fig. 10.14. Free energy W of the formation of a crystal nucleus

where B is the average number of collisions per unit time between a nucleus and its basic atoms or molecules in the liquid.

According to thermodynamics, we have

$$\frac{\partial \Delta G_v}{\partial T} = \Delta S, \tag{10.5.5}$$

where ΔS is the difference of entropy per unit volume between liquid and solid. On the other hand, the entropy change is expressed in terms of the molar latent heat L_m of melting and the molar volume V as $\Delta S \approx L_m/VT_m$. Therefore, ΔG_v is given approximately by

$$\Delta G_v \approx -\frac{(T_m - T) L_m}{VT_m}, \tag{10.5.6}$$

and the insertion of (10.5.6) into (10.5.3) leads to an expression

$$A \approx \frac{4\alpha^3 \gamma^3}{27} \left[\frac{V}{L_m \theta}\right]^2 \quad \left(\theta \equiv \frac{T_m - T}{T_m}\right). \tag{10.5.7}$$

We can see from (10.5.7) that A is quite large just below the melting temperature and, therefore, from (10.5.4) that the rate of nucleus growth is very slow. The growing rate of the nucleus is also very low at extremely low temperatures because the number of atoms whose energy surmounts the potential barrier is reduced. For liquids of complex molecules, B is supposed to be very small, and cooling of these liquids does not produce a crystal but a noncrystalline solid (glassy state) where the viscosity is very high and few atoms diffuse.

Among others, the following three types are typical examples of materials whose noncrystalline state can be easily obtained: oxides $A_m O_n$ such as SiO_2, B_2O_3, GeO_2, P_2O_3, As_2O_5 etc, where three or four oxygen atoms are bound to atom A to make AO_3 type triangles or AO_4 type tetrahedrons. These triangles or tetrahedrons are connected to adjacent ones through at least three corner atoms to form a three-dimensional continuous network and hence the viscosity is still very high at the melting temperature; chain molecules of chalcogenides S and Se, organic high polymers or others; hydrogen-bonded materials such as

HPO_3, in which the viscosity is high due to the hydrogen bonds and a regular array of molecules is hardly ever achieved.

It takes a finite time in general for a crystal to grow. Therefore, as a liquid is cooled at a fixed pressure, a crystal does not show up immediately after the temperature passes the melting point T_m and remains in the form of super-cooled liquid, unless the cooling rate is sufficiently slow. The situation is shown schematically in Fig. 10.15. When the temperature is lowered further, however, all materials which potentially become crystals undergo a jump in volume at a certain temperature to move on the line ef in Fig. 10.15, whilst the volume of those materials mentioned above, which can become noncrystalline solids, keep contracting along the liquid line $a \rightarrow b \rightarrow c$ of the volume versus temperature plot. Eventually, at a certain point c, the slope of the line changes sharply, and again the volume decreases along with $c \rightarrow d$ parallel to the crystalline line $e \rightarrow f$. In the latter case, c is called a glass transition point and the temperature at c is denoted by T_g. The state corresponding to $c \rightarrow d$ is known as the glassy state. As materials in the glassy state perform the transition from the glassy state to super-cooled liquid, they show a sharp increase of the specific heat around T_g as shown in Fig. 10.16. Therefore, if we measure the specific heat of non-crystalline materials during heating, we can judge whether the material in question was in the glassy state or in the super-cooled liquid state according to whether the observed specific heat shows a jump or not. Along with the specific heat, the viscosity exhibits a drastic change at the glass transition, namely, while it is about 10^{14} P in the vicinity of T_g, it decreases quickly above T_g and becomes that of the liquid at much higher temperatures. The glass transition temperature T_g depends strongly on the cooling rate. For a fast cooling, the volume contrac-

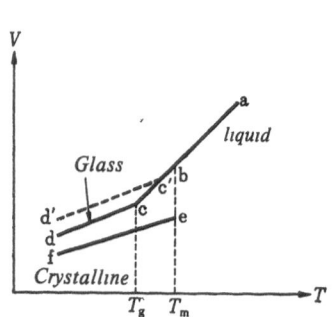

Fig. 10.15. Schematic plot of volume versus temperature for liquids, glassy states and crystals

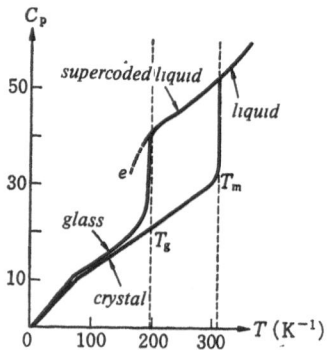

Fig. 10.16. Example of the change of the specific heat around T_g: specific heat of glycerine. [10.8] (e) the expected curve for super-cooled liquid in the quasi-equilibrium state

tion takes place along the line $a \to c' \to d'$ in Fig. 10.15 and the transition temperature will be higher than that for the slow cooling.

From the above discussion, we may conclude that the molecular configuration in the glassy state is regarded as a quenched state of the super-cooled liquid at T_g and the quenching is supposed to dissolve sharply as the temperature is increased above T_g. According to (10.5.4), the growing rate of a crystal nucleus becomes vanishingly slow as $T \to 0$, and hence all materials will show the quenching mentioned above if the cooling is carried out sufficiently fast. In fact, a computer simulation by molecular dynamics on classical particles confined in a box predicts that the existence of the glassy state is common. For example, Fig. 10.17 shows the dependence of the diffusion constant on the density obtained by the computer simulation, where the behavior of particles whose pair potential obeys $\phi(r) = C/r^{12}$ (C is a positive constant and r is the interatomic

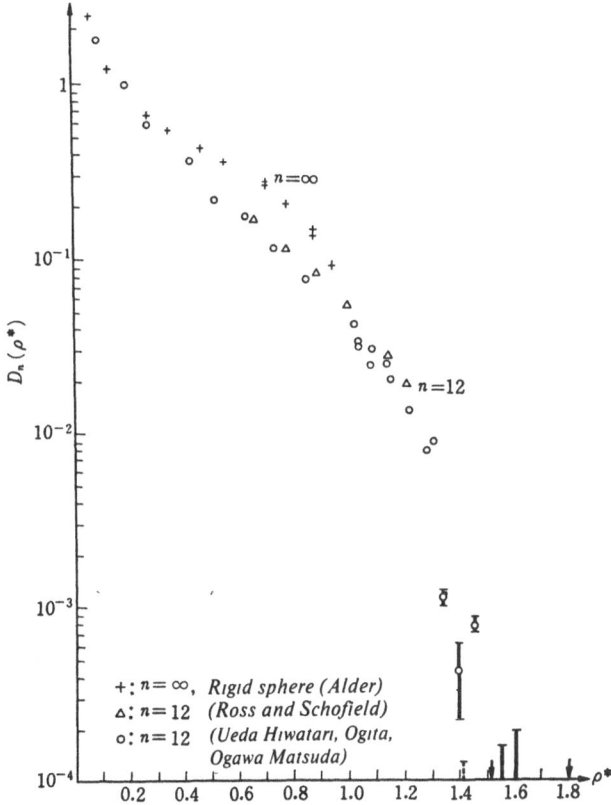

Fig. 10.17. Self-diffusion coefficients of a classical particle system with pair potential $\phi_n(r) = C/r^n$. Note the sharp change near $\rho \approx 1.3$. The abscissa denotes $\rho = (C/kT)^{3/n}\rho$ (ρ being the number density) and the ordinate denotes $D_n(\rho^*) = \sqrt{m}C^{-1/n}(kT)^{(2-n)/2n}D$ (D being the diffusion coefficient and m is the particle mass). $D_n(\rho^*)$ is shown to be a function only of ρ^* with a parameter n [10.9]

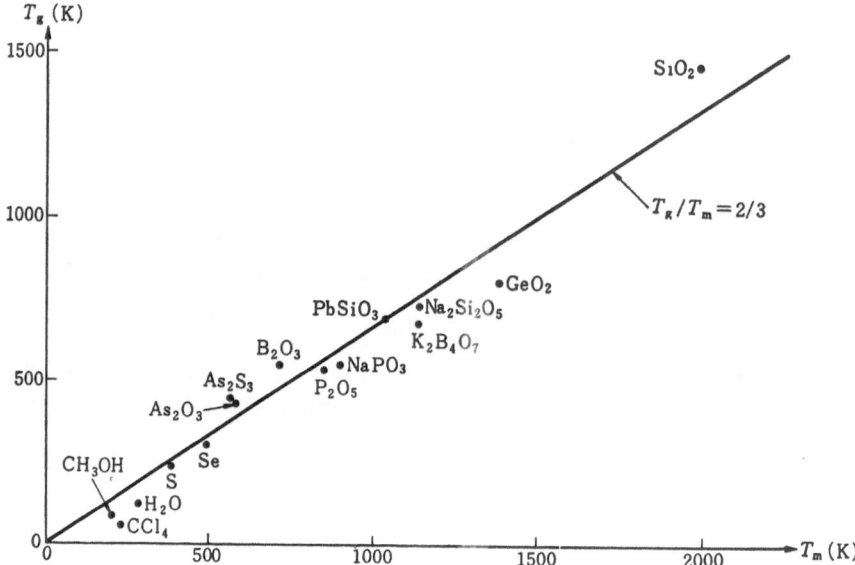

Fig. 10.18. T_g and T_m of various materials. [10.10]

distance) is examined when a sudden cooling from a liquid state is carried out. As one can see in Fig. 10.17, the drastic change in the diffusion constant implies that below a certain temperature the atomic diffusion almost disappears and only the vibrational modes survive in the atomic motion. Therefore, the glassy state is suggested to exist commonly, though the stability of the glassy state will depend heavily on the interatomic interaction.

So far, the glassy state has not yet been obtained for inert gases, which are monatomic substances whose interatomic interaction is almost isotropic, and for metals except for Fe. However, the glass transition temperature T_g has been measured for simple organic molecules such as propene, chloroform, carbon tetrachloride, methyl alcohol, for inorganic molecule H_2O and for metal Fe. As is shown in Fig. 10.18, $T_g/T_m \approx 2/3$ holds in most materials, but the ratio for those simple molecules is somewhat smaller than 2/3.

The characteristic molecular motion in the glassy state is reflected in the specific heat. If the molecular motion in the glassy state is restricted only to small oscillations, the vibrations of long wavelength are dominantly excited at low temperatures, and hence Debye's theory of specific heat applies, regardless of the molecular arrangement of the system. The specific heat is then expected to be proportional to T^3. Recently, however, *Zeller* and *Pohl* [10.11] found a linear dependence of the specific heat on temperature for various glasses at very low temperatures (Fig. 10.19).

Anderson et al. [10.12] gave the following account for these interesting experimental results. They first assumed that atoms (or atomic groups) which have two adjacent minima of potential energy exist. If one expresses the position of an atom by a configuration coordinate x, the potential energy $E(x)$ has two minima as a function of x as shown in Fig. 10.20. Let ΔE be the energy difference between two minima and let two minima be separated from each other by an energy barrier of height V. In order that the motion of atom between two minima contributes to the specific heat, V must be small so that the transition can really occur within the observation time. But, we assume that V is properly large enough so that we can neglect the resonant tunneling effect between x_1 and x_2. Thus the energy change associated with the transition is considered to be about ΔE. Let $n(\Delta E)$ denote the number per unit energy difference of a pair of these minima which satisfy the above restriction. We assume that $n(\Delta E)$ is continuous in the vicinity of $\Delta E = 0$. At sufficiently low temperatures where $n(kT) \approx n(0)$ holds, the specific heat due to the jump of atoms (or atomic groups) between two minima is given by

$$
C = k \int_0^\infty n(\Delta E) \left[\left(\frac{\Delta E}{kT} \right)^2 \frac{e^{-\Delta E/kT}}{(1 + e^{-\Delta E/kT})^2} \right] d(\Delta E)
$$

$$
\approx \frac{\pi^2}{6} k^2 T n(0) , \tag{10.5.8}
$$

and thus we have a term proportional to T.

Though a further investigation will be necessary to see if this model describes the actual glassy states in every detail, the study along this line is quite important for exploring the characteristics of glassy states. Future development is to be expected.

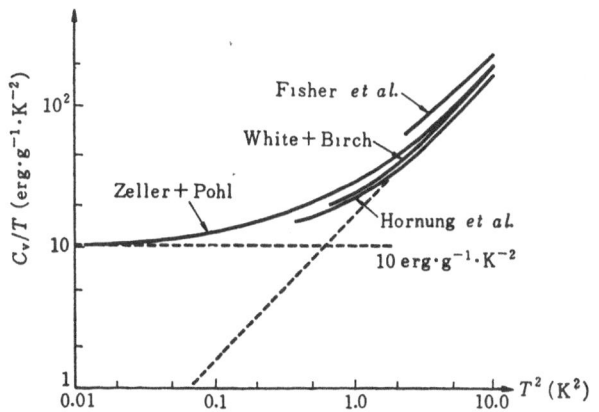

Fig. 10.19. Specific heat of glassy SiO_2 measured by several authors [10.11]

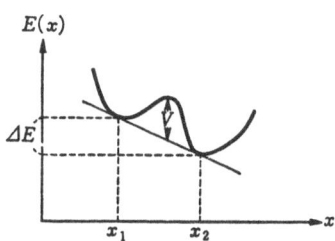

Fig. 10.20. Model potential energy curve of atoms in the glassy state from *Anderson* et al. [10.12]

10.6 Liquid Metals and Electronic and Vibrational States of Random Systems

Metals are still good conductors in the liquid phase, where the regular atomic array in the crystalline phase is destroyed by melting, since they have enough free electrons as in the solid phase. As we have discussed in Chap. 5, the deviation of atoms from their equilibrium position, which forms a periodic lattice of crystal, is caused mainly by lattice vibrations. The lattice vibration is described as phonons and the electric conductivity is determined by the scattering of electrons by phonons. In liquid metals, however, phonons with definite wave numbers as in crystals do not describe atomic vibrations properly because of the random configuration, and thus it is generally difficult to calculate the conductivity in a reliable manner. Nevertheless, if the scattering cross section of electrons due to atoms is small, and hence the electron scattering due to the pseudopotential of each atom can be calculated by the Born approximation, then the problem can be easily handled. Moreover, as *Ziman* [10.13] has pointed out, the electric conductivity of many liquid metals can be roughly explained on the basis of the above approximation; we will employ this method in the following discussion.

Let m, e and v be the electron mass, the electron charge and its velocity vector, respectively, and E denote the strength of the electric field. Now, if one assumes that when an electron moves in the material a friction proportional to its velocity acts on the electron in the opposite sense of its velocity, then the equation of motion for electrons reads as

$$m\frac{dv}{dt} = eE - \frac{mv}{\tau}.$$ (10.6.1)

Here, τ is a constant called the relaxation time. Now, the steady state solution of the equation is given by $v = (e\tau/m)E$, since $dv/dt = 0$ in the steady state. Hence, the electric conductivity is written as

$$\sigma = \frac{e^2\tau n}{m},$$ (10.6.2)

where n is the number density of the free electrons.

To estimate the relaxation time τ in (10.6.1), we must consider a scattering process of electrons. Suppose that electrons are scattered elastically by N_i atoms in a volume V. Let $u(r - R_j)$ be the pseudopotential of an atom at R_j exerting on an electron at r and w be the sum of the pseudopotential of all atoms $w = \sum_{j=1}^{N_i} u$ $(r - R_j)$. In the Born approximation, the scattering probability per unit time for a plane wave of wave vector k into that of wave vector k' is expressed as

$$P_{kk'} = \frac{2\pi}{\hbar} |<k|w|k'>|^2 \delta(E_k - E_{k'})$$

$$= \frac{2\pi}{\hbar} \sum_{i,j} \exp[i q(R_i - R_j)] \left|\frac{1}{V} \int u(r) \, e^{i q \cdot r} d^3 r\right|^2 \delta(E_k - E_{k'}). \tag{10.6.3}$$

Here,

$$q = k' - k, \quad |k'| = |k|, \quad E_k = \frac{\hbar^2}{2m} k^2 . \tag{10.6.4}$$

Now, let $(N_t/V)P(R)$ be the probability of finding other atoms in a unit volume at a distance R from a given atom. Assuming $N_t \gg 1$, we have

$$\sum_{i,j} \exp[i q(R_i - R_j)] \approx N_t \left[1 + \frac{N_t}{V} \int e^{i q \cdot R} P(R) d^3 R\right]$$

$$= N_t \left\{1 + \frac{N_t}{V} \int_0^\infty [P(R) - 1] \frac{\sin qR}{qR} 4\pi R^2 dR\right\}, \tag{10.6.5}$$

and hence the scattering probability is expressed in terms of

$$u(q) \equiv \int u(r) \, e^{i q \cdot R} d^3 r , \tag{10.6.6}$$

$$a(q) \equiv 1 + \frac{N_t}{V} \int_0^\infty [P(R) - 1] \frac{\sin qR}{qR} 4\pi R^2 dR \tag{10.6.7}$$

as

$$P_{kk'} = \frac{2\pi}{\hbar} \frac{N_t}{V^2} |u(q)|^2 a(q) \delta(E_k - E_{k'}) . \tag{10.6.8}$$

Now, for free electrons, $k = mv/\hbar$ and hence it is easy to see that the reduction of velocity v per unit time due to a scattering $k \to k'$ takes a form v $(1 - \cos\theta)P_{kk'}$, where θ is the angle between k and k'. Only the electrons with Fermi energy ε_F are vulnerable to the scattering and hence we can put $E_k = \varepsilon_F$. Finally, the reduction of velocity v per unit time due to the elastic scattering into various directions is written as

$$\frac{v}{\tau} = v \int_{-\infty}^\infty \int_0^\pi (1 - \cos\theta) P_{kk'} \rho(\varepsilon_F) \frac{\sin\theta}{2} d\theta \, dE_{k'} . \tag{10.6.9}$$

Here, $\rho(\varepsilon_F)$ denotes the density of electronic states for a given spin direction at the Fermi surface

$$\rho(\varepsilon_F) = \frac{3}{2} \frac{(N/2)}{\varepsilon_F} . \tag{10.6.10}$$

It follows that

$$\frac{1}{\tau} = \left[\frac{2\pi}{\hbar}\right] \frac{3}{8} \frac{N_i N}{V^2} \frac{1}{\varepsilon_F} \int_0^\pi (1 - \cos\theta) |u(q)|^2 a(q) \sin\theta \, d\theta , \tag{10.6.11}$$

where

$$q = 2k_F \sin\frac{\theta}{2} , \tag{10.6.12}$$

$$\varepsilon_F = \frac{\hbar^2}{2m} k_F^2 . \tag{10.6.13}$$

Using (10.6.12, 13), (10.6.11) is rewritten as

$$\frac{1}{\tau} = \frac{1}{4\pi} \frac{m}{(\hbar k_F)^3} \frac{N_i}{V} \int_0^{2k_F} |u(q)|^2 a(q) q^3 dq \tag{10.6.14}$$

and insertion of (10.6.14) into (10.6.2) yields the electric conductivity. Here, $a(q)$ is obtained by x-ray scattering experiments and $u(q)$ is evaluated by making use of various assumptions. Figure 10.21 shows the schematic behavior of $a(q)$ and $u(q)$.

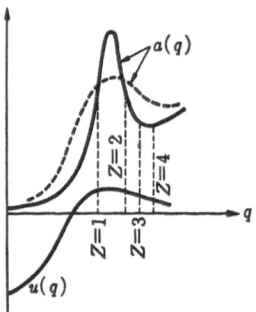

Fig. 10.21. Dependence of $a(q)$ and $u(q)$ on q and the Fermi wave number

The electric conductivity thus calculated is in good agreement with experiments. This is remarkable if one notices that the theoretical calculations of the high temperature conductivity for solids have failed to reproduce even the magnitude of the conductivity. Furthermore, the temperature dependence of the electric conductivity of monovalent and divalent metals, namely, experiments where the conductivity increases with temperature for monovalent metals whilst the converse being the case for divalent metals, can be explained qualitatively by the present simple model. The temperature dependence of the conductivity is attributed to that of $a(q)$, which is expected at higher temperatures to behave as the broken curve in Fig. 10.21. As we can see in Fig. 10.21, $a(q)$ is increased for lower values of q and decreased near the peak of $a(q)$ with temperature. For monovalent metals, the Fermi wave number is located far below the peak of $a(q)$ and hence mostly smaller values of q contribute to the integral in (10.6.14). On the other hand, the contribution near the peak of $a(q)$ to the integral is dominant for divalent metals. Thus, this effect provides a qualitative account for the peculiar temperature dependence of monovalent and divalent metals. The perturbation approach such as the Born approximation thus works well in some cases in predicting the gross nature of electronic states in random systems. But, it has been also pointed out that electronic and vibrational states in random systems have characteristics which can be hardly understood by a perturbative approach to a periodic system.

For instance, let $a_j(t)$ be the amplitude of wave function at position j and time t in the tight binding model for the electronic states in solids. The time variation of $a_i(t)$ obeys the Schrödinger equation

$$i\hbar \frac{da_j}{dt} = E_j a_j + \sum_k V_{jk} a_k , \tag{10.6.15}$$

where E_j is the energy of electrons localized on site j and V_{jk} is the energy related to a transfer of electrons between sites j and k. If an electron was localized at site 0 at time $t = 0$, the initial condition is given by $a_j(0) = \delta_{j0}$. Now, if all sites are on periodic lattice points and V_{jk} is a function only of the distance between sites j and k, and if E_j is a constant independent of site j, then the eigenfunctions of the Hamiltonian are spatially extended as we have seen in Chap. 5. Therefore, $\lim_{t\to 0} a_0(t) = 0$ holds in an infinite system, and hence electrons diffuse away infinitely after an infinite time. On the other hand, *Anderson* [10.14] first predicted that an absence of diffusion takes place in a certain case where E_j is not a constant but obeys a random distribution. Explicitly, he concluded that if a ratio W/V, where W is the width of the distribution of E_j and V is the band width of electron energy due to transfer energy V_{jk}, exceeds a critical value specific to the lattice structure, then $\lim_{t\to\infty} a_0(t)$ does not vanish, provided that the interactions with other freedoms are negligible so that (10.6.15) is valid. This fact suggests that a spatially varying $\{E_j\}$ produces localized eigenfunctions, and the diffusion is absent in the states expressed as a linear combination of such localized eigenfunctions. If the one electron eigenfunction near the Fermi surface is localized such that $\langle (x - \langle x \rangle)^2 \rangle$ remains finite as the system size is increased infinitely, the static conductivity at 0 K is proved to vanish (*Halperin* [10.15]). Here, x is the position of an electron and $\langle \cdots \rangle$ denotes an average over energy eigenfunctions. Therefore, even if the density of states at the Fermi surface is nonzero, the static conductivity at 0 K vanishes when the localization takes place. When the energy of an electron is located inside the energy band and the electron mobility is zero, the energy is said to be in the mobility gap. In fact, the existence of such a mobility gap has been experimentally suggested in some amorphous semiconductors. From the theoretical point of view, however, it is not clear at present whether the absence of diffusion and localization of eigenfunctions really take place or not within the simple model (10.6.15) and what the condition is for the mobility gap to exist. The exact results proved so far are only that the absence of diffusion occurs in one-dimensional random systems. In 1972, *Ishii* [10.16] showed that the absence of diffusion in a weak sense

$$\int_0^\infty |a_0(t)|^2 dt \to \infty \tag{10.6.16}$$

takes place, and in 1978, *Molchanov* [10.17] proved that $\lim_{t \to \infty} |a_0(t)|^2 > 0$ holds.

A harmonic oscillator obeys a similar equation of motion to (10.6.15):

$$m_j \frac{d^2 u_j}{dt^2} = \sum_k K_{jk}(u_k - u_j) . \tag{10.6.17}$$

Here, u_j is the displacement of the jth atom from its equilibrium position, m_j is the mass of the jth atom and K_{jk} is a spring constant matrix. The stationary solution obeying a time dependence $u_j \propto \exp(i\omega t)$ satisfies

$$(\sum_k K_{jk} - m_j \omega^2) u_j = \sum_k K_{jk} u_k . \tag{10.6.18}$$

On the other hand, a stationary solution of (10.6.15) having $a_j \propto \exp(i\omega t)$ obeys

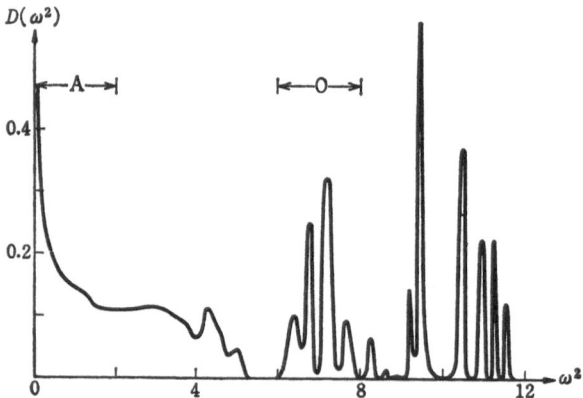

Fig. 10.22. Distribution of normal frequencies of vibration in random one-dimensional chain. $D(\omega^2)$ is the density of squared frequency. The result is shown for the random chain where isotopes with mass ratio 3:1 are distributed randomly. A and O denote the frequency range of the acoustical and optical modes of a diatomic linear lattice of these two isotopes

$$-(E_j + \hbar\omega) \, a_j = \sum_k V_{jk}a_k \, . \tag{10.6.19}$$

Comparing these two equations, we notice that (10.6.18) with a random distribution of m_j has a close relationship with (10.6.19) with random E_j. In recent years, numerical analyses of finite systems using computers have made the properties of the distribution of allowed angular frequency or normal angular frequency ω of the system obeying (10.6.18) remarkably clear. In random systems, in particular in one-dimensional random chains, the distribution of normal frequencies is found in many cases to have a fine structure and forbidden frequencies where the density of states is always zero, as shown in Fig. 10.22. The various conditions for a given frequency range to be in the forbidden gap have been theoretically obtained. On the other hand, when the singular behaviors mentioned above do not appear, the coherent potential approximation has been shown to be very useful in approximately deriving the behavior of aperiodic systems. In the coherent potential approximation, an actual random system is simulated by a virtual effective periodic system with periodic potential which is introduced to reduce the effect of perturbation due to aperiodicity. A detailed discussion of the coherent potential approximation will be given in Chap. 11.

11. Coherent Potential Approximation (CPA)

This final chapter is an introduction to the coherent potential approximation (CPA). This is now regarded as one of the standard theoretical methods to treat random systems and so various aspects of CPA are discussed along with examples of applications.

11.1 Survey

Over the last 50 years since quantum mechanics was invented, solid-state physics has made remarkable progress. Thereby the Bloch theorem has played a central role. Various subjects of solid-state physics are discussed in the main text of this volume. The Bloch theorem is established on the basis of the periodic structures of crystals. It is regarded that the essential theoretical foundation of solid-state physics for these materials with long-range order or periodic structures has already been laid. On the other hand, for systems without periodic structures or for so-called "disordered systems", a fundamental principle such as the Bloch theorem in the case of periodic crystals has not yet been attained. Besides, it is important to realize that almost all real materials in this world contain some degree of deviation from periodic structures.

At an international symposium on "Amorphous Magnetism" in 1972, *Krumhansl* [11.1] gave an interesting review talk about the history of research on disordered systems. For his talk, he chose an attractive title: "It's a random world". Maybe, he might have wanted to say: "Just try and look around yourself carefully. Then you will notice that not only physical systems but also chemical systems, biological systems, or even geographical phenomena and social phenomena are substantially characterized by the random nature of the constituent elements or factors held in these systems or phenomena."—You see, the world is full of disorder.

Anyway, we have not yet conquered this random world. When we restrict our interest to physical systems, we notice that disordered alloys, impurity semiconductors, amorphous metals, amorphous semiconductors, amorphous magnets, liquids, etc., are classified as random or disordered systems. Recently, problems of these disordered systems have become increasingly important in various senses, that is, from a theoretical, experimental and practical point of view. In this situation, it is one of the most significant and urgent tasks remaining in the field of solid-state physics to construct a systematic theoretical scheme for the purpose of generally understanding various properties of disordered systems. In view of this, the coherent potential approximation (CPA) which we are going to discuss in this chapter is considered to have served as a milestone in the process towards this ambitious goal of attaining a unified view of disordered systems.

When a system as an object of solid-state physics is called "disordered", that indicates that the microscopic structure of constituent atoms in the system does not have a complete long-range periodicity. The way deviations from a complete periodicity occur varies from material to material. Therefore, it is important to define types of disorder in the atomic structure before we start any discussion at all on disordered systems. An example of classifying types of disorder is schematically shown in Fig. 11.1. Figure 11.1a represents a complete crystal in general while in Fig. 11.1b, a covalently-bonded crystal is illustrated with a special emphasis on the covalent bonds (three bonds per atom in the case of Fig. 11.1b).

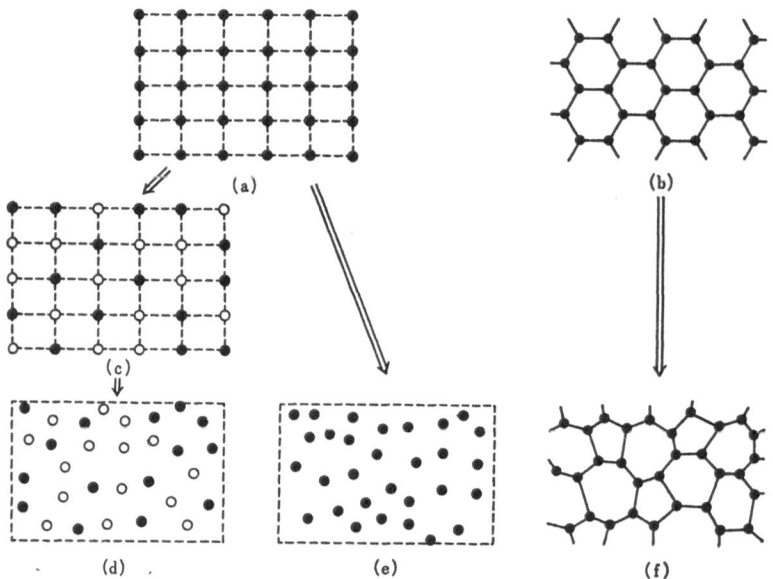

Fig. 11.1. A schematic illustration of various kinds of disordered systems

A disordered system as denoted by Fig. 11.1c is composed of atoms placed on a periodic lattice with translational invariance. Two kinds of atoms are randomly distributed on the lattice points. Although a binary system is shown in this figure where two kinds of atoms are described by open and filled circles, respectively, multicomponent systems are naturally included in this category. Disorder of this type is usually called "substitutional disorder" (or sometimes "cellular disorder", "quantitative disorder", etc.). This sort of structural model for disordered systems is applied to disordered binary alloys, mixed crystals and so on.

On the other hand, structural models (e) and (d) are obtained from models (a) and (c), respectively, by breaking the complete periodicity in the atomic positions in (a) and (c). Disorder of this type is usually called "structural disorder". Model (e) is used to describe liquids or liquid metals composed of single species while (d) serves as a structural model for multispecies liquids or liquid metals. (This kind of "structural disorder" is also referred to as "qualitative disorder" or "positional disorder".)

Structural models such as denoted by Fig. 11.1f have originally been proposed to apply to amorphous semiconductors. As is apparent from the figure, the long-range order in the atomic distribution is completely broken while the short-range order is maintained in the sense that the coordination number of each atom remains the same as in the case of a corresponding ordered crystal. In the example shown in Fig. 11.1, each atom is bonded to its nearest-neighbor atoms by three covalent bonds both in a crystal (b) and in a disordered system (f), although bond lengths and angles in the latter system fluctuate. Essentially, this type of disorder is considered to belong to the category of structural disorder. However, in view of the fact that networks formed by the covalent bonds are regarded as playing a significant role in determining various physical properties of amorphous semiconductors, the type of disorder illustrated by Fig. 11.1f is distinguished from more general "structural disorder" as represented by Figs. 11.1d, e, and is specifically called "topological disorder".

In Table 11.1, some examples of actual physical materials are listed to show to which categories of disordered systems these real materials belong. It must be emphasized, however, that actual disordered materials in this world cannot be necessarily sorted out in such a simple way as explained in the above. For instance, even when we confine ourselves to the examples shown in Fig. 11.1, model (d) contains the two aspects represented both by (c) and (e). If we take doped semiconductors as another example, they are, by nature, classified into substitutionally disordered systems as can be seen from Table 11.1. But owing to the relatively large effective Bohr radii of doped impurity atoms compared to

Table 11.1. An example of classifications of disordered system

Types of disorder	Correspondence to disordered systems illustrated in Fig. 11.1.	Actual materials
(1) Substitutional disorder /Cellular disorder \ \Quantitative disorder/	(c)	Disordered alloys Mixed crystals Doped semiconductors
(2) Structural disorder /Positional disorder \ \Qualitative disorder/	(d), (e)	Non-metal liquids Liquid metals Amorphous metals etc.
(3) Topological disorder	(f)	} Liquid semiconductors Amorphous semiconductors

the average distance between them, it is often more convenient and practically more adequate to treat doped semiconductors as structurally disordered systems in the study of their physical properties. In fact, the microscopic structure of an individual disordered system shows the mixed character of several types of disorder classified above, holding a certain degree of features attributable to each different kind of disorder. Actually, it would not be too much of an exaggeration if we say that as many types of disorder exist in this world as the number of actual disordered materials. An important point to be made, therefore, is to know which aspects or which factors of the microscopically disordered structure dominate others in characterizing some macroscopic properties of the system.

The CPA which is to be explained in this chapter was originally proposed to study one-particle properties of various elementary excitations in substitutionally-disordered systems as illustrated by Fig. 11.1c. The idea of the original CPA has been extended to evaluate other physical properties in addition to just one-particle properties. The idea has also been modified to treat systems with other types of disorder.

11.2 A Few Examples of Experimental Results

Among observed physical properties of actual disordered materials, there are quite a number of quantities which definitely show features characteristic of disordered structures. In other words, by no means can they be understood nor explained from simple extensions of theories for ordered systems to small perturbations and distortions. This fact means that we need theoretical methods other than naive perturbation theories so that we can deal with properties far from ordered nature; and this is actually our purpose in this chapter. In order to make our motivation and aim clearer, it would be instructive to show at the outset some experimental results showing interesting behaviour typical of disordered systems. Some examples are shown in the present section where we confine ourselves to the experimental results associated with one-particle properties of quasiparticles in substitutionally-disordered binary alloys.

11.2.1 Optical Absorption Spectra in Mixed Ionic Crystals

The effects of foreign alkali metal ions or foreign halogen ions on the excitation absorption in some mixed alkali halides have been investigated. An outstanding feature of the optical absorption is that (a) some experiments show two-mode behaviour in the sense that two structures corresponding to the two constituent substances do exist and (b) others show one-mode behaviour in the sense that two structures from respective constituent substances are amalgamated in a single one. The former category is classified as a persistence type while the latter an amalgamation type. In general, mixed crystals of the form MX_A–MX_B tend to show two-mode behaviour, M denoting alkali ions and X_A and X_B denoting two different kinds of halogen ions. The optical absorption spectra of KCl–KI solid solutions at liquid N_2 temperature are shown in Fig. 11.2. The peak at about 5.8 eV, which is characteristic of KI, remains rather appreciable at nearly the same position for a wide range of the I concentration. On the other hand, mixed crystals of M_AX–M_BX type seem to show one-mode behaviour where M_A and M_B denote two different kinds of alkali metals. In Fig. 11.3, peak energies of the excitation doublet measured at the temperature of liquid air are plotted against the concentration in mole percent for the mixed system KBr–RbBr. Peaks shift almost linearly with the change of composition of the mixture.

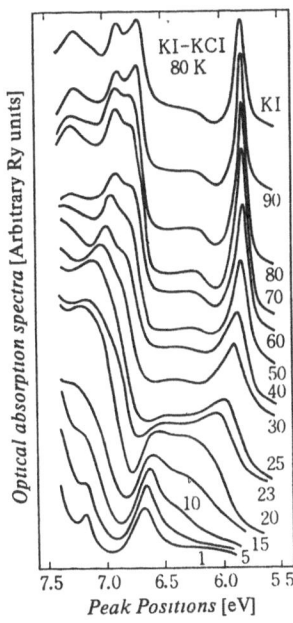

Fig. 11.2. Optical absorption spectra of KCl–KI mixed crystals. Numbers on the right-hand side denote the mole percentages of KI in KCl–KI. [11.2]

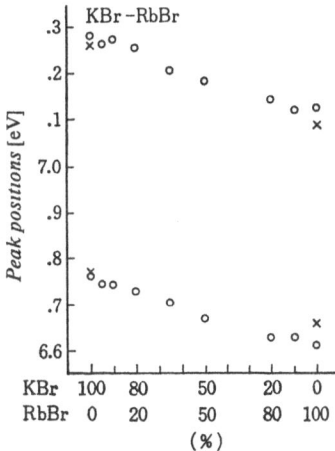

Fig. 11.3. Peak positions of absorption maxima of KBr–RbBr mixed crystals as functions of the mole percentages of the constituent components. [11.3]

According to a simplistic interpolation scheme, it is natural to expect one-mode behaviour. The experimental results, however, do not always favour this simplistic scheme, and for this reason, this problem of excitation absorption in some mixed ionic crystals attracted considerable attention by many workers from the fairly early stage of research in this field.

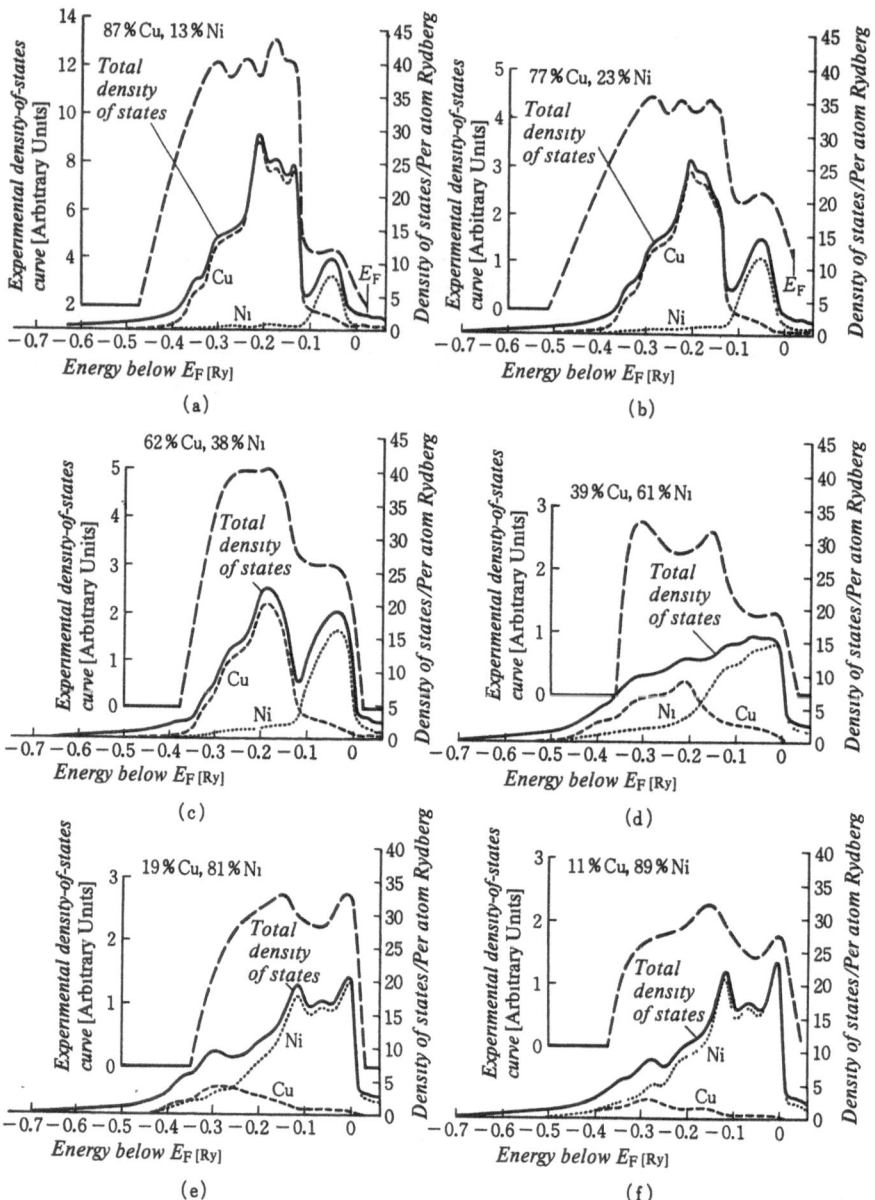

Fig. 11.4. Densities of states of paramagnetic Cu–Ni Alloys. The upper long dashed curves denote the experimental optical densities of states (ODS) [11.4]. The solid curves represent the calculated total densities of states while the lower two curves (dots and dashes) are the partial densities of states of Ni and Cu, respectively. [11.5]

11.2.2 Photoemission of Transition Metal Alloys and Noble Metal Alloys

The behavior of the electronic density of states of transition metal alloys and noble metal alloys are evaluated through appropriate analysis of photoemission data from these materials. Let us examine the example of Cu–Ni alloys. In Fig. 11.4, the densities of states calculated from experimental photoemission data are given by the long dashed curves in the upper scale of each figure, the concentration of Ni increasing from (a) to (f). As is apparent from the figures, the main features of these experimental data are summarized as follows. For Cu–Ni alloys at high Cu concentrations, we can see (I) that a subsidiary emission peak appears between the structure originating from the d bands of Cu and the Fermi energy and (II) that the high-energy emission edge ε_{edge} of the Cu d states does not change its position with respect to the Fermi energy ε_F as the Ni concentration is increased, i.e., $(\varepsilon_{edge}\text{-}\varepsilon_F)$ is nearly constant. Although it is suggested from an analysis of experimental data that the "subsidiary peak" mentioned in (I) is due to the d level of Ni, no traditional theories could explain the appearance of this peak. For Ni rich Ni–Cu alloys, on the other hand, it is known from experiments of photoemission and seen from Figs. 11.4e, f that two persistent peaks appear at almost the same positions, irrespective of the Cu concentration, that the splitting between these peaks is therefore nearly constant and that the second peak at the lower energy grows with respect to the first peak at the higher energy when the Cu concentration is increased.

In order to see the situation stated in (II) above, more clearly, $(\varepsilon_{edge} - \varepsilon_F)$ is given in Fig. 11.5 as a function of the Cu concentration. For the sake of comparison, the results of calculations based upon the fixed band model and the (lowest-order) virtual crystal model are presented in the same figure. The former describes a model in which both the shape and the position in energy of the band is kept unchanged for the whole concentrations of an alloy while the Fermi energy is shifted according to the number of electrons for each concentration of a constituent element. The latter model corresponds to the "rigid band model" which will be introduced in a later section. These are both simple approximate

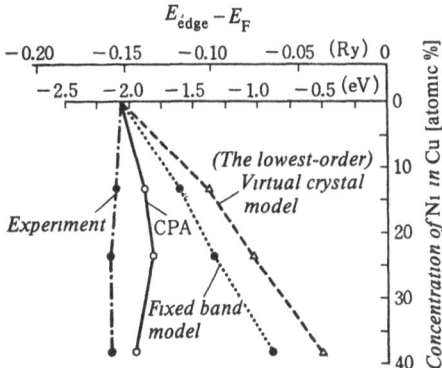

Fig. 11.5. The energy at its half-height of the high-energy edge in the Cu subband measured relative to the Fermi energy. The curves CPA, rigid-band, and (the lowest-order) virtual crystal were obtained from the calculated densities of states while the experimental curve was taken from energy-distribution curves at an incident photon energy of 10.2 eV. The experimental curve has been moved uniformly by 0.06 eV such that theory and experiment coincide for pure Cu. [11.5]

theoretical models based on the naive interpolation concept. It is readily seen from Fig. 11.5 that neither the fixed band model nor the rigid band model can explain the observed feature even qualitatively. Although it is known that there are some alloys for which these simple approximate theories are applicable, the experimental data of Cu–Ni alloys indicate that the characteristic features of the alloys cannot be understood within the simple theories.

11.2.3 Magnetic Moments and Electron Specific Heat of Ferromagnetic Alloys

Since ferromagnetic alloys show interesting characteristic features depending on the kinds of constituent elements, they have been studied from various points of view. Several approximate theories including the rigid band model have been established at an early stage to deal with ferromagnetic alloys, and it has been shown that these earlier theories are useful in explaining physical properties of some ferromagnetic alloys. In this section, however, let us show a couple of examples of experimental data which cannot be explained by these earlier approximate theories e.g., (I) the magnetic moments at a Ni site and at an Fe site, respectively, in Ni–Fe alloys and (II) the specific heat of Ni–Fe alloys. In Figs. 11.14, 15 the observed magnetic moments and the specific heat are given as functions of the Fe concentration, respectively. From Fig. 11.14, we can see that the magnetic moments at a Ni site and at an Fe site both remain nearly unchanged even in alloys, taking almost the same values as the respective values for pure Ni and pure Fe. This feature can hardly be understood by the itinerant electron picture where electrons form a band. Another example is the experimental data of the electron specific heat shown by a dashed curve in Fig. 11.15. It is also easy to imagine that the experimental curve cannot be obtained from any simple theories.

Finally, let us remark that the examples described in this section are only a small portion of the whole abundant experimental results which seem to reflect the complexity and wide variety of disordered systems.

11.3 Computer Simulations and Exact Solutions

Among various theoretical methods of investigating disordered systems there are some remarkably successful methods, which include a method of carrying out computer experiments of given model systems to evaluate appropriate physical properties of these systems or a method of obtaining rigorous solutions of some simplified model systems. Attempts to study disordered systems by means of these methods have been made from fairly early days in the history of the research of disordered systems. The characteristic feature common to both computer experiments and rigorous solutions is that these methods are free from any ambiguity or arbitrariness due to approximations, and accordingly, the obtained results are exact within the scope of the models. These methods are generally used to see what is happening in realistic disordered systems. They can be made more realistic by choosing models more carefully so as to embrace the essential aspects of the corresponding realistic systems. These methods are advantageous in the sense that they can be used to examine the validity of an arbitrary approximation. That is to say, by applying an approximation to the model used in computer work or in a rigorous treatment and testing the results against the exact results through computer work or rigorous solutions, we can discuss the validity and the limit of applications of the approximation under consideration.

In this section, let us show some examples of computer experiments and rigorous solutions which are often used to examine the validities of the CPA and other approximations before the CPA.

11.3.1 Results of Computer Experiments

Owing to the remarkable progress of computers in recent years, computer experiments are used fully from various aspects.

Figure 10.22 shows the phonon spectra $D(\omega^2)$ obtained from the computer simulation of a disordered linear chain of oscillators where ω is the normal frequency. Similarly, the phonon spectra of an isotopically disordered three-dimensional system is also calculated from a computer experiment (see the histograms in Figs. 11.6a,b).

11.3.2 Exact Results

Most of the models for which exact solutions are obtained are one-dimensional because, in one-dimensional models, problems can be solved either analytically or numerically by making the best use of mathematics characteristic of one dimension. One famous example is the Saxon–Hutner theorem illustrated in Fig. 11.7. Consider an alloy composed of two different types of atoms, denoted by A and B. The bandwidths of the A crystal and of the B crystal are represented by two hatched bands in the lower part of the figure. The Saxon–Hutner theorem states that the energy region where the density of states of the AB alloy is finite is within these regions where either the A crystal or the B crystal has a non-zero density of states, the A crystal being a pure crystal composed of A atoms alone and the B crystal of B atoms alone. The theorem can also be described as follows: an alloy A_cB_{1-c} with the arbitrary concentration c of A atoms can have no states in a region of energy simultaneously forbidden in pure crystals of either type. Accordingly, when the pure crystals of A and B have bands which do not overlap if superposed, then the density of states of the AB alloy consists of two isolated parts separated by an energy gap in between. This theorem was rigorously proven only for one-dimensional systems (for two and three-dimensional systems, somewhat similar theorems exist).

Apart from this theorem, several methods have been proposed to calculate the exact density of states of one-dimensional systems. One of them is the so-

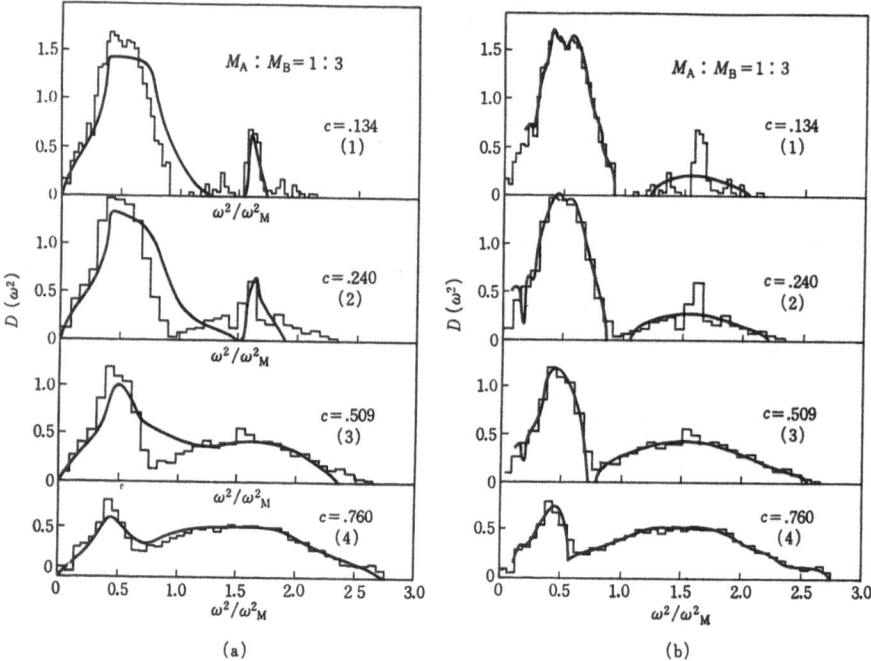

(a)   (b)

Fig. 11.6a, b. The phonon density of states $D(\omega^2)$ vs ω / ω_M^2 for substitutionally-disordered $(A_c B_{1-c})$ simple cubic lattices with $M_B = 3 M_A$ at four concentrations c of A atoms. ω_M denotes the maximum frequency in the system composed of only B atoms. The histograms represent machine calculations [11.6]. The solid curves in (a) are the calculated phonon spectra on the basis of the ATA [11.7] while the solid curves in (b) show the calculated phonon spectra obtained through CPA. [11.8]

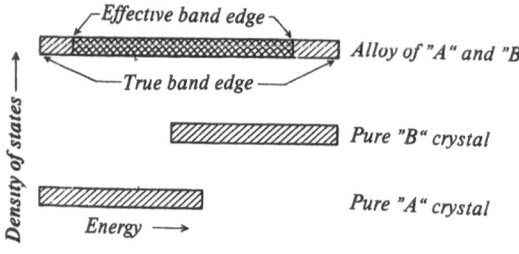

Fig. 11.7. A schematic representation of the band of a one-dimensional alloy of δ-function potentials. The different types of atoms are denoted by "A" and "B". [11.9]

called Schmidt method. An example of the integrated density of states $N(E)$ $\equiv \int_{-\infty}^{E} \rho(E) dE$ calculated by means of the Schmidt method is shown in Fig. 11.8. The calculation has been performed for a disordered binary alloy composed of A and B atoms, each having δ-function potential where the amplitudes of the δ-functions for A and B atoms are different from each other. The open circles in Figs. 11.8a, b denote the results obtained from the Schmidt method. Two arrows in each figure mark the positions of the localized levels of two δ-function poten-

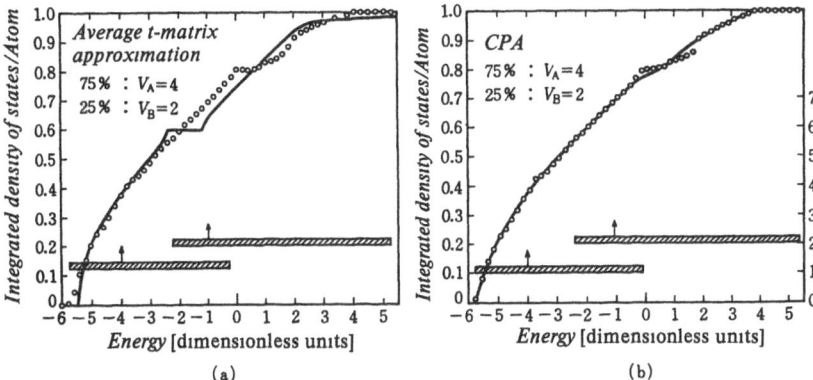

Fig. 11.8a, b. The cumulative densities of states for a substitutionally-disordered one-dimensional AB alloy; the circles denote the results of the numerical calculations and the solid curve in (a) is obtained in the average t-matrix approximation (ATA) while the solid curve in (b) shows the CPA result. The energy is measured in units of the square of the inverse of the lattice constant and the amplitudes of the δ function in units of the inverse lattice constant. The bands of pure crystals of the two potentials (at the same lattice spacing) are indicated by the bars, while the eigenvalues of the isolated potentials are given by the vertical arrows. [11.9]

tials due to A and B atoms, respectively. Two hatched bars represent the band widths of the pure A crystal and the B crystal, respectively. As is expected from the Saxon–Hutner theorem illustrated in Fig. 11.7, the density of states of the alloy is finite over almost the whole region of energy where either the A band or the B band exists. However, the following two facts are noticeable in the curve of $N(E)$, i.e., first, that $N(E)$ is nearly flat in the energy region near $E = 0$ where the density of states, $\rho(E)$, tends to vanish, and secondly, that a similar behaviour is observed at the top of the band. Although the true band width of the alloy would cover the whole energy region where either the A band or the B band is nonzero as is stated by the Saxon–Hutner theorem or as is illustrated in Fig. 11.7, the "effective band width" would be narrower than the true band width. There, we define the effective band width as the energy region where the density of states is large enough to be observable (the cross-hatched region on the energy band of the AB alloy in Fig. 11.7). Outside this effective band, the density of states is extremely small though not zero. An important point concerning the calculated results of the Schmidt method in Fig. 11.8 is the fact that the effective band widths cover the two localized levels of the δ-functions due to A and B atoms (denoted by arrows).

11.4 Model Hamiltonian

Electrons, Frenkel excitons, phonons and magnons are among elementary excitations which are discussed very frequently in problems of solid-state physics. In the present section, we first give an explicit form of the Hamiltonian (or the equation of motion) for these elementary excitations in disordered systems. The Hamiltonian corresponding to each kind of elementary excitation can be reduced to the Hamiltonian of the same mathematical structure when the physical meanings of notations in each case are appropriately interpreted. As a result, the same approximations and the same mathematical treatments are available *mutatis mutandis* to various kinds of elementary excitations.

First of all, let us choose electrons as our prototype. When the disturbance due to each impurity atom or defect is spatially localized in the neighborhood of that atom or defect, then it is convenient to describe the behaviour of elementary excitations in the site representation. In the case of electrons, this corresponds to writing the Hamiltonian in the tight-binding representation. Within the one-band scheme, the one-electron Hamiltonian operator is expressed as

$$\boldsymbol{H} = \sum_{j} |j\rangle E_j \langle j| + \sum_{j \neq l} |j\rangle V_{jl} \langle l|$$
$$\equiv \boldsymbol{D} + \boldsymbol{W}, \tag{11.4.1}$$

where R_j is the position of the jth atom and $|j\rangle$ is the Wannier function associated with site j, E_j is the eigen-energy of the eigenfunction $|j\rangle$ and V_{jl} is the transfer energy from the lth atom at $\boldsymbol{R}_l$ to the jth atom at $\boldsymbol{R}_j$. The summations $\sum_j$, etc., are taken over N atoms. The Hamilton operator (11.4.1) can also be expressed in the second quantized form as

$$\boldsymbol{H} = \sum_{j} E_j a_j^\dagger a_j + \sum_{j \neq l} V_{jl} a_j^\dagger a_l, \tag{11.4.2}$$

where a_j^+ and a_j are the creation and annihilation operators, respectively. Since we shall concern ourselves mainly with one-particle properties, the one-particle (one-electron in the present case) Hamiltonian defined by (11.4.1) or (11.4.2) is sufficient. When the interaction between particles must be taken into account, it is necessary to include such terms as $a_j^+ a_i^+ a_m a_n$ in addition to the terms appearing on the right-hand side of (11.4.2). Note that both (11.4.1, 2) carry the same physical meaning within the one-particle scheme.

As mentioned in the above, the Hamiltonian (11.4.1) or (11.4.2) can be interpreted as describing the behaviour of elementary excitations other than electrons. Typical examples are shown in Table 11.2.

In the case of electrons, the elementary excitations are Fermions, and a_j^+ and a_j in (11.4.2) obey the Fermi statistics. On the other hand, Frenkel excitons shown in Table 11.2 are an example in which the elementary excitations behave as Bosons. The new interpretation of the notations is as follows: $a_j^+(a_j)$ denotes the creation (annihilation) operator of an exciton on the jth atom at $\boldsymbol{R}_j$; E_j is the atomic excitation energy associated with the jth atom and V_{jl} is the transfer energy between excitons at $\boldsymbol{R}_l$ and $\boldsymbol{R}_j$.

Table 11.2. Various kinds of elementary excitations as described by the Hamiltonian (11.4.1) or (11.4.2). Physical meanings of E_j and V_{jl} for each case are given in the table

Elementary excitations	E_j	V_{jl}
(1) electron	atomic (or Wannier) level of an electron around the jth atom	transfer energy of an electron from the lth atom to the jth atom
(2) Frenkel exciton	excitation energy at the jth atom	transfer energy of an exciton from the lth atom to the jth atom
(3) phonon	related to $M_j \omega^2$ where M_j is the mass of the jth atom	related to the coupling constant between the jth and lth atoms
(4) magnon	related to the localized (or atomic) level and electron correlations associated with an electron at the jth atom	related to transfer energy and exchange of an electron between the jth and lth atoms

Phonons are also discussed in Sect. 10. 6. The lattice vibrations of a disordered system are described by the equation of motion for a system of harmonic oscillators. The stationary solution having time dependence of the form $\exp(i\omega t)$ is expressed by (10.6.18). It can be shown that (10.6.18) has the same mathematical structure as (10.6.15, 19) for electrons in the tight-binding model.

Elementary excitations in magnetic materials are spin waves or magnons. In general, the interaction between spins is described by

$$H = \sum_{j \neq l} \{I(j, l)\, S_z(j)\, S_z(l) + J(j, l)[S_x(j)\, S_x(l) + S_y(j)\, S_y(l)]\}, \qquad (11.4.3)$$

where j and l denote the jth and lth atoms with spins, respectively. When $J = 0$, (11.4.3) represents the Ising model while it corresponds to the Heisenberg model when $I = J$. In this case, we can cast (11.4.3) into the form of the Hamiltonian with the same mathematical structure as in the cases of electrons, excitons and phonons described above.

As a next step, let us discuss the types of disorder. Various types of disorder classified according to the categories shown in Fig. 11.1 or in Table 11.1 can be formally expressed by a common Hamiltonian as in the form of (11.4.1) or (11.4.2) when appropriate consideration is given to the way the matrix elements E_j and V_{jl} of H do depend on the kinds of atoms at R_j and R_l or to the form of the many-atom distribution for N atoms at $\{R_j\} \equiv \{R_1, R_2, ..., R_N\}$. The situation is summarized in Table 11.3. When $\{R_j\}$ form a regular crystal lattice and the matrix elements E_j and V_{jl} are not random, then, apparently, the Hamiltonian describes a crystal. Distinctions between substitutionally-disordered

Table 11.3. How various kinds of disorder are described by the Hamiltonian (11.4.1) or (11.4.2)

		Diagonal terms (E_J)	Off-diagonal terms (V_{Jl})	Positions of atoms $\{R_J\}$
(0) Crystals		$E_J \equiv E_0 = $ constant	function of R_{Jl} alone	
(1) Substitutionally disordered systems	(a) diagonal disorder	random (dependent on the kinds of the Jth atom)	function of R_{Jl} alone	ordered lattice points
	(b) Off-diagonal disorder	$E_J \equiv E_0 = $ constant	random (dependent on the kinds of the Jth and lth atoms)	
	(c) Realistic case	random		
(2) Structurally disordered systems				
(3) Topologically disordered systems		see (1)	see (1)	structurally disordered distribution

The three types of disorder (a), (b) and (c) can be realized for both cases (2) and (3).

systems and structurally (or topologically) disordered systems are made by the types of the many-atom distribution functions for $\{R_J\}$.

For each of substitutional, structural and topological disorders, we can further define site-diagonal disorder and off-diagonal disorder. Site-diagonal disorder is realized when we can assume that the effects of disorder are embodied only in the diagonal terms of a model Hamiltonian. This corresponds to the case where E_J depends only on the kind of atom at R_J while V_{Jl} is independent of the kinds of atoms at R_J and R_l as well as of the atomic configuration in the neighborhood. This assumption of diagonal randomness is guaranteed when the spatial extension of each Wannier function is sufficiently small. In actual cases, this assumption is valid when the band structure of a pure crystal composed of one constituent element of an alloy is similar to that of a pure crystal composed of another constituent. It is worth mentioning here that the Anderson model used in the discussion of electron localization in Sect. 10.6 is one typical example of a system with diagonal disorder alone. For phonon systems, the assumption of diagonal randomness is appropriate when systems are only isotopically disordered (i.e., when disorder of mass can be regarded as most significant).

On the other hand, a picture of off-diagonal disorder is acceptable for the exchange interaction between spins or for the problem of phonons in the case where the coupling (or spring) constant between atoms depends on the kinds of atoms, etc.

In general, however, the effects of disorder show up both in site-diagonal and off-diagonal terms in the model. Therefore, either pure site-diagonal disorder or off-diagonal disorder is a simplified model. When disorder in diagonal terms and that in off-diagonal terms are equally important and if they are correlated, then it is not easy to solve the problem. The theoretical treatments developed so far mainly concern themselves with situations where only diagonal terms are dominant, or where only off-diagonal terms are dominant, or where both terms are not correlated. The CPA was originally proposed for disordered systems which belong to category (1a) in Table 11.3, i.e., to substitutionally disordered systems with diagonal randomness alone. In the succeeding three sections (Sects. 11.5–7), we confine ourselves to this type of disorder. We also treat only those cases in which there is only one atom per lattice site and regard $\sum_j$ as the summation over all-lattice sites rather than the summation over all the atoms.

11.5 One-Particle Green's Function

In the present and succeeding sections, all formulations are given for the case of electrons as defined by (11.4.1). It is possible to apply completely analogous treatments to elementary excitations or quasiparticles other than electrons. In later sections, therefore, we will show numerical results concerning other elementary excitations without giving further explanations about the formulations specific to the elementary excitations under consideration.

The first term D in (11.4.1) is diagonal in the Wannier representation while the second term W is off-diagonal. Since it is assumed that only the diagonal term is random, W is not random and therefore carries the characteristic features of the ordered lattice. In other words, W possesses the translational invariance of the crystal lattice. Accordingly, W is diagonal in the Bloch representation which is defined by $|k\rangle = N^{-1/2} \sum_l \exp(ik \cdot R_l)|l\rangle$, as shown in the following equation

$$\langle k|W|k'\rangle = \delta_{kk'} \sum_l \exp(ik \cdot R_l) V_{0l} \equiv \delta_{kk'}\varepsilon(k) , \qquad (11.5.1)$$

in which $\varepsilon(k)$ is a function of k which determines the energy band structure or the dispersion relation.

Among macroscopic physical quantities reflecting the effects of randomness in the microscopic configurations of atoms in disordered systems, the simplest and most essential quantities are one-particle (one-electron) properties of elementary excitations which are derived from the ensemble average of the one-particle Green's function $G(z) = (z1-H)^{-1}$ in the form

$$G_{av}(z) \equiv \langle G(z)\rangle \equiv (z1 - H_{eff})^{-1} , \qquad (11.5.2)$$

where $\langle \cdots \rangle$ denotes the ensemble average over all possible microscopic configurations of atoms. Notation z is used to denote complex energies; 1 is a unit operator which is defined as $1 = \sum_j |j\rangle\langle j|$ in the Wannier representation, or as $1 = \sum_k |k\rangle\langle k|$ in the Bloch representation. H is the one-electron Hamiltonian operator corresponding to "a given configuration of atoms" and has the form (11.4.1). H_{eff} is an "effective Hamiltonian operator" which is related to $G_{av}(z)$ by the definition (11.5.2). In our model, the microscopic configurations of atoms are defined by all possible ways of distributing different kinds of atoms (each having a given concentration) on a regular crystal lattice. The ensemble in our case therefore, consists of all different microscopic configurations of atoms which correspond to the same observable quantities ("the same" indicates "macroscopically undistinguishable").

The most typical example of one-particle properties to be derived from $G_{av}(z)$ is the electronic density of states per atom given by

$$\rho(E) = N^{-1} \operatorname{Tr}\{\langle\delta(E1 - H)\rangle\}$$
$$= -(N\pi)^{-1} \operatorname{Im}\{\operatorname{Tr}\{G_{av}(E + i0)\}\}. \tag{11.5.3}$$

The exciton density of states, the phonon spectra $D(\omega^2)$, the spectral function $\mathscr{A}(z; k) = -\pi^{-1} \operatorname{Im}\{\langle k | G_{eff}(z) | k\rangle\}$ and the optical absorption spectra $\mathscr{A}(z; k = 0)$ are included in the same category. Now, let us describe by $\rho_0(E) \equiv N^{-1} \sum_k \delta[E - \varepsilon(k)]$ the density of states for the case where $D = 0$ in the Hamiltonian (11.4.1). The density of states $\rho_0(E)$, or more fundamentally, the dispersion relation $\varepsilon(k)$, can be evaluated through the traditional band theory calculations for a given crystal structure and V_{jl}. On the other hand, it is easy to show that $-N^{-1} \operatorname{Tr}\{G_{av}(z)\}$ is expressed as a function of complex energy z in the form $F_0 \equiv N^{-1} \sum_k [z - \varepsilon(k)]^{-1}$, which is related to $\rho_0(E)$ by

$$F_0(z) = N^{-1} \sum_k [z - \varepsilon(k)]^{-1} = \int_{-\infty}^{\infty} dE(z - E)^{-1} \rho_0(E). \tag{11.5.4}$$

Since the translational invariance of the crystal is totally recovered when the ensemble average is carried out over all random configurations in (11.5.2), both $G_{av}(z)$ and H_{eff} also carry the translational invariance of the crystal, and accordingly become diagonal in the Bloch representation. This means that we can write $\langle k | G_{av}(z) | k'\rangle \equiv \delta_{kk'} G_{av}(z; k)$. The effective Hamiltonian is written as

$$\langle k | H_{eff}(z) | k'\rangle \equiv \langle k | [W + \Sigma_t(z)] | k'\rangle$$
$$\equiv \delta_{kk'} [\varepsilon(k) + \Sigma_t(z; k)]. \tag{11.5.5}$$

This equation defines $\Sigma_t(z)$ which is usually called the total self-energy. As can be seen from the definition, $\Sigma_t(z)$ carries all information concerning the effects of scattering by the random configurations of atoms. Therefore, our problem is reduced to the problem of calculating $\Sigma_t(z)$ by some appropriate methods. In what follows, we investigate how $\Sigma_t(z)$ can be evaluated.

In $\langle G(z) \rangle$ of (11.5.2) defining $G_{av}(z)$, randomness in the atomic configurations is included in H which appears in the inversed form $(z1 - H)^{-1}$ so that it is normally rather difficult to take the ensemble average in this form. A conventional procedure to deal with this kind of situation is to calculate $G_{av}(z)$ through the following three steps.

(I)   Divide H into an unperturbed part K (which does not include randomness) and the perturbation $(H - K)$ (which has random terms); then expand $G(z) = [z1 - H]^{-1} = [(z1 - K) - (H - K)]^{-1}$ in the perturbation series where the unperturbed term is chosen to be $P(z) \equiv [z1 - K]^{-1}$ and the perturbation to be $(H - K)$.

(II)   Take the ensemble average of each term in the perturbation series.

(III) Take the summation of the ensemble-averaged terms.

In this procedure, the quantities which must be averaged are all of the form $(H - K)^n$, n being 1, 2, ..., and naturally the ensemble average becomes simpler compared to the case in which averages must be taken of the quantity in the inversed form. Note that the way of dividing H into two parts as stated in (I) is not determined uniquely. As will be explained in later sections, K can be chosen in such a manner as is most convenient in each case. To start with, let us consider the case where K is chosen in the following form:

$$K = \sum_{j} |j\rangle u(z) \langle j| + \sum_{j \neq l} |j\rangle V_{jl}\langle l|. \tag{11.5.6}$$

The trace $F(z)$ of $P(z)$, the Green's function for K, is given by

$$F(z) = N^{-1} \operatorname{Tr}\{P(z)\} = \langle 0|P(z)|0\rangle = F_0[z - u(z)], \tag{11.5.7}$$

where use has been made of (11.5.4) and of the fact that K is translationally invariant. We write the basis $|l = 0\rangle$ in the Wannier representation as $|0\rangle$. It is readily seen that the density of states for the unperturbed Hamiltonian K can be evaluated by using $F(z)$ in (11.5.7). This function $F(z)$ is also used in the equation which defines the t-matrix of scattering as will be shown below. Furthermore, it will turn out that $F(z)$ is generally very useful in the formulations in what follows. Hereafter, the variable z in operators will be omitted unless it causes confusion. Using (11.5.6), H and G are rearranged in the following forms:

$$\begin{aligned} H &= K + \sum_{j} |j\rangle [E_j - u(z)]\langle j| \\ &\equiv K + \sum_{j}|j\rangle v_j\langle j| \equiv K + \sum_{j} v_j, \end{aligned} \tag{11.5.8}$$

$$G = P + P \sum_{j} v_j P + P \sum_{j} v_j P \sum_{l} v_l P + P \sum_{j} v_j P \sum_{l} v_l P \sum_{m} v_m P + \cdots. \tag{11.5.9}$$

If we introduce the notation t_j to denote the effects of multiple scattering by the same lattice site, we can further rearrange G as

$$G = P + P \sum_j t_j P + P \sum_j t_j P \sum_{l(\neq j)} t_l P + P \sum_j t_j P \sum_{l(\neq j)} t_l P \sum_{m(\neq l)} t_m P + \cdots .$$

$$(11.5.10)$$

In the above equation, the summations over lattice sites j, l, m, etc., are carried out with the restriction that the successive parameters of each pair (j and l, or l and m for instance) are not equal to each other. The operator t_j is given by

$$t_j \equiv v_j[1 - Pv_j]^{-1} = |j\rangle v_j[1 - v_j F_0(z - u(z))]^{-1}\langle j| \equiv |j\rangle t_j \langle j| . \quad (11.5.11)$$

The c number, t_j, which appears in the last equation of (11.5.11) is given by $t_j = v_j[1 - v_j F_0(z - u(z))]^{-1}$ which is nothing but the scattering t-matrix (usually called the Slater–Koster t-matrix) due to a single impurity atom at R_j specified by a potential v_j which is located in the hypothetical crystal characterized by the Hamiltonian K.

The t-matrix, T, for G is normally defined by $G = P + PTP$; then the average G_{av} is related to the average $\langle T \rangle$ by $G_{av} = P + P\langle T \rangle P$. It is easy to see from (11.5.10) that T is expressed by the perturbation series

$$T = \sum_j t_j + \sum_j t_j P \sum_{l(\neq j)} t_l + \sum_j t_j P \sum_{l(\neq j)} t_l P \sum_{m(\neq l)} t_m + \cdots . \quad (11.5.12)$$

If we divide the total self-energy $\Sigma_t(z)$ as defined by (11.5.5) into two terms, i.e., into the diagonal operator $u(z)\mathbf{1}$ and the remaining part [denoted by $\Sigma(z)$], then we write $\Sigma_t(z) = u(z)\mathbf{1} + \Sigma(z)$. Now, it is readily seen that $\Sigma(z)$ corresponds to the total self-energy defined for the case where the unperturbed system is so chosen that it is described by K. Then, the average Green's function satisfies the relation

$$G_{av} = P + P \Sigma G_{av} \quad (11.5.13)$$

which gives the relationship between Σ and $\langle T \rangle$ in the form

$$\Sigma = \langle T \rangle[\mathbf{1} + P\langle T \rangle]^{-1} . \quad (11.5.14)$$

To summarize, (11.5.14) suggests that Σ can be obtained when $\langle T \rangle$ is calculated, and then the total self-energy Σ_t can be derived from Σ. Step (I) is the easiest of the three steps described above and can always be performed in principle.

Let us study Step (II) and see how we can perform the statistical average. For this purpose, let us take as an example a disordered binary alloy denoted by $A_c B_{1-c}$, where c is the normalized concentration of A atoms. When N_A is the number of A atoms and N_B that of B atoms, then c is given by $c \equiv N_A/(N_A + N_B) \equiv N_A/N$, N being the total number of atoms. In this chapter, we shall be concerned with only completely random distributions of atoms in the sense that there is no correlations in the ways of distributing A and B atoms on a lattice. To be more precise, a completely random distribution indicates that a probability of finding an A or B atom on an arbitrary site is, respectively, c or $(1 - c)$, irrespective of the atomic configuration in the neighborhood. As a result, the whole statistical average is replaced by a product of mutually-independent statistical averages, each at every lattice site. When this replacement is allowed, then it is convenient to introduce new random variables defined by

$$\xi_j = \begin{cases} 1 \ (\text{when the site } j \text{ is occupied by an atom A}), \\ 0 \ (\text{when the site } j \text{ is occupied by an atom B}). \end{cases} \tag{11.5.15}$$

Using ξ_j, (11.5.8) is rewritten in the form

$$\boldsymbol{H} = \boldsymbol{K} + \sum_j |j\rangle[\xi_j(E_A - u(z)) + [(1 - \xi_j)(E_B - u(z))]\langle j| \, . \tag{11.5.16}$$

Note that $\langle \xi_j \rangle = c$. The average $\langle \xi_j \xi_l \rangle$ for $j = l$ is calculated as $\langle \xi_j \xi_l \rangle = \langle \xi_j^2 \rangle$ $= \langle \xi_j \rangle = c$ while $\langle \xi_j \xi_l \rangle$ for $j \neq l$ is reduced to $\langle \xi_j \xi_l \rangle = \langle \xi_j \rangle \langle \xi_l \rangle = c^2$. Therefore, when we take the average of the third term in (11.5.9), we must distinguish the case where $j = l$ from the case where $j \neq l$. Similarly, the fourth term of (11.5.10) yields different average values for $j = m$ and for $j \neq m$. In this way, the averages of the higher-order terms in the perturbation series (11.5.9) or (11.5.10) become more complicated when the terms become higher.

However, it is still possible, in principle, to take the statistical averages of the terms in the perturbation series since the replacement of the total average by the product of individual averages has simplified the problem enormously. Thus, Step (II) is essentially within our reach. The most serious difficulty arises in Step (III). In general, it is impossible to sum up in some compact forms all the averaged terms in the perturbation series up to the infinite order because the infinite perturbation expansion when averaged becomes far more complicated than the pre-average perturbation series, the latter being known to be a geometric series which can be neatly summed up. This means that G_{av} cannot be evaluated rigorously except for some idealized cases. Therefore, G_{av} is usually calculated by some approximate methods. Namely, on the basis of some physical arguments, we choose the most dominant terms out of the post-average infinite perturbation series and neglect all other terms in order to take appropriate partial summations, which leads to approximate G_{av}. The characteristics and validity of an approximation are examined by asking which partial summation has been chosen to derive that approximation and what were the criteria leading to it.

The CPA, which is the theme of this chapter, is the best single-site approximation for calculating G_{av} among various approximations proposed so far, the best in the sense that its physical ground is confirmed, its mathematical significance is verified and its actual mathematical structure is reasonably simple. What is more, the CPA has provided really ample results in various aspects which can hardly be expected from its simplicity. The CPA has succeeded in explaining systematically and beautifully a vast variety of experimental data which were not understood before. It is no exaggeration to say that the epoch-making success of the CPA is regarded as one strong support to our belief that "nature" is after all simple and is governed by simple rules, and therefore it must be explained by some simple concepts.

11.6 Approximations before CPA

In this section, a short review is given to see what kinds of approximations were proposed before the CPA and in what respects they were not successful.

Generally speaking, an approximation is judged by seeing whether or not that approximation satisfies some minimum requirements which appropriate approximations ought to satisfy. For example, the minimum requirements an approximation for a disordered binary alloy A_cB_{1-c} should satisfy may be stated as follows:

1) Is the physical ground of the approximation confirmed?
2) Is the mathematical meaning of the approximation acceptable?
3) Is the approximation applicable to the whole concentration region, i.e., to the whole value of c $(0 \leq c \leq 1)$? Analogously, is the dual symmetry satisfied?
4) Is the behaviour in various limits reproduced within the scope of the approximation?
5) Is it possible to attain both amalgamation and persistence types from the same formulations when appropriate parameters are changed?
6) Is the Green's function analytic?
7) Is the sum rule fulfilled?

The Hamiltonian for our binary alloy A_cB_{1-c} is expressed by (11.4.1) or (11.4.2) in which E_j is E_A or E_B according to whether the atom on the jth site is of type A or B. The dual symmetry mentioned in Item 3) means that the formulations obtained from an approximation are symmetric with respect to the exchange of $c_A \equiv c$ by $c_B \equiv 1 - c$ and E_A by E_B. Concerning Item 5), the appearance of both amalgamation and persistence types has been suggested from the experiments of the exciton absorption spectra of mixed ionic crystals. This can also be expected from the Saxon–Hutner theorem. In our model, the situation is stated as follows. Consider a parameter $\delta \equiv |\Delta/2w|$ where $\Delta \equiv E_A - E_B$ and $2w$ is the bandwidth (which in our model is common to the pure A crystal and the pure B crystal). From naive physical consideration, a one-mode spectra (the amalgamation type) is expected when $\delta \ll 1$. On the other hand, a two-mode spectra (two bands extending around E_A and E_B, respectively, and referred to as the persistence type) is expected when $\delta \gg 1$. The sum rule required in Item 7) asserts that, in the case of the two-mode type, the total density of states in the band around E_A should be c while that in the band around E_B should $(1 - c)$.

With these requirements in mind, let us study the approximate methods proposed in the pre-CPA period. Description of an approximation is made by clarifying (I) what do we choose for the unperturbed Hamiltonian K and (II) what partial summation do we choose out of the infinite perturbation series. The approximations to be discussed in this section are limited to those which belong to the category of single-site approximations (SSA). The SSA indicates that, from the perturbation series in (11.5.9) or (11.5.10), we take into account only those terms which can be reorganized in the products of terms with "essentially single-site" nature. A clearer picture of the SSA would be obtained from the analysis of each approximation in the following.

As explained in the preceding section, our task is reduced to evaluating the self-energy. Talking of the self-energy, the SSA is characterized by a self-energy which is diagonal (and naturally independent of site) in the Wannier representation, i.e., $\Sigma(z) = \sum_j |j\rangle \Sigma^{(1)}(z)\langle j|$. This indicates that $\Sigma(z, k) = \Sigma^{(1)}(z)$ and thus the self-energy is diagonal and independent of k (As a matter of fact, the self-energy in this form is diagonal and independent of each basis function in any representation.). Accordingly, the total self-energy $\langle k|\Sigma_t(z)|k'\rangle$ becomes diagonal and independent of k. Expressing this total self-energy by $\Sigma_t(z)$, we have

$$\Sigma_t(z) = u(z) + \Sigma^{(1)}(z). \tag{11.6.1}$$

In this case, the function $\mathscr{F}(z)$ related to the trace of G_{av} is written in terms of the function $F_0(z)$ defined by (11.5.4). That is to say

$$\mathscr{F}(z) = N^{-1}\operatorname{Tr}\{G_{av}(z)\} = N^{-1}\sum_k [z - \Sigma_t(z) - \varepsilon(k)]^{-1}$$
$$= F_0[z - \Sigma_t(z)]. \tag{11.6.2}$$

The density of states is given by $\rho(E) = -\pi^{-1}\operatorname{Im}\{\mathscr{F}(E + i0)\}$.

The limitation of the SSA will be discussed in a later section.

11.6.1 Rigid Band Model (the Lowest-Order Virtual Crystal Model)

To start with, let us study the case where $K = W$, i.e., $u(z) = 0$. Out of the terms in the perturbation expansion of (11.5.9), we take into account only those terms in which the concerned sites, j, l, m, etc., are different from one another. Then the average of $v_j v_l v_m$, etc., can be taken independently for each v_j, v_l, v_m, etc., and we have

$$\langle G \rangle = P + P\sum_j \langle v_j \rangle P + P\sum_j \langle v_j \rangle P \sum_{l(\neq j)} \langle v_l \rangle P$$
$$+ P\sum_j \langle v_j \rangle P \sum_{l(\neq j)} \langle v_l \rangle P \sum_{\substack{m(\neq l) \\ (\neq j)}} \langle v_m \rangle P + \cdots. \tag{11.6.3}$$

Then, if we ignore the restrictions on the summations, the whole terms are of the form of an infinite-order geometric series. Therefore, the summation can be performed and the self-energy is obtained as $\Sigma^{(1)}(z) = \langle v_j \rangle = cE_A + (1 - c)$ E_B. This is the arithmetic mean of E_A and E_B with the respective weights c and $(1 - c)$. Let us denote this average by $\bar{v}$. Thus, the approximation described in the above corresponds to taking a virtual crystal with a potential of magnitude $\bar{v}$ on each lattice site. The density of states can be evaluated from $\mathscr{F}(z) = F_0$ $(z - \bar{v})$. As a result, the shape and width of the band is unchanged when c is increased from 0 to 1, but the centre of gravity of the band shifts linearly in c from E_B to E_A. Apparently, the requirement stated in Item 5) at the beginning of this section is not fulfilled in this approximation. This, however, is a natural consequence of the assumption used in approximating (11.5.9) by (11.6.3). The assumption corresponds to neglecting $\langle v_j P v_j \rangle$, $\langle v_j P v_j P v_j \rangle$, etc., in comparison with $\langle v_j \rangle$ which is valid only in the weak-scattering limit, or in other words, only in the case of $|E_A - E_B|/2w = \delta \ll 1$.

11.6.2 Virtual-Crystal Model

In order to improve the rigid band model discussed in the preceding section, let us take $u(z) = \bar{v}$. Out of the terms in the expansion of (11.5.9), all those terms appearing in (11.6.3) have already been included in this approximation. Now, out of the terms which have been discarded in the approximation from (11.5.9) to (11.6.3), we now take into account those terms which can be expressed by products of some $\langle v_j \rangle$ and some $\langle v_l \cdot v_l \rangle$. For instance, terms included are $\langle v_j \cdot v_j \rangle$, $\langle v_j \rangle \langle v_l \cdot v_l \rangle$, $\langle v_j \cdot v_j \rangle \langle v_l \rangle$, $\langle v_j \cdot v_j \rangle \langle v_l \cdot v_l \rangle$, $\langle v_j \rangle \langle v_l \rangle \langle v_m \cdot v_m \rangle$, $\langle v_j \rangle \langle v_l$ $\cdot v_l \rangle \langle v_m \rangle$, $\langle v_j \cdot v_j \rangle \langle v_l \rangle \langle v_m \rangle$, etc. It is an easy task to see that, when the restrictions on summation are ignored, the whole terms are rearranged in the form of an infinite geometric series and, accordingly, the summation can be evaluated. The self-energy is derived as

$$\Sigma^{(1)}(z) = \langle v_j^2 \rangle F_0(z - \bar{v}) = c(1 - c) \Delta^2 F_0(z - \bar{v}). \tag{11.6.4}$$

The calculated density of states for the binary alloy based on this virtual-crystal approximation is given in Fig. 11.9a for various values of δ where c is fixed. The calculation is performed for the case where $\rho_0(E)$ has the semi-elliptic form. The band width becomes larger as δ increases. From the figure, we can conclude that this virtual-crystal model is an improvement over the rigid band model in the sense that the atomic levels $E_A = 2w\delta$ and $E_B = 0$ are definitely covered by the alloy band for any value of δ. However, the situation where the amalgamated one-mode band is always achieved even for large values of δ is not improved. This result also originates from the same cause as before, that is, $\langle v_j P v_j P v_j \rangle$ and higher-order terms are discarded compared with $\langle v_j \rangle$ and $\langle v_j P v_j \rangle$.

11.6.3 Dilute Limit

In this section, we study the case where a small number of A atoms are doped as impurity atoms in the host B crystal. The best unperturbed system in this case is

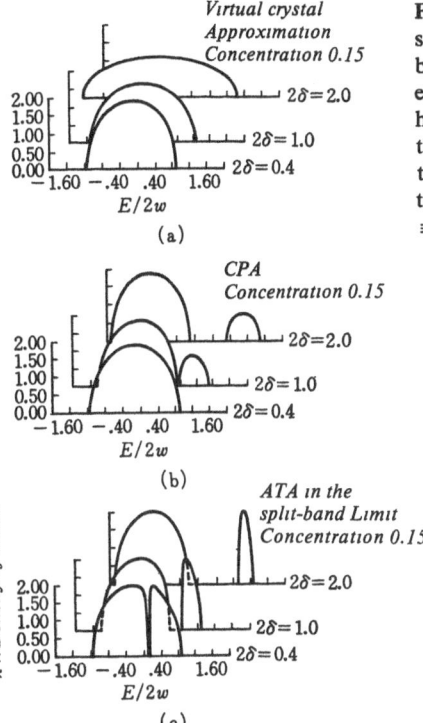

Fig. 11.9a–c. Comparison of the densities of states of a substitutionally disordered AB binary alloy as calculated in (a) the self-consistent virtual crystal approximation, (b) the coherent potential approximation (CPA), and (c) the average t-matrix approximation (ATA) in the split-band limit. In each case, the concentration c of A atoms is 0.15. The values of $2\delta \equiv |E_A - E_B|/w$ are 0.4, 1.0 and 2.0. [11.10]

naturally the pure B crystal. Therefore, we have $u(z) = E_B$. Since c is very small, the effects of various impurity (A) atoms can be regarded as independent of one another. In order for the approximation to be valid even when δ is large, let us start from (11.5.10) rather than from (11.5.9). If we apply the same procedure to (11.5.9) as we have used to approximate (11.5.9) by (11.6.3), we obtain

$$\langle G \rangle = P + P \sum_j \langle t_j \rangle P + P \sum_j \langle t_j \rangle P \sum_{l(\neq j)} \langle t_l \rangle P$$
$$+ P \sum_j \langle t_j \rangle P \sum_{l(\neq j)} \langle t_l \rangle P \sum_{m\binom{\neq l}{\neq j}} \langle t_m \rangle P + \cdots \tag{11.6.5}$$

which has the same structure as (11.6.3) with v_j, etc., replaced by t_j, etc. If we again ignore the restrictions on summation, the self-energy is written as

$$\Sigma^{(1)}(z) = \langle t_j \rangle = \frac{c\Delta}{1 - F_0(z - E_B)\Delta}, \tag{11.6.6}$$

where we have used (11.5.11). This is equal to the result of multiplying the Slater-Koster scattering t-matrix of a single A impurity by c with the scattering potential $\Delta = E_A - E_B$ in the medium described by $u(z) = E_B$. Therefore, this approximation corresponds to treating N_A independent impurity atoms in an exact manner, under the condition that the correlations between the impurity atoms are completely switched off. This condition is plausible when c is small enough. The density of states calculated on the basis of this approximation gives an isolated impurity level off the B band in the limit of $c \to 0$. The level corresponds to a localized mode due to an A atom. The level is broadened into an impurity band for a small but finite value of c. However, this impurity band shows some unphysical features such as the divergence of the density of states. As can be expected from the assumptions of independent impurity atoms described above, the applicability of this approximation is limited. Naturally, the dual symmetry is not satisfied.

11.6.4 The Average t-Matrix Approximation (ATA)

The idea of the approximation explained in this section is valid irrespective of the choice of the unperturbed system. Therefore, we present, in what follows, the formalism with $u(z)$ left unspecified. The procedure here follows that of the preceding section up to the point where (11.5.10) is approximated by (11.6.5). The difference enters in the way of treating the restrictions on summation. In the dilute limit we have ignored the restrictions completely, while we include the effects of the restrictions partially. The most dominant influence of the restrictions appears in the fact that the diagonal terms of some of the P's in (11.6.5) are automatically excluded from the summation. For instance, the diagonal terms of the second P in the third term in (11.6.5) are excluded because of the restriction $l = j$. Similarly, the diagonal terms of the second and third P's in the fourth term are excluded. In the dilute limit, this influence has safely been ignored since c is small. Now, in order to make the formulation more tractable when this influence is taken into account, we express (11.6.5) in terms of the matrix representation. By $\hat{P}$, we denote a matrix whose (j, l) element is $\langle j|P|l\rangle$, i.e., $\hat{P} = \{\langle j|P|l\rangle\}$. In the same way, we define $\hat{t} \equiv \delta_{jl}t_j$ and $\hat{G} = \{\langle\{j|G|l\rangle\}$. Then, the matrix for (11.6.5) is expressed as

$$\langle\hat{G}\rangle = \hat{P} + \hat{P}\langle\hat{t}\rangle\hat{P} + \hat{P}\langle\hat{t}\rangle\hat{P}'\langle\hat{t}\rangle\hat{P} + \hat{P}\langle\hat{t}\rangle\hat{P}'\langle\hat{t}\rangle\hat{P}'\langle\hat{t}\rangle\hat{P} + \cdots$$
$$= \hat{P} + \hat{P}[\langle\hat{t}\rangle(\hat{1} - \hat{P}'\langle\hat{t}\rangle)^{-1}]\hat{P}, \tag{11.6.7}$$

where $\hat{1}$ denotes the unit matrix and $\hat{P}' \equiv (\langle j|P|l\rangle - \delta_{jl}\langle 0|P|0\rangle)$. From (11.6.7), it is readily seen that the matrix of the approximate t-matrix $\langle T\rangle$ is given by $[\langle\hat{t}\rangle(\hat{1} - \hat{P}'\langle\hat{t}\rangle)^{-1}]$. By inserting this into the matrix of $\langle T\rangle$ in (11.5.14), we obtain the self-energy in the form

$$\Sigma^{(1)}(z) = \frac{\langle t_j\rangle}{1 + \langle t_j\rangle\langle 0|P|0\rangle}. \tag{11.6.8}$$

Equation (11.5.7) gives $\langle 0|P|0\rangle = F_0[z - u(z)]$. It is easy to show that $\langle t_j\rangle$ is given by

$$\langle t_j \rangle = \frac{c[E_A - u(z)]}{1 - [E_A - u(z)]F_0[z - u(z)]} + \frac{(1 - c)[E_B - u(z)]}{1 - [E_B - u(z)]F_0[z - u(z)]} \quad (11.6.9)$$

Equations (11.6.8, 9) are the general equations for the ATA.

In the first place, let us apply our result for the ATA to the case where the unperturbed system is the pure B crystal as in the case of the dilute limit. Then, $u(z) = E_B$. The second term of (11.6.9) vanishes, and $\langle t_j \rangle$ has the same form as in the case of the dilute limit, i.e., $\langle t_j \rangle = c\varDelta/[1 - F_0(z - E_B)\varDelta]$. Inserting this in (11.6.8), we can evaluate the self-energy as

$$\Sigma^{(1)}(z) = \frac{c\varDelta}{1 - (1 - c)F_0(z - E_B)\varDelta} \,. \quad (11.6.10)$$

The only difference between (11.6.6, 10) is the factor $(1 - c)$ which appears in the denominator of (11.6.10). This factor is shown to remove some of the unphysical properties of the impurity band calculated from (11.6.6), and the approximation is applicable to a wider range of c (though c is still small) than the dilute limit approximation in the preceding subsection. In the limit of $c \to 1$, the approximation approaches the virtual crystal approximation. Therefore, the requirement stated in Item 5) is formally fulfilled although the dual symmetry is not yet attained.

The dual symmetry is realized by taking the rigid band model as an unperturbed system. This indicates that $u(z) = \bar{v}$. It is easily noticed from (11.6.8, 9) that the dual symmetry is achieved in this case. Therefore, this approximation yields the formulations which are the reasonable interpolations over the whole range of c. The phonon density of states $D(\omega^2)$ calculated by this approximation is shown by the solid curves in Fig. 11.6a for four values of c. As a whole, the approximate results show satisfactory agreement with the results of the computer experiment. Definitely, the agreement is remarkably better than those of any previous methods.

The weakest point of this approximation, as is obvious from Fig. 11.6a, is its failure to reproduce the band edges correctly. Another point noticeable from the figure is that the results of the approximation deviate from the results of the computer work most seriously near the upper band edge of the pure B crystal, i.e., near $\omega^2/\omega_M^2 \sim 1$. It is also noted that the approximation is better for the cases with large c values than for the cases with small c values.

Before concluding this section, let us study another application of the ATA equations (11.6.8, 9). We replace $F_0[z - u(z)]$ in (11.6.9) by $[z - u(z)]^{-1}$ where $u(z)$ is left unspecified. As can be derived from (11.5.4), this replacement is exact in the limit of $\varepsilon(\boldsymbol{k}) \to 0$. The self-energy in (11.6.8) then becomes

$$\Sigma^{(1)}(z) = \frac{z\bar{v} - E_A E_B}{z - [cE_B + (1 - c)E_A]} - u(z) \,. \quad (11.6.11)$$

The integrated density of states for a disordered chain is calculated by this approximation and shown by the solid curve in Fig. 11.8a. This approximation yields a clear-cut band gap over the energy region between -2.5 and -1. Since the localized level due to the delta-function potential of a B atom appears at $E = -1$, the energy gap between -2.5 and -1 is obviously attributed to the inadequacy of the approximation. The density of states for a three-dimensional binary alloy calculated on the basis of this approximation is shown in Fig. 11.9c for various values of δ. A most remarkable feature of the calculated density of states is that the energy gap always appears, no matter how small c may be. This corresponds to the opposite extreme of the virtual crystal approximation whose results always give the one-mode band as shown in Fig. 11.9a. The appearance of the gap can be detected from the form of the self-energy. The density of states is related to the function $\mathscr{F}(z)$ defined by $\mathscr{F}(z) = N^{-1}\mathrm{Tr}\{G_{\mathrm{av}}(z)\}$ which, on the other hand, is given by $\mathscr{F}(z) = F_0[z - u(z) - \Sigma^{(1)}(z)]$. Inserting the self-energy (11.6.11) into $F_0[z - u(z) - \Sigma^{(1)}(z)]$ and remembering the definition of $F_0(z)$ given by (11.5.4), we can readily show that $\mathscr{F}(z)$ is always zero at $z = cE_B + (1 - c)E_A$. Therefore, the density of states for the alloy always vanishes at $E = cE_B + (1 - c)E_A$ irrespective of the value of δ. Because of this fact, this approximation is sometimes called the approximation in the split-band limit.

It is interesting to note that (11.6.11) can be rearranged in the following form

$$\frac{1}{z - \Sigma_t(z)} = \frac{c}{z - E_A} + \frac{1 - c}{z - E_B}, \tag{11.6.12}$$

where $\Sigma_t(z) = u(z) + \Sigma^{(1)}(z)$. This equation is suggestive if we introduce the following interpretation. Suppose the average Green's function can be approximated by the arithmetic mean of the Green's function $G_A(z) = (z1 - H_A)^{-1}$ for the pure A crystal and the Green's function $G_B(z) = (z1 - H_B)^{-1}$ for the pure B crystal with the respective weight c and $(1 - c)$. That is to say

$$G_{\mathrm{av}}(z) = cG_A(z) + (1 - c)\, G_B(z). \tag{11.6.13}$$

Here, $H_A = E_A 1 + W$ and $H_B = E_B 1 + W$. The trace of (11.6.13) is given by

$$F_0[z - \Sigma_t(z)] = cF_0(z - E_A) + (1 - c)\, F_0(z - E_B). \tag{11.6.14}$$

When we take the same limit $\varepsilon(\mathbf{k}) \to 0$ as we have used to obtain (11.6.11), then (11.6.14) reduces to (11.6.12). Therefore, (11.6.12) defines the self-energy which is an improvement over the atomic limit while the latter requires $\varepsilon(\mathbf{k}) = 0$ throughout the formulation. This clarifies the meaning of the approximation (11.6.11), and gives an explanation for the split band.

The pre-CPA methods presented in this section are applicable only in some limited regions of parameters. In view of this situation, the development of a unified theory which was valid in the wide range of parameters including the whole range of the concentration c and the magnitude of scattering (described by

δ) was desired. The CPA was proposed as an example of such a unified theory. But, as we have learned in this section, we really had to go a long way before we could reach the CPA.

11.7 Derivation and the Characteristic Features of the CPA

First, let us recall the fundamental principle for evaluating G_{av}. The closer the unperturbed Hamiltonian K is chosen to the effective Hamiltonian H_{eff}, the smaller the effects of the perturbation terms become. Actually, the best choice of K is to take $K = H_{eff}$. Note, however, that K in this case cannot be expressed in the simple form of (11.5.6) because H_{eff} has the off-diagonal terms other than just V_{jl}. If $K = H_{eff}$, the Green's function P for K is equal to $G_{av} \equiv \langle G \rangle$. Therefore, the average of the equation

$$G = P + PTP \tag{11.7.1}$$

which is written as

$$G_{av} = \langle G \rangle = P + P\langle T \rangle P, \tag{11.7.2}$$

indicates $\langle T \rangle = 0$ because $P = \langle G \rangle = G_{av}$. On the other hand, it is straightforward to show from (11.5.13, 14) that $\Sigma = 0$ when $P = G_{av}$.

For a general K, G_{av} is evaluated if $\langle T \rangle$ is obtained while $\langle T \rangle$ is calculated by taking the average of the total t-matrix which is usually expressed in the form of the perturbation series (11.5.12). Here again, the average t-matrix $\langle T \rangle$ thus calculated will turn out to be zero if K is chosen to be H_{eff}. This fact can be made use of as follows. Since, for a given K, $\langle T \rangle$ is calculated as a function of K, $\langle T \rangle \equiv \langle T(K) \rangle = 0$ may be regarded as an equation defining K. Then by solving this equation, we can find an ideal K, i.e., we can find $K = H_{eff}$. In practice, however, we have learned in the preceding section that, in general cases, it is almost impossible to calculate the average of T exactly because T is given in the form of (11.5.12).

An idea, therefore, is to evaluate some approximate form of $\langle T \rangle$ (let us denote this approximation by $\langle T \rangle_{app}$) for some unknown K and use $\langle T(K) \rangle_{app} = 0$ as an equation to define K. As an example, let us apply this idea to the approximation for $\langle T \rangle$ as expressed in (11.6.5), when P is now regarded as the Green's function for unknown K. It is obvious from (11.6.5) that $\langle T \rangle_{app} = 0$ if $\langle t_j \rangle = 0$. Since t_j is associated with site j alone, t_j is expressed as $t_j \equiv |j\rangle t_j \langle j|$. This indicates that the condition $\langle t_j \rangle = 0$ is equivalent to the condition $\langle t_j \rangle = 0$. In addition, it is seen that this approximation is an SSA as explained in Sect. 11.6. This means that the self-energy is a function of z alone and that K can be expressed in the form (11.5.6). All these reduce to the condition $\langle t_j \rangle = 0$, $\langle t_j \rangle$ being given by (11.6.9). This approximation is nothing but the CPA.

The physical meaning of the condition $\langle t_j \rangle = 0$ is illustrated in the following way. Using (11.6.9), we can rewrite this condition as

$$\langle t_j \rangle \equiv c t_A + (1 - c) t_B = 0, \tag{11.7.3}$$

where t_A is the Slater–Koster t-matrix for a single impurity atom specified by the scattering potential $E_A - u(z)$ and located at R_j [or at the origin] in the effective medium defined by the Hamiltonian $K = \sum_j |j\rangle u(z)\langle j| + W$. Similarly, t_B is the corresponding quantity for a single B impurity atom. The CPA equation (11.7.3) is illustrated in Fig. 11.10, in which every lattice site except for the central site is occupied by a pseudoatom characterized by potential $u(z)$ (which we call the "coherent potential"). The pseudoatoms in the figure are represented by the hatched circles. The central site is occupied either by an A' atom characterized by potential $E'_A \equiv E_A - u(z)$ or by a B' atom characterized by potential $E'_B \equiv E_B - u(z)$. The probability of finding an A' atom or a B' atom on the central site is, respectively, $c_A \equiv c$ or $c_B \equiv 1 - c$. Thus, disorder is maintained only on the central site. Fluctuation $E'_A = E_A - u(z)$ or $E'_B = E_B - u(z)$ is regarded as an excess scattering potential which acts on electrons. The requirement that the central site be on the average equivalent to any other site on the lattice is rephrased by the requirement that the effect of the t-matrix due to the excess scattering potential on the central site be zero on the average. The latter requirement defines $u(z)$.

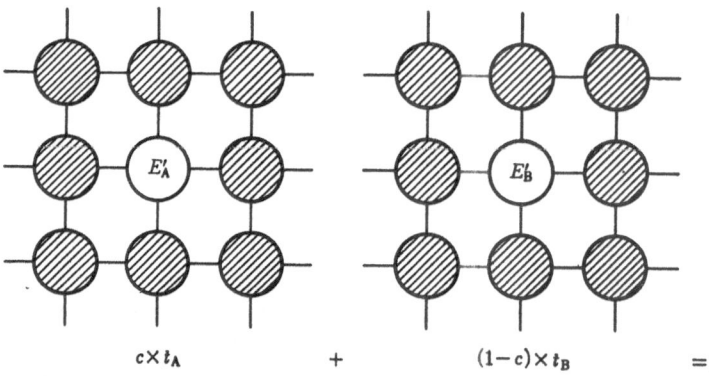

$$c \times t_A \qquad + \qquad (1-c) \times t_B \qquad = 0$$

Fig. 11.10. A schematic illustration of the fundamental concept of CPA in the case of a substitutionally-disordered binary $(A_c B_{1-c})$ alloy

The idea illustrated by Fig. 11.10 is generally known as an effective medium approximation and has been used in various fields of physics rather independently. The outline of the approximation is stated as follows. Normally, it is very difficult or nearly impossible to obtain the exact average of some quantity over the ensemble of all microscopic configurations of atoms in a disordered system.

This is mainly because disorder or fluctuation is not spatially local, which means that there is a finite probability of finding fluctuation in any part of the system. To deal with this situation, the principle of the effective medium approximation is to assume that the exact average can be approximated by the average of a hypothetical disordered system in which disorder is maintained only in a restricted region Ω_0 in the system specified as an effective medium. The region Ω_0 is chosen to be the central site in our problem as shown in Fig. 11.10. The effective medium outside this region is defined by requiring that the chosen region Ω_0 on the average must behave exactly in a similar manner to any arbitrary part of the effective medium. Obviously, this theory is equivalent to the above-described idea of the CPA for which the fluctuation or the deviation in Ω_0 from the average value must vanish on the average. As mentioned, this idea has been widely used not only for problems due to spatial disorder of the static type, such as those discussed in this section, but also for problems due to disorder of the dynamical type such as thermal disorder. An example included in the former category is the effective medium approximation for continuous media while the molecular field approximation in the problem of phase transitions is considered a typical example of the latter class. In both these cases, it is well known that the idea is very successful. This confirms that the CPA is physically grounded, and thus the condition stated in Item 1) of Sect. 11.6 is guaranteed.

The CPA is named after the "coherent potential" denoting the potential of the "pseudoatom". It is important to notice that the potential $u(z)$ to be determined self-consistently through the procedure explained above is not a real value but a complex value.

The mathematical ground of the CPA as required in Item 2) will be studied in Sect. 11.6 and it will be shown that the CPA is mathematically verified as well. As will be mentioned in Sect. 11.9, the CPA has been achieved independently through various different methods by several research groups. This fact seems to lend support to the validity and universality of the CPA. It is also obvious from (11.7.3) that the CPA satisfies the dual symmetry as stated in Item 3).

The behaviour of the CPA in various limits as mentioned in Item 3) can be examined by methods presented below. In the discussions to follow, we use the fact that (11.7.3) can be rearranged in the form

$$u(z) - E_{\mathrm{B}} = \frac{c\Delta}{1 - \mathscr{F}(z)\{\Delta - [u(z) - E_{\mathrm{B}}]\}} \ . \tag{11.7.4}$$

On the other hand, the CPA condition in (11.7.3) can also be rearranged as

$$\mathscr{F}(z) = \frac{c}{\mathscr{F}(z)^{-1} + u(z) - E_{\mathrm{A}}} + \frac{1 - c}{\mathscr{F}(z)^{-1} + u(z) - E_{\mathrm{B}}} \ , \tag{11.7.5}$$

where (11.6.2) has been used.

11.7.1 Rigid Band Limit

The rigid band limit corresponds to the limit of $\delta = |\Delta/2w| \ll 1$. In the CPA equation (11.7.4), the sum of the terms collected in the square brackets appearing in the denominator is of order Δ since $[u(z) - E_B]$ is of order Δ. Because the product of Δ and $\mathscr{F}(z)$ is of order δ, it is small compared to unity. Accordingly, within the approximation which neglects terms $O(\delta)$ in comparison with $O(1)$, (11.7.4) reduces to $[u(z) - E_B] = c\Delta$, which yields $u(z)$ in the form $u(z) = cE_A + (1 - c)E_B = \bar{v}$. Therefore, the CPA in the limit of $\delta \ll 1$ reproduces the result of the rigid band which has been introduced in Sect. 11.6.1.

11.7.2 Virtual-Crystal Limit

Now, let us include the next, higher-order terms in δ. The terms in the square brackets in the denominator of (11.7.4) are written as $c\Delta[1 + O(\delta)]$. If we drop the second term $O(\delta)$ in the parentheses on the basis that it is much smaller than 1, then we can write (11.7.4) as

$$u(z) - E_B = c\Delta[1 + (1 - c)\Delta\mathscr{F}(z) + O(\delta^2)]$$

$$\therefore \quad u(z) = \bar{v} + c(1 - c)\Delta^2\mathscr{F}(z) + \text{higher-order terms.} \qquad (11.7.6)$$

In (11.7.6), $\mathscr{F}(z)$ in the second term is given by $\mathscr{F}(z) = F_0[z - u(z)]$. Instead of determining $u(z)$ self-consistency, we approximate $u(z)$ appearing in $F_0[z - u(z)]$ by $u(z) = \bar{v}$. This is consistent with the policy of the present approximation that $O(\delta^3)$ and higher-order terms are neglected compared to $O(1)$, $O(\delta)$ and $O(\delta^2)$. Substituting $\mathscr{F}(z)$ by $F_0(z - \bar{v})$ and dropping the third term of (11.7.6), we obtain the equation identical to (11.6.4). Thus the CPA has been proved to reproduce the virtual-crystal model.

11.7.3 Dilute Limit

The dilute limit corresponds to the limit of $c \to 0$. It is obvious from (11.7.4) or (11.7.6) that the lowest order in c of $[u(z) - E_B]$ is $O(c)$. Therefore, concerning the terms gathered in the square brackets in the denominator of (11.7.4), $[u(z) - E_B]$ can be neglected compared to Δ, which gives (11.7.4) in the form

$$u(z) - E_B = \frac{c\Delta}{1 - \mathscr{F}(z)\Delta}.$$

As was the case for (11.7.6), here again the equation for $u(z)$ contains $\mathscr{F}(z)$ which is defined as $\mathscr{F}(z) = F_0[z - u(z)]$. For $u(z)$ in $F_0[z - u(z)]$, we employ $u(z) = E_B$, which is obtained from the above equation by discarding terms of order δ or higher-order terms compared to unity. Substituting $\mathscr{F}(z)$ in the denominator by $F_0(z - E_B)$, we have $u(z) = E_B + c\Delta/[1 - F_0(z - E_B)\Delta]$ which is identical with

(11.6.6) for the dilute limit [note that $\Sigma_t(z) = u(z) + E_B$]. Thus this limit is also included in the CPA.

11.7.4 Split-Band Limit

The split-band limit corresponds to the limit of $w \ll |\Delta|$, or $\delta \gg 1$. When $w = 0$, $\varepsilon(\boldsymbol{k})$ is also zero. In this case, (11.6.2) yields $\mathscr{F}(z) = [z - \Sigma_t(z)^{-1}]$. If we take $\Sigma_t(z) = u(z)$, we have $\mathscr{F}(z) = [z - u(z)]^{-1}$ which, inserted into $\mathscr{F}(z)$ in (11.7.5), yields the split-band limit described by (11.6.12). This proves that the CPA also holds the split-band limit.

Thus in the above, the CPA equation, as described by (11.7.3, 4) or (11.7.5), has been shown to comprehend the previous theories, each of which is applicable only in some limit as explained in the preceding section. Therefore, the requirement stated in Item 4) is also satisfied by the CPA. As for the requirement that a desirable approximation must explain both aspects of amalgamation and persistence, a detailed discussion will be given in a later section. On the other hand, analyticity of the CPA Green's function on the complex energy (z) plane is also proved although the actual demonstration of the proof is omitted from this text to avoid too detailed mathematics. Let us just mention that this proof confirms the fulfillment of the requirement Item 5).

As for the sum rule as stated in Item 7), the proof will be given in the following way. We introduce a new variable ξ which is defined by

$$\xi \equiv \langle 0|G_{av}|0\rangle^{-1} + u(z) = \mathscr{F}(z)^{-1} + u(z)$$
$$= F_0[z - u(z)]^{-1} + u(z),\tag{11.7.7}$$

where we have used $\boldsymbol{P} = G_{av}$. Equations (11.5.7, 6.2) have also been made use of to obtain (11.7.7). Using ξ, (11.7.5) can be rewritten as

$$\mathscr{F}(z) = \frac{c}{\xi - E_A} + \frac{1 - c}{\xi - E_B}.\tag{11.7.8}$$

Now, the density of states per atom is given by $-(N\pi)^{-1}\mathrm{Im}\{\mathrm{Tr}\{G_{av}\}\} = -\pi^{-1}$. $\mathrm{Im}\{\mathscr{F}(E^+)\} = +(2\pi i)^{-1}[\mathscr{F}(E^-) - \mathscr{F}(E^+)]$, where $E^{\pm} = E \pm i0_+$. The total number of states per atom contained in a given band can be calculated by integrating the density of states over the energy region covered by the band. Let us consider the case where the density of states of a disordered binary alloy $A_c B_{1-c}$ is of a two-mode type, i.e., the case of the persistence type. Then, the total number of states per atom contained in the sub-band due to A atoms is given by

$$-(N\pi)^{-1}\int_{E_1}^{E_2}\mathrm{Im}\{\mathrm{Tr}\{G_{av}\}\{dE = (2\pi i)^{-1}\left[\int_{E_1}^{E_2}\mathscr{F}(E^-)dE + \int_{E_2}^{E_1}\mathscr{F}(E^+)dE\right]$$

$$\tag{11.7.9}$$

where E_1 and E_2 denote the energies of the bottom and top of this sub-band. It is shown that the one-to-one mapping is possible between a path of the integral on the complex energy (z) plane and that on the complex ξ plane. The path of the integral in (11.7.9) from E_1 to E_2 on $E^- = E - iO_+$ together with that from E_2 to E_1 on $E^+ = E + iO_+$ is mapped on the closed path c_A around $\xi = E_A$ on the complex ξ plane. The point $\xi = E_B$ is excluded from the area circled by the path c_A. Therefore, using (11.7.8, 9), we obtain

$$-(N\pi)^{-1} \int_{E_1}^{E_2} \text{Im} \{\text{Tr} \{\boldsymbol{G}_{av}\}\} \, dE = (2\pi i)^{-1} \oint_{c_A} \left[\frac{c}{\xi - E_A} + \frac{1-c}{\xi - E_B} \right] d\xi = c$$

$$(11.7.10)$$

and the total number of states per atom in the sub-band corresponding to A atoms is shown to be c. Similarly, we can show that the total number of states per atom in the B sub-band is given by $(1 - c)$. This completes the proof of the sum rule stated in Item 7).

11.8 Applications of the CPA

In this section, we present several examples of applications of the CPA to the calculations of various physical properties of a disordered binary alloy.

11.8.1 A One-Dimensional AB Alloy

In Fig. 11.8b, the solid curve represents the result of the CPA calculation of the integrated density of states for a one-dimensional AB alloy by *Soven* [11.9]. When this result of the CPA is compared with that of the ATA split-band limit approximation given by the solid curve in Fig. 11.8a, it is obvious that both qualitatively and quantitatively the CPA is a definite improvement over the ATA. A most remarkable point is that the unphysical band gap characteristic of the results in Fig. 11.8a does not appear in the results of the CPA.

11.8.2 Phonon Density of States of an Isotopically Disordered Model System

In Fig. 11.6b, the solid curve denotes the phonon density of states $D(\omega^2)$ of an isotopically disordered binary alloy calculated by the CPA. When we compare this result with that of the ATA given in Fig. 11.6a, we can see that the agreement with the computer work is largely improved. The correct bandwidths are obtained and the difficulty of the ATA giving the incorrect behaviour near $\omega^2/\omega_M^2 \sim 1$ has been eliminated. In particular, it is very noticeable that, for large values of c, the fine structures obtained from the computer experiment are beautifully reproduced in a coarse-grained manner.

The reason why the CPA gives less satisfactory results for small values of c than for large values of c may be explained as follows. (This tendency is not characteristic of the CPA alone, but can also exist in the results of the ATA and actually in most SSAs). In the limit of small c, almost all lattice sites are occupied by B atoms and only a small portion of sites are occupied by A atoms. In this situation, sites occupied by A atoms are isolated from one another and there is only an infinitesimally small probability of finding more than one neighboring site being occupied by A atoms. Since A atoms are lighter in our model, they move more easily than the host B atoms and yield the local mode with higher frequency than those for the B band. When c becomes finite and increases, the probability of finding clusters of two A atoms, of three A atoms, etc., also increases. These clusters contribute to respective local modes. When c is increased further, but is still not very large, the size of these clusters also increase. When c reaches a threshold value c_0, a cluster of infinite size appears. This threshold value c_0 is called "the critical percolation concentration" in the percolation theory. In the case of a simple cubic lattice, $c_0 \sim 0.28$. For $c < c_0$, the local modes due to clusters of A atoms are localized. In this case, the system is very inhomogeneous in the microscopic scale, and the effective medium approximation such as CPA, becomes inappropriate. In addition, the single-site nature of the CPA makes it difficult to reproduce those modes originating from two-atom clusters, three-atom clusters, etc.

11.8.3 Density of States of a Three-Dimensional Alloy

In Fig. 11.9b, the densities of states calculated on the basis of the CPA for various values of δ are shown, where the unperturbed band is taken to be of a semi-elliptic form. We can see from the figure that the amalgamation-persistence behaviour is beautifully reproduced. That is, when δ is small, the amalgamation-type one-mode band is given; when δ is increased, the band splits and a gap appears which grows wider when δ is increased further.

11.8.4 Optical Absorption Spectra of Mixed Crystals

Let us first study alkali-halide ionic mixed crystals as presented in Figs. 11.2, 3. In general, it is suggested from the split of the peaks that, in mixed crystals denoted by $MX_A - MX_B$ (M indicating an alkali atom and X a halogen atom), $\Delta = |E_A - E_B|$ is as large as 1 eV or so. On the other hand, in mixed crystals denoted by $M_A X - M_B X$, Δ is approximately 0.1 eV and much smaller than that for $MX_A - MX_B$. Since the width of each peak is about $0.1 \sim 0.15$ eV on the average, $MX_A - MX_B$ gives the situation $\delta > 1$ while $M_A X - M_B X$ provides the situation $\delta < 1$. From experiments for these mixed crystals, we have learned in Sect. 11.2 that the materials of the former type $MX_A - MX_B$ generally show the persistence behaviour as illustrated in Fig. 11.2, while the materials of the latter type $M_A X - M_B X$ tend to have the amalgamation behaviour as shown in

Fig. 11.3. Combining all the information described above, we may assert that the experimental results for mixed ionic crystals can be explained by the analysis due to the CPA as shown in Fig. 11.9b.

Analyses similar to Fig. 11.9b have also been performed for various concentrations. From the analyses, it has been shown that, when δ is small enough, the resultant spectra are always of the one-mode type or in other words, of the amalgamation type over the whole range of c. On the other hand, when δ is large enough, then the spectra are always of the two-mode type, or of the persistence type over the whole range of c. In Fig. 11.11, the excitonic density of states and the optical absorption spectra evaluated on the basis of a CPA model calculation for $c = 0.5$ are shown. Here again, the same behaviour as shown in Fig. 11.9b is observed, i.e., the obtained spectra are either of the persistence type or of the amalgamation type according to whether δ is large or not. Within the scope of the CPA, a given binary mixed crystal (or a binary alloy) can be classified either as the persistence type or the amalgamation type when the values of c and δ are given. The phase diagram is given in Fig. 11.12 where the calculations have been performed using the semi-elliptic model band for the unperturbed system.

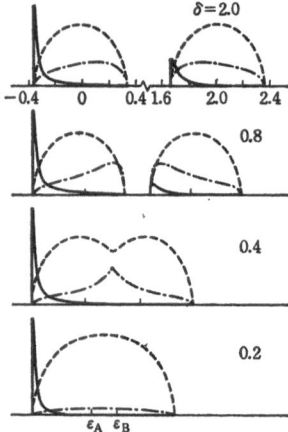

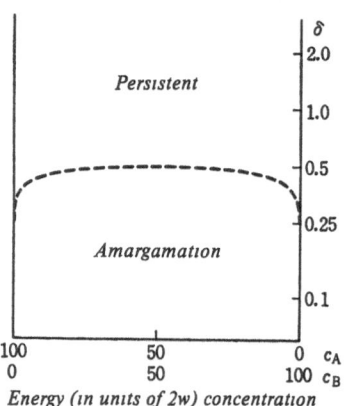

Fig. 11.11. The densities of states, etc., of excitons in a substitutionally-disordered binary (A_cB_{1-c}) mixed crystal as calculated in CPA for $c = 0.5$. The densities of states are represented by the dashed curves, the imaginary parts of the self-energies by the dot-dash curves and the optical absorption spectra by the solid curves. Energy is measured by taking $2w$ as a unit. [11.11]

Fig. 11.12. The phase diagram in the δ-c space which shows a region for the amalgamation type and a region for the persistent type. The calculation was carried out in the CPA using the semi-elliptic model band. [11.11]

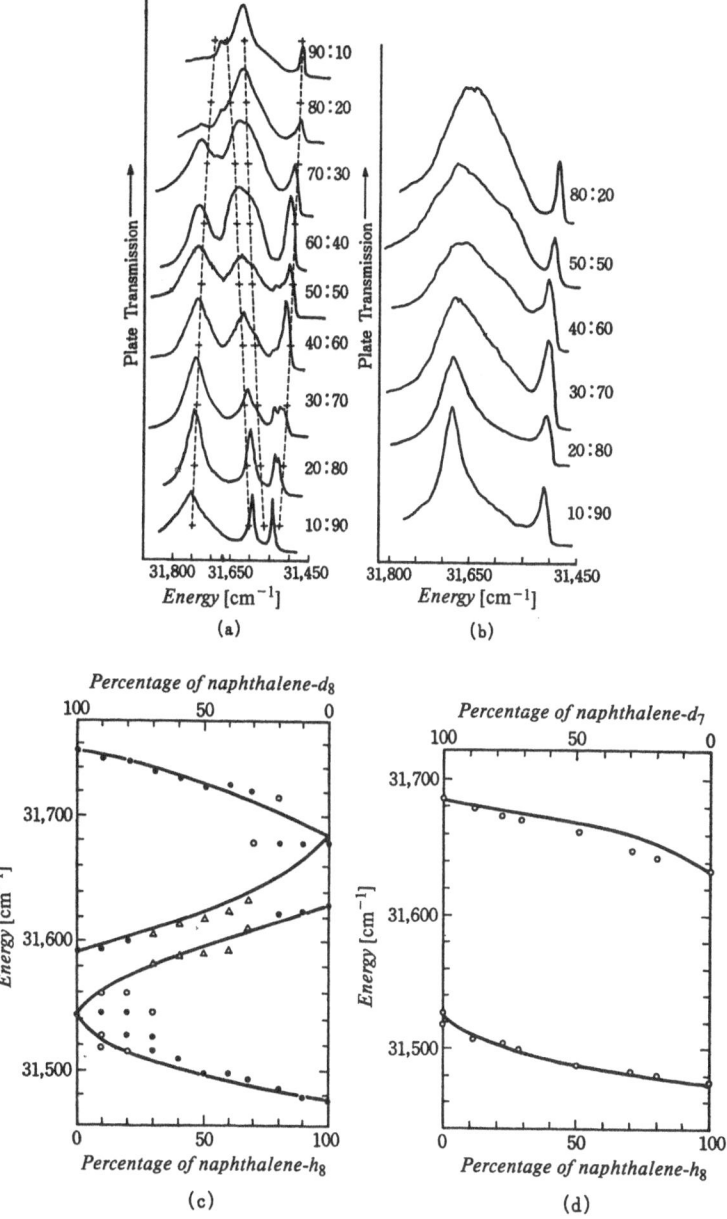

Fig. 11.13. (a) Absorption spectra of mixed crystals of naphthalene h_8 and d_8. (b) Absorption spectra of mixed crystals of naphthalene h_8 and d_4. (c) Observed peak positions for mixed crystals described in (a) the solid curves representing CPA calculations. (d) Positions as stated in (c) for mixed crystals described in (b). [11.12]

Similar experimental results concerning the absorption spectra have been obtained for organic crystals with hydrogeneous compounds such as naphthalenes and benzenes. In these materials, deuteration of the molecule changes the zero-point energy. In naphthalene, the shift due to the replacement of 8 protons (h_8) by 4 protons and 4 deuterons (d_4) is about $E(h_8) - E(d_4) \sim 50$ cm^{-1} while the shift due to the replacement of 8 protons (h_8) by 8 deuterons (d_8) is $E(h_8) - E(d_8) \sim 115$ cm^{-1}. The exciton bandwidth is ~ 100 cm^{-1}. The most detailed results of the optical spectra show two-mode behaviour in the $h_8 - d_8$ mixtures as given in Fig. 11.13a and one-mode behaviour in the $h_8 - d_4$ mixtures as given in Fig. 11.13b. The peak positions, as given in Fig. 11.13a, b are represented by triangles and circles in Figs. 11.13c,d as functions of concentration. The solid curves in Figs. 11.13c, d are the results of the CPA calculations. As can be derived from the numbers given in the above, δ is 1.15 in the case of ($h_8 - d_8$) while δ is 0.5 in the case of ($h_8 - d_4$). This example, therefore, gives another proof of the success of the CPA. In Figs. 11.9b, and 11.12 we can see how successful the CPA is in explaining qualitative behaviour, but the quantitative agreement shown in Figs. 11.13c, d demonstrates the undoubtedly superior feature of the CPA

11.8.5 Density of States of Transition and Noble Metal Alloys

The density of states of transition and noble metal alloys is derived through some appropriate analysis of experimental data of photoemission. Theoretical investigations of these alloys are usually made by studying the two-band model Hamiltonian of the form $H = H_{ss} + H_{sd} + H_{dd}$, where H_{ss} denotes the Hamiltonian for s electrons, H_{dd} for d electrons and H_{sd} represents the s-d hybridization term. When it is assumed that the effects of disorder are reflected only in the site-diagonal terms in H_{dd}, then the CPA formulation becomes applicable to H. By fitting the parameters in the Hamiltonian so that they can explain several experiments consistently, the density of states can be calculated using the CPA formulation. From the results of these calculations, it has been shown that, for alloys such as Cu–Ni and Au–Ag, the CPA densities of states are in good agreement with those evaluated from the photoemission data. In this section, let us study the case of Cu–Ni alloys, about which some results are shown in Fig. 11.4.

First, let us study the Cu–rich Cu–Ni alloys which correspond to Figs. 11.4a, b and c. Using the CPA, the total density of states $\rho(E)$ are evaluated as well as the partial densities of states $\rho_{Cu}^d(E)$ and $\rho_{Ni}^d(E)$ associated with a Cu site and a Ni site, respectively. They are shown in the lower scale in Fig. 11.4. By examining each partial density of states carefully, we can see that the Lorentzian sub-band results almost entirely from Ni sites while the main sub-band results from Cu sites. This lends support to the conjecture deduced from the analysis of the experimental data as mentioned in the preceding section. As the concentration of Ni is increased, the Ni sub-band grows while the Cu sub-band becomes smaller, but the Cu sub-band changes its position very little with respect to the Fermi energy of the alloy. These results of the CPA theory reproduce the characteristic feature of the experimental data, as stated in (II) in Sect. 11.2.2. This situation is illustrated in Fig. 11.5 where we present the CPA results for the energy ε_{edge} at the half height of the high-energy edge in the Cu sub-band measured relative to the Fermi energy ε_F. The CPA results are compared with experiments and with the calculations based upon the rigid band theory and the virtual-crystal approximation. It is apparent that the CPA gives semiquantitative agreement with experimental data while the other two methods (two examples of the pre-CPA methods) completely fail to explain the experimental results.

In the next place, let us study the Ni-rich Cu–Ni alloys (as shown in Fig. 11.4d–f). The CPA calculation reproduces the characteristic features of the experimental data which show that, even when the concentration is changed, the positions of the main peaks hardly shift although the shapes and heights of these peaks are modified. Thus, the behaviour of some features in connection with the concentration dependence are explained theoretically when the CPA is applied.

11.8.6 Ferromagnetic Alloys

For quite a long time, (I) the dilute limit (as described in Sect. 11.7.3) and (II) the rigid band model (as discussed in Sect. 11.2) have been the central theoretical approaches in the field of ferromagnetic alloys. In particular, the latter has been widely studied and shown to be successful in that this approximation can give the reasonable ferromagnetic-paramagnetic transition when the parameters are suitably chosen. This approximation, however, has some deficiencies such as its incapability of reproducing the results of the single impurity theory in the dilute limit and its well-known possibility of yielding unphysical results even for alloys composed of neighbouring elements in the periodic table. With a view to improving this situation, the CPA has been applied to ferromagnetic alloys in general and to Ni–Fe alloys specifically. The CPA has been proved to be free from the above-described difficulties inherent in the rigid band approximation. The CPA is also shown to be successful in various aspects as follows: (I) the magnetic moments of Ni and Fe in a $Ni_{1-c}Fe_c$ can be calculated independently and the calculated results are in good agreement with the experimental data (Fig. 11.14), (II) from the analysis of the concentration dependence of the density of states at the Fermi level, the experimental specific heat can be satisfactorily explained (Fig. 11.15), and (III) the instability of the paramagnetic state at finite temperatures can be identified from the analysis of the calculated susceptibility and as a result the transition temperature from the paramagnetic state to the ferromagnetic state can be derived. When the one-electron picture is valid and the assumption of the second-order phase transition is verified, then this transition temperature obtained from the CPA is regarded as corresponding to the Curie temperature. Note that the solid curve in the middle of Fig. 11.14 represents the average magnetic moment of the Ni-Fe alloy. From the curve, we can see that the average magnetic moment increases linearly in c for $c \leq 0.5$ and q is qualitatively in agreement with the Slater–Pauling curve.

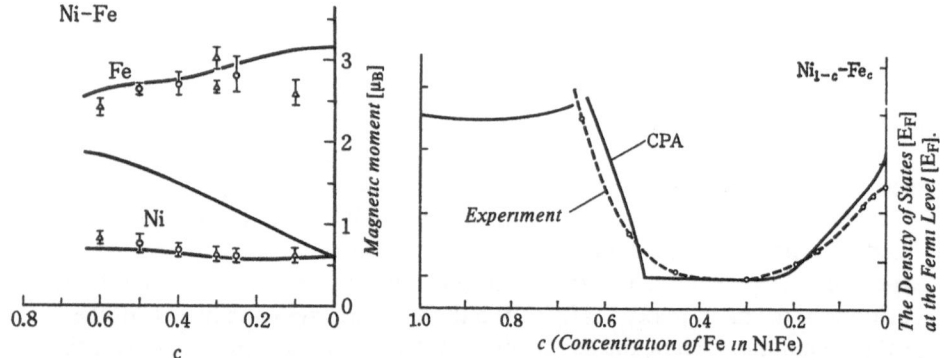

Fig. 11.14   **Fig. 11.15**

Fig. 11.14. A comparison of the calculated magnetic moments in CPA (the upper and lower solid curves) of Ni and Fe substitutionally-disordered ferromagnetic NiFe alloys to the experimental data obtained from the neutron diffraction (the circles and triangles with error bars). The comparison is made for $0 \leq c \leq 0.6$ where c is the concentration of Fe. The solid curve in the middle is the arithmetic mean of the magnetic moments of Ni and Fe. [11.13]

Fig. 11.15. A comparison of the calculated density of states in the CPA of substitutionally-disordered $Ni_{1-c}Fe_c$ alloys to the values (the dashed) obtained from the experimental data of the electronic specific heat. The calculated results are scaled so that they agree with the experimental results at the flat part, i.e., for $0.3 \leq c \leq 0.4$. The solid curve for $c \geq 0.65$ corresponds to the paramagnetic case. [11.13]

Recalling the fact that CPA comprises the dilute limit ($c \to 0$) as well as the weak scattering limit ($\delta \to 0$), we can conclude that CPA is also useful for those ferromagnetic alloys to which the dilute limit approximation and the rigid band approximation have been applied successfully as stated in (I) and (II) above. In this way, CPA has turned out to be advantageous in ferromagnetic alloys as well in the sense that this approximation gives remarkably good results in the whole regions of c and δ.

In this section, we have shown some examples of results by computer simulations and of experimental data for various realistic disordered systems, which can be beautifully explained by CPA. Before concluding this section, we would like to emphasize that the success of CPA is extremely general and overwhelming and the successful applications of CPA are not restricted to those examples which have been demonstrated in this and previous sections but can be extended to many other examples which have been omitted because of the limited space allowed.

11.9 Universality of the CPA

We have shown in the preceding section that CPA can serve as a unified theory which can systematically explain a wide variety of experimental data. This fact can be regarded as a proof or guarantee for the validity of CPA from a physical point of view. The CPA condition $\langle t_j \rangle = 0$ as discussed in the last section rather directly appeals to our physical instinct and therefore it is readily expected that this condition is physically reasonable. On the other hand, for the purpose of properly appreciating this approximation, it is equally important to clarify the mathematical meaning of the condition $\langle t_j \rangle = 0$. This is achieved by referring to the standard method of evaluating G_{av}, as discussed in Sect. 11.6, and examining which partial summation in the averaged perturbation series is equivalent to the approximation $\langle t_j \rangle = 0$. This problem has already been solved at an early stage of the history of CPA. Before presenting this mathematical analysis, let us briefly give a historical review of the period when CPA was first proposed.

Although CPA was proposed in 1967 and 1968, the importance of this research field of disordered systems had been realized much earlier and some serious work had been carried out towards the end of 1950's and at the beginning of 1960's. It was in this period when that famous paper on localization of electrons by *Anderson* [11.14] was published (introduced in Chap. 10), the Ziman theory on liquid metals was established and the Edwards theory was proposed to show that we can use the Green's function method for the study of electrons in disordered systems, as explained in Sect. 11.6. One of the strongest motivations for the intensive theoretical study of disordered systems is the fact that a remarkable amount of interesting experimental data had been accumulated by that time concerning typical disordered systems such as liquid metals, doped semiconductors and disordered alloys. Some reasonable theories which could explain these experimental results were seriously needed. Besides these conventional theoretical approaches as mentioned above, a completely new approach was introduced in 1960's, i.e., an approach by means of computers. This new approach is totally due to the fact that computers were made available for scientific research. *Dean* [11.15] was one of the earliest theoreticians who carried out computer experiments. He studied a disordered linear chain. Since then, an enormous amount of work has been performed using methods of computer simulations in almost all fields of physics including the problems of disordered systems. Owing to the strikingly rapid progress of the computer techniques, this approach of computer experiments or computer simulations has even reached the extent that it is now established as one commonplace tool for dealing with difficult problems which cannot be solved by analytical theories. Side by side with the computer work, theoretical research has also made considerable progress. Rigorous solutions of one-dimensional problems have provided significant suggestions concerning disordered systems in general. The Green's function method has been studied from various viewpoints and several approximate theories have been developed in connection with the Green's function method. All these attempts, however, were neither made systematically nor in the most efficient manner. Similar attempts and theories were independently proposed to deal with different systems. Since systems to which these attempts and theories were applied were different from one another according to the interest of the individual group, it was only after CPA was widely acknowledged that the interrelationship of these independently proposed theories was noticed. CPA itself was invented independently and repeatedly by various groups.

To be more precise, the condition $\langle t_j \rangle = 0$ was proposed through various methods. For instance, this condition is equivalent to taking all the self-contained and self-consistent single-site terms as the partial summation in the perturbation expansion series of the Green's function. This is achieved by rewriting the average of (11.5.9) by the cumulant expansion series and by using the diagram representation of the equations [11.16]. A detailed explanation of this method will be given below. Another method is to evaluate the interpolation of the self-energy between the weak scattering limit ($\delta \to 0$) and the split band limit ($\delta \to \infty$), and the obtained interpolation form has turned out to be identical with the CPA self-energy. This method naturally guaranteed the validity of CPA in the mentioned limits.

For the purpose of showing the mathematical ground of CPA, let us study the cumulant method as mentioned above. We consider the case where the unperturbed system is taken to be a pure B crystal, i.e., $u(z) = E_B$. We can write (11.5.16) as

$$H = H_B + \sum_j v_j \tag{11.9.1}$$

$$v_j = \sum_j |j\rangle \xi_j \Delta \langle j| \equiv \sum_j |j\rangle v_j \langle j| . \tag{11.9.2}$$

Then (11.5.9) becomes

$$\begin{aligned} G = G_B &+ G_B \sum_j v_j G_B + G_B \sum_j v_j G_B \sum_l v_l G_B \\ &+ G_B \sum_j v_j G_B \sum_l v_l G_B \sum_m v_m G_B \\ &+ G_B \sum_j v_j G_B \sum_l v_l G_B \sum_m v_m G_B \sum_n v_n G_B + \cdots . \end{aligned} \tag{11.9.3}$$

Disorder in the microscopic configuration of atoms now appears in ξ_j, ξ_l, etc., included in v_j, v_l, etc. The ensemble average is carried out by taking the averages of ξ_j, ξ_l, etc. Note that $\xi_j^2 = \xi_j$ as stated before, and the following relations hold,

$$\langle \xi_j \xi_l \rangle = \begin{cases} \langle \xi_j^2 \rangle = \langle \xi_j \rangle = c & (j = l) \\ \langle \xi_j \rangle \langle \xi_l \rangle = c^2 & (j \neq l) . \end{cases} \tag{11.9.4}$$

The suffices j, l, m, etc., in (11.9.3) denote the positions of atoms, or the lattice sites. Equation (11.9.4) asserts that, in carrying out the ensemble averages, proper care must be taken when some of the suffices in the same term coincide. Since it is too clumsy to write down all the distinct terms arising from the coincidence of some of the suffices, the diagram representation of equations is one popular way of eliminating this difficulty. Diagrams are convenient because they provide the visual illustrations of equations to make it easier to understand the meanings of the terms. Prescriptions are given in Fig. 11.16 for expressing variables and operators by means of diagrams. Using these rules, the leading four terms of (11.9.3) are expressed by diagrams given in Fig. 11.17. In this

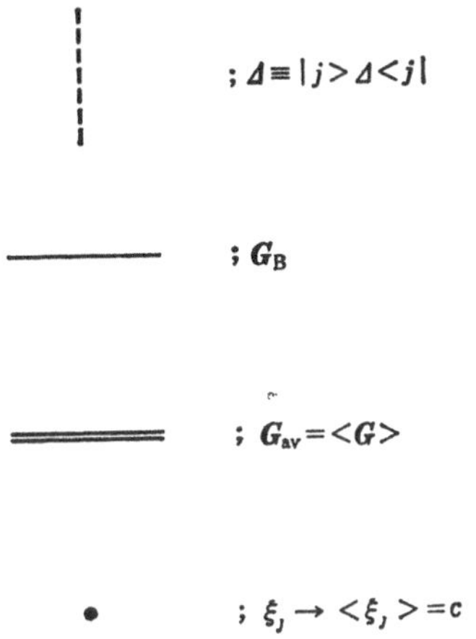

$; \varDelta \equiv |j> \varDelta <j|$

$; G_B$

$; G_{av} = <G>$

$; \xi_j \rightarrow <\xi_j> = c$

Fig. 11.16. Representation of the Green's function equations by means of diagrams

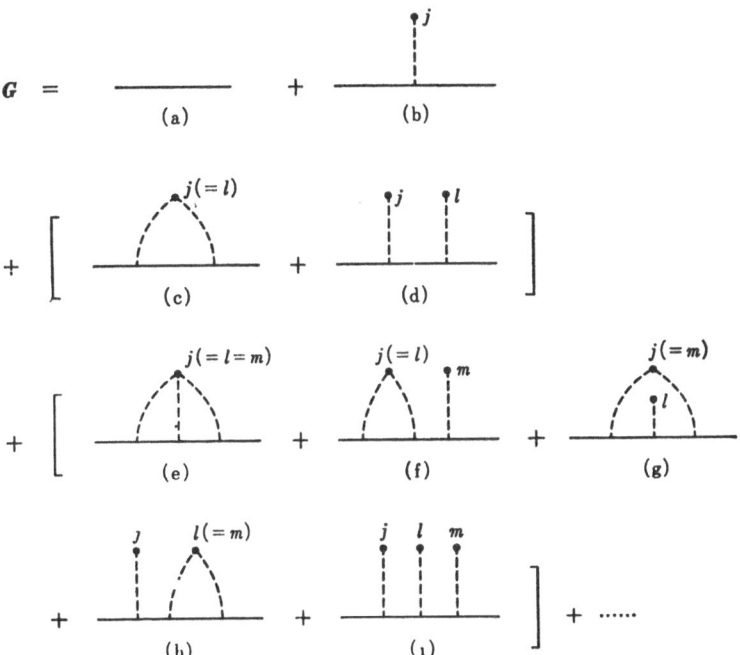

Fig. 11.17. Diagrams corresponding to the leading terms of G

figure, different filled circles correspond to different lattice sites as in the diagram (Fig. 11.17d) where $j \neq l$. Therefore, the averages can be carried out independently for each site represented by each filled circles. After carrying out the average, $\langle \xi_j \rangle = c$ is associated with each filled circle. An important point here is that the averaging procedure does not alter the restrictions included in the summations over j, l, etc. Namely, the summations appearing in connection with the diagram (Fig. 11.17d) must be carried out under the restriction that $j \neq l$. The significance of these restrictions over the summation indices would be understood better in the following paragraph.

By definition, the averaged Green's function $G_{av} = \langle G \rangle$ is expressed by using the self-energy Σ in the form

$$G_{av} = G_B + G_B \Sigma G_{av} , \tag{11.9.5}$$

or equivalently in the form

$$G_{av} = G_B + G_B \Sigma G_B + G_B \Sigma G_B \Sigma G_B + \cdots . \tag{11.9.6}$$

Equation (11.9.6) indicates that the self-energy must be the sum of all "proper" diagrams alone, where a "proper" diagram is the kind of diagram in the expansion series illustrated by Fig. 11.17 that cannot be separated into two parts by cutting one of the inner G_B lines only once. In Fig. 11.17, diagrams (b), (c), (e) and (g) are proper diagrams, while diagrams (d), (f) and (i) are not "proper" diagrams. It must be noted however that the restriction $j \neq l$ in Fig. 11.17d, for instance, prevents the two scattering lines l and j from being separated into two completely independent terms. This difficulty is overcome by a technique called the cumulant method, which is explained as follows by taking the term corresponding to Fig. 11.17d as an example. The restriction $j \neq l$ can be removed by adding a similar term with $j = l$. In order to keep the whole summation unchanged, we subtract the same term added to Fig. 11.17d from the term corresponding Fig. 11.17c. Noting that the sum of (c) and (d) is called the second moment and expressed as $G_B M_2 G_B$, the adding-subtracting procedure described above is written as

$$M_2 = c \sum_j \varDelta G_B \varDelta + c^2 \sum_{j \neq l} \sum \varDelta G_B \varDelta$$
$$= (c - c^2) \sum_j \varDelta G_B \varDelta + c^2 \sum_j \sum_l \varDelta G_B \varDelta . \tag{11.9.7}$$

The first equation is a direct interpretation of Fig. 11.17c, d. By adding the term $c^2 \sum_j \varDelta G_B \varDelta$ to the second term, the restriction $j = l$ on the summation can be eliminated. On the other hand, the same term is subtracted from the first term so that the equation holds. Thus the factor in front of the first term changes from c to $(c - c^2)$, which is called "the second-order cumulant". The restrictions on the summation indices can be removed in a completely similar manner in terms corresponding to Fig. 11.17f–i in higher-order diagrams.

(a)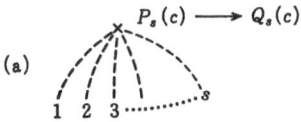

$$P_s(c) \longrightarrow Q_s(c)$$

(b) G_{av} $=$ G_B $+$ $G_B \Sigma^{(1)} G_{av}$

$$\equiv \; = \; \text{---} \; + \; \bigcirc{\Sigma^{(1)}} \, \equiv$$

(c) $\left(\Sigma^{(1)}\right)$ $=$ $\stackrel{\times}{\vdots}$ $+$ $\triangle$ $+$ $\bigtriangleup$ $+$ $\bigtriangleup$ $+$ $\cdots$

(d)

(e)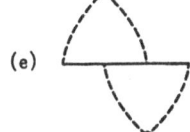

Fig. 11.18. The diagram method for evaluating the proper self-energy $\Sigma^{(1)}$

Aftert this rearrangement, which replaces all the moments by the cumulants, the factor to be associated with each filled circle in the diagrams is not c but a quantity called a cumulant. In order to avoid confusion, let us denote lattice sites by crosses as illustrated by Fig. 11.18a. With a vertex having s interaction lines ($s\Delta$ lines), an sth-order cumulant $P_s(c)$ is associated. The cumulant $P_s(c)$ can be evaluated by using the method of statistics.

After these considerations have been taken, the self-energy is now rigorously defined by the sum of all proper diagrams. As has been mentioned many times, it is almost impossible to sum up all proper diagrams to obtain the rigorous self-energy. One example of the approximations is to evaluate a partial summation by taking into account all essentially single-site diagrams as defined by diagrams given in Fig. 11.18c. The double horizontal line represents the total Green's function G_{av}, while a single horizontal line denotes the unperturbed Green's function as defined before. It is straightforward to show by expanding all the inner double lines that $\Sigma^{(1)}$ includes all "essentially single-site" diagrams.

Now, according to our previous consideration, a cumulant $P_s(c)$ is associated with each cross. It has been proved, however, that a further consideration is necessary. Namely, it has been asserted that $P_s(c)$ includes excess corrections in the following sense. According to the procedure described in the above, the correction due to a term corresponding to Fig. 11.18e is included to modify the factor associated with the filled circle in Fig. 11.18d. Therefore, $P_4(c)$ includes the correction due to Fig. 11.18d. However, it is important to notice here that the term itself corresponding to Fig. 11.18d is excluded from the single-site self-energy $\Sigma^{(1)}$ as is obvious from Fig. 11.18d. This means that, if we employ $P_s(c)$ as defined before, we have also included the corrections due to the diagrams which are not taken into account and the treatment is not self-contained. If we pay proper attention to this point, the factor to be associated with each cross in the figure becomes $Q_s(c)$ instead of $P_s(c)$ where $Q_s(c)$ is a modified single-site cumulant. Using the explicit forms for $Q_s(c)$ obtained from a mathematical method of statistics, we can show that the summation $\Sigma^{(1)}$ becomes identical with the CPA self-energy. This completes the proof that CPA is the best single-site approximation which takes into account all essentially single-site diagrams and in addition includes all relevant corrections.

It has been pointed out that the Hubbard solution in the problem of electron correlation is essentially identical with CPA. Although the problem of electron correlation is not directly related to the problem of disordered systems, the potential acting on an electron in the Hubbard model is shown to be analogous to that in a disordered binary alloy. The situation is expressed in terms of the Hamiltonian. The original Hubbard Hamiltonian is written in the tight-binding form which includes the terms corresponding to (11.4.2) plus the terms like $a_i^\dagger a_i^\dagger a_j a_l$ which describe the electron correlation. This means that the Hamiltonian is the two-particle Hamiltonian and therefore the corresponding Green's function must be treated by the method for double-time Green's functions. When the approximation for the double-time Green's function is not self-consistent, the obtained density of states always has a gap between two separated bands irrespective of parameters. As a result, this approximation gives a conclusion that a system with one electron per atom is always an insulator no matter how weak or strong the electron correlation may be, which is definitely unphysical. Actually, we can see that this approximation corresponds to the split-band limit explained in Sect. 11.5. When the approximation for the double-time Green's function is improved and the treatment is made to be self-consistent, then the obtained density of states is an amalgamation type one-mode band when the electron correlation is weak. Therefore, the system becomes metallic while the persistence-type two-mode band is obtained when the electron correlation is strong, and therefore, the system is shown to be insulating. In our problem, this corresponds to the capability of CPA in producing both the amalgamation and persistence bands according to the small and large values of δ, respectively.

The requirements stated at the beginning of Sect. 11.6 have thus been shown to be fulfilled by CPA. Although CPA is a reasonable approximation in various aspects, as described in the preceding and present sections, it has also some weak points because CPA is after all, an approximation but not an exact solution. For instance, CPA is not successful in explaining the fine structures of the phonon

Table 11.4. How far is the CPA successful in treating the vast variety of disordered systems.

Disordered systems		Types of disorder	Approximations	Distribution of E_J or/and V_{jl}	Has the CPA accomplished?
(1) Substitutionally-disordered systems		(a) Diagonal disorder	(I) single-site	completely random	◎
				correlated	○
			(II) multi-site	completely random	○
				correlated	○
		(b) Off-diagonal disorder	(I) single-bond	completely random	○
				correlated	○
			(II) multi-bond		○
		(c) Both diagonal and off-diagonal disorder are included			○
(2) Structurally-disordered systems	composed of single species	V_{jl} is a function of R_{jl} alone	(I) single-site	completely random	○
				correlated	○
			(II) multi-site		
(3) Topologically-disordered systems		(a) Diagonal disorder E_J = random V_{jl}=a function of R_{jl} alone			△
	mixed systems	(b) Off-diagonal disorder E_J = const V_{jl} depends on R_{jl} as well as on kinds of the jth and lth atoms			
		(c) Both diagonal and off-diagonal disorder are included			

density of states for $c < c_0$. This is owing to the fact that CPA is essentially a single-site approximation. When we refer to the classification of disorder, as shown in Table 11.4, CPA has originally been proposed for substitutionally-disordered systems in category (1) with diagonal disorder (a) where the configuration of E_j is assumed to be mutually independent and the single-site terms are taken into account. (Thus the original CPA is intended for the category marked by a double circle.) The applicability of CPA is therefore limited by these conditions used in the process of evaluating CPA. Improvements and extensions of CPA are realized by eliminating these conditions. We introduce some of the attempts in the following, and the summary is given in Table 11.4. Circles denote that the original philosophy of CPA is maintained while triangles indicate that the original philosophy of CPA is modified.

11.9.1 Cluster Effects (Multi-Site Approximations)

An extension of CPA to include the cluster effects has been made for substitutionally-disordered systems with site-diagonal disorder. According to the original CPA idea, the cluster CPA equations are derived by requiring that the t-matrix with two-site terms, three-site terms, etc., be zero on the average. This extension, however, introduces more unknown variables and the simplicity of the CPA formulations is spoiled (in other words, the CPA self-energy is the only one unknown variable to be determined self-consistently for a given energy z, while the cluster CPA contains more than one unknown variable and the problem becomes too complicated to be solved by the ordinary methods). In addition, the analyticity of the Green's function in many cases is not guaranteed even when the cluster CPA equation for more than one variable is solved. Nonanalyticity of the Green's function means that the approximation is physically meaningless. To avoid this difficulty of nonanalyticity, a method has been proposed in which the total Hamiltonian is divided into a sum of homomorphic cluster Hamiltonians and the CPA procedure is applied to one of these homomorphic clusters. This method is called the homomorphic cluster CPA (HCPA) [11.17]. It has been shown rigorously that the HCPA Green's function is always analytic. Other alternatives are to employ the molecular CPA or to solve the problem by using only a partial information given by the total cluster CPA equations.

11.9.2 Extension to Systems with Off-Diagonal Disorder

To substitutionally-disordered systems with off-diagonal disorder [in Category (1b)], the cluster CPA equations developed for systems with site-diagonal disorder alone can be applied. Accordingly, the same difficulty of nonanalyticity as in the case of diagonal disorder may take place. This can also be eliminated by using the HCPA. In addition, it has been shown that the HCPA for off-diagonal disorder leads to several interesting results including the reproduction of the results due to the computer experiments of the bond-type percolation problem.

11.9.3 Extension to Systems with Both Diagonal and Off-Diagonal Disorder

Now, let us consider those substitutionally-disordered systems where both E_j and V_{jl} are random. When the distribution of E_j is assumed to be independent of that of V_{jl}, we can apply the methods discussed in (a)–(i) and (b)–(i) to each of them, respectively, and this is a simple superposition of two independent procedures. When both distributions are correlated, as is the case in realistic systems, the HCPA is still available and the analyticity of the approximation can be proved in a similar manner.

11.9.4 Extension to Structurally-Disordered Systems Composed of Single Species

It is easy to extend the CPA to systems where the distribution of atoms is completely random. As explained in Sect. 11.1, the complete random distribution of atoms serves as a reasonable model of doped semiconductors. On the other hand, although V_{jl} is the function of R_{jl} alone in single-species liquid metals, the distribution of R_{jl} itself is random, but usually not completely random in the sense that correlation exists. In general, the n-atomic distribution function $g^{(n)}(R_1, R_2, ..., R_n)$ is defined as

$$\rho^n g^{(n)}(R_1, R_2, ..., R_n)$$
$$= \frac{N!}{(N-n)!} \frac{\int \exp\left[-U(R_1, R_2, ..., R_N)/kT\right] dR_{n+1} ... dR_N}{\int \exp\left[-U(R_1, R_2, ..., R_N)/kT\right] dR_1 dR_2 ... dR_N}, \qquad (11.9.8)$$

where ρ is the number density and $U(R_1, R_2, ..., R_N)$ is the potential energy for a given configuration of $\{R_n\}$. In order to calculate G_{av}, the average of (11.5.9) or (11.5.10) must be taken using the distribution function $g^{(n)}$ for $2 \leqq n \leqq N$. In practice, $g^{(n)}$ has neither been obtained experimentally nor theoretically for $n \geqq 3$. Even if $g^{(n)}$ for $n \geqq 3$ is given somehow and the averages of the terms in (11.5.9) or in (11.5.10) can be calculated, it is impossible to sum up all these terms or to sum even appropriate partial terms. Therefore, $g^{(n)}$ is usually approximated by some appropriate functionals of pair distribution functions $g^{(2)}(R_{jl}) \doteq g(R_{jl})$. A commonly used approximation to $g^{(n)}$ is to employ the superposition approximation in which $g^{(n)}$ is approximated by the product of the $n(n-1)/2$ pair distribution functions $g(R_{12}), g(R_{13}), ..., g(R_{n-1,n})$.

A self-consistent single-site approximation has been proposed on the basis of the superposition approximation and the cumulant method explained at the beginning of this section. When the treatment is made self-consistent and self-contained, the obtained approximation is the best single-site approximation and proved to be equivalent to CPA for substitutionally-disordered systems. This approximation is called "the effective medium approximation (EMA)." The density of states and other properties of liquid metals have been calculated by EMA and shown to be improved both qualitatively and quantitatively, com-

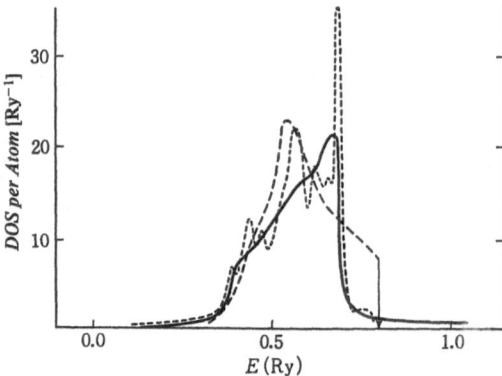

Fig. 11.19. The densities of states of liquid Ni in a muffin-tin model as calculated in the approximation corresponding to the average t-matrix approximation (ATA) (the long dashed curve), and in CPA (the solid curve). Those results are compared to the density of states (the short dashed curve) for a crystal Ni. [11.18]

pared to the results of the previous approximations. In Fig. 11.19 one example of the numerical results based on EMA is shown. The solid curve represents the density of states of liquid Ni calculated by EMA while for the sake of comparison, the density of states by the average t-matrix approximation (ATA) is given by the long dash curve. The short dash curve represents the band for the solid Ni. The ATA density of states becomes negative in the higher energy region which means that the analyticity of the Green's function is broken in the ATA. It is obvious that EMA (the CPA-equivalent approximation) gives the analytic result.

11.9.5 Extension to the Multi-Band Hamiltonian

So far, we have carried out discussions based on the Hamiltonian (11.4.1) or (11.4.2). This Hamiltonian is a so-called one-band model in which there is only one kind of Wannier function associated with each site. In realistic materials, however, it is generally necessary to take into account the effects of the degenerate multi-orbitals or of the multi-bands (actually, the results shown in Fig. 11.19 have been obtained from the calculations which include one s orbital, three p orbitals and five d orbitals). When there is more than one orbital associated with site j, we take the Wannier functions $|j\mu\rangle$ as the basis where μ denotes the μth orbital at the jth site. The Hamiltonian operator in this case is written as follows:

$$H = \sum_{j\mu} |j\mu\rangle E_j^{\mu} \langle j\mu| + \sum_{j\neq l} \sum_{\mu\nu} |j\mu\rangle V_{jl}^{\mu\nu} \langle l\nu| , \qquad (11.9.9)$$

where E_j^{μ} is the localized level associated with the orbital $|j\mu\rangle$ and $V_{jl}^{\mu\nu}$ denotes the transfer probability of an electron from the νth state at the lth site to the μth state at the jth site. All the formulations in Sects. 11.4–9 developed for the one-band Hamiltonian (11.4.1) can be transferred to the formulations for the multi-band Hamiltonian (11.9.9) by replacing the scalar quantities in the former by the matrix quantity with the (μ, ν) matrix elements.

11.9.6 Extension of CPA to the Calculations of Other Physical Quantities

So far in this chapter we have been concerned with one-particle properties which are evaluated from $\langle G \rangle$. The transport properties, such as the electric conductivity and the Hall coefficients on the other hand, are derived from $\langle GG \rangle$ or $\langle GGG \rangle$. It has been shown that for a one-s-band model, the treatment consistent with CPA leads to the relations $\langle G \rangle = \langle G \rangle \langle G \rangle$ and $\langle GGG \rangle = \langle G \rangle \langle G \rangle \langle G \rangle$. These relations do not hold when a band other than an s band is considered or when the effects of the multiband are taken into account. It is expected even for a one-s-band that the region of parameters (c and $\varDelta$), for which CPA is reasonable, is smaller in the case of $\langle GG \rangle$ and $\langle GGG \rangle$ than in the case of $\langle G \rangle$.

Before concluding, let us mention the effective medium approximation for a continuous medium composed of two constituent materials at a macroscopic level.

Suppose the two kinds of materials with the dielectric constants ϵ_A and ϵ_B, respectively, are mixed in a completely random manner. Now we examine how the effective dielectric constant ϵ_m of the mixture is evaluated. The idea of this approximation is schematically shown in Fig. 11.10. The essential point of this approximation is to replace the exact average by the average over the systems in which disorder is maintained in a sphere with radius a placed in an effective medium. When this sphere is made of material A, the induced polarization P_A at the centre of the sphere is given by $P_A = Ea^3(\epsilon_A - \epsilon_m)/(\epsilon_A + 2\epsilon_m)$, where E is the electric field. When the sphere is made of material B, the polarization P_B at the centre is $P_B = Ea^3(\epsilon_B - \epsilon_m)/(\epsilon_B + 2\epsilon_m)$. An equation to define ϵ_m is obtained by requiring that the polarization at the centre of the sphere should be zero on the average. That is to say

$$x_A \frac{\epsilon_A - \epsilon_m}{\epsilon_A + 2\epsilon_m} + x_B \frac{\epsilon_B - \epsilon_m}{\epsilon_B + 2\epsilon_m} = 0, \tag{11.9.10}$$

where x_A and x_B are the volume fractions of A and B materials, respectively. Apparently, we have $x_A + x_B = 1$. It is easily understood that (11.9.10) corresponds to our CPA equation (11.7.3). (This point is self-evident when we compare Fig. 11.10 with Fig. 11.20.) Moreover, a detailed analysis of the mathematical structure of (11.9.10) makes it possible to show that (11.9.10) is the best single-site approximation and rigorously equivalent to the CPA. It is interesting to note that (11.9.10) is also applicable to the evaluation of the effective conductivity σ_m of a mixture by replacing ϵ in (11.9.10) by the conductivity σ. In practice, (11.9.10) has been used in various problems such as the calculation of the dielectric constant in heterogeneous media and the evaluation of the conductivity in the percolation theory.

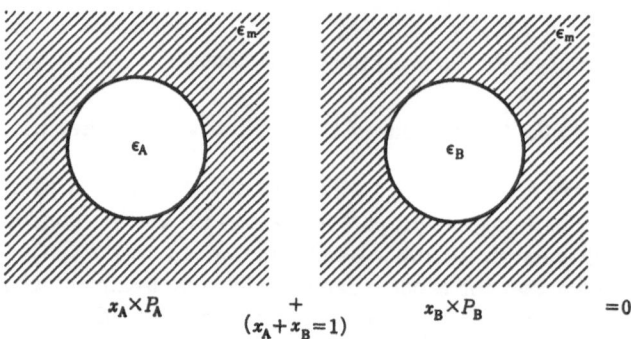

$$x_A \times P_A \qquad + \qquad x_B \times P_B \qquad = 0$$
$$(x_A + x_B = 1)$$

Fig. 11.20. A schematic illustration of the effective medium approximation for continuous binary systems

To summarize, several independent approaches for different materials based on different philosophies have all reached the CPA-equivalent approximations. This fact itself is regarded as indicating that CPA is a stable approximation. The stability of CPA is also suggested by the fact that most of the attempts to extend CPA have somehow failed. For instance, the extension of CPA to clusters or to off-diagonal disorder introduces extremely complicated numerical calculations or yields the breakdown of analyticity.

So far, the discussions have been given to see the advantageous points of CPA both from physical and mathematical points of view. However, it is difficult to describe the significance of CPA fully. We may be able to estimate the popularity or universality of CPA and the wide variety of its applications if we know how many papers have been written in connection with CPA for the last 10 years or so since the invention of CPA. *Elliott* et al. [11.19] have quite properly described the situation in one of their papers saying that, owing to the CPA, even experimentalists no longer despise the Green's functions.

Even now, attempts are being made daily somewhere in the world to analyze experimental data by CPA in enormously wide fields. This way, various new aspects are added day by day to the history of CPA. And yet, we have not yet revealed the complete range of the possibilities of CPA. We believe that epoch-making progress to the next step in statistical mechanics of random systems will become possible when the whole scope of CPA is brought to light.

References

Chapter 1

1.1 F. Herman, S. Skillman: *Atomic Structure Calculations* (Prentice Hall, Reading, MA 1963)
1.2 L.J. Schiff: *Quantum Mechanics* (McGraw-Hill, New York 1949)
1.3 J.C. Slater: *Quantum Theory of Matter*, 2nd ed., (McGraw-Hill, New York 1968)

Additional Reading

Condon, E.U., Shortley, G.H.: *The Theory of Atomic Spectra* (University Press, Cambridge 1935)
Coulson, C.A.: *Valence* (Clarendon, Oxford 1952)
Lindgren, I., Morrison, J.: *Atomic Many-Body Theory*, Springer Ser. Chem. Phys., Vol.13 (Springer, Berlin, Heidelberg, New York 1981)
Sobelman, I.I.: *Atomic Spectra and Radiative Transitions*, Springer Ser. Chem. Phys., Vol.1 (Springer, Berlin, Heidelberg, New York 1979)
Sobelman, I.I., Vainshtein, L.A., Yukov, E.A.: *Excitation of Atoms and Broadening of Spectral Lines*, Springer Ser. Chem. Phys., Vol.7 (Springer, Berlin, Heidelberg, New York 1981)

Chapter 2

2.1 D. Pines, P. Nozieres: *The Theory of Quantum Liquids*, Vol.1 (Benjamin, New York 1966) Chap.3

Additional Reading

Slater, J.C.: *Quantum Theory of Molecules and Solids*, Vols.1-3 (McGraw-Hill, New York 1963, 1965, 1967)

Chapter 3

3.1 R.A. Cowley, A.D.B. Wood: Can. J. Phys. *49*, 177 (1971)
3.2 W.M. Fairbank, M.J. Buckingham, C.F. Kellers: Proceedings of the Fifth International Conference on Low Temperature Physics and Chemistry, ed. by J.R. Dillinger, University of Wisconsin (1958)
3.3 A.L. Thomson, H. Meyer, E.D. Adams: Phys. Rev. *128*, 509 (1962)
3.4 W.R. Abel, A.C. Anderson, J.C. Wheatley: Phys. Rev. Lett. *17*, 74 (1966)
3.5 R.A. Webb, T.J. Greytak, R.T. Johnson, J.C. Wheatley: Phys. Rev. Lett. *30*, 210 (1973)

Additional Reading

Armitager, J.G.M., Farguhar, I.E. (eds.): *The Helium Liquids* (Academic, New York 1975)
Benneman, K.H., Ketterson, J.B. (eds.): *The Physics of Liquid and Solid Helium Helium*, I,II (Wiley, New York 1976)
Brewer, D.F. (ed.): *Quantum Fluids* (North-Holland, Amsterdam 1966)
Donnelly, P.J.: *Experimental Superfluidity* (University of Chicago Press, Chicago 1967)
Keller, W.E.: *Helium 3 and Helium 4* (Plenum, New York 1969)
Khalatnikov, I.M.: *Introduction to the Theory of Superfluidity* (Benjamin, New York 1965)
London, F.: *Superfluids*, Vol.2 (Wiley, New York 1954)
Wilks, J.: *The Properties of Liquid and Solid Helium* (Clarendon, Oxford 1967)

Chapter 4

4.1 D.N. Paulson, R.T. Johnson, J.C. Wheatley: Phys. Rev. Lett. *30*, 829 (1973)
4.2 D.D. Oscheroff, W.J. Gully, R.C. Richardson, D.M. Lee: Phys. Rev. Lett.*29*, 920 (1972)
4.3 J.D. Reppy: Physics B and C *90*, 64 (1977)
4.4 R.A. Webb, R.T. Kleinberg, J.C. Wheatley: Phys. Lett. A*48*, 421 (1974);
4.5 K. Maki, T. Tsuneto: Prog. Theor. Phys. *52*, 773 (1974)

Additional Reading

Leggett, A.J.: A theoretical description of the new phases of liquid ^{3}He. Rev. Mod. Phys. *47*, 331 (1975)
Wheatley, J.C.: Experimental properties of superfluid ^{3}He. Rev. Mod. Phys. *47*, 415 (1975)

Chapter 5

5.1 J.M. Ziman: "The Band Structure Problem", in *Solid State Physics*, Vol.26, ed. by F. Seitz, D. Turnbull, H. Ehrenreich (Academic, New York 1971) p.1
5.2 B. Alder, S. Fernbach, M. Rotenberg: "Energy Bands in Solids", in *Methods of Computational Physics*, Vol.8 (Academic, New York 1968)
5.3 J.O. Dimock: "Calculations of Electronic Energy Bands by the APW Methods", in *Solid State Physics*, Vol.25, ed. by E. Seitz, D. Turnbull, H. Ehrenreich (Academic, New York 1971) p.103
5.4 L.M. Falikov: In *Energy Bands in Metals and Alloys*, ed. by L.H. Bennett, J.T. Waber (Gordon and Breach, New York 1968)
5.5a J.M. Ziman (ed.): *The Physics of Metals* (University Press, Cambridge 1969)
5.5b D. Shoenberg: "Electronic Structure: The Experimental Results", ibid., p.62
5.5c A.B. Pippard: "Metallic Electrons in a Magnetic Field", ibid., p.113
5.5d R.G. Chambers: "Transport Properties: Surface and Size Effects", ibid., p.175
5.6 W.A. Harrison: *Pseudopotentials in the Theory of Metals* (Benjamin, New York 1966);
V. Heine: "The Pseudopotential Concept", in *Solid State Physics*, Vol.24, ed. by F. Seitz, D. Turnbull, H. Ehrenreich (Academic, New York 1970) p.1
5.7 A.O. Animalu, V. Heine: Phil. Mag. *12*, 1249 (1965)
5.8 M.L. Cohen, V. Heine: "The Fitting of Pseudopotential to Experimental Data and Their Subsequent Application", in *Solid State Physics*, Vol.24, ed. by F. Seitz, D. Turnbull, H. Ehrenreich (Academic, New York 1970) p.38

5.9   V. Heine, D. Weaire: "Pseudopotential Theory of Cohesion and Structure", in *Solid State Physics*, Vol.24, ed. by F. Seitz, D. Turnbull, H. Ehren-reich (Academic, New York 1970) p.248
5.10 D.G. Pettifor: J. Phys. C*3*, 366 (1970)
5.11 J. Friedel: In [Ref.5.5a, p.340]
5.12 H. Sato, R.S. Toth: Phys. Rev. *124*, 1833 (1961)
5.13 S. Nakajima, Y. Toyozawa, R. Abe: *The Physics of Elementary Excitations*, Springer Ser. Solid-State Sci., Vol.12 (Springer, Berlin, Heidelberg, New York 1980)
5.14 W.L. McMillan: Phys. Rev. *167*, 331 (1968)

Additional Reading

Friedel, J., Guinier, A. (eds.): *Metallic Solid Solutions* (Benjamin, New York 1963)
Jones, H.: *The Theory of Brillouin Zones and Electronic States in Crystals* (North-Holland, Amsterdam 1960)
Massalski, T.B. (ed.): *Alloying Behavior and Effect in Concentrated Solid Solution* (Gordon and Breach, New York 1965)
Mott, N.F., Jones, H.: *Theory of the Properties of Metals and Alloys* (University Press, Oxford 1936)
Parks, R.O. (ed.): *Superconductivity*, Vols.1,2 (Dekker, New York 1969)
Rudman, P.S., Stringer, J., Jaffee, R.I.: *Phase Stability in Metals and Alloys* (McGraw-Hill, New York 1966)
Vonsovsky, S.V., Izyumov, Y.A., Kurmaev, E.Z.: *Superconductivity of Transition Metals*, Springer Ser. Solid-State Sci., Vol.27 (Springer, Berlin, Heidel-berg, New York 1982)
Ziman, J.M.: *Electrons and Phonons* (Clarendon, Oxford 1966)

Chapter 6

6.1   S. Nakajima, Y. Toyozawa, R. Abe: *The Physics of Elementary Excitations*, Springer Ser. Solid-State Sci., Vol.12 (Springer, Berlin, Heidelberg, New York 1980)
6.2   J. Sherman: Chem. Rev. *11*, 93 (1932)
6.3   L. Pauling: *The Nature of Chemical Bond* (University Press, Cornell 1960)
6.4   R. Landshoff: Z. Phys. *12*, 201 (1936); Phys. Rev. *52*, 246 (1937)
6.5   L.P. Howland: Phys. Rev. *109*, 1927 (1958)
6.6   P.O. Löwdin: Adv. Phys. *5*, 1 (1956)
6.7   F. Herman: Phys. Rev. *95*, 847 (1954)
6.8   D. Brust: Phys. Rev. *109*, 1337 (1964)
6.9   V. Heine, R.O. Jones: J. Phys. C*2*, 719 (1969)
6.10 K. Okada, H. Sugie: J. Phys. Soc. Jpn. *25*, 1128 (1968)
6.11 J.C. Slater: J. Chem. Phys. *9*, 16 (1941)
6.12 M.L. Cohen, T.K. Bergstresser: Phys. Rev. *141*, 117 (1966)
6.13 D.R. Penn: Phys. Rev. *128*, 2093 (1962)
6.14 J.A. Van Vechten: Phys. Rev. *182*, 891 (1967)
6.15 P.J. Lin, W. Saslow, M.L. Cohen: Solid State Commun. *5*, 893 (1967)
6.16 Y. Onodera: Solid State Commun. *11*, 1397 (1972)
6.17 M.H. Cohen, L.M. Falikov, S. Golin: IBM J. Res. Dev. *8*, 215 (1964)
6.18 B.M. Riggleman, H.G. Drickamer: J. Chem. Phys. *34*, 446 (1962)
6.19 T.E. Slykhouse, H.G. Drickamer: Phys. Chem. Solids *7*, 210 (1958)
6.20 A.L. Edwards, H.G. Drickamer: Phys. Rev. *122*, 1149 (1961)
6.21 S. Minomura, H.G. Drickamer: Phys. Chem. Solids *23*, 451 (1963)

Additional Reading

Drickamer, H.G.: "The Effect of High Pressure on the Electronic Structure of Solids", in *Solid State Physics*, Vol.17, ed. by F. Seitz, D. Turnbull, H. Ehrenreich (Academic, New York 1965) p.1
Fridkin, V.M.: *Photoferroelectrics*, Springer Ser. Solid-State Phys., Vol.9 (Springer, Berlin, Heidelberg, New York 1979)
Luis, J., Amoros, M.: *Molecular Crystals* (Wiley, New York 1969)
Madelung, O.: *Physics of III–V Compounds* (Wiley, New York 1964)
Mayer, L.: Phase transitions in Van der Waals lattice. Adv. Chem. Phys. *16*, 343 (1969)
Mitsui, T., Tatsuzaki, I., Nakamura, E.: *An Introduction to the Physics of Ferroelectrics* (Gordon and Breach, New York 1976)
Mott, N.F., Gurney, R.W.: *Electronic Process in Ionic Crystals* (University Press, Oxford 1940)
Phillips, J.C.: *Bonds and Bands in Semiconductors* (Academic, New York 1973)
Phillips, J.C.: "The Fundamental Optical Spectra of Solids", in *Solid State Physics*, Vol.18, ed. by F. Seitz, D. Turnbull, H. Ehrenreich (Academic, New York 1966) p.55
Seeger, K.: *Semiconductor Physics*, Springer Ser. Solid-State Sci., Vol.40, 2nd ed. (Springer, Berlin, Heidelberg, New York 1982)
Silinsh, E.A.: *Organic Molecular Crystals*, Springer Ser. Solid-State Sci., Vol.16 (Springer, Berlin, Heidelberg, New York 1980)
Thomas, D.G. (ed.): *II–VI Semiconducting Compounds* (Benjamin, New York 1969)
Tosi, M.P.: "Cohesion of Ionic Solids", in *Solid State Physics*, Vol.16, ed. by F. Seitz, D. Turnbull, H. Ehrenreich (Academic, New York 1964) p.1
Willardson, R.K., Beer, A.C. (eds.): *Semiconductors and Semimetals*, Vols.1–4, (Academic, New York 1966–1968)

Chapter 7

7.1 D. Weaire, N.F. Thorpe: Phys. Rev. B*4*, 2508, 3518 (1971)
7.2 W.A. Harrison: Phys. Rev. B*8*, 4487 (1973);
    W.A. Harrison, S. Ciraci: Phys. Rev. B*10*, 1516 (1973)
7.3 W.A. Harrison, J.C. Phillips: Phys. Rev. Lett. *33*, 410 (1974);
    R.K. Sundfors: Phys. Rev. B*10*, 4244 (1974)
7.4 D.J. Chadi, W.A. Harrison: Phys. Rev. Lett. *35*, 1372 (1975)
7.5 D.J. Chadi, R.M. White, W.A. Harrison: Phys. Rev. Lett. *35*, 1372 (1975)
7.6 J.C. Phillips: Rev. Mod. Phys. *42*, 317 (1970)
7.7 S.T. Pantelids, W.A. Harrison: Phys. Rev. B*11*, 3006 (1975)

Additional Reading

Harrison, W.A.: *Electronic Structure and the Properties of Solids* (Freeman, San Francisco 1980)
Pantelids, S.T., Harrison, W.A.: Structures of the valence bands of zinc-blende-type semiconductors. Phys. Rev. B*13*, 2667 (1976)

Chapter 8

8.1 J.H. Van Vleck: *The Theory of Electric and Magnetic Susceptibilities* (University Press, Oxford 1932)
8.2 R. Kubo: J. Phys. Soc. Jpn. *12*, 149 (1957)
8.3 S. Kachi, K. Kosuge, H. Okinaka: J. Solid State Chem. *6*, 258 (1973)

8.4 J. Smart: In *Magnetism*, Vol.3, ed. by G.T. Rado, H. Suhl (Academic, New York 1963) p.63
8.5 J. Kanamori: *Magnetism* (Baifukan, Tokyo 1969) p.74 [in Japanese]
8.6 P.G. de Gennes: Phys. Rev. *118*, 141 (1960)
8.7 R.M. Bozorth: *Ferromagnetism* (Van Nostrand, New York 1951) p.441
8.8 J.W.D. Connolly: Phys. Rev. *159*, 415 (1967)
8.9 T. Kasuya: In *Magnetism*, Vol.2B, ed. by G.T. Rado, H. Suhl (Academic, New York 1966) p.215
8.10 S.C. Keeton, T.L. Loucks: Phys. Rev. *168*, 672 (1968)
8.11 S. Nakajima, Y. Toyozawa, R. Abe: *The Physics of Elementary Excitations*, Springer Ser. Solid-State Sci., Vol.12 (Springer, Berlin, Heidelberg, New York 1980)
8.12 L. Néel: Ann. Phys. Paris *3*, 137 (1948)
8.13 T. Moriya: J. Magn. Magn. Mater. *14*, 1 (1979)
8.14 T. Moriya (ed.): *Electron Correlation and Magnetism in Narrow-Band Systems*, Springer Ser. Solid-State Sci., Vol.29 (Springer, Berlin, Heidelberg, New York 1981)
8.15 A.M. Clogstone, B.T. Matthias, M. Peter, H.J. Williams, E. Corenzwit, R.C. Sherwood: Phys. Rev. *125*, 541 (1962)

Additional Reading

Abragam, A.: *The Principle of Nuclear Magnetism* (University Press, Oxford 1961)
Abragam, A., Bleaney, B.: *Electron Paramagnetic Resonance of Transition Ions* (Clarendon, Oxford 1970)
Adler, D.: "Insulating and Metallic States in Transition Metal Oxides", in *Solid State Physics*, Vol.21, ed. by F. Seitz, D. Turnbull, H. Ehrenreich (Academic, New York 1968) p.1
Anderson, P.W.: "Theory of Magnetic Exchange Interactions", in *Solid State Physics*, Vol.14, ed. by F. Seitz, D. Turnbull, H. Ehrenreich (Academic, New York) p.99
Brout, R.: *Phase Transition* (Benjamin, New York 1965)
De Gennes, P.G.: "Theory of Neutron Scattering by Magnetic Crystals", in *Magnetism*, Vol.3, ed. by G.T. Rado, H. Suhl (Academic, New York 1963) p.115
Dyson, J.F.: General theory of spin-wave interactions. Phys. Rev. *102*, 1217 (1956)
Elliott, R.J., Harley, R.T., Hays, W., Smith, S.R.P.: Raman scattering and theoretical studies of Jahn-Teller induced phase transitions in some rare-earth compounds. Proc. Roy. Soc. London A*328*, 217 (1972)
Fischer, Ø., Maple, B. (eds.): *Superconductivity in Ternary Compounds I*, Topics Curr. Phys., Vol.32 (Springer, Berlin, Heidelberg, New York 1982)
Frauenfelder, H.: *Mössbauer Effect* (Benjamin, New York 1962)
Freeman, A.J., Watson, R.E.: "Hyperfine Interactions in Magnetic Materials", in *Magnetism*, Vol.2A, ed. by G.T. Rado, H. Suhl (Academic, New York 1962)
Goodenough, J.B.: *Magnetism and Chemical Bond* (Interscience, New York 1963)
Griffith, J.S.: *The Theory of Transition-Metal Ions* (University Press, Cambridge 1964)
Halperin, B.: "The Excitonic State at the Semiconductor-Semimetal Transition," in *Solid State Physics*, Vol.21, ed. by F. Seitz, D. Turnbull, H. Ehrenreich (Academic, New York 1968) p.115
Herring, C.: "Exchange Interactions among Itinerant Electrons", in *Magnetism*, Vol.4, ed. by G.T. Rado, H. Suhl (Academic, New York 1966) p.1
Kataoka, M., Kanamori, J.: A theory of the cooperative Jahn-Teller effect. J. Phys. Soc. Jpn. *32*, 113 (1972)
Kittel, C.: *Quantum Theory of Solids* (Wiley, New York 1963)
Kondo, J.: "Theory of Dilute Magnetic Alloys", in *Solid State Physics*, Vol.23, ed. by F. Seitz, D. Turnbull, H. Ehrenreich (Academic, New York 1969) p.183
Kosterlitz, J.M., Thouless, D.J.: Ordering, metastability and phase transitions in two-dimensional systems. J. Phys. C*6*, 1181 (1973)

Kosterlitz, J.M.: The critical properties of the two-dimensional xy model.
    J. Phys. C7, 1046 (1974)
Lieb, E.H., Mattis, D.C.: *Mathematical Physics in One Dimension* (Academic,
    New York 1966)
Maple, B., Fischer, Ø. (eds.): *Superconductivity in Ternary Compounds II*,
    Topics Curr. Phys., Vol.34 (Springer, Berlin, Heidelberg, New York 1982)
Mattis, D.C.: *The Theory of Magnetism I*, Springer Ser. Solid-State Sci., Vol.17
    (Springer, Berlin, Heidelberg, New York 1981)
Mott, N.F.: *Metal-Insulator Transition* (Taylor and Francis, London 1974)
Nagamiya, T., Yosida, K., Kubo, R.: Antiferromagnetism. Adv. Phys. 4, 1 (1955)
Nagamiya, T.: "Helical Spin Ordering-I Theory of Helical Spin Configuration",
    in *Solid State Physics*, Vol.20, ed. by F. Seitz, D. Turnbull, H. Ehrenreich
    (Academic, New York 1967) p.305
Opechowski, W.: "Magnetic Symmetry", in *Magnetism*, Vol.2A, ed. by G.T. Rado,
    H. Suhl (Academic, New York 1965)
Slichter, C.P.: *Principles of Magnetic Resonances*, Springer Ser. Solid-State
    Sci., Vol.1, 2nd ed. (Springer, Berlin, Heidelberg, New York 1980)
Stanley, H.E.: *Introduction to Phase Transition and Critical Phenomena*
    (Clarendon, Oxford 1971)
Sturge, M.D.: "The Jahn-Teller Effect in Solids", in *Solid State Physics*,
    Vol.20, ed. by F. Seitz, D. Turnbull, H. Ehrenreich (Academic, New York
    1967) p.91
Sugano, S., Tanabe, Y., Kamimura, H.: *Multiplets of Transition-Metal Ions in
    Crystals* (Academic, New York 1970)
Tyablikov, S.V.: *Methods in the Quantum Theory of Magnetism* (Plenum, New York
    1967)
Vonsovsky, S.V., Izyumov, Y.A., Kurmaev, E.Z.: *Superconductivity of Transition
    Metals*, Springer Ser. Solid-State Sci., Vol.27 (Springer, Berlin, Heidel-
    berg, New York 1982)
White, R.M.: *Quantum Theory of Magnetism*, Springer Ser. Solid-State Sci.,
    Vol.32 (Springer, Berlin, Heidelberg, New York 1982)

Chapter 9

9.1   K. Yosida, A. Yoshimori: In *Magnetism*, Vol.5, ed. by G.T. Rado, H. Suhl
      (Academic, New York 1973) p.253
9.2   K.G. Wilson: Rev. Mod. Phys. 47, 773 (1975)
9.3   K. Yamada: Prog. Theor. Phys. 53, 970 (1975)
9.4   K. Yosida, K. Yamada: Prog. Theor. Phys. 53, 1286 (1975)
9.5   K. Kume: Nippon Butsuri-Gakkai-shi 32, 562 (1977) [in Japanese]
9.6   P. Steiner: Phys. Rev. 10, 4704 (1974)
9.7   K. Kume, K. Mizoguchi: J. Phys. F7, 1555 (1977)
9.8   B.B. Triplett, N.E. Phillips: Phys. Rev. Lett. 27, 1001 (1971)
9.9   C. Rizzuto: Rep. Prog. Phys. 37, 147 (1974)
9.10 A. Yoshimori: Prog. Theor. Phys. 55, 67 (1976)
9.11 K. Yosida: Nippon Butsuri-Gakkai-shi 31, 116 (1976) [in Japanese]
9.12 H.R. Krishna-Murthy, J.W. Wilkins, K.G. Wilson: Phys. Rev. B21, 1003,
      1044 (1980)
9.13 F.D.M. Haldane: Phys. Rev. Lett. 40, 416 (1978)

Additional Reading

Noziere, P.: Fermi-liquid description of the Kondo problem at low temperature.
    J. Low Temp. Phys. 17, 51 (1974)
Suhl, H. (ed.): *Magnetism*, Vol.5 (Academic, New York 1973)

White, R.M.: *Quantum Theory of Magnetism*, Springer Ser. Solid-State Sci.,
    Vol.32 (Springer, Berlin, Heidelberg, New York 1982)
Wiegmann, P.B.: Towards an exact solution of the Anderson model. Phys. Lett.
    A*80*, 163 (1980)

Chapter 10

10.1   F.A. Lindeman: Phys. Z. *11*, 609 (1910)
10.2   A.K. Singh, P.K. Sharma: Can. J. Phys. *46*, 1677 (1968)
10.3   J.N. Shapiro: Phys. Rev. B*1*, 3982 (1970)
10.4   B.J. Alder, T.E. Wainwright: J. Chem. Phys. *27*, 1208 (1957)
10.5   B.J. Alder, W.G. Hoover: In *Physics of Simple Liquids*, ed. by H.N.V.
       Temperley, J.S. Powlinsor, G.S. Rushbrooke (North-Holland, Amsterdam
       1968) Chap.4
10.6   J.E. Lennard-Jones, A.F. Devonshire: Proc. Roy. Soc. London A*163*, 53
       (1937); *170*, 464 (1939)
10.7   D. Turnbull, J.C. Fisher: J. Chem. Phys. *17*, 71 (1949)
10.8   F.E. Simon, F. Lange: Z. Phys. *38*, 227 (1926)
10.9   M. Ross, P. Schofield: J. Phys. C*4*, L306 (1971);
       B.J. Alder, T.E. Wainwright: J. Chem. Phys. *53*, 3813 (1970);
       Y. Hiwatari, H. Matsuda, T. Ogawa, N. Ogita, A. Ueda: Prog. Theor.
       Phys. *52*, 1105 (1974)
10.10 S. Sakka, J.D. Mackenzie: J. Non Cryst. Solids *6*, 145 (1971)
10.11 R.C. Zeller, R.O. Pohl: Phys. Rev. B*4*, 2029 (1971)
10.12 P.W. Anderson, B.I. Halperin, C.M. Verma: Phil. Mag. *25*, 1 (1972)
10.13 J.M. Ziman: Phil. Mag. *6*, 1013 (1961)
10.14 P.W. Anderson: Phys. Rev. *109*, 1492 (1958)
10.15 B.I. Halperin: Adv. Chem. Phys. *13*, 123 (1967)
10.16 K. Ishii: Prog. Theor. Phys. Suppl. *53*, 77 (1973)
10.17 S.A. Molchanov: Izv. Akad. Nauk SSSR Ser. Mat. *42*, 70 (1978)

Additional Reading

Bourgoin, J., Lannoo, M.: *Point Defects in Semiconductors II*, Springer Ser.
    Solid-State Sci., Vol.35 (Springer, Berlin, Heidelberg, New York 1982)
Davison, S.G., Levine, J.D.: "Surfaca States", in *Solid State Physics*, Vol.25,
    ed. by F. Seitz, D. Turnbull, H. Ehrenreich (Academic, New York 1970) p.1
Dean, P.: The vibrational properties of disordered systems: numerical studies.
    Rev. Mod. Phys. *44*, 127 (1972)
Egelstaff, P.M.: *An Introduction to the Liquid State* (Academic, New York 1967)
Faber, T.E.: *An Introduction to the Theory of Liquid Metals* (University Press,
    Cambridge 1972)
Friedel, J.: *Dislocations* (Addison-Wesley, Reading, MA 1964)
Guttman, L.: "Order-Disorder Phenomena in Metals", in *Solid State Physics*,
    Vol.3, ed. by F. Seitz, D. Turnbull, H. Ehrenreich (Academic, New York
    1956) p.145
Kawai, N., Inokuti, Y.: Low melting of elements under high pressure and its
    progression in the periodic table. Jpn. J. Appl. Phys. *7*, 989 (1968)
Landau, L.D., Lifshitz, E.M.: *Statistical Physics* (Pergamon, London 1958)
Lannoo, M., Bourgoin, J.: *Point Defects in Semiconductors I*, Springer Ser.
    Solid-State Sci., Vol.22 (Springer, Berlin, Heidelberg, New York 1981)
Leibfried, G., Breuer, N.: *Point Defects in Metals I*, Springer Tracts Mod.
    Phys., Vol.81 (Springer, Berlin, Heidelberg, New York 1980)
Maradudin, A.A.: "Theoretical and Experimental Aspects of the Effects of Point
    Defects and Disorder on the Vibrations of Crystals", in *Solid State Physics*,
    Vols.18,19, ed. by F. Seitz, D. Turnbull, H. Ehrenreich (Academic, New York
    1966) p.273; p.1

Maradudin, A.A., Montroll, E.W., Weiss, G.H.: "Theory of Lattice Dynamics in
the Harmonic Approximation", in *Solid State Physics*, Suppl.3, ed. by
F. Seitz, D. Turnbull, H. Ehrenreich (Academic, New York 1963) Chap.6
Markham, J.J.: "F-Centers in Alkali Hilides", in *Solid State Physics*, Suppl.8,
ed. by F. Seitz, D. Turnbull, H. Ehrenreich (Academic, New York 1966)
Mott, N.F.: Electrons in disordered structures. Adv. Phys. *16*, 49 (1967)
Peterson, N.L.: *Diffusion in Solids* (McGraw-Hill, New York 1963)
Powler, W.: *Physics of Color Centers* (Academic, New York 1968)
Rosenberger, F.: *Fundamentals of Crystal Growth I*, Springer Ser. Solid-State
Sci., Vol.5 (Springer, Berlin, Heidelberg, New York 1981)
Shewmon, P.G.: *Diffusion in Solids* (McGraw-Hill, New York 1963)
Shockley, W., Holloman, J.H., Maurer, R., Seitz, F. (eds.): *Imperfections in
Nearly Perfect Crystals* (Wiley, New York 1952)
Stishow, S.M.: Melting at high pressures. Sov. Phys. Usp. *11*, 816 (1969)
Ubbelohde, A.R.: *Melting and Crystal Structure* (Clarendon, Oxford 1965)
Vainshtein, B.K., Fridkin, V.M., Indenbom, V.L.: *Modern Crystallography II,
Structure of Crystals*, Springer Ser. Solid-State Sci., Vol.21 (Springer,
Berlin, Heidelberg, New York 1982)
Weertman, J., Weertman, J.R.: *Elementary Dislocation Theory* (McMillan,
New York 1964)

Chapter 11

11.1   J.A. Krumhansl: In *Amorphous Magnetism*, ed. by H.O. Mooper, A.M. deGraaf
(Plenum, New York 1973) p.15
11.2   N. Nagasawa, N. Nakagawa, Y. Nakai: J. Phys. Soc. Jpn. *24*, 1403 (1968)
11.3   Y. Nakai, T. Murata, K. Nakamura: Jpn. J. Appl. Phys. Suppl. *1*, 616
(1965)
11.4   D.H. Seib, W.E. Spicer: Phys. Rev. Lett. *20*, 1441 (1968); Phys. Rev.
B*2*, 1676, 1694 (1970)
11.5   G.M. Stocks, R.W. Williams, J.S. Faulkner: Phys. Rev. Lett. *26*, 253
(1971); Phys. Rev. B*4*, 4390 (1971)
11.6   D.N. Payton, W.M. Visscher: Phys. Rev. *154*, 802 (1967); *156*, 1032 (1967)
11.7   P.L. Leath, B. Goodmann: Phys. Rev. *181*, 1062 (1969)
11.8   D.W. Taylor: Phys. Rev. *156*, 1017 (1967)
11.9   P. Soven: Phys. Rev. *156*, 809 (1967)
11.10 B. Velicky, S. Kirkpatrick, H. Ehrenreich: Phys. Rev. *175*, 747 (1968)
11.11 Y. Onodera, Y. Toyozawa: J. Phys. Soc. Jpn. *24*, 341 (1968)
11.12 G.W. Robinsen, H.K. Hong: J. Chem. Phys. *54*, 1369 (1971)
11.13 H. Hasegawa, J. Kanamori: J. Phys. Soc. Jpn. *31*, 382 (1971)
11.14 P.W. Anderson: Phys. Rev. *109*, 1492 (1958)
11.15 P. Dean: Proc. Roy. Soc. London A*254*, 507 (1960)
11.16 F. Yonezawa: Prog. Theor. Phys. *40*, 734 (1968)
11.17 F. Yonezawa, T. Odagaki: Solid State Commun. *27*, 1199, 1203 (1978)
11.18 S. Asano, F. Yonezawa: J. Phys. F*10*, 75 (1980)
11.19 R.J. Elliott, J.A. Krumhansl, P.L. Leath: Rev. Mod. Phys. *46*, 465 (1974)

Additional Reading

Jones, G.O.: *Glass* (Wiley, New York 1965)
Parker, R.L.: "Crystal Growth Mechanisms", in *Solid State Physics*, Vol.25,
ed. by F. Seitz, D. Turnbull, H. Ehrenreich (Academic, New York 1970) p.152
Prins, A.J.: *Physics of Non-Crystalline Solids* (North-Holland, Amsterdam 1965)

Subject Index

M. A. van Hove, S. Y. Tong

Surface Crystallography by LEED

Theory, Computation and Structural Results

1979. 19 figures, 2 tables. IX, 286 pages.
(Springer Series in Chemical Physics,
Volume 2). ISBN 3-540-09194-7

"...This is an excellent book for anyone
seriously interested in LEED who would like
to be able to perform his own sophisticated
calculations. It is clearly and carefully written,
and all the nuances of the techniques are
thoroughly covered. If any book will make
the practical calculation of LEED intensity
profiles widely available, this is the one."
American Scientist

"...The extensive experience of the authors, the
welltested nature of the programs, and the
many practical suggestions in the text make this
book an essential reference for workers
studying surface structure by LEED and a
valuable reference for quantitative surface
theory generally."
Contemporary Physics

Hydrogen in Metals I

Basic Properties

Editors: G. Alefeld, J. Völkl
With contributions by numerous experts

1978. 178 figures, 31 tables. XVI, 426 pages.
(Topics in Applied Physics, Volume 28)
ISBN 3-540-08705-2

Contents:
Elastic Interaction and Phase Transition in
Coherent Metal-Hydrogen Alloys. – Lattice
Strains due to Hydrogen in Metals. – Investi-
gation of Vibrations in Metal Hydrides by
Neutron Spectroscopy. – The Change in Elec-
tronic Properties on Hydrogen Alloying and
Hydride Formation. – Mössbauer Studies of
Metal-Hydrogen Systems. – Magnetic Proper-
ties of Metal Hydrides and Hydrogenated Inter-
metallic Compounds. – Theory of the Diffu-
sion of Hydrogen in Metals. – Nuclear Magne-
tic Resonance on Metal-Hydrogen Systems. –
Quasielastic Neutron Scattering Studies of
Metal Hydrides. – Magnetic Aftereffects of
Hydrogen Isotopes in Ferromagnetic Metals
and Alloys. – Diffusion of Hydrogen in
Metals. – Positive Muons as Light Isotopes
of Hydrogen.

Photoemission in Solids I

General Principles

Editors: M. Cardona, L. Ley

1978. 90 figures, 17 tables. XI, 290 pages.
(Topics in Applied Physics, Volume 26)
ISBN 3-540-08685-4

Contents:
M. Cardona, L. Ley: Introduction. –
W. L. Schaich: Theory of Photoemission:
Independent Particle Model. – *S. T. Hanson:*
The Calculation of Photoionization Cross
Sections: An Atomic View. – *D. A. Shirley:*
Many-Electron and Final-State Effects:
Beyond the One-Electron Picture – *G. K. Wert-
heim, P. H. Citrin:* Fermi Surface Excitations in
X-Ray Photoemission Line Shapes from
Metals. – *N. V. Smith:* Angular Dependent
Photoemission. – Appendix.

Springer-Verlag
Berlin
Heidelberg
New York

Light Scattering in Solids II

Basic Concepts and Instrumentation

Editors: M. Cardona, G. Güntherodt

1982. 88 figures. Approx. 275 pages. (Topics in Applied Physics, Volume 50)
ISBN 3-540-11380-0

Contents:
M. Cardona, G. Güntherodt: Introduction. – *M. Cardona:* Resonance Phenomena. – *R. K. Chang, M. B. Long:* Optical Multichannel Detection. – *H. Vogt:* Coherent and Hyper-Raman Techniques. – Subject Index.

Structural Phase Transitions I

Editors: K. A. Müller, H. Thomas

1981. 61 figures. IX, 190 pages. (Topics in Current Physics, Volume 23)
ISBN 3-540-10329-5

Contents:
K. A. Müller: Introduction. – *P. A. Fleury, K. Lyons:* Optical Studies of Structural Phase Transitions. – *B. Dorner:* Investigation of Structural Phase Transformations by Inelastic Neutron Scattering. – *B. Lüthi, W. Rehwald:* Ultrasonic Studies Near Structural Phase Transitions.

Real-Space Renormalization

Editors: T. W. Burkhardt, J. M. J. van Leeuwen

1982. 60 figures, approx. 16 tables. Approx. 300 pages. (Topics in Current Physics, Volume 30). ISBN 3-540-11459-9

Contents:
T. W. Burkhardt, J. M. J. van Leeuwen: Progress and Problems in Real-Space Renormalization. – *T. W. Burkhardt:* Bond-Moving and Variational Methods in Real-Space Renormalization. – *R. H. Swendsen:* Monte Carlo Renormalization. – *G. F. Mazenko, O. T. Valls:* The Real Space Dynamic Renormalization Group. – *P. Pfeuty, R. Jullien, K. A. Penson:* Renormalization for Quantum Systems. – *M. Schick:* Application of the Real-Space Renormalization to Adsorbed Systems. – *H. E. Stanley, P. J. Reynolds, S. Redner, F. Family:* Position-Space Renormalization Group for Models of Linear Polymers, Branched Polymers, and Gels.

B. Dorner

Coherent Inelastic Neutron Scattering in Lattice Dynamics

1982. 47 figures. VIII, 96 pages. (Springer Tracts in Modern Physics, Volume 93)
ISBN 3-540-11049-6

Contents:
Introduction. – Experimental Technique with Three-Axis Spectrometers. – The Scattering Function and Symmetry Operations in the Crystal. – Lattice Dynamical Models. – Analysis of Phonon Intensities. – Analysis of Phonon Line Shapes. – Final Remarks. – References. – Subject Index.

Springer-Verlag
Berlin
Heidelberg
New York

Springer Series in Solid-State Sciences

Editors: M. Cardona P. Fulde H.-J. Queisser